# Advances in Geographical and Environmental Sciences

**Series Editor**

R. B. Singh, University of Delhi, Delhi, India

Advances in Geographical and Environmental Sciences synthesizes series diagnostigation and prognostication of earth environment, incorporating challenging interactive areas within ecological envelope of geosphere, biosphere, hydrosphere, atmosphere and cryosphere. It deals with land use land cover change (LUCC), urbanization, energy flux, land-ocean fluxes, climate, food security, ecohydrology, biodiversity, natural hazards and disasters, human health and their mutual interaction and feedback mechanism in order to contribute towards sustainable future. The geosciences methods range from traditional field techniques and conventional data collection, use of remote sensing and geographical information system, computer aided technique to advance geostatistical and dynamic modeling.

The series integrate past, present and future of geospheric attributes incorporating biophysical and human dimensions in spatio-temporal perspectives. The geosciences, encompassing land-ocean-atmosphere interaction is considered as a vital component in the context of environmental issues, especially in observation and prediction of air and water pollution, global warming and urban heat islands. It is important to communicate the advances in geosciences to increase resilience of society through capacity building for mitigating the impact of natural hazards and disasters. Sustainability of human society depends strongly on the earth environment, and thus the development of geosciences is critical for a better understanding of our living environment, and its sustainable development.

Geoscience also has the responsibility to not confine itself to addressing current problems but it is also developing a framework to address future issues. In order to build a 'Future Earth Model' for understanding and predicting the functioning of the whole climatic system, collaboration of experts in the traditional earth disciplines as well as in ecology, information technology, instrumentation and complex system is essential, through initiatives from human geoscientists. Thus human geosceince is emerging as key policy science for contributing towards sustainability/survivality science together with future earth initiative.

Advances in Geographical and Environmental Sciences series publishes books that contain novel approaches in tackling issues of human geoscience in its broadest sense—books in the series should focus on true progress in a particular area or region. The series includes monographs and edited volumes without any limitations in the page numbers.

More information about this series at https://link.springer.com/bookseries/13113

Pankaj Kumar · Gaurav Kant Nigam ·
Manish Kumar Sinha · Anju Singh
Editors

# Water Resources Management and Sustainability

Springer

*Editors*
Pankaj Kumar
Department of Geography
Delhi School of Economics
University of Delhi
New Delhi, Delhi, India

Manish Kumar Sinha
Department of Environmental
and Water Resources Engineering
Chhattisgarh Swami Vivekanand
Technical University
Bhilai, Chhattisgarh, India

Gaurav Kant Nigam
Krishi Vigyan Kendra
Korba, Chhattisgarh, India

Anju Singh
Department of Geography
Aditi Mahavidyalya
University of Delhi
New Delhi, Delhi, India

ISSN 2198-3542 ISSN 2198-3550 (electronic)
Advances in Geographical and Environmental Sciences
ISBN 978-981-16-6572-1 ISBN 978-981-16-6573-8 (eBook)
https://doi.org/10.1007/978-981-16-6573-8



This Springer imprint is published by the registered company Springer Nature Singapore Pte Ltd.
The registered company address is: 152 Beach Road, #21-01/04 Gateway East, Singapore 189721, Singapore

# A Homage to the Memory of Our Mentor–Master Prof. R. B. Singh (b. 3 February 1955 ~ d. 22 July 2021)

R. B. Singh: watching the rural landscape near
Bishkek, Kyrgyzstan, 24 July 2017

वासांसजिीर्णानयिथावहािय, नवानगिृह्णातनिरोऽपराणि।
तथाशरीराणविहाियजीर्णा, न्यन्यानसिंयातनिवानदिेही ॥

*vāsānsijīrṇāniyathāvihāya, navānigṛihṇātinaro 'parāṇi* |
*tathāśharīrāṇivihāyajīrṇānya, nyānisanyātinavānidehī.* ||

*As a person sheds worn-out garments and wears new ones,*
*Likewise, at the time of death, the soul casts off its worn-out body*
*and enters a new one*
—*ShrīmadaBhāgavatGītā* (2.22)

# Foreword and Appraising the Contributions of Prof. R. B. Singh

The *Mahābhārata* (12.182.14–19), a circa-fourteenth-century BCE text, states that the Supreme God created Primordial Man who first made sky; from sky, water was then made; and the seed of water, together with fire and air, made the earth; hence, in a metaphysical sense, these elements are not separated from each other (Singh 2000, p. 81). According to the *BhāgavataPurāṇa* (1.3.2–5), an eighth-century text, Primordial Man was lying down in the water of the universe. Not only are these elements related to one another by their intrinsic nature, but they foster bonds among creatures. In fact, "water plays a cardinal role in most creation myths, frequently associated with female element, in reciprocal relationship with the male elements of sky and earth" (Buttimer 1989, p. 265). The *Rig Veda* (10.90. 11–14), a circa-fifteenth-century BCE text, describes water as a unifying fluid between sky/heaven and earth (Singh 1996, p. 87). In a part of the *Rig Veda*, the *ShatapathaBrāhmaṇa* (1.8.1.1–6), the protector of the Hindu Trinity, Vishnu, has a fish incarnation that symbolizes the origin of life in water. In a fish form, he saved organic life-seeds from the great cosmic flood (Singh 1992, p. 77). The water is regarded as the primary materialization of Vishnu's *māyā* (energy) and is therefore a visible manifestation of the divine essence (Zimmer 1991, p. 34). Metaphorically and metaphysically, the ancient mythologies refer to water as the container of life, strength, and eternity, that is how preservation and conservation within the frame of sustainability the issue of water becomes vital. Following this line of thought, scholars and seekers pay their attention for "Water Resource Management and Sustainability". The present anthology is also an addition in this relevant and contemporary concern, taking multifaced approach and multidisciplinary outline, as reflected by scholars from different field.

The idea of this book was firstly initiated by (*late*) Prof. R. B. Singh (b. 3 February 1955–d. 22 July 2021) to proceed under the series initiated and edited by him, i.e. Advances in Geographical and Environmental Sciences (Springer Nature Pte Ltd., Singapore). The present volume is a sequel to the similar volume published by his associates, and dedicated to him—under Sustainable Development Goals Series of the Springer Nature, which was also conceptualized and started by him (see, Pandey and Anand 2021); this book deals with the importance of water resources for socio-economic and ecological development including geomorphic and ecological environments. In fact, "potentiality of becoming water as critically scarce resource in the coming years is increasing continuously due to various factors. Looking the importance of water from local to global level, its integrated, appropriate and long-term strategies are much needed for sustainable water resource management" (ibid., p. 5). However, the present book laid emphasis on the mapping, modelling and management of water resources and attempt to cover geo-engineering and scientific problems, illustrated with case studies within the spectrum of sustainability.

The serene environment of the Sarayu riverfront, the sacred groves and gardens in the surroundings, small monastery of a Vaishnavite saint in the vicinity and beautiful lush trees and shrubs in the surroundings, and also the all-around scattered green fertile fields in the countryside where R. B. Singh was born, altogether made an environmental setting where his innocent childhood mind had developed deep attachment to the spirit of place (*genius loci*) and the village ecology that flourish and sustained throughout his life journey (Singh 2021, p. 163). The deep sense of his place attachment and ecological understanding has got momentum in his master's class with the collaboration of his teacher and elder brother (myself), as illustrated with a study of their native village, Majhanpura (Bihar), which resulted in a monograph, viz. Singh and Singh (1981), *Changing Frontiers of Indian Village Ecology*. In his Foreword, Prof. Walther Manshard (Secretary General IGU) has remarked, "I hope that this study will reflect a new direction for an interdisciplinary approach and inspire fellow workers for the detailed investigation of the micro-level rural habitat and ecology. With this objective, I wish for its success and wide dissemination" (ibid., p. 9). This publication has also opened door for his involvement and associated activities in the International Geographical Union (IGU). This study has further been enhanced in his doctoral dissertation (period 1977–1981), emphasising the ecological perspectives of rural development, illustrated with a case study of Siwan district (Bihar), awarded in 1981 and published in 1986. Additionally, in his first co-authored research paper, this frame is analysed (Singh and Singh 1979).

The training R. B. Singh had received in his master's studies (1975–1977) at Banaras Hindu University (Varanasi, India) was slightly different from the contemporary geography centres involved in making a geographer as part of academic pigeonholing; instead, it opened the mind to put geography anywhere and create that beautiful. Here for the first time, he learned the use of technology and technocratic approach in understanding landscape and culture; however, he finally put himself into the ocean of empiricism and technocratic narration. What is in the contemporary world is the common scenario (Singh 2021, p. 165).

Carrying background training at master and doctoral level at Banaras Hindu University, and the childhood experiences, R. B. Singh had advanced and exposed his understanding and interpretation of interfaces between water resources and sustainability as one of the major concerns of his researches, direction of various projects, guidance of doctoral students and of course publications.

The gamut of publications by R. B. Singh consists of variety and distinct topics, crossing all the traditional border and themes of geography, and under different categories, which include: books; edited volumes; papers—peer reviewed; papers in Indian journals/books; papers in foreign books and proceedings; and papers in Indian books and proceedings (see, Singh 2021, Appendix 1, pp. 179–197). Altogether themes and subject matter of his publications could be broadly categorized under 13 categories, viz. environmental studies, geoecology; land resources, land use/land cover; water issues, hydrology; disaster, natural hazard; quality of life, livelihood; climatic change, air pollution study; urban environment, health, well-being; SDGs, environmental monitoring; geography, development studies-rural and urban; mountain studies, forestry, tourism; RS, GIS, recent trends appraisal; regional science and SDGs; and INSA geography report, and related. Out of his total 260 research papers, 31 (i.e. only 11.8%, see Appendix A at the end) deal directly on the problems of water resources and hydrology. Additionally, he has also co-edited a proceedings based volume on the topic landscape ecology and water management (Singh et al. 2014). This volume focuses on holistic natural resource-based spatio-temporal planning, development and management and considers them as essential to save the degraded ecosystem for sustainable resource management. Contributions are compiled in two volumes: (i) climate change and biodiversity and (ii) landscape ecology and water management. Moreover, geoinformatics along with its tools such as remote sensing and geographical information systems (GIS) have been used in assessing the results of various environmental problems both physical and social.

All the thematic papers (independent and co-authored) by R. B. Singh are categorized into four groups by himself and followed here the same for review and appraisal (see, Appendix A). The first group consists of papers in peer-reviewed journals (**A**), which recorded the frequency of 15 during the period 1997–2018. With reference to water resource management, hydrological parameters are tested in the Mahanadi River Basin of the Himalaya, and attributes of environmental sustainability are delineated and compared (1997a). The followed-up study in the same year deals with regional water management, using case of agriculturally developed region of the Indo-Gangetic Plains (2002) and further extended to the overall appraisal of water resource management in the vulnerable Indian Environment (2009). In a project-based study, using spatial analysis of water quality of the Yamuna River has been vividly narrated (2011a) and further extended to link with climatic variability modes (2011b), and also the problem of declining water sources in the traditional water bodies in Delhi (2013).

Similar modes and models are further implied to analyse low streamflow of Paranaiba River of Brazil (2014a). The impact of monsoon rainfall—its variability and consequences in the Western Himalaya—is another representative of the study carried out in this period (2014b). Under the purview of urban sustainability, the issues of drinking water, sanitation and health are examined in case of Kolkata Metropolitan City (2015a). The problem of water scarcity in the dryland of Rajasthan represents another study of applied approach and policy implications (2015b). The two other follow-up studies discuss the hydrological and groundwater conditions in spatio-temporal frame, illustrated with Marathwada region of Maharashtra, western India (2018a and 2018b).

The second group consists of papers in the Indian journals and books (**B**), which includes seven papers, published during 2001–2017. With a case study of the Himalayan Mountain Region, a study reveals the context of management of water-related disasters and finally purposes sustainable strategies for mitigation and policy making (2001). As an extension of his earlier studies, a paper on impact of climate change on water resources sustainability in the Himalayan-Gangetic Region was published (2004). Another study refers to hydro-informatics for the integrated watershed management, where interfaces and linkages between science and community participation are dealt in making a rational policy (2006). The problems of living with flood and sustainable livelihood development in lower Brahmaputra River Basin of Assam is the first-hand study of its kind in this area (2014a); this study is extended to the analyse the scale of adaptation strategies for flood risk mitigation in the same area through integrated river basin management (2014b) and also measuring the extent and magnitude of flood due to soil erosion (2014c). The last study in this group is a review appraisal that discusses ensuring clean water accessibility and mitigating water-induced disasters as a base for contribution towards sustainable development goals, SDGs (2017).

The third group consists of papers in foreign books and proceedings (**C**), which include four papers, published during 1983–2015. All these four papers in this group deal with the four themes, of course interconnected, viz. ecological approach of irrigation system: potentiality constrains and management in traditional society (1983), local knowledge and community participation in water management for poverty alleviation: towards integrating water science and ethics in India (2003), review and appraisal of the studies on water scarcity (2014), and drinking water, sanitation and health problems in Kolkata metropolitan city—implications towards urban sustainability (2015).

The fourth group consists of papers in the Indian books and proceedings (**D**), which includes five papers, published during 1996–2008. The five papers cover the analysis and perspectives on the issues of managing rural water resources, as observed in a study of the Ganga-Gomati Doab (1996); the interfaces between dam and water resource development: myth or reality, as visualized in the case of Hirakud Dam(1998); critical appraisal and evaluation of influence of water availability on poverty in India (2005); examination of integrated watershed management for livelihood security, with a case study of Uppalgaon Watershed, Uttaranchal Himalaya (2006); and geographical analysis of water logging induced land degradation in Karnal District of Haryana (2008).

Similarly, the pathway what R. B. Singh had dealt with taken in this anthology and framed to prepare the present anthology, which consists of 23 chapters. The essays can easily be grouped under four broad themes.

The first is Environmental and Water quality management, which addresses water quality-related issues and sediment deposition in the river basin. The chapter covers sediment depositions in the river basin of Lippe river, NRW, Germany due to Mining Related PCB, analysis of textural characteristics and depositional environment of sediments in Gumti River, Tripura, India, and estimation of sediment rate through stage-discharge rating curve for two mountain streams in Sikkim, India. Water quality-related issues were addressed in the Mahanadi river system in India using soft computing techniques. A similar GIS-based technique was applied for site suitability analysis for water harvesting structures in the Tapti river in India. A uniquely IoT-based machine learning application for a remotely operated wetland siphon system during hurricanes was designed in the United States, Florida and Wastewater treatment plants to combat climate change and help sustainable water management in Kumamoto city discussed in this section.

The second section addresses the application of remote sensing and GIS environment in water resources and climate change implications. Detailed application of parametric and non-parametric statistical approaches and Budyko framework was demonstrated for northern and central Indian regions. The results demonstrated a significant relationship between temperature and rainfall extremes in the basins for shorter and long-term rainfall series. Four case studies using land-use land-cover analysis in reveals the floodwater harvesting for groundwater development in the upper delta of Cauvery River Basin, root causes for drying of Kanari River system, morphometric analysis and geohydrological inference of Bhilangna river basin and geospatial analysis of Kosi River course was showcase the importance and applicability of open source remote sensing products.

The third section demonstrates the in-depth hydraulic and hydrologic modeling and simulation using open-source software/programs ex.: EPA-SWMM, MODFLOW, HEC-HMS, HEC-RAS, SWAT, EPANET, WATERNETGEN. This study covers the various hydraulics and hydrologic analysis methods in cities and basins of Indian regions.

The last and fourth section covers water resources management for society and communities in the Philippines, West Africa, and India. A community-based water-related problem and management strategy were proposed for the Wassa Amenfi East District of the Western Region, Ghana and Rural Desert Communities of Jaisalmer, India. Similarly, potential groundwater zoning was done using maximum entropy theory in the Philippines and irrigation planning using fuzzy parametric programming approach was used in irrigation fields of the central Indian regions.

This volume would open a new door of interdisciplinary understanding using similar modeling techniques and assessing the potentials and prospects of water resources in different regions. The conclusion will further be used for further investigations in similar conditions and landscapes. *Reverence*—the deeper vision of the sanctity of life; *responsibility*—the connecting link between ethics and rationality; *frugality*—grace without waste; and *ecojustice* all form the minimal core of intrinsic values for right conservation and preservation of the spirit of sustainability (Skolimowski 1990, pp. 101–102). This book will help to make good march and awakening on the various contexts of water resource management and linkages with sustainability.

Rana P. B. Singh
(Former) Professor and Head—Department of Geography, Banaras Hindu University (BHU)

President
ACLA Asian Cultural Landscape Association (Korea-India)
Varanasi, India

## References

Buttimer A (1989) Nature, water symbols and the human quest for wholeness. In: Seamon D, Mugerauer R (eds) Dwelling, place and environment. Columbia University Press, New York, pp 257–280

Pandey BW, Anand S (eds) (2021) Water science and sustainability. Sustainable Development Goals Series (ISSN 2523-3084). Springer Nature Switzerland AG, Cham. https://doi.org/10.1007/978-3-030-57488-8

Singh M, Singh RB, Hasan MI (eds) (2014) Landscape ecology and water management. Springer Japan, Tokyo

Singh RB (1986) Geography of rural development (a case study of Siwan district, Bihar). Inter India Publisher, New Delhi. [Revised doctoral dissertation, awarded in 1981].

Singh RPB (1996) The Gaṅgā River and the spirit of sustainability in Hinduism. In: Swan J, Swan R (eds) Dialogues with the living earth. New Ideas from Spirit of Place from Designers, Architects and Innovators. Quest Books, Wheaton, IL, pp 86–107

Singh, RPB (2020) Sacrality and waterfront sacred places in India: myths and making of place. In: Ray C (ed) Sacred waters: a cross-cultural compendium of hallowed springs and holy wells. Routledge, London & New York, pp 80–94. https://doi.org/10.1007/978-100-30-1014-2_6

Singh, RPB (2021) Professor R. B. Singh (1955 ~ 2021), an icon of Indian geography: a passage on the path of lineage, legacy and liminality. Space and Culture, India (ACCB Publishing, England, ISSN: 2052-8396, Scopus indexed), vol 9(2), September, pp 161–200. https://doi.org/10.20896/saci.v9i2.1204

Singh, RPB, Singh, RB (1979) Ecological perspectives of rural development in North Indian plain: a case of Siwan district. Geographica (Singapore), vol 13, pp 49–57

Singh, RPB, Singh RB (1981) Changing frontiers of Indian village ecology: case of a Bihari village, Majhanpura. Foreword: Prof. Walther Manshard, Secretary-General IGU. Varanasi: National Geographical Society of India, Banaras Hindu University, Res. Pub. No. 27; 88-pp.: 20-tables, 17-figures. [A model study of their own native village, under IGU WG on 'Transformation of Rural Habitat in Developing Countries'].

Singh, RPB (1992) Nature and cosmic integrity: a search in Hindu geographic thought. GeoJournal 26(2):139–147

Skolimowski H (1990) Reverence for life. In: Ronald Engel J, Engel JG (eds) Ethics of environment and development. University of Arizona Press, Tuscon, AZ.

Zimmer H (1991) Myths and symbols in Indian art and civilization. Princeton University Press, Princeton

# Preface

Water resource is life's elixir and essential component for sustainable development. Previously, the availability of water was thought to be an unlimited resource and entirely renewable natural resource in nature, but over the last 20 years, there has been immense pressure on these valuable resources, primarily as a result of rapid industrialization and increase in human population, which has resulted in increased in demand of water resources for irrigation and agricultural needs. Though agricultural technology has advanced significantly, poor irrigation management has resulted in significant groundwater depletion, degraded soils quality and availability and degradation in water quality, making future water supply highly uncertain. In the light of the imminent lack of usable water supplies in the foreseeable future, it has become critical for water scientists and planners to measure and conjunctively control the available water resources for their judicial use.

Sustainable Water Resources Management is a key environmental challenge of the twenty-first century. It is a concept that emphasizes geographical and environmental scientists the need to consider the long-term future as well as the present need of water management practices and challenges. Under this avenue, many engineering and scientific problems have been addressed to cope with the current situation of water management problems in the world. The most current thinking on the environment, earth system, climate change, land use and social dimensions indicates to sustainably managing the water demand at local, regional or basin levels. The twin challenges of ensuring an adequate water demand and the optimal allocation for different uses are driven by changes in climate, land use, demographic patterns, water demand and water availability.

This book covers the engineering and scientific problems, case study and sustainable solutions in water resources management domain. The chapters of the book provide an in-depth coverage of such topics as integrated water resources management, hydrological modelling, remote sensing and geographical information system applications in geosciences, hydro-climatology in water resources, modelling and analysis of environmental problems related to water management. This book advances the scientific understanding, development and technologies related to spatial dimensions of surface and groundwater resource planning and management in changing hydrological system.

This practical, state-of-the-art reference book is a valuable resource for students, designers of water resources systems, researchers and scientists interested in sustainable water resources management. Apart, this book will be very useful in understanding uses and applications of Earth Observation (EO) open data sources, platforms, freeware GIS applications for watershed assessment and management, hydro-climatology and water resources modelling to achieve environmental sustainability.

New Delhi, India — Pankaj Kumar
Korba, India — Gaurav Kant Nigam
Bhilai, India — Manish Kumar Sinha
New Delhi, India — Anju Singh

# Memoirs of Academic and Educational Contribution of Late Professor R. B. Singh

**Keywords** Disaster · Water resource · Land-use change · Glocal environmental education

## Introduction

22 July 2021 was a sad day for the world geographic community. Professor R. B. Singh, Delhi School of Economics, Delhi, India, who had been International Geographical Union (IGU), Vice President (August 2012–August 2018) and Secretary General and Treasurer (August 2018), suddenly deceased on that day from illness. I worked with him closely for around a quarter of a century, first in IGU-LUCC (IGU Commission on Land Use/Cover Change) since its birth in 1996, then in a number of projects, meetings and publications related with land-use/cover change, sustainability, disasters, environmental education, etc., as well as in the IGU Executive Committee since he joined it in 2012, which was two years after my joining it. This article is memoirs of academic and educational contribution of Professor Singh viewed from me as a geographer and his long-time collaborator and colleague aiming together at enhancing global to local sustainability.

## Studies of Land-Use and Land Cover Change

Professor Singh's long-time involvement in the study of land-use and land cover change strongly reflects its particular importance in India where such pressing issues as food production, population, biodiversity, environmental conservation, water and land management, human health and welfare are closely related with it. I initiated IGU-LUCC (IGU Commission on Land Use and Land Cover Change) in 1996 in order to support the LUCC Programme, which was jointly established by (International Geosphere-Biosphere Programme (IGBP) and International Human

Dimensions Programme for Global Environmental Change (IHDP), and chaired it till 2004, and he was Steering Committee Member of it through that period.

As a contribution to IGU-LUCC, he took the role of chief editor of *Land Use and Cover Change*, a book published by Science Publishers in 2001, which contained articles selected from the papers presented at the IGU-LUCC conference at Honolulu in 1999 (Singh et al. 2001). The book covers a wide range of geographic regions and technical/modelling applications in the LUCC context.

*Land-Use Changes in Comparative Perspective*, a second book of IGU-LUCC commercially published, was edited by Himiyama et al. (2002), containing papers presented at the IGU-LUCC summer workshops held in Japan and Korea in 2000 with focus on comparative case studies of land-use/cover changes. It includes a paper written by Prof. Singh entitled "Urbanization and Challenges to Sustainable Rural Land Use in India—Comparative Case Study" (Singh 2002). In this paper, he compares four carefully selected villages near Delhi, where he conducted interviews to as many as 172 persons and concluded that in a situation where urban impact was both beneficial and otherwise, there was an urgent need to categorize and prioritize rural and agricultural land use and to plan for sustainable development of the rural–urban fringe.

Professor Singh wrote on forest-cover mapping in India (Singh and Mishira 2005), demonstrating the use of GIS and remote sensing in forest cover mapping by identifying some districts having recorded decline in forest cover and drawing attention to them for forest conservation and renewal in *Understanding Land-Use and Land-Cover Change in Global and Regional Context* edited by Milanova et al. (2005). The book is an outcome of a workshop on "Global and Regional Land-Use/Land-Cover Change" and a session on "Land-use dynamics and sustainable development" held in Moscow-Barnaul, Russia, in July 2003.

IGBP/IHDP LUCC merged with Global Change and Terrestrial Ecosystems (GCTE) and together formed Global Land Project (GLP) in 2006. While keeping support to GLP, IGU-LUCC did not re-establish formal cooperative relation with it and continued its own mission of promoting research on land-use/cover change. It is noted that in many of the geographic conferences of international significance held in India since the early 2000s, Professor Singh seemed to have suggested the local organizers to include IGU-LUCC as a supporting commission. The consequence of his long-time initiative in promoting land-use studies in India has been impressive, as seen in the publication of *Land Use–Reflection on Spatial Informatics, Agriculture & Development* (Jha and Singh 2008). In this book, he wrote on urban land use and its impact on groundwater quality in the Delhi Metropolitan Region with his daughter Dr. Anju Singh (Singh and Singh 2008).

I directed Sustainable Land Use for Asia (SLUAS) Project funded by Japan Society for Promotion of Sciences for 2009–2014. Major issues of the SLUAS project included sustainability, urbanization, rural development, land-related problems such as food problems and land-related disasters in Monsoon Asia, as well as reconstruction of historical land-use/cover changes. One of its outcomes was publication of *Exploring Sustainable Land Use in Monsoon Asia* (Himiyama 2018). It includes two papers written by Professor Singh and his colleagues, namely

"Low Carbon Resilient Delhi Megacity for Sustainable Future Earth" (Singh et al. 2018) and "Dynamics of Land Use and Climate Change in Subhumid Region of Rajasthan, India" (Singh and Kumar 2018). It is noted that Monsoon Asia, with its huge and increasing population and rapid socio-economic changes, is a major hot spot of global change in general and of land-use change in particular. Land-use change has been an essential driving force of environmental change, and at the same time a result of socio-economic and environmental changes, and is a major environmental change itself.

## Disaster Studies

Before joining IGU-LUCC, Professor Singh was active in IGU Study Group on Development Issues in Marginal Regions. He coordinated "International Conference on Disasters, Environment and Development" in December 1994 at University of Delhi, with support from the study group, UN International Decade of Natural Disaster Reduction, UN Development Programme, Ministry of Agriculture, etc., edited its 53 proceedings papers and published the book under the title *Disasters, Development and Development* (Singh 1996a). He himself contributed three papers, namely on disaster monitoring and mitigation (Singh 1996b), hazard zone mapping and risk assessment analysis (Singh and Pandey 1996c) and environmental degradation and its impact on livelihood strategies in the urban fringe of Delhi (Druijven and Singh 1996).

Thus, Prof. Singh was an eminent scholar of not only geographical but also environmental, human-geoscientific sciences and beyond. In May 2012, he was invited to the annual meeting of Japan Geoscience Union (JpGU) held at Makuhari, Japan, as a guest keynote speaker. His lecture was on "Spatial Information Technology for Diagnosticating Urban Landscape and Prognosticating Heat Islands in Mega Cities of India". He concluded that: (a) intensive anthropogenic activities in mega cities are largely responsible for UHIs and micro-climate change; (b) it is imperative to examine spatial and temporal dynamics and dimensions of urban heat island; (c) the micro-climate also influences the long range monsoon system; (d) forest and vegetation plantation can have an influence on carbon sequestration; (e) carbon resilience measures can play an effective role in mitigating climate change by adapting afforestation, non-conventional energy, enhancing energy efficiency and promoting green buildings and neighbourhoods together with good governance. His lecture impressed a large number of audience and attracted many questions and comments.

As SLUAS project included land-related problems such as geo disasters from its start, it could easily incorporate the Great East Japan Disaster, which was triggered by the M 9.0 earthquake of 11 March 2011. SLUAS carried out a number of field surveys in East Japan, observing the extremely uneasy situation of the tsunami-stricken areas and the hopelessly battered land and people's lives in the areas contaminated by the radioactive substances ejected from the Tokyo Electric Company Fukushima Daiichi Nuclear Power Plant.

In June 2013, Prof. Singh joined a field trip of SLUAS to the tsunami-stricken areas in East Japan. Figure 1 was taken in front of Hotel Kanyo (Ocean View) with him on the left, Mrs. Noriko Abe, the landlady, in the middle, and me to the left. Kanyo is the largest hotel in the area hit by the gigantic tsunami. It survived the tsunami, saved many hundreds of sufferers, and has become a major learning core of disaster and its prevention in Japan. In the photographs, Mrs. Abe is explaining how her father's three-story house in Kesennuma City, which had outside spiral stairs to the rooftop, saved the lives of tens of neighbours while the house itself was severely damaged by the tsunami. Professor Singh took many photographs during the field trip along the tsunami-stricken coastal areas and took notes on the papers he had in fine letters.

In fact I was invited, through Prof. Singh, to "International Seminar on Global Environment & Disaster Management: Law & Society", as guest speaker on the Great East Japan Disaster, in July 2011, i.e. only four months after the disaster was triggered. The seminar was held at Vigyan Bhawan, New Delhi, and was opened by the President and closed by the Prime Minister of India. Figure 2 was taken soon after the closing of the seminar, with me and Professor Singh standing side by side in front. It was a memorable moment realized by the trust and respect Professor Singh enjoyed among the political leaders of India.

**Fig. 1** Interviewing at Hotel Kanyo in Minami Sanriku Town, which was hit by the gigantic East Japan earthquake/tsunami of 11 March 2011

**Fig. 2** International Seminar on Global Environment & Disaster Management: Law & Society

## Glocal Environmental Education

Professor Singh's commitment to global–local environmental education was also great. Global environmental problems, such as global warming, resource problems and population problems, have been becoming increasingly serious in recent decades. G8 Hokkaido-Toyako Summit held in July 2008 highlighted these issues, but a deep gap existed between the serious global environmental situation and the state of environmental education in the world. Environmental education is often influenced by local to national conditions or political will, and global environmental problems, global–local linkages and responsibility as global citizens may have been treated only marginally at schools. Thus, Hokkaido University of Education located in Hokkaido decided to organize "International Symposium on Glocal Environmental Education 2008" on that occasion. Glocal is a term meaning global to local, and it emphasizes multi-scale views in addressing global to local problems or activities.

Professor Singh was one of the guest speakers of the Symposium, and played a major role in writing the "Sapporo Declaration on Glocal Environmental Education" towards the end of the symposium. The declaration stated that we needed to re-orientate our approach to environmental education through: *improving learning opportunities*; *strengthening existing good practices and overcoming the barriers and gaps that hinder their spread and development*; and *enhancing research-led teaching*. The full text of the declaration can be found in *Glocal Environmental Education*, a book edited by Himiyama, Singh and two others, as outcome of the Symposium (Himiyama et al. 2010). It is noted that the well-known Bonn Declaration of the UNESCO World Conference on Education for Sustainable Development issued in April 2009 shared many things with the Sapporo Declaration issued one year earlier. In this book, Singh wrote an article entitled "Designing and shaping global to local environmental education: challenges and opportunities—the Indian experience" (Singh 2010). In this article, he analysed the linkages between global to local environmental problems within the environmental education system in India. He said that the introduction of global to local environmental education for the people leads to the opening up of a vast pool of opportunities and that it helps in the promotion of awareness, knowledge, positive attitudes, skill building of the individual and increases participation level at local, regional, national and global levels. He emphasized that this education system should focus on priority areas, such as food, energy, water, biodiversity and health and that key thematic issues to be incorporated in glocal environmental education in India should include land-use and land cover change, carbon/energy, food security, water, biodiversity, and processes affecting health.

Figure 3 shows some of the participants of the International Year of Global Understanding (IYGU) Regional Action Centre for South Asia Conference on "South Asia Initiative towards IYGU 2016—A step towards education, health care, gender sensitivity and women empowerment", 17 March 2016. IYGU is an international sustainability programme promoted by the IGU with the support by International Council for Philosophy and Human Sciences (CIPSH) and UNESCO

**Fig. 3** IYGU Regional Action Centre for South Asia Conference on: "South Asia Initiative Towards International Year of Global Understanding 2016"

(Werlen 2015). In fact what is really meant by global understanding is glocal (global to local) understanding, and Professor Singh was deeply involved in promoting IYGU in India. Figure 3 includes some academics including Professor Harsh Gupta, a former International Union of Geodesy and Geophysics (IUGC) President, sitting at the centre, Professor Singh second to the left and Himiyama to the right, and other active geographers, together with leading school teachers, students and practitioners, young and senior, men and women.

## Conclusions

Professor Singh and I had very similar interests in research, education and duties as scientists, and we helped each other, learned from each other and worked together for quarter of a century in research, education and promotion of geography. His a little too early passing in the middle of research, education, writing and editing of a number of articles and the work as Secretary General and Treasurer of the International Geographical Union made a big hole in my and many other's mind. It is hoped that this short memoir is of use to those who are trying to fill such hole by knowing some aspects of Professor Singh which may have been overlooked.

## References

Druijven PCJ, Singh RB (1996) Environmental degradation and its impact on livelihood strategies in the urban fringe of Delhi—some theoretical reflections. In: Singh RB (ed) Disasters, environment and development. Oxford & IBH Publishing Co., New Delhi, India, pp 355–368

Himiyama Y (ed) (2018) Exploring sustainable land use in monsoon Asia. Springer, Singapore

Himiyama Y, Hwang M, Ichinose T (eds) (2002) Land-use changes in comparative perspective. Science Publishers, Enfield, USA

Himiyama Y, Singh RB, Kanda F, Hindson J (eds) (2010) Glocal environmental education. Rawat Publications, Jaipur, India, p 246

Jha MM, Singh RB (eds) (2008) Land use—reflection on spatial informatics, agriculture & development. Concept, New Delhi, India
Milanova E, Himiyama Y, Bicik I (eds) (2005) Understanding land-use and land-cover change in global and regional context. Science Publishers, Enfield, USA
Singh A, Singh RB (2008) Urban land use and its impact on groundwater quality in the Delhi metropolitan region. In: Jha MM, Singh RB (eds) Land use—reflection on spatial informatics, agriculture & development. Concept, New Delhi, India, pp 185–194
Singh RB (1996a) Disasters, environment and development. Oxford & IBH Publishing Co., New Delhi, India
Singh RB (1996b) Disaster monitoring and mitigation. In: Singh RB (ed) Disasters, environment and development. Oxford & IBH Publishing Co., New Delhi, India, pp 31–62
Singh RB (2002) Urbanization and challenges to sustainable rural land use in India—comparative case study. In: Himiyama Y, Hwang M, Ichinose T (eds) Land-use changes in comparative perspective. Science Publishers, Enfield, USA, pp 53–62
Singh RB (2010) Designing and shaping global to local environmental education: challenges and opportunities—the Indian experience. In: Himiyama Y, Singh RB, Kanda F, Hindson J (eds) Glocal environmental education. Rawat Publications, Jaipur, India, pp 62–80
Singh RB, Kumar A (2018) Dynamics of land use and climate change in subhumid region of Rajasthan, India. In: Himiyama Y (ed) Exploring Sustainable Land Use in Monsoon Asia. Springer, Singapore, pp 137–156
Singh RB, Pandey BW (1996c) Hazard zone mapping and risk assessment analysis in the upper Beas Basin: cida-sici experience. In: Singh RB (ed) Disasters, environment and development. Oxford & IBH Publishing Co., New Delhi, India, pp 121–136
Singh RB, Fox J, Himiyama Y (eds) (2001) Land use and cover change. Science Publishers, Enfield, USA
Singh RB, Mishra P (2005) Forest-cover mapping of Uttaranchal State, India. In: Milanova E, Himiyama Y, Bicik I (eds) Understanding land-use and land-cover change in global and regional context. Science Publishers, Enfield, USA, pp 249–258
Singh RB, Anand S, Saluja V (2018) Low carbon resilient Delhi megacity for sustainable future earth. In: Himiyama Y (ed) Exploring sustainable land use in monsoon Asia. Springer, Singapore, pp 137–156
Werlen B (ed) (2015) Global sustainability—cultural perspectives and challenges for transdisciplinary integrated research. Springer, Switzerland

Yukio Himiyama
Hokkaido University of Education
Sapporo, Japan
himiyamay@kkd.biglobe.ne.jp

# Contents

# Chapter 1
# Water Resources, Livelihood Vulnerability and Management in Rural Desert Communities of Jaisalmer, India

**Pankaj Kumar, Shreya Ojha, Gaurav Kant Nigam, Anju Singh, and Manish Kumar Sinha**

**Abstract** Water is the lifeblood of natural life on our planet. Climate change and augmented use of the resource has goaded it to scarcity. Given its dearth, it is imperative to understand how water can be managed in order to sustain human populations. The concept, need and various approaches to manage water sustainably have been incorporated in this chapter. An empirical case study using mixed methods of research has been delivered to illustrate the Livelihood Vulnerability Index (LVI) of rural communities in the desert region of Jaisalmer, with special attention to vulnerability of people in terms of water. The LVI is constructed by using IPCC's 3 major dimensions: exposure, sensitivity and adaptive capacity. Seven major components and 42 sub-components have been included to conduct this research. Various local water management practices applied by the rural desert communities to sustain water have been briefly covered. Primary data collection through household and group surveys indicates how vulnerable these communities are present which indicates how their condition could worsen in the coming decades. Suggestions have been drawn to help ease the plight.

P. Kumar · S. Ojha (✉)
Department of Geography, Delhi School of Economics, University of Delhi, New Delhi, Delhi 110007, India
e-mail: sojha@ksu.edu

P. Kumar
e-mail: pkumar@geography.du.ac.in

G. K. Nigam
Subject Matter Specialist, Krishi Vigyan Kendra–Korba, Chhattisgarh, India

A. Singh
Department of Geography, Aditi Mahavidlaya, University of Delhi, New Delhi, Delhi 110039, India

M. K. Sinha
Department of Environmental & Water Resources Engineering, Chhattisgarh Swami Vivekanand Technical University Bhilai, Bhilai 491107, India
e-mail: sinha@lih.rwth-aachen.de

Department of Engineering Geology and Hydrogeology, RWTH Aachen University, 52074 Aachen, Germany

 
P. Kumar et al. (eds.), *Water Resources Management and Sustainability*, Advances in Geographical and Environmental Sciences,
https://doi.org/10.1007/978-981-16-6573-8_1

**Keywords** Vulnerability index · Climate change · Rural adaptation · Sustainable development · Water management · India

## 1.1 Introduction

Water is one of the most essential resources on our planet. We cannot imagine life thriving without this resource. It is an indispensable component of our ecosystem and the essence of our life. The very beginning of human civilization began across water bodies like the river valleys of Indus, Nile, Euphrates, Tigris, etc. and this pattern continues even today as several major urban areas are located on coastal areas, across water bodies. Water resource has acted as a base for cultural, economic and social development. The belief that water is abundant as it covers 70% of the earth's surface is covered by water. This is false, as only 2.5% of all water is freshwater and only 1% is accessible. The major source of freshwater is rainfall which occurs as a result of 'hydrological cycle' and the other sources of freshwater include underground water and water from the melting of glaciers. There are several uses of water apart from the drinking purpose. An average human uses about 40 L of water per day in rural areas and 150 L of water in urban areas.

Water is no more an unlimited resource. According to the United Nations, water use has increased more than twice the rate of population increase in the last century. It is estimated that about 3.9 billion (which would be 40% of the world's population in 2050) will live in severely stressed water basins and the limited resource will need to support about 9.7 billion people by 2050. However, it is estimated that population pressure is not the only cause of scarceness of water. It is the excessive use, wastage and lack of conservation that is driving the world towards extreme stresses. It is expected that from now to 2050, the demand for water will increase by 400% from the manufacturing sector and 130% from the household sector.

This water insecurity will be exacerbated by recurring draughts, climate change, lack of proper diversion and conservation of flood and deluge water. Due to the inaccessibility and lack of availability of fresh water, optimizing it remains a challenge in different locations of the world. With increased vulnerabilities of global climate change and irrational use of water, managing and conserving water for our future generations will become increasingly difficult. The future holds unprecedented challenges which are yet to be encountered and this calls for alternate and sustainable strategies to manage water.

## 1.2 Concepts and Need of Water Management

The importance of water in twenty-first century is analogous to the importance that oil had in the twentieth century. However, there have been several alternatives to oil like natural gas, shale gas, solar energy, wind energy, etc. but, there is no alternative to water. The issue is not its lack of adequate water for all, but its ludicrous and reckless overuse. Managing water resources is the key to a water-secure future. There is a need for systemic change in planning and water management.

Water resource management refers to the 'planning, development, review, assessment, decision making, distribution and optimum use of the resource keeping in mind the present limitation of the water and changing priorities of human use'. It is also referred to as a sub-set of water cycle management.

Water management is the management of water under a set of regulations and policies. Water was once an abundant resource available free of cost from natural sources, now it has become highly valuable due to increased droughts and overuse, to the level that it has become a priced precious commodity.

Water management lays a particular focus on how managing the resource lays an effect on the quality of life of both present and future generations. Managing water resources is inherently complex. It involves hydrological cycle, climate, humans, various ecosystems, plants, animals, etc. all these components are dynamic and interrelated. A change in any one of them may have a long-term impact which might be irreversible. Thus, several considerations need to be made while managing water.

In addition to the resource and the natural components, there are several other factors like various stakeholders and their interests, policies, politics, geographical boundaries and economic implications which need to be managed. Figure 1.1 shows the components involved in managing water resources.

Each one of us has a right to live in an environment of quality and have access to clean water for various purposes. The developed countries of the world have already developed a system of water management at all levels- national, provincial, local and domestic. However, in developing countries, this has to be done both through

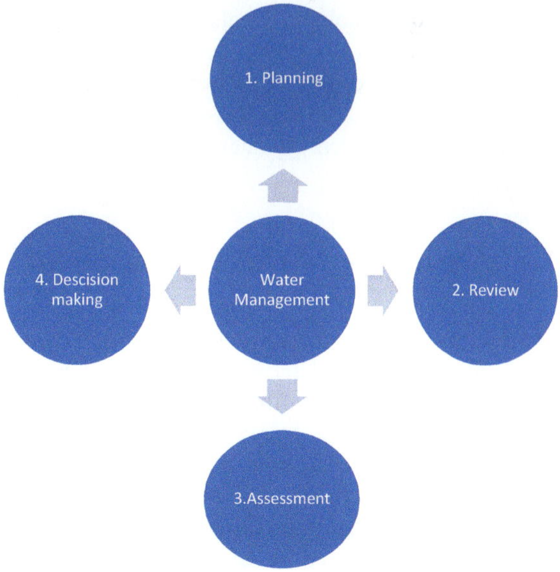


**Fig. 1.1** Components of Water Management. *Source* Prepared by author

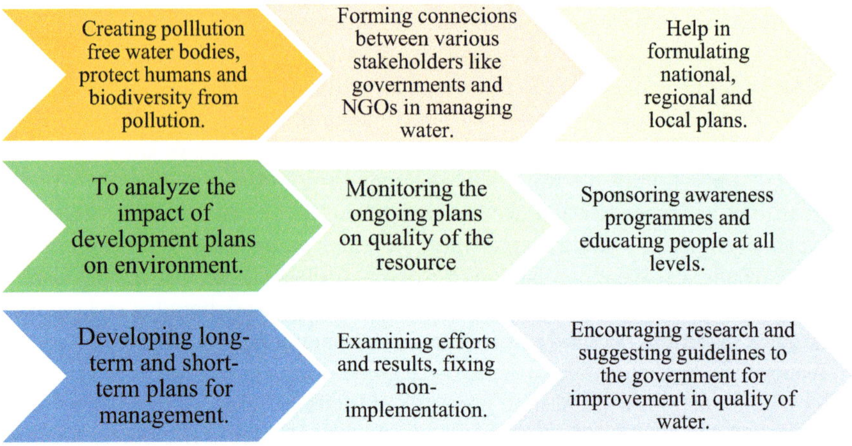


**Fig. 1.2** Diagram representing various aspects (steps and methods) to be considered in water management. *Source* Prepared by author

legislation as well as people's participation. Even today, large number of people in India directly depend on nature and its resources to sustain their livelihood.

Therefore, it is essential to understand the functions and interaction of physical and societal elements of the environment to apply this knowledge in sound management programmes while conserving water and culture. "Although humans have made efforts to better comprehend the resource, their efforts to manage it has fallen short of preventing its deterioration and exhaustion, due to their continuous indulgence in exploitation for immediate gains. The complex nature of problems that could be caused when the planet runs dry of its water, not only need imaginative and innovative specialists and technology, but also the people's participation. A rigorous process of 'planning, review, assessment, decision making and likewise, which is necessary for real-life situations of limited resources and changing priorities is what is meant by management".

The essential aspects of water management are depicted in Fig. 1.2.

## 1.3 Approaches in Water Management

The increase in the urgency of managing water has become a priority all over the world due to the imposition of serious consequences which have or could be experienced by humankind. "Any plan that envisages the management of the resource must be based on the fact that each area or region has certain characteristics and can tolerate a range of physical conditions to a limited extent. Each human activity affects the resource favourably or unfavourably".

We can make use of our surroundings by two methods, either by modifying an area or region to change its capabilities or by adopting our needs and demands to the capability of each region or area. The first method is much more intensive and is thus called, Intensive Management as it controls the physical features of the environment (example: changing the course of a river, constructing dams, channelling river water by modifications, etc.). Whereas, adopting our needs is a more sustainable, thoughtful, conscious and eco-friendly method of mutual co-existence while respecting nature for providing the resource (ex: waste disposal without overloading the water, rainwater harvesting, reuse of purified sludge water, storing rainwater runoff, etc.).

The other two approaches according to the World Conservation Strategy (1980) were: (i) Preservative Approach, according to this, humans should not disturb the natural system and should adjust according to it. However, since this is not applicable everywhere, imbalances of varied intensities could arise. (ii) Conservative Approach advocates that there should not be over-exploitation of the resources and conserving them is essential for sustainable development. Proper utilization and conservation are the prime objectives of water management. Figures 1.3 and 1.4 summarize different kinds of approaches used in water management.

Doxiadis (1977) has developed a science of planning settlements in balance with nature approach and has named it *ekistics*.

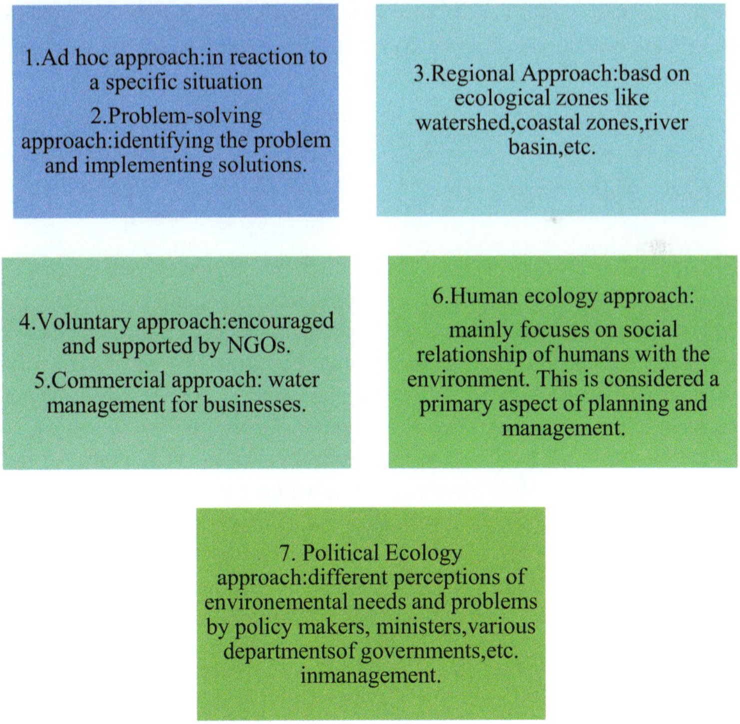


**Fig. 1.3** Approaches in Water Management. *Source* Prepared by author

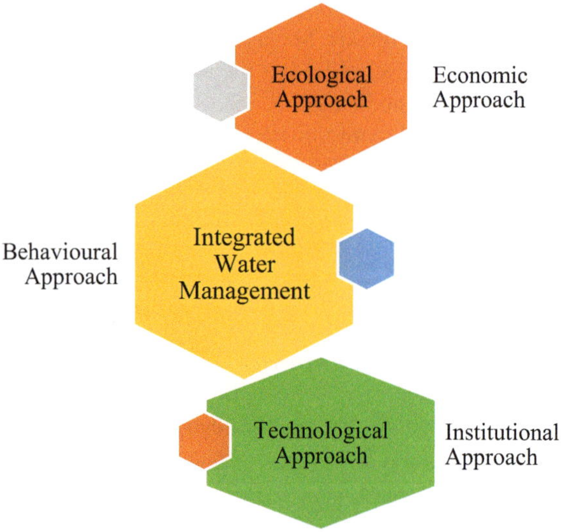


**Fig. 1.4** Various approaches in Management of Water. *Source* Prepared by author

Water management systems is a means to assist different bodies and industries to ensure that the quality and quantity of water is maintained according to policies and objectives. Wate management needs to be an integrated and proactive approach to ensure a balanced view across all functions.

### *1.3.1 Ecological Approach*

The ecological approach gained importance after Barrows proposed human ecology as a focus in geography in 1993. It is one of the new ways/approaches of managing natural resources that takes into account the entire ecosystem. This approach focuses on natural sciences and empirical methods. It studies how the environment influences the abundance and availability of a given resource and is a shared approach amongst hydrologists, geomorphologists, ecologists and biogeographers.

It stresses nature and limits of resources but is silent on human implications. It prioritizes ecological values over economic values. It balances recreational use, economic development and conservation of natural resources in a way that all the needs are met in a sustainable manner (ex: purification of water).

### *1.3.2 Economic Approach*

"The economics of natural resources and the environment provides a set of theories and conceptual tools to monitor, analyze, evaluate and regulate". Economics enables the creation of models for sustainable use of renewable natural resources such as 'water' which can be managed by economic tools of taxation, grants, subsidies, standards, permits and market rights.

### 1.3.3 Behavioural Approach

A behavioural approach for studying reactions to natural hazards and adaptation to floods was introduced by Gilbert White in 1945. This approach gained popularity as it added the 'human' dimension of 'behaviour' (decision making, practices and human activities in environment), to theories of interaction with environment. The behaviour of humans acts as a connecting link between ecological and social (human) systems. Thus, making it an essential component in managing natural resources.

Recognizing how the customs, actions, judgements and practices of humans are interlinked and affect the environment is the foremost stride in promoting sustainable resource management as it calls for attempts in remarkably changing behaviour (example: individual, community, regional, national and global efforts to conserve and manage water) and maintaining other behaviour patterns (example: application of traditional methods in conserving water, judicious use of water at individual, household and community level).

Factors that influence behaviour are knowledge, values, social norms, sociocultural factors, skills, economics, laws, policies and gender. The tools and methods of this approach are direct behavioural observation survey, interviews, community meetings, focus groups, matrices and contrastive analysis.

Behavioural interventions have proven effective in the field of resource conservation especially in water, energy resources. Mullainathan and World Bank argue that there is considerable scope for using behavioural insights to policy issues in the developing world, including those related to climate change and resource use.

### 1.3.4 Integrated Water Management

Integrated Water Resource Management is the proactive, holistic and system-based approach that integrated different aspect for resource management. Cairns and Crawford's (1991) definition: "Coordinated control, direction or influence of all human activities in a defined environmental system to achieve and balance the broadest possible range of short- and long-term objectives". Integrated Water Resource Management is better than other approaches in management because it is more inclusive, interconnected, goal oriented and strategic. Figure 1.5 is a representation of the dimensions included in Integrated Water Management.

**Fig. 1.5** Represents the various dimensions included while managing water through the IWRM method

### *1.3.5 Institutional Approach*

Institutional approach refers to an approach, typically used by an organization. It is an approach that is applied and proven to be effective in achieving goals of resource management. An Institution is the backbone of any resource management.

Institutions are responsible for regulating, maintaining policies and programmes regarding any resource. They play a role in regulating and implementing policies. It focuses on the analysis of formation of institutional arrangements in response to the changing needs and is biased to economic, political law and business administration institutions which emphasize the role of social political and economic organizations in determining economic events. Hence, the success and failure of water management are tied up with institutional structure, the pattern of agencies, laws and policies which pertain to water resource issues. (Ex: Upper Yamuna River Board, Central Water Commission New Delhi.)

### *1.3.6 Technological Approach*

Information technology mainly counts tools of data science, remote sensing and GIS, in natural resource management which have great potential in decision making. Resource data are to analyse and present the information needed to develop sustainable resource management programmes in different sectors. Remote sensing helps in data acquisition. GIS application has enlarged the horizon of data assimilation.

## 1.4 Case Study

Livelihood Vulnerability in Rural Desert Communities of Jaisalmer, India.

### *1.4.1 Study Area*

Jaisalmer district is the western most region of Rajasthan. It is bordered in the north and west by Pakistan, in the east by Jodhpur district and in south by Barmer district. It stretches between 25° 58′ 12.47″ to 28° 04′ 45.94″ north latitude and 69° 25′ 31.11″ to 72° 22′ 35.22″ east longitude and covers an area of approximately 38,487.2 $km^2$. Major part of this district does not have a systematic drainage system, so the whole district is part of an 'Outside' Basin. Jaisalmer district is divided into three blocks and three divisions, administratively. It has 639 towns and villages, out of which three are block headquarters. The three divisions are Pokhran, Fatehgarh and Jaisalmer. Rural area in this study is considered to be the countryside, away from the main city, where maximum population depended majorly on agricultural and animal rearing income and where services like health, education and sanitation were appalling (Plate 1.1).

### *1.4.2 Source of Data Collection*

The study is based on the primary data collected from field surveys, conducted in Jaisalmer, Pokhran and Fathegarh divisions of Jaisalmer district. Several villages like Amar Sagar, Deva, Baramsar, Ramdevri, Pokhran village, were included. Secondary data was collected from several literatures, University libraries and related government departments. The data was analysed and is presented in the form of tables, charts and graphs. For cartographic work, ArcGIS 10.5 was used. Livelihood Vulnerability Index (LVI) was used.

This index has been delineated to dispense a heuristic tool that can be applied by policy makers, government and non-government development organizations and social welfare practitioners to comprehend several social factors like health, demography, occupation and access to natural, economic and social resources, useful in analysing climate vulnerability at community or district level. The tool is pliable such that it can be modified according to the area of focus and the appropriate desideratum of the geographical area. Both, composite index and sectoral vulnerability scores can be separated to intervene by picking the potential dimensions, as required.

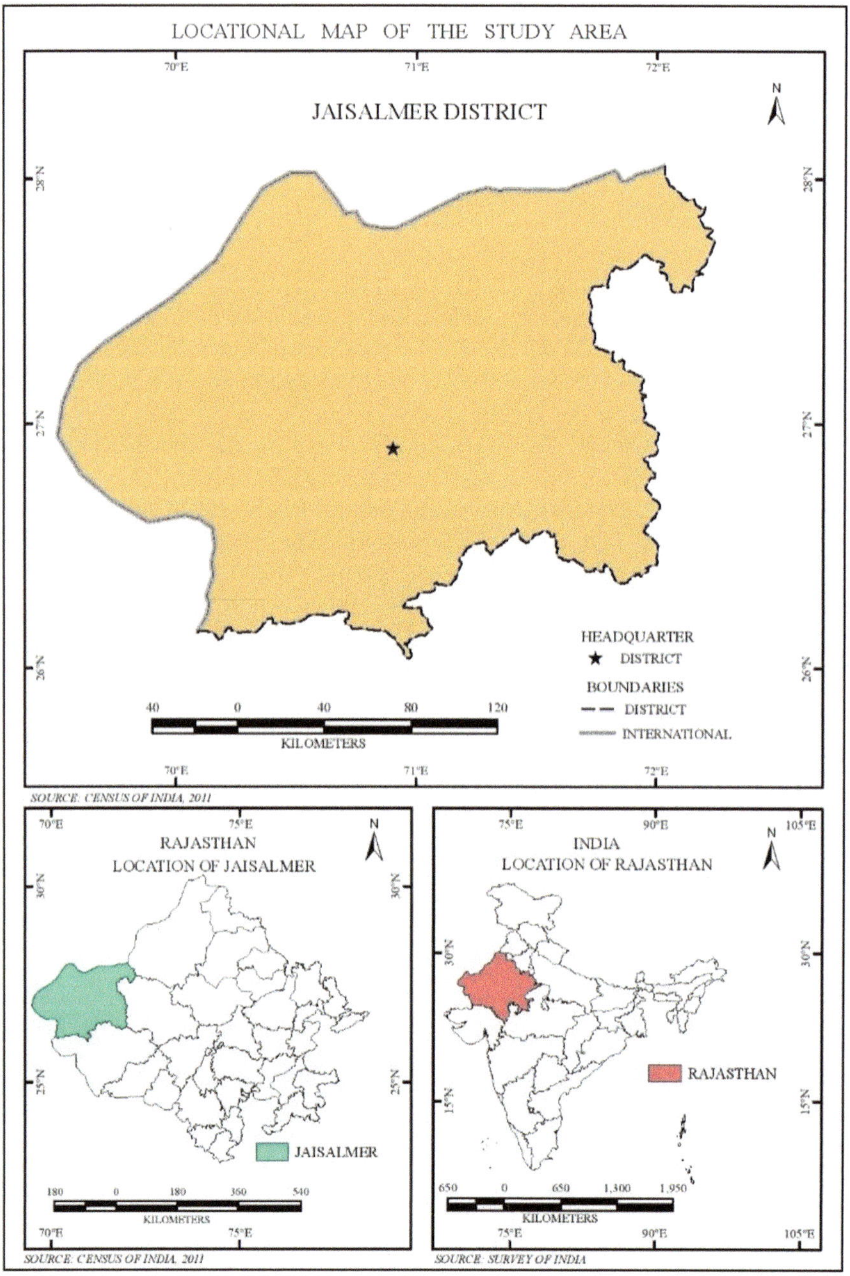


**Plate 1.1** Map representing location of the Study Area. *Source* Prepared by author using ArcGIS software

#### 1.4.2.1 Primary Data

Survey of 36 households was conducted to collect data on socio-demographics, livelihoods, social networks, health, food and water security, natural disasters and climate variability. It was then aggregated using a composite index and differential vulnerabilities were compared. This pragmatic approach can be used to monitor vulnerability, programme resources for assistance and/or evaluate potential programme/policy effectiveness in data-scarce regions by introducing scenarios into the LVI model for baseline comparison. A variety of methods were used for primary data collection such as questionnaire survey, focussed group interviews, tourist interviews and administrative individual interviews so as to obtain first-hand information, a general picture of the area. Later, proceedings towards specific issues of water, sanitation, health and other aspects were obtained. It was difficult to cover all villages due to paucity of time and financial constraints.

#### 1.4.2.2 Secondary Data

Vulnerability analysis involves the inclusion of varied techniques to interlink and assesses human-ecological interactions with their social and biophysical surroundings. Such assessments have been employed in several contexts, including "USAID's Famine Early Warning System (FEWS-NET) (USAID 2007), World Food Programme's Vulnerability Analysis Mapping tool for targeting food aid (World Food Programme 2007) and other geographic analyses that include data on biodiversity, health, poverty and globalization" (O'Brien et al. 2004; UNEP 2004).

This study attempts to quantify multidimensional concerns by substituting these issues in the form of indicators which are combined to create a composite index to include a diversified range of variables. Various methods have been applied to combine indicators of the Human Development Index like health, education, standard of living and life expectancy to depict the well-being of the communities of the surveyed region.

Livelihood Vulnerability Index (LVI) comprises exposure, sensitivity and adaptive capacity as its three major dimensions and it can be represented as a "single value index or broken down into its three major dimensions".

To evaluate the deviation of water allocation and use from a pre-established standard for Water Poverty Index, the Gap Method by Sullivan et al. (2002, p. 1204) was used. HDI and WPI are samples of the composite indices which were calculated using weighted averages of each indicator. Most of the indicators reckon on the IPCC's way of "defining vulnerability as a function of exposure, sensitivity and adaptive capacity" (IPCC 2001).

"Exposure is defined as the magnitude and duration of the climate-related exposure such as a drought or change in precipitation. Sensitivity is recognized to be the degree to which the system is affected by the exposure and adaptive capacity is the system's ability to withstand or recover from the exposure" (IPCC 2001).

Formulating the indicators involved the use of two approaches. Expressing LVI as a composite index comprising of seven major components was the first approach and aggregating the seven components into IPCC's 3 major dimensions (exposure, sensitivity and adaptive capacity) was the second approach. These approaches called for the necessity of collecting primary data through questionnaire surveys at household level in order to build the index. This approach of constructing the index can be used to present a framework in order to cluster and classify the indicators on the district level in order to make it explanatory for planning adaptation and developmental activities by policy makers.

Adoption of household data is advantageous as it steers away from the perils of using secondary data and fills a gap in knowledge as most vulnerability and climate models are enforced and exhibited at large scales. This approach could be essential to present precise projections at various levels that would be beneficial for community development IPCC (2007a, p. 443).

### *1.4.3 Techniques and Analysis*

Calculating the LVI: composite index approach:

LVI incorporates seven major components that are: "socio-demographic profile, livelihood strategies, social networks, health, food, water, water vulnerability, natural disasters and climate variability". Each of these components encompasses various sub-components or indicators.

"Even though the major components comprise of different number of sub-components, a balanced weighted average approach is used where each sub-component contributes equally to the overall index" (Sullivan et al. 2002). A straightforward, 'easy to apply' and 'comprehend' approach of applying equal weights to all major components has been done as we intend to develop an 'assessment tool' which can be accessible to a diverse set of users in resource deficient settings. The weighing scheme is simple and can be adjusted by future users as desired.

"Since each sub-component was measured on a different scale, it was essential to standardize each as an index. The equation utilized for the conversion was adapted from the one used in Human Development Index to calculate life expectancy index, which is a ratio of difference of actual life expectancy, a pre-selected minimum and range of pre-determined maximum and minimum life expectancy".

$$\text{Index}_{sd} = \frac{s_d - s_{\min}}{{}^1/_4 s_{\max} - s_{\min}} \tag{1.1}$$

where $s_d$ is the original sub-component for district $d$ and $s_{\min}$ and $s_{\max}$ are the minimum 1and maximum values, respectively, for each sub-component determined using data from both districts. For example, the 'average time to travel to reach the primary water source' sub-component ranged from 5 to 945 min in the district we surveyed. These minimum and maximum values were used to transform this indicator into a standardized index so it could be integrated into the water component of the LVI. For variables that measure frequencies such as the 'percent of households reporting having heard about conflicts over water resources in their community,' the minimum value was set at zero and the maximum at 100.

$$\text{Index}_{sv} = \frac{S_v - S_{\min}}{S_{\max} - S_{\min}},$$

Will be enabled if the value of indicators has a positive relationship with vulnerability.

$$\text{Index}_{sv} = \frac{S_{\max} - S_v}{S_{\max} - S_{\min}},$$

Max–min approach will be enabled if the above condition is not met. Where $S_v$ is the actual value of indicator in the series. $S_{\min}$ and $S_{\max}$ are the minimum and maximum values of indicators in the series. After each was standardized, the sub-components were averaged using Eq. 1.2 to calculate the value of each major component.

$$\frac{M_d{}^1\!/_4 \times \mathrm{P} ni^1\!/_4 1 index_s di}{N} \tag{1.2}$$

where, $M_d$ = one of the seven major components for district $d$ [Socio-Demographic Profile (SDP), Livelihood Strategies (LS), Social Networks (SN), Health (H), Food (F), Water (W), or Natural Disasters and Climate Variability (NDCV).

$\text{index}_{sd}i$ represents the sub-components, indexed by $i$, that make up each major component.

$n$ is the number of sub-components in each major component. Once values for each of the seven major components for a district were calculated, they were averaged using Eq. 1.3 to obtain the district level LVI:

$$\text{LVI}(\text{I}_d) = \frac{{}_d\text{¼}\; i\text{¼}1\; w M_i M_d i w_{SDP} SDP_d \text{þ}\; w_{LS} LS_d \text{þ}\; w_{SN} SN_d \text{þ}\; w_H H_d \text{þ}\; w_F F_d \text{þ} w_W W_d \text{þ} w_N NDCV_d}{wSDP \text{ þ } wLS \text{ þ } wH \text{ þ } wSN \text{ þ } wF \text{ þ } wW \text{þ} wNDC} \tag{1.3}$$

where $\text{LVI}_d$, the Livelihood Vulnerability Index for district $d$, equals the weighted average of the seven major components. The weights of each major component, $w_{Mi}$, are determined by the number of sub-components that make up each major component and are included to ensure that all sub-components contribute equally to the overall LVI (Sullivan et al. 2002). In this study, the LVI is scaled from 0 (least vulnerable) to 0.5 (most vulnerable).

### 1.4.4 Calculating the LVI-IPCC: IPCC Framework Approach

Exposure of the population studied is measured by the preparedness and awareness the rural communities received by the government authorities and the media, while climate variability is measured by the average standard deviation of the maximum and minimum monthly temperatures and monthly precipitation over a 6-year period. Adaptive capacity is quantified by the demographic profile of a district (ex: dependency ratio, average monthly income of households), the types of livelihood strategies employed (e.g. predominately agricultural, livestock diversification) and the strength of social networks (e.g. percent of residents assisting neighbours with chores, dependency on banking facilities, number of households attending social events). Last, sensitivity is measured by assessing the current state of a district's food, water security and health status. A special section of vulnerability of households to water has also been included. The same sub- components are outlined in Table 1.1 as well as Eqs. (1.1)–(1.3) were used to calculate the LVI-IPCC. The LVI-IPCC diverges from the LVI when the major components are combined. Rather than merge the major components into the LVI in one step, they are first combined according to the categorization scheme in Tables, using the following equation:

$$\mathrm{CF}_d ¼ n i ¼ 1 w_{Mi} M_{di} / w_M$$

where $\mathrm{CF}_d$ is an IPCC-defined contributing factor (exposure, sensitivity, or adaptive capacity) for district *d*, $M_{di}$ is the major component for district *d* Indexed by *i*, $w_{Mi}$ is the weight of each major component and '*n*' is the number of major components in each contributing factor. Once exposure, sensitivity and adaptive capacity were calculated, the three contributing factors were combined using the following equation:

$$\mathrm{LVI—IPCC}_d ¼ ð e_d — a_d Þ \, \mathrm{m} s_{d\ (FINAL\ CALCLUATION)} \qquad (1.4)$$

where $\mathrm{LVI–IPCC}_d$ is the LVI for district '*d*' expressed using the IPCC vulnerability framework, *e* is the calculated exposure score for district *d* (equivalent to the Natural Disaster and Climate Variability major component), '*a*' is the calculated adaptive capacity score for district *d* (weighted average of the Socio-Demographic, Livelihood Strategies and Social Networks major components) and '*s*' is the calculated sensitivity score for district *d* (weighted average of the Health, Food and Water major components). We scaled the LVI-IPCC from −0 (least vulnerable) to 1 (most vulnerable).

**Table 1.1** Major components and sub-components comprising the Livelihood Vulnerability Index (LVI) developed for Jaisalmer district

| Major components | Sub-components | Explanation of sub-components | Survey question | Source | Potential limitations |
|---|---|---|---|---|---|
| Socio-demographic profile | Dependency ratio | Proportion of the populace under 15 and more than 65 years old to the populace somewhere in the range of 19 and 64 years old | Could you please list the ages and sexes of every person who eats and sleeps in this house? | Adapted from Domestic Household Survey (DHS) (2006). Measure DHS: Model Questionnaire with Commentary | Large extended families; Confusion about who is a member of the household; Lack of birth certificates |
| | Percent of households where head of household has not attended school | Level of family units where the leader of the family reports that they have gone to below 10 long periods of school | What is your educational qualification? | | |
| | Average Monthly income in rupees | Average income less than Rs. 10,000 is recorded | What is your average monthly income? | | |
| Livelihood | Number of households with family members working in a different community | Average number of families that report at any rate one relative who works outside of the network for their essential work action | How many people in your family go to a different community to work? | Adapted from World Bank (1997). Household Questionnaire: Survey of Living Conditions, Uttar Pradesh and Bihar | Confusion regarding who is a member of the family; Does not count members of the family who previously worked outside of community; Confusion about what is "outside of the community." |

(continued)

**Table 1.1** (continued)

| Major components | Sub-components | Explanation of sub-components | Survey question | Source | Potential limitations |
|---|---|---|---|---|---|
| | Percent of households dependent solely on agriculture as a source of income | Percent of family units subordinate exclusively on agriculture as a wellspring of salary | Do you or someone else in your household raise animals?<br>Do you or someone else in your household grow crops? | Adapted from World Bank (1997) | Survey only asked about the three primary sources of income for families in the area |
| | Average Agricultural Livelihood Diversification Index (range: 0.20–1) | The inverse of (the number of agricultural livelihood activities +1) reported by a household, e.g. A household that farms, raises animals and collects natural resources will have a Livelihood Diversification Index = $1/(3 + 1) = 0.25$ | Same as above | Adapted from DHS (2006) | |
| | Percentage of unirrigated land | Percentage of land that depends on rainfall for irrigation | Is the land rainfed? | Adapted from DHS (2006) | |

(continued)

**Table 1.1** (continued)

| Major components | Sub-components | Explanation of sub-components | Survey question | Source | Potential limitations |
|---|---|---|---|---|---|
| Social Networks | Percent of households that receive assistance | Percentage of households that received help | In the past month, did relatives or friends help you and your family: (e.g. Get medical care or medicines, Sell animal products or other goods produced by family) | Adapted from DHS (2006) | Confusion about who is family (immediate) and who is a relative (extended); Reliance on self-reported types of help/support |
| | Percent households that attend social events | Percentage of number of households that visit their relatives (last 12 months) | Have you attended any social events in past 12 months? | Adapted from World Bank (1997) | |
| | Percent of households that have not gone to their local government for assistance in the past 12 months | Percentage of households that reported that they have not asked their local government for any assistance in the past 12 months | In the past 12 months, have you or someone in your family gone to your community leader or bank for help? | Adapted from WHO/RBM (2003) | Reliance on self-reported money exchanges; Does not consider exchange of non-monetary goods Reliance on self-reported visits to government; Recall bias (more likely to remember going to government for dire issues |
| Migration | Percent of households where people have migrated | Percentage of households that have at least one migrant | Has anyone from your family migrated for employment or education? | Developed for the purposes of this questionnaire | |

(continued)

**Table 1.1** (continued)

| Major components | Sub-components | Explanation of sub-components | Survey question | Source | Potential limitations |
|---|---|---|---|---|---|
| | Percent of households where people migrate for employment | Percentage of people who could not find job in their home town and had to move elsewhere | Same as above | Developed for the purposes of this questionnaire | |
| | Percent of households where people migrate for education | Percent of households where at least one member has gone out of village to study | Same as above | Developed for the purposes of this questionnaire | |
| | Monthly remittances received from migrant workers (less than Rs. 10,000) | Percentage of people who received remittances | Do you receive any remittances? How much? | Developed for the purposes of this questionnaire | Hesitation in disclosing the correct amount |
| Health | Average time to reach hospital (minutes) | Average time it takes the households to get to the nearest health facility | How long does it take you to get to a health facility? | Developed for the purposes of this questionnaire | No watches; Subjective estimates of travel time |
| | Percent of households with family members with chronic illness | Percentage of households that report at least one family member with chronic illness. Chronic illness was defined subjectively by respondent | Is anybody in your family chronically ill (they get sick very often)? | Developed for the purposes of this questionnaire | "Chronically ill" was subjectively defined by respondent |

(continued)

**Table 1.1** (continued)

| Major components | Sub-components | Explanation of sub-components | Survey question | Source | Potential limitations |
|---|---|---|---|---|---|
| | Percentage of households with family members going to doctor for allopathic treatment | Percentage of households where all family members take allopathic treatment | Do all family members resort to allopathic treatment or still believe in taking homoeopathy/Traditional medication? | Developed for the purposes of this questionnaire | |
| | Average households who do not have proper toilet facilities | Average number of households that do not have a well-built pucca toilet and who practice open defecation | Does your house have a toilet inside its premises? | Developed for the purposes of this questionnaire | |
| Food | Percent of households dependent on family farm for food | Percentage of households that get their food primarily from their personal farms | Where does your family get most of its food? | Adapted from World Bank(1997) | No specification regarding the year in question |
| | Average Crop Diversity Index (range: >0 to 1) | The inverse of (the number of crops grown by a household +1). For, e.g. A household that grows guar, bajra, jow and millet will have a Crop Diversity. Index = 1/(4 + 1) = 0.20 | What kind of crops does your household grow? | Developed for the purposes of this questionnaire | |

(continued)

**Table 1.1** (continued)

| Major components | Sub-components | Explanation of sub-components | Survey question | Source | Potential limitations |
|---|---|---|---|---|---|
| | Percent of households that do not save crops | Percentage of households that do not save crops from each harvest | Does your family save some of the crops you harvest to eat during a different time of year? | Developed for the purposes of this questionnaire | |
| | Percent of households that do not save seeds | Percentage of households that do not have seeds from year to year | Does your family save seeds to grow the next year? | Developed for the purposes of this questionnaire | No specification regarding the year in question |
| | Average number of months households struggle to find food (range: 0–12) | Average number of months households struggle to obtain food for their family | Does your family have adequate food the whole year, or are there times during the year that your family does not have enough food? How many months a year does your family have trouble getting enough food? | Developed for the purposes of this questionnaire | |
| | Percent of households that depend on government subsidy and agriculture for food | Percentage of households where the family is solely dependent on what they cultivate and on government's subsidy for food | Do you consume what you grow and depend on government subsidies? | Adapted from DHS (2006) | |

(continued)

**Table 1.1** (continued)

| Major components | Sub-components | Explanation of sub-components | Survey question | Source | Potential limitations |
|---|---|---|---|---|---|
| | Percentage of people who consume vegetarian food only | Percentage of people who are unable to afford non-vegetarian food to fulfil their protein requirement | Do you prefer vegetarian food over non-vegetarian food? | Adapted from DHS (2006) | |
| Water | Percent of households that utilize a natural water source | Percentage of households that report a well, an oasis, or hole as their primary water source | Where do you collect your water from? | Adapted from World Bank (1997) | Confusion when families have multiple water sources |
| | Average time to water source (minutes) | Average time it takes the households to travel to their primary water source | How long does it take to get to your water source? | Developed for the purpose of this questionnaire | No watches; Subjective estimates of travel time; Different family members collect water |
| | Percent of households that do not have a consistent water supply | Percent of households that receive irregular supply of water | Do you receive water every day? | | Recall bias (more likely to remember several consecutive days of water shortage) |
| | Inverse of the average number of litres of water stored per household (range: >0 to 1) | Average amount of water saved by households | What containers do you usually store water in? How many litres? | | Lack of understanding of containers |

(continued)

**Table 1.1** (continued)

| Major components | Sub-components | Explanation of sub-components | Survey question | Source | Potential limitations |
|---|---|---|---|---|---|
| Natural disasters and climate variability | Percent of households that did not receive a warning about the pending natural disasters | Percentage of households that did not receive a warning about the most severe flood, drought and cyclone event in the past 6 years | Did you receive a warning about the flood/cyclone/drought before it happened? | Adapted from Williamsburg Emergency Management (2004) | Subjective definition of "warning." |
| | Percent of households with an injury or death as a result of the most severe natural disaster in the past 6 years | Percentage of households that reported either an injury to or death of one of their family members as a result of the most severe flood, drought, or cyclone in the past 6 years | Was anyone in your family injured in the flood/cyclone drought? Did anyone in your family die during the flood/cyclone/drought? | Household Natural Hazards Preparedness Questionnaire | Recall bias (severe injuries are most likely to be remembered) |
| | Mean standard deviation of the daily average maximum temperature by month | Standard deviation of the average daily maximum temperature by month between 1998 and 2003 was averaged for each province | District data; weather station based in the state capital | Adapted from Williamsburg Emergency Management (2004) | Reliance on average data; Short time period |

(continued)

**Table 1.1** (continued)

| Major components | Sub-components | Explanation of sub-components | Survey question | Source | Potential limitations |
|---|---|---|---|---|---|
| | Mean standard deviation of the daily average minimum temperature by month | Standard deviation of the average daily minimum temperature by month between 1998 and 2003 was averaged for each province | | Developed for the purposes of this questionnaire | Reliance on average data; Short time period |
| | Mean standard deviation of average precipitation by month | Standard deviation of the average monthly precipitation between 1998 and 2003 was averaged for each province | | Developed for the purposes of this questionnaire | Reliance on average data; Short time period |
| Water Vulnerability | Percentage of households dependent on community wells | Percentage of households who have only community well as their source of drinking water | What is the source of your drinking water? | Instituto Nacional de E statistica (2007) | Confusion due to recall bias |
| | Percentage of households dependent on pipeline (tapped water supply) | Percentage of households that receive regular tap water | Do you receive regular tap water? | Developed for the purpose of this questionnaire | |
| | Percentage of households dependent on water tanker for drinking purpose | Percentage of households that have the capacity to buy water delivered by tankers | Do you purchase tank water for drinking? | | |

(continued)

**Table 1.1** (continued)

| Major components | Sub-components | Explanation of sub-components | Survey question | Source | Potential limitations |
|---|---|---|---|---|---|
| | Percentage of households dependent on water only from Indira Gandhi Canal | Percentage of households that are situated near the IGC and use its water on daily basis | Do you receive water from Indira Gandhi Canal? Do you use this water source or depend on other sources? | | |
| | Percentage of households receiving water from more than two sources | Percentage of households where people receive water from more than two water sources available. (ex: one for drinking and other for domestic purposes) | Do you receive water from multiple sources? | | |
| | Average distance from water source (in kilometres) | Average distance the people from households need to travel in order to fetch water | How far do you travel in search of water? | | |
| | Percentage of households where women and children travel to water sources | Percentage of households where women and their children travel in groups to fetch water for the daily requirements of living | Who goes to fetch water from your family? | Developed for the purpose of this questionnaire | Men travelled in certain areas, however mostly it was women and children. Hesitant in replying |

(continued)

**Table 1.1** (continued)

| Major components | Sub-components | Explanation of sub-components | Survey question | Source | Potential limitations |
|---|---|---|---|---|---|
| | Average months of water scarcity | Average number of months during which households face maximum scarcity of water | Do you face water scarcity? If yes, for how many months? | Developed for the purpose of this questionnaire | Exaggerated responses |
| | Percentage of households that do not have water storing facilities | Percentage of households that do not have an underground or rooftop tank to save water and only save in drums | Do you have underground/rooftop water tank to save water? If not, how do you store it? | Developed for the purpose of this questionnaire | |
| | Percentage of households that drink impure water | Percentage of people who store rainwater in tanks and consume it directly without purifying it | Do you boil/purify water before drinking? | Developed for the purpose of this questionnaire | |
| | Average decline in groundwater level | Percentage of households that complained about increased water table depth | Has the depth of underground water availability increased? | Developed for the purpose of this questionnaire | |

(continued)

**Table 1.1** (continued)

| Major components | Sub-components | Explanation of sub-components | Survey question | Source | Potential limitations |
|---|---|---|---|---|---|
| | Percentage of households that cannot afford to purchase Jar/Jerry can water | Percentage of households that cannot afford to purchase jar water during social event like festivals or marriages | Do you purchase jar water in social gatherings? | Developed for the purpose of this questionnaire | |
| | Percentage of households that cannot pay for clean water for all activities | Percentage of households that cannot afford to pay for a clean supply of water | Do you pay for the water you use? | Developed for the purpose of this questionnaire | |
| | Percent of households who struggle to get clean water by paying | Percentage of households who somehow manage to pay for a clean supply of water | How easy is it to get a clean supply of water by paying for it? | Developed for the purpose of this questionnaire | |
| | Percentage of households where people suffer from water borne health problems | Percentage of households where family members face health problems due to consumption of impure water | Has the consumption of impure water caused any health issues to any family member? | Developed for the purpose of this questionnaire | Subjective definition of health problems |
| | Percent of households that received water polluted with fluoride/nitrate/arsenic | Percentage of households that receive water infested with a high amount of chemicals | Does the water you consume contain any chemicals? | Developed for the purpose of this questionnaire | |

*Source* Prepared by author

**Table 1.2** Categorization of significant segments into contributing elements from the IPCC (Intergovernmental Panel on Climate Change) vulnerability definition for computation of LVI-IPCC

| IPCC contributing factors to vulnerability | Major components |
|---|---|
| Exposure | Natural disaster and climate variability |
| Adaptive capacity | Socio-demographic profile, Livelihood strategies, Social Networks |
| Sensitivity | Migration, food, health, water security |

### *1.4.5 Results*

Each interview lasted 30 min on average and was carried out in Hindi language as it was the language common to the households of the region. Respondents were natives of the villages of Amar Sagar, Deva, Baramsaar and Ramdevri (Pokhran) in Jaisalmer. Interviewer was a native speaker of Hindi language and is trained in sample design, survey technique and confidentiality protocol.

Table 1.2 describes the major components and contributing factors of vulnerability.

Table 1.3 shows the maximum and minimum calculated LVI based on the sub-components of the major components described in Table 1.2.

Table 1.4 represents indexed major components and sub-components of the contributing factors.

Table 1.5 depicts the overall LVI of Jaisalmer.

### *1.4.6 Inference of the Tabular Data*

#### 1.4.6.1 Adaptive Capacity

**Socio-demographic profile**: the dependency ratio of the rural communities of the desert region of Jaisalmer was high (0.715) and it showed a higher proportion of young children below the age of 15 as compared to older population (above 65). Rural households where the population is not educated above 10th standard was very high (0.86) which shows high vulnerability in terms of education level. Most of the respondents had not completed high school and dropped out after 8 years of formal schooling. The illiteracy was as high as 86.11% as compared to the state average of approximately 27.01%. The average reported age of the respondents was 50 years old (males) and 45 years old (females). The average monthly income of most of the households was less than Rs. 10,000 and it shows a high vulnerability (0.805). Most of the families were joint in nature with about 10 members in one household.

**Table 1.3** Livelihood Vulnerability Index (LVI) sub-components values and minimum and maximum sub-component values for Jaisalmer district, Rajasthan, India

| Major Component | Sub-component | Units | Jaisalmer | Maximum value in Jaisalmer | Minimum value in Jaisalmer |
|---|---|---|---|---|---|
| Socio-demographic profile | Dependency ratio | Ratio | 0.479 | 421 | 202 |
| | Percentage of households where head of the house is not educated till high school | Percent | 86.11 | 100 | 0 |
| | Percentage of households where average income is less than Rs. 10,000 | Percent | 80.5 | 100 | 0 |
| Livelihood strategies | Average number of households with family members working in a different community | Percent | 0.6136 | 100 | 0 |
| | Percent of households dependent solely on agriculture as a source of income | Percent | 52.77 | 100 | 0 |
| | Average agricultural livelihood diversification index (livestock) | 1/#Livelihoods | 0.78 | 12 | 9.361 |
| | Percentage of unirrigated land | Percent | 74.07 | 100 | 0 |
| Social networks | Percentage of households that received assistance | Percent | 19.44 | 100 | 0 |
| | Percentage of households that do have not attended any social events in 12 months | Percent | 30.33 | 100 | 0 |
| | Percent of households that do not go to the government for assistance | Percent | **77.77** | 100 | 0 |
| Migration | Percent of households where people have migrated | Percent | 19.44 | 100 | 0 |

(continued)

**Table 1.3** (continued)

| Major Component | Sub-component | Units | Jaisalmer | Maximum value in Jaisalmer | Minimum value in Jaisalmer |
|---|---|---|---|---|---|
| | Percent of households where people migrate for employment | Percent | 16.66 | 100 | 0 |
| | Percent of households where people migrate for education | Percent | 3 | 100 | 0 |
| | Monthly remittances received from migrant workers (less than Rs. 10,000) | Percent | 16.66 | 100 | 0 |
| Health | Average time to reach hospital (minutes) | Minutes | 1240 | 34.44 | 5 |
| | Percent of households with family members with chronic illness | Percent | 11 | 100 | 0 |
| | Percentage of households with family members going to doctor for allopathic treatment | Percent | 72 | 100 | 0 |
| | Average households who do not have proper toilet facilities | 1/# toilets | 25 | 100 | 0 |
| Food | Percent of households dependent on family farm for food | Percent | 41.66 | 100 | 0 |
| | Average Crop Diversity Index (range: >0 to 1) | 1/#crops | 0.221 | 1 | 0.16 |
| | Percent of households that do not save crops | Percent | 36.11 | 100 | 0 |
| | Percent of households that do not save seeds | Percent | 30.5 | 100 | 0 |
| | Average number of months households struggle to find food (range: 0–12) | Months | 0.51 | 9 | 3 |
| | Percent of households that depend on government subsidy for food | Percent | 30.55 | 100 | 0 |

(continued)

Table 1.3 (continued)

| Major Component | Sub-component | Units | Jaisalmer | Maximum value in Jaisalmer | Minimum value in Jaisalmer |
|---|---|---|---|---|---|
| | Percentage of people who consume vegetarian food only | Percent | 50 | 100 | 0 |
| Water | Percent of households that utilize a natural water source | Percent | 33.33 | 100 | 0 |
| | Average time taken to reach water source | Minutes | 945 | 26.25 | 5 |
| | Percent of households that do not have a consistent supply of water | Percent | 61.11 | 100 | 0 |
| | Inverse of the average number of litres of water stored per household | 1/Litres | 37.5 | 40 | 10 |
| Water vulnerability | Percentage of households dependent on community wells | Percent | 13.8 | 100 | 0 |
| | Percentage of households dependent on pipeline (tapped water supply) | Percent | 16.66 | 100 | 0 |
| | Percentage of households dependent on water tanker for drinking purpose | Percent | 22.22 | 100 | 0 |
| | Percentage of households dependent on water only from Indira Gandhi Canal | Percent | 8 | 100 | 0 |
| | Percentage of households receiving water from more than two sources | Percent | 41.66 | 100 | 0 |
| | Average distance from water source (in kilometres) | 1/#distance | 1.5 | 2 | 0 |
| | Percentage of households where women and children travel to water sources | Percentage | 68.9 | 100 | 0 |

(continued)

**Table 1.3** (continued)

| Major Component | Sub-component | Units | Jaisalmer | Maximum value in Jaisalmer | Minimum value in Jaisalmer |
|---|---|---|---|---|---|
| | Average months of water scarcity | 1/#months | 2.08 | 3 | 1 |
| | Percentage of households that do not have water storing facilities | Percent | 16.66 | 100 | 0 |
| | Percentage of households that drink impure water | Percent | 83.33 | 100 | 0 |
| | Average decline in groundwater level | 1/#groundwater level | 11.91 | 15 | 9 |
| | Percentage of households that cannot afford to purchase Jar/Jerry can water | Percent | 66.6 | 100 | 0 |
| | Percentage of households that cannot pay for clean water for all activities | Percent | 38.88 | 100 | 0 |
| | Percent of households who struggle to get clean water by paying | Percent | 61.11 | 100 | 0 |
| | Percentage of households where people suffer from water borne health problems | Percent | 52.7 | 100 | 0 |
| | Percent of households that received water polluted with fluoride/nitrate/arsenic | Percent | 58.33 | 100 | 0 |
| Natural disasters and climatic variability | Percentage of households that did not receive a warning about an upcoming natural disaster | Percent | 72.2 | 100 | 0 |
| | Percent of households with an injury or death due to natural disaster | Percent | 13.8 | 100 | 0 |

(continued)

**Table 1.3** (continued)

| Major Component | Sub-component | Units | Jaisalmer | Maximum value in Jaisalmer | Minimum value in Jaisalmer |
|---|---|---|---|---|---|
| | Mean standard deviation of monthly average of maximum temperature (2011–2019) | Celsius | 1.29 | 1.54 | 1.01 |
| | Mean standard deviation of monthly average minimum temperature (2011–2019) | Celsius | 0.99 | 1.46 | 0.64 |
| | Mean standard deviation of monthly average precipitation (2011–2019) | Millimetres | 19.92 | 73.76 | 3.82 |
| | Mean standard deviation of monthly wind speed (2011–2019) | km/h | 0.59 | 0.93 | 0.33 |

*Source* Made by author. Source: prepared by author on the basis of primary survey data

**Livelihood Strategies**: the rural households showed a high vulnerability on livelihood strategies component (0.666). 52.27% households depend solely on agricultural income and most of them depend on both agriculture and livestock. The rural households owned a large variety of domestic animals like cattle, sheep, dogs, goats and a few households owned camels as well which shows a diversification index of 0.78. Every household owned at least nine animals. A large number of households owned land of up to 10 bighas. Large land owners owned about 52 bigha land and small land owners had about two bigha land.

The agricultural land is mostly rainfed in this region and people lack the means to irrigate their fields due to scarcity of water in the arid region. This makes them highly dependent on rainfall for watering their fields (0.747) only 25.94% households have regularly well irrigated farms. Apart from them, a major chunk of population (61.3%) comprised tenants as well who owned no land and worked on the agricultural farms of others. Most of their income was generated through agriculture related work for which they worked in different communities.

**Social Networks**: The social networks of the rural households seem to be strong with an overall vulnerability of 0.424. A large number of households depend on informal means of receiving and lending money/assistance. The households have a well-knit network wherein they help each other in times of need in medication, lending money, food, cattle. The vulnerability in this aspect is low at 0.194. The households are very

**Table 1.4** Indexed major components and sub-components and overall LVI of Jaisalmer

| Sub-components | Jaisalmer | Major components | Jaisalmer |
|---|---|---|---|
| Dependency ratio | 0.479 | Socio-demographic profile | 0.715 |
| Percent of households where head of household has not attended school | 0.861 | | |
| Percent of households where monthly income is less than Rs. 10,000 | 0.805 | | |
| Percent of households with family members working in a different community | 0.613 | Livelihood strategies | 0.666 |
| Percent of households dependent solely on agriculture as a source of income | 0.527 | | |
| Average agricultural Livelihood Diversification Index (livestock) | 0.78 | | |
| Percentage of unirrigated land | 0.747 | | |
| Percentage of households that receive assistance | 0.194 | Social networks | 0.425 |
| Percentage of households that have not attended any social events in 12 months | 0.305 | | |
| Percent of households that have not gone to their local government for assistance in 12 months | 0.777 | | |
| Percent of households where people have migrated | 0.194 | Migration | 0.139 |
| Percent of households where people migrate for employment | 0.166 | | |
| Percent of households where people migrate for education | 0.03 | | |
| Monthly remittances received from migrant workers (less than Rs. 10,000) | 0.166 | | |
| Average time to reach health facility | 0.84 | Health | 0.669 |
| Percent of households with family members with chronic illness | 0.305 | | |
| Percentage of households with family members going to doctor for allopathic treatment | 0.72 | | |

(continued)

**Table 1.4** (continued)

| Sub-components | Jaisalmer | Major components | Jaisalmer |
|---|---|---|---|
| Average households who do not have proper toilet facilities | 0.25 | | |
| Percent of households dependent solely on family farm for food | 0.416 | Food | 0.268 |
| Average number of months households struggle to find food | 0.513 | | |
| Average Crop Diversity Index | 0.341 | | |
| Percent of households that do not save crops | 0.366 | | |
| Percent of households that do not save seeds | 0.55 | | |
| Percent of households that depend on government subsidies for food | 0.305 | | |
| Percent of households that consume only vegetarian food | 0.5 | | |
| Percent of households that utilize natural sources of water | 0.33 | Water | 0.676 |
| Average time taken to reach water source | 0.85 | | |
| Percent of households that do not have a consistent water | 0.611 | | |
| Average water stored in litres | 0.916 | | |
| Percent of households that did not receive warning about an upcoming disaster | 0.722 | Natural disasters and climate variability | |
| Percent of households with an injury or death as a result of natural disaster | 0.138 | | |
| Mean standard deviation of monthly average of average maximum daily temperature (years: 2011–2019) | 0.528 | | |
| Mean standard deviation of monthly average of average minimum average temperature (years: 2011–2019) | 0.426 | | |
| Mean standard deviation of monthly average precipitation | 0.230 | | |
| Mean standard division of monthly wind speed | 0.433 | | |

(continued)

**Table 1.4** (continued)

| Sub-components | Jaisalmer | Major components | Jaisalmer |
|---|---|---|---|
| Percentage of households dependent on community wells | 0.138 | Water Vulnerability | 0.452 |
| Percentage of households dependent on pipeline (tapped water supply) | 0.166 | | |
| Percentage of households dependent on water tanker for drinking purpose | 0.222 | | |
| Percentage of households dependent on water only from Indira Gandhi Canal | 0.08 | | |
| Percentage of households receiving water from more than two sources | 0.416 | | |
| Average distance from water source (in km) | 0.726 | | |
| Average months of water scarcity | 0.541 | | |
| Percentage of households where women and children travel to water sources | 0.689 | | |
| Percentage of households that do not have water storing facilities | 0.166 | | |
| Percentage of households that drink impure water | 0.833 | | |
| Average decline in groundwater level | 0.486 | | |
| Percentage of households that cannot afford to purchase Jar/Jerry can water | 0.666 | | |
| Percentage of households that cannot pay for clean water for all activities | 0.388 | | |
| Percent of households who struggle to get clean water by paying | 0.611 | | |
| Percentage of households where people suffer from water borne health problems | 0.527 | | |
| Percent of households that received water polluted with fluoride/nitrate/arsenic | 0.583 | | |

*Source* Prepared by author

**Table 1.5** Calculations showing LVI-IPCC for Jaisalmer district

| Contributing factors | Major components for Jaisalmer district | Major component values for Jaisalmer district | Number of sub-components per major component | Contributing factor values | LVI-IPCC VALUE for Jaisalmer |
|---|---|---|---|---|---|
| Adaptive capacity | Socio-demographic profile | 0.715 | 3 | 0.607 | |
| | Livelihood strategies | 0.665 | 4 | | |
| | Social Networks | 0.425 | 3 | | |
| | Migration | 0.139 | 4 | 0.430 | |
| Sensitivity | Health | 0.480 | 4 | | 0.083 |
| | Food | 0.374 | 7 | | |
| | Water | 0.676 | 4 | | |
| | Water vulnerability | 0.452 | 16 | | |
| Exposure | Natural disasters and climate variability | 0.413 | 6 | 0.412 | |

*Source* Prepared by author

social and regularly attend all festivals, marriages and auspicious events together. The ones who have not attended any functions are low 0.305.

These sub-categories show that the people are not left on their own terms when they are in need. Households have mostly resorted to informal means of borrowing and lending money and have avoided the formal banking credit and loan facilities even though 63% of the respondents owned bank accounts, they do not use it. They depend on their immediate relatives or friends for money due to which they show a high vulnerability of 0.777 due to lack of possession of a formal bank account. They have not approached their local government for assistance and borrow and receive in-kind assistance. Total adaptive capacity values are represented in Table 1.6.

**Table 1.6** LVI-IPCC contributing factors to vulnerability for rural communities of Jaisalmer

| IPCC contributing factors to vulnerability | Jaisalmer |
|---|---|
| Exposure | 0.412 |
| Adaptive capacity | 0.607 |
| Sensitivity | 0.430 |
| LVI-IPCC | 0.083 |

*Source* Prepared by author
The LVI-IPCC is on a scale from −1 (least vulnerable) to 1 (most vulnerable)

#### 1.4.6.2 Sensitivity

**Migration**: the rural households of Jaisalmer have reported low numbers of people migrating to other districts for work and education. The overall vulnerability is 0.139. The people who migrated for work show low vulnerability as only 19.44% people have moved out of the village. 16.6% people migrated for job opportunities and 3% for education. The monthly remittances they received was very less and only a few of them received them on a regular basis.

**Health**: Jaisalmer households reported travelling an average of 35 min and a high of 50 min to reach health care facilities. Adding all indicators, the vulnerability of the households in terms of health is high at 0.667 vulnerability of average time taken to reach health facility is 0.84. 61% of the households reach their nearest community hospital within 30 min. 27% people take about 45 min to 1 h and only 11.11% take over an hour.

Not many households have reported chronic illness which makes their vulnerability a medium of 0.305. Only 30.5% of the respondents had at least one family member who was chronically ill. However, the members of the household who were suffering from diabetes, high blood pressure and joint pains were over 65 years of age and had age-related health issues. Only one female member from a household in Baramsar village reported TB (tuberculosis). Cases of malaria have been reported in areas through which Indira Gandhi Canal passes. Most of the people resorted to allopathic means of medication. About 25% of the households surveyed did not have a toilet/ proper sanitation facility inside their house. This shows that the goal of hundred percent open defecation free has not been achieved, however, there has been a drastic improvement in the number of households with toilets inside their premises.

**Food**: the rural households of Jaisalmer reported struggling to find adequate food for their families (0.513). The struggle was maximum after the cropping season ended. About 41.6% families rely solely on agriculture for income. The average cropping season in Jaisalmer lasts for 3 months during the monsoon season after which the families tend to consume what they grow or find alternative means of food such as meat, meat products. However, most of the households depend on vegetarian food (0.5). Farmers grow about an average of three crops at least, some grow five. The most common crops found in this region are guar, pearl millets (bajra), sorghum (jowar), sesamum, groundnuts, aloe vera, wheat and barley are rarely grown in rabi season. Therefore, the vulnerability in terms of crop diversification is low 0.341 which means that the farmers grow at least three types of crops. 36.6% percentage of farmers do not save crops and 55% do not save seeds for the next season which makes them vulnerable more as compared to those who save. This shows that they buy new seeds for the next season and do not consume crops that they produce. There is a certain class of households, the landless and the tenants who only depend on the government's subsidy for food (0.035). They receive rice at a price of Rs. 3/kg and wheat at Rs. 2/kg. The sensitivity of the rural communities is represented in Table 1.6.

#### 1.4.6.3 Exposure

Natural Disaster and Climate Variability: Jaisalmer has reported several dust storms. The drought-like condition prevails all year round. The percentage of households reported to receive no warning about any upcoming dust storm or drought was 72.22%. There were very few cases of disaster related injuries or deaths (13.8%). Households reported that they faced extremely strong summer months and biting cold winters for over 3–4 years continuously and the temperature has increased every year. The skies are clear and sunshine is received for all 365 days of the year. They are more vulnerable to climate change impacts (0.412). The vulnerability in high and low temperatures months was 0.52 and 0.426. The average wind speed calculated for 12 months over 10 years is 43 kmph and the average precipitation is 23 cm. Table 1.6 shows the exposure of communities to changing climate conditions.

### *1.4.7 Water Availability*

Water is an extremely precious resource in Jaisalmer. About 33% of the rural households depend on open sources of water to meet their daily needs. Open sources include water from oases and wells. 61.11% of those who receive piped water, complain of irregular supply. The members of the household travel an average of about 26.25 min to reach the water source, the shortest duration being 5 min and the maximum being 60 min (to and fro). 30.5% of the respondents travel more than an hour in search of water. Jaisalmer households reported storing 37.5 L of water every day and about 91.6% of the households save water, out of this, 38.88% of the households save 20–40 L of water per day, 30.5% households save less than 20 L of water. Only a small portion save above 40 L.

**Table 1.7** The Ground water resources as per 31.03.2013 data

| Block | Allocation for domestic and industrial requirement | Net G.W. availability for future irrigation Development | Stage of G.W. Development | Category of Block |
|---|---|---|---|---|
| Jaisalmer | 11.2878 | 3.8811 | 332.71 | Over exploited |
| Sam | 8.2513 | 15.2119 | 125.33 | Over exploited |
| Sankra Pokhhran | 6.071 | 4.8415 | 331.66 | Over exploited |

*Source* CGWB, Jaipur

**Table 1.8** The Ground water resources as per 31.03.2013 data

| Block | Area of block | Total annual ground water recharge | Natural discharge during non-monsoon season | Net annual ground water availability | Existing gross ground water for irrigation | Existing gross G.W. for domestic and industrial use | Existing gross ground water draft for all uses |
|---|---|---|---|---|---|---|---|
| Jaisalmer | 115,951 | 21.8423 | 1.8582 | 19.984 | 53.2718 | 13.2167 | 66.4885 |
| Sam | 21,194.8 | 28.483 | 2.8483 | 25.6347 | 19.5125 | 12.6146 | 32.1271 |
| Sankra Pokhran | 5615.2 | 19.8406 | 1.8395 | 18.0011 | 52.8013 | 6.901 | 59.7023 |

*Source* CGWB, Jaipur

**Table 1.9** Categorization of aquifers of different blocks on the basis of Stage of Development

| Categorization on the basis of stage of development of ground water | Block name |
|---|---|
| Critical | Sam |
| Over exploited | Sankra Pokhran, Jaisalmer |

*Source* CGWB, Jaipur
Basis for categorization: Ground water development $\leq$ 100%—Critical and > 100%—Over Exploited

Table 1.7 provides information on net ground water availability for future irrigation development and status of ground water development in the three blocks of Jaisalmer, i.e. Sam, Pokhran and Jaisalmer.

Natural discharge during non-monsoon season, existing gross ground water for irrigation, industrial use and other uses on the basis of 2013 data is shown in Table 1.8.

Categorization on the basis of stage of development of ground water (critical and over exploited) is represented in Table 1.9.

## *1.4.8 Water Vulnerability*

Water woes remain a common thread in the lives of people of Jaisalmer. Struggling to access drinking water has become a norm for people in Thar Desert, irrespective of the season. From red-tapeism to environmental clearances, various factors have kept the efficiency of water projects low. Rajasthan's relationship with summer is not a pleasant one. The shortage of water in the region only adds to the misery of the people. 13.8% of the people depend only on open sources of water to meet their daily needs. They depend on oasis, *beris (*traditional percolation tanks that get recharged at night with freshwater), open wells and take multiple rounds to fetch. 16.6% of the households have a tapped water pipeline. The water from this source is salty and cannot be used for drinking purposes, similar is the case with the water from Indira Gandhi Canal which serves the water needs for the villagers of Ramdevri, Pokhran (8%). They too complain of the same problem that the water cannot be used for drinking due to the high amount of fluoride and salts in it. 22.22% households depend on supplies from the water tanker of PHED which delivers water from surrounding districts. This tanker supplies drinking water to each village at a gap of about one week due to which the households have to rely on multiple sources of water and this, in turn, increases their vulnerability. About 41.6% of the households depend on stored rainwater, tank water and water from wells and beris for drinking and tap water and water from Indira Gandhi Canal for other domestic purposes. Women and children are the ones who travel long distances of about 3 to 4 km every day in order to fetch water for the family in every season (Figs. 1.12, 1.13).

"*I have seen my mother and other family members struggle to get drinking water in my childhood. Now, I see my wife and daughters-in-law go through the same trouble. Nothing has changed*," says 75-year-old Hemant Ram, a resident of Baramsar village in Jaisalmer block.

Women, accompanied by their children, in this region continue to go through the ordeal of walking long distances to access potable water, even after years of political promises and pretence to improve availability of water to this region. Surprised to see a herd of cows out of which only three cows belonged to Manisha Devi, I asked why she was drawing water for animals that belonged to other households. "*It's a common practice here. Whoever comes to the **Beri** around this time provides water to the animals gathered here*," she said. The strong sense of community Manisha displays is not uncommon in this arid region.

The struggle to find water is the same in 3–7 °C or 45–50 °C. However, maximum scarcity is felt during summer (May to July). The vulnerability touches 0.541.

"*We avoid taking bath for 10–15 days and don't wash clothes for around 20 days," says 55-year-old Kamla Devi. "We depend on the water tanker*," says Sonu. "*Since it costs around Rs. 800 to Rs. 1,000, we call it once in 10 days. The animals are also given less water than what they require*," she says, adding that they have lost several livestocks to harsh summers.

Maximum households store drinking water in underground tanks which have a capacity of about 1000 L which lasts them an entire year (Fig. 1.14 provides more details). About 16.66% people do not have the infrastructure or the ability to store water on such a large scale so they try saving certain amounts on a daily basis. 66.66% of the villagers cannot afford to purchase jerry cans or jar water for providing drinking water to guests in marriages and social events.

61% of the people directly consume water without purifying it. They lack resources and are hardly able to meet their daily needs so boiling water and purifying it prior to drinking, is out of question. Due to this reason, maximum population (especially the old) suffer from problems such as knee and joint pain, yellowing of teeth, haziness in vision (cataract in almost all old people) and falling of hair. This is due to the high amount of fluoride in the water. About 58.3% of the households complained of impurities in water. Most of the older population has developed a hunch back and their joints have become stiff, young people start to look old sooner due to the high amount of salinity in water. Increased content of fluoride is due to high rate of exploitation of groundwater.

### 1.4.9 Inference of the Spider Web Chart

The results of major component calculations are presented collectively in a spider diagram. The scale of the diagram ranges from 0 (less vulnerable) at the centre of the web, increasing to 0.5 (more vulnerable) and to 1 (most vulnerable). Figure 1.6 represents the vulnerabilities of each major component which in turn comprises several sub-components to present an overall web. The households of Jaisalmer are most vulnerable in terms of dependency ratio (0.715) which is followed by water availability (0.676) and their livelihood strategies (0.665).

In terms of various aspects of water vulnerability, the households show that they are affected by scarcity and are vulnerable at a higher side (0.452). They are more vulnerable in getting access to health facilities (0.480) and they have a well-knit system of social networks. However, since maximum households surveyed do not hold bank accounts, the average vulnerability increased to (0.425). The households are very prone to facing natural hazards (0.413) such as dust storms and all year round dry conditions, the monthly average of maximum and minimum temperate is also at extremes. They have also seen moderate floods in 1992, 1994, 1996, 1998, 2007 and 2008. The rural communities are not informed by the government regarding an

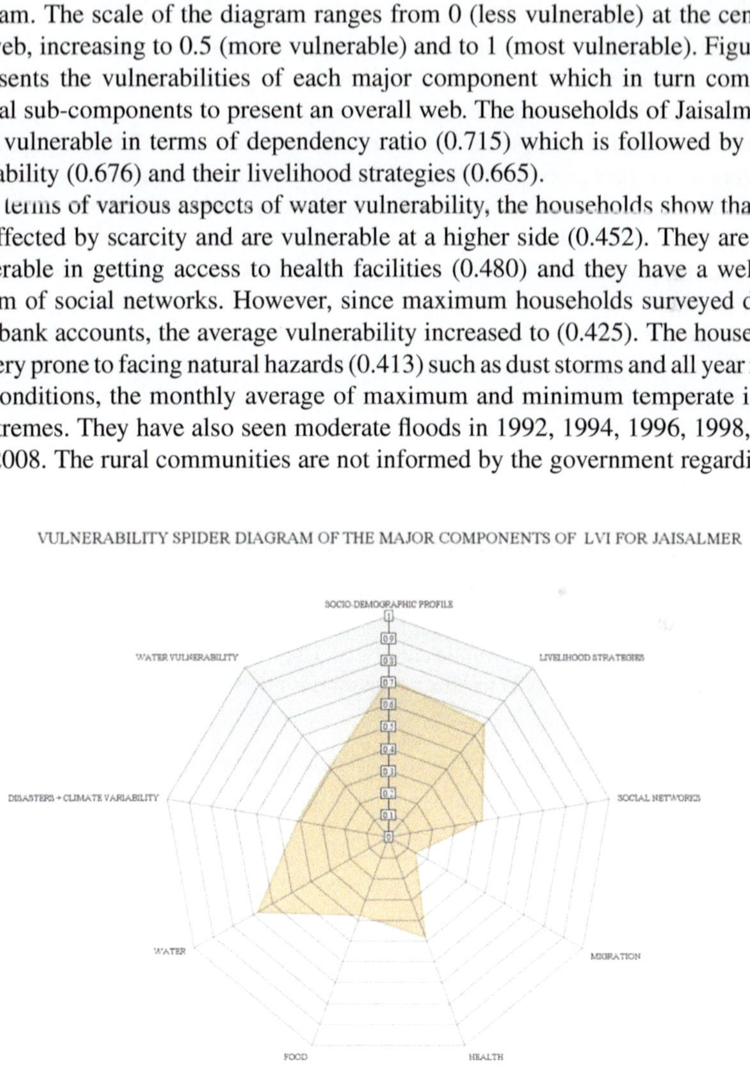


**Fig. 1.6** Spider web chart diagram of the major components of LVI-IPCC for Jaisalmer district. *Source* Prepared by author

upcoming storm and have to save their own lives. The casualty rate is low due to a strong interlinked community.

Maximum households of the village households are unaware of any water related schemes of the government and complain about lack of effectiveness in management of water by the authorities. The residents say that the water that is delivered to them through tankers is clean, water from other sources looks clean, has no odour but the people have shown signs of water related health problems.

The major problem that they complain of is travelling long distances to collect water multiple times a day (Fig. 1.10 shows the time taken (in %) to reach water source). The ones who have a tapped water supply complain of impurity in water and it cannot be consumed for drinking so they too are highly vulnerable. The people who buy water through tankers complain of irregularity.

In terms of food, the households are moderately secure. 37.4% of the population is vulnerable as their crop diversification index is low, they are unable to save crops or seeds for the next season, they are major tenants and work on the farms of big landlords. They have a hand-to-mouth earning thus, they are vulnerable. Lack of water makes the cropping season short to about only 4 months after which the communities depend on their livestock. Thus, those who do not own land, at least own livestock to insure themselves in tough times. 75% of the households' own cows, 38.8% own goats, 8% own sheep and 0.2% own camels. Thus, the rural communities of Jaisalmer are more vulnerable and exposed to the vagaries of nature and show a high degree of vulnerability.

### *1.4.10 Inference of the Vulnerability Triangle*

Figure 1.7 shows a vulnerability triangle, which plots the contributing factor scores for exposure, adaptive capacity and sensitivity. The triangle illustrates that Jaisalmer may be more exposed (0.412) to climate change impacts. accounting for the current

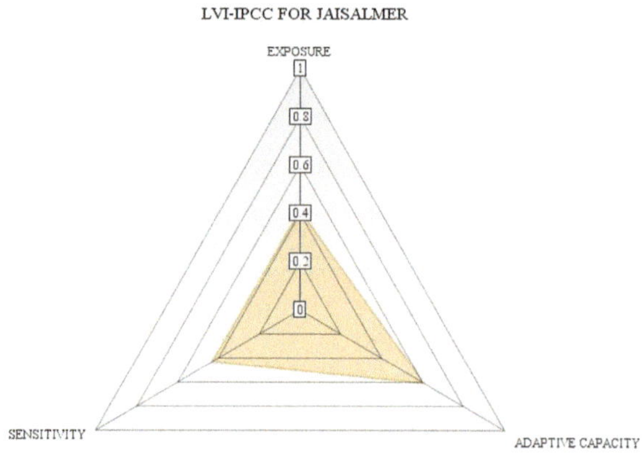


**Fig. 1.7** Final IPCC contributing factors to vulnerability for Jaisalmer. *Source* Prepared by author

health status as well as food and water security, the rural households are more sensitive to climate change impacts (0.430). Based on demographics, livelihoods and social networks, Jaisalmer showed a higher adaptive capacity (0.715). The overall LVI-IPCC scores indicate that households are very vulnerable (0.083) but have shown tendencies to adapt.

## 1.5 Water Management Techniques in Jaisalmer

**Water Conservation and Artificial Recharge**
In Jaisalmer, water is harvested in two ways such as: surface water harvesting and groundwater harvesting. Within surface water, one can identify two different modes such as those which are harvesting the rainwater in the form of roof top harvesting and those which collect the surface runoff at suitable sites from well-defined at times well maintained catchment areas. Storing and harvesting of water is controlled by physical conditions such as climatic factors and surface morphology. Figure 1.9 depicts various means of water sources used by households (village communities) in Jaisalmer.

1. Talaab/Talai

Talabs are reservoirs. They may be natural, such as the ponds (pokhariyan). A repository territory of under-five bighas is known as a talai; a medium estimated lake is known as a bandhi or talab; greater lakes are called sagar or samand. They fill water systems and drinking needs. These were not just built by the leaders of the little realm yet in addition by the dealers and Banjaras. A portion of these Talaabs (Image 1.1) are made by certain networks to take care of the drinking water issue. For instance, 'Gomat' a small village near Pokaran, in Jaisalmer district has its own source of water for domestic utilization.

2. Kueen, Kuan and other similar structures

All these structures are the means of groundwater harvesting. Some of these structures are examples of subsurface water harvesting. The term subsurface water denotes shallow groundwater, availability of which largely depends on rainfall conditions of previous monsoon season. For instance, Kueen, Beri and Kuan are the structures used to get percolated water.

'Kueen' is a small form of dug wells. Old Kueen is now lined with stone walls. In old days Kueen was lined only with interlocking stone blocks. In other types of lining the Kueen has rope lining. This is done by 'Chelwanji' the people who are skilled persons to do rope tinning in the Kueen up to 30–60 m deep and only 1.5–2.5 m in diameter. Kueen (Image 1.2) is particularly observed in the areas of interdunal depressions.

Water availability through Kueen is controlled by the rainfall amount received and the seepage of surface water. Kueens are observed in groups and generally located

**Image 1.1** A Talaab with Panghat in Pokhran

**Image 1.2** Kueen in Deva village, Jaisalmer

outside the settlement. People believe that the greater number of Kueens in the area do not affect the subsurface water availability. In fact, they dug more Kueens to harvest groundwater at its optimum and believe that the subsurface water can be harvested only when there is enough rainfall in the monsoon season. The rainwater percolates down up to subsurface below level that can be harvested, however further downward the movement of water becomes slow and potentials of getting water beyond 50–60 m greatly reduced.

3. Beri

Beri is the smallest structure among the family of dug wells. Beri is observed in very few areas of Rajasthan like, 'Thaat' village off Pokaran-Devikot road in Pokaran tahsil, of Jaisalmer district. This area is only along the ephemeral streams and in the areas where subsurface stratum is more impervious than the overlying stratum. Beris are also dug in the reservoir of the Johad or Medhbandi (soil bund used to divert surface runoff in depressions or in the Johad reservoir). In the areas of very high temperature like Jaisalmer, when water from reservoirs dries out, the percolated water can be used through the Beri.

4. Tanka and Kund

Roof top rainwater harvesting is a traditional system in many parts of Jaisalmer. Wherever possible the people of Jaisalmer have tried to collect the rainwater. Tanka is the most popular rainwater harvesting structure in Jaisalmer. This structure is constructed below the ground. The size of Tanka depends on the demand of users. When the Tanka is used for family then it is small in size and is generally called 'Kund'. When it is for any community, the size of such Tanka is bigger. Some Kunds that are smaller in size are also called Kundi. The catchments for these structures are either the roofs of houses or paved surfaces or unpaved surfaces particularly demarcated for the purpose. Many such Kunds are observed in villages of Jaisalmer. Figure 1.12 provides a percentage wise estimate of methods of water storage by households in the surveyed area (Images 1.3, 1.4, 1.5 and 1.6).

## 1.6 Practical Implications of LVI of the Surveyed Area

The major vulnerability components shown in the spider web chart provides information about which characteristics contribute most to climate vulnerability in Jaisalmer. These in turn might be programmed for community assistance. For example, although Jaisalmer suffers from recurrent droughts and an all-time dryness, we observed that many households have adapted by storing water in plastic/metal drums which have a capacity of about 200 L. Similarly, they use traditional means of storing water like kueen, beri (underground water storage methods) and johad, kund and talab which are used surface water harvesting. Underground tanks which hold up to 1000 L of rainwater are a common sight in most of the households that are better off than the ones that are landless.

**Image 1.3** Kund at Sam block

**Image 1.4** Tanka at Deva village

**Image 1.5** Matka and drum

**Image 1.6** Interior of a Kuaan/Water well

These practices have likely decreased the vulnerability of travelling long distances in search of water. Similarly, although the households reported struggles to find food for 5 months or more per year, 64% households store crops which helps them sustain their need for food for the rest of the months.

This shows that the households are engaged in sustainable food management practices (Fig. 1.8). About 45% of the people store seeds for the next season which suggests that education on seed preservation and seed diversification would constitute an appropriate intervention for Jaisalmer households, despite having a secure status in terms of food. The region needs to invest in programmes aiming to increase livestock production and enhance livelihood support in rural areas.

**Migration**: Migration is seen in need of immediate medical conditions or income needs. the percentage of people who migrated remains low (19.4%). Those who fail

**Fig. 1.8** Pie chart representing variety of livestock domesticated by the rural households. *Source* Prepared by author with primary survey data

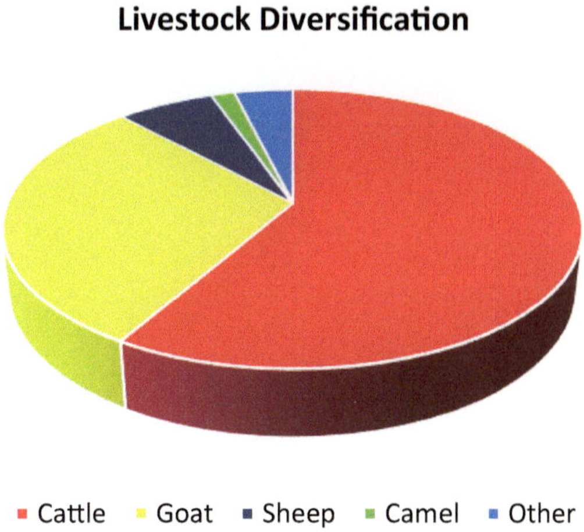

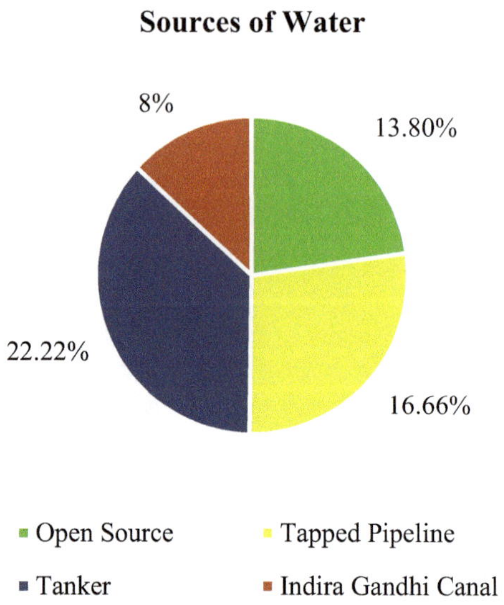


**Fig. 1.9** Pie chart representing the sources of water used by rural households. *Source* Prepared by author with primary survey data

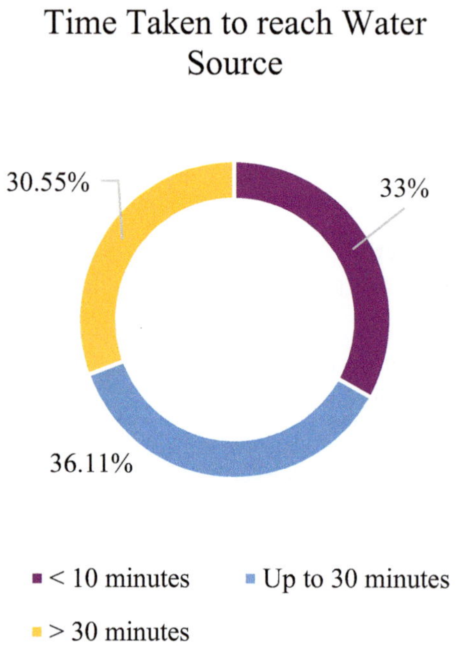


**Fig. 1.10** Donut pie chart showing time taken (in %) to reach water source. Source Prepared by author with primary survey data

to find work in the village have adapted by sending the male members and children for better employment and education to meet the livelihood stability.

**Health**: The rural communities have shown resilience in recovering from and protecting themselves from diseases and infections after the construction of primary health care centres. They are also aware of saving the girl child and the ill-effects of defecating in the open due to awareness created by the government. They have thus resorted to keep sanitation a priority.

**Social assistance**: The borrow-lend money and receive-give assistance averages were created to measure the degree to which households rely on family and friends for financial assistance and in-kind help. It was assumed that a household that receives money or in-kind assistance often but offers little assistance to others is more insecure and vulnerable compared to those with excess money and time to help others.

Since the familial structure was joint in nature and comprised of a greater number of people with immediate family and friends, these living arrangements may have influenced the way in which residents judged 'helping' versus 'obligation'. Community bonds and high-level trust among households are important for decreasing vulnerability to climate change impacts. However, these social characteristics can be more difficult to measure than measuring food security and health indicators.

Other measures of social capital include a household's range of contacts or access to formal government structures, access to information, agricultural and technical support, degree of gender equity as well as the number of social groups to which a household belongs. Despite the challenges in quantifying social networks, their inclusion in climate vulnerability assessment is essential as many adaptation behaviours depend on collective insurance mechanisms such as agricultural cooperatives.

Water scenarios can be summarized through charts prepared by the researcher in Figs. 1.9, 1.10, 1.11, 1.12, 1.13 and 1.14 through data received during primary survey of the region. Their in-line descriptions have been covered in the previous sections.

## 1.7 Conclusion

All blocks of the district have over exploited groundwater, thus, leaving a very limited scope of further groundwater development for consumption. The area is devoid of sustained surface water bodies. Villages of Jaisalmer block could be attributable to better means of livelihood as compared to the other regions due to close proximity to the city and thus witnessed higher commutation from the villages to the city.

Villages of Pokhran block were found to be more vulnerable and exposed in terms of water and climate change and variability. Unawareness regarding pending natural disasters, fluctuation in precipitation and temperature, illiteracy, large family sizes and inadequate access to medical care are some of the dimensions in which the people of this block are vulnerable. There was an increase in the vulnerabilities of farming communities in the region towards extreme climatic events.

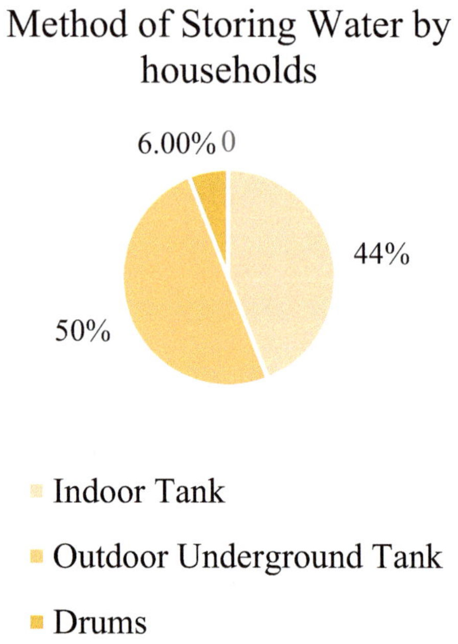


**Fig. 1.11** Pie chart depicting water storage methods by rural communities of Jaisalmer. *Source* Prepared by author with primary survey data

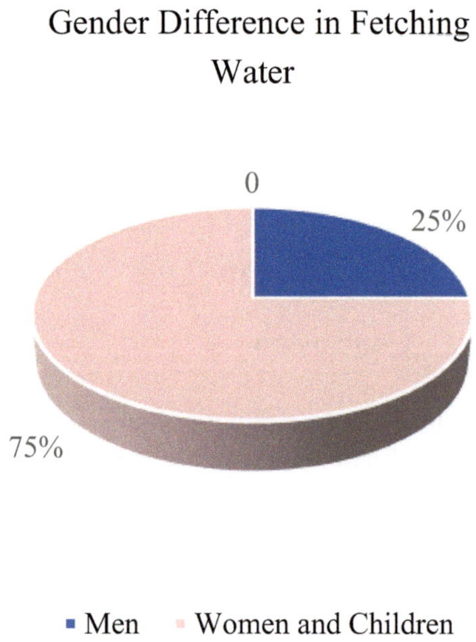


**Fig. 1.12** Gender disparity in fetching water is represented. *Source* Prepared by author with primary survey data

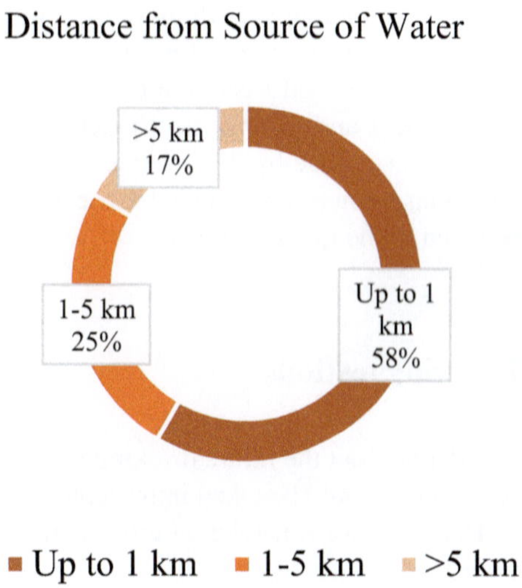


**Fig. 1.13** Donut pie representing distance travelled by people to fetch water. Source Prepared by author with primary survey data

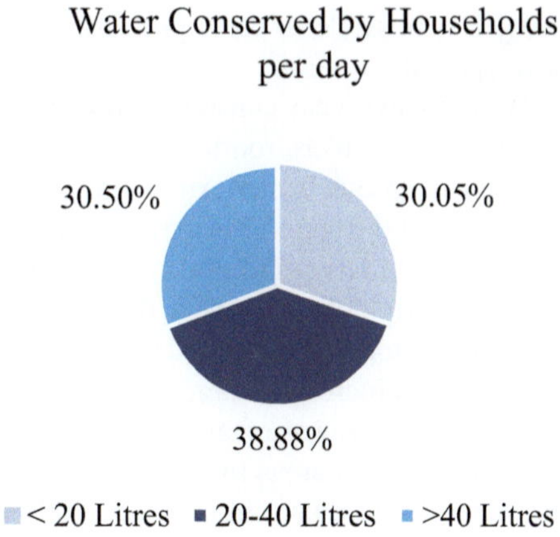


**Fig. 1.14** Pie chart showing percentage of households preserving different quantities of water per day. *Source* Prepared by author with primary survey data

Sam region was highly vulnerable in terms of natural phenomena like dust storms, water availability, health and food resources. This region was found to be the most sensitive region to extreme climatic events probably because of the region's inadequate access to immediate resources. The findings of this study have important policy relevance that could enable the rural communities of the desert region to better adapt to the effects of climate change and variability.

The westernmost region and the interior deserts regions should be given priority by both government and donors in terms of income generation, water sources and food security projects in order to reduce the water and livelihood. The government has done an impressive job in providing employment to people through MNREGS to clean wells. However, there is the need to improve water supply in the region by undertaking measures such as construction of tankas and kunds in order to reduce the duration to get to far off water sources like beris.

If a hypothetical assumption regarding the continuation of the present kind of development and management practices, then by the year 2045, the irrigated area will further reduce by 28% whereas, the population would increase by 88%. This highlights the need for sustainable use of water resources and specific attention should be given to the quality of water.

## 1.8 Suggestions

In order to meet the future drinking water and irrigation requirements the following suggestions could be taken into account:

Halophilic or salt-tolerant crops could be grown in areas with brackish groundwater.

Deep groundwater exploration can be done in areas that fall under a 'relatively' safe category like, in the Sam region, expanses beneath the tertiary and Lathi formations, it is possible to carry out further groundwater development exploration for development.

Water for everyday consumption should be used from conservation and harvesting structures like tankas, rooftops and underground water storage which would help in reducing the groundwater extraction. This is the need of the hour since large number of aquifers of the region has the potential to recharge by surface water from the IGNP (Indira Gandhi Nahar Project) system and rainwater discharge during the rainy years.

Construction of small check dams and earthen dams could be done at appropriate locations to store rainwater and increase groundwater recharge which will, as a result, replenish and increase water levels in wells.

Heavy withdrawal of groundwater for agriculture and drinking purpose, from potential zones in areas like Chandan-Bhairawa and Lathi need to be controlled as the water development has crossed 100%. Awareness programmes to educate the urban and rural communities regarding conservation of groundwater resources and training in rainwater harvesting will be beneficial to check decline in water level and make sure of its justified use. Financial assistance for groundwater development in over exploited, critical and semi-critical areas should not be encouraged. Use of water saving devices such as sprinklers and close field distribution channels should be promoted.

Reduction in the wastage of water and controlling irrigation by reducing the pumping hours according to the minimum requirement of water per field, according to the crop sown could be helpful in optimizing and effectively managing the requirements of modern agricultural techniques. Crops that require low amount of water and are cost-effective need to be selected for the desert regions rather than cultivating rice which is a water intensive crop. Agriculture extension services could go a long way in providing knowledge and skills to farmers to adopt alternate crops that demand lesser water.

Traditional methods of water harvesting in Rajasthan are structures that were designed back in the fifteenth century. They were constructed not merely with experience but were scientific in their architecture. This can be witnessed all along Western Rajasthan where the science can still be explained by the local community. Each drop of water is saved such that the rural networks of Jaisalmer are not devoid of the resource. This is the most impressive adaptation method based on sustainable scientific techniques that have helped in the continuance of existence of people in the region. However, now, increasing the 'ideas of development' had caused a spike in salinity, fluoride, nitrate and chloride content in groundwater. Detailed scientific research is required by geologists, hydrologists, engineers, environmentalists and social scientists to study the rocks of the region, groundwater flow, depth of water table and continuously monitor them to suggest sustainable groundwater management which has been polluted due to industrialization and pollution.

This study is proof of compelling changes in the climate, water availability and adaptation techniques utilized by the people of Jaisalmer. Change is the only factor that is persistent in this tough environment. Therefore, there is immense need to conserve, preserve and defend this environment, its biodiversity and its people.

Community participation in managing cross-scale and multi-level concerns could be an effective way of protecting the region from losing its resources. Climatic vulnerability, social vulnerability and livelihood vulnerability studies need to be carried out extensively along with monitoring their exposure, sensitivity, adapting capacity and resilience to perturbations with the support of the government, NGOs and community level action demands attention be included in formulating policies and programmes which target the specific region of Jaisalmer according to the social, environmental, economic and political needs of this region.

This case study, thus, shows us the methods of using traditional knowledge in conserving, storing and managing water through sustainable means, without harming the environment. It also shows us the use of a well-developed integrated and ecological approach in water management.

## References

Aalst V, Cannon MK, Burton T (2008) Community level adaptation to climate change: the potential role of participatory community risk assessment. Glob Environ Change 18(1):165–179

Adger WN (2003) Social aspects of adaptive capacity. In: Smith JB, Klein RJT, Huq S (eds) Climate change, adaptive capacity and development. Imperial College Press, London, pp 29–49

Adger WN, Brooks NN, Bentham G, Agnew M, Eriksen S (2004) New indicators of vulnerability and adaptive capacity. Tyndall Centre Technical paper 7.

Adger WN, Arnell NW, Tompkins EL (2005) Successful adaptation to climate change across scales. Glob Environ Change 15(2):77–86

Adger WN, Agrawala S, Mirza MMQ, Conde C, O'Brien K, Pulhin J, Pulwarty R, Smit B, Takahashi K (2007) Assessment of adaptation practices, options, constraints and capacity. In: Parry ML, Canziani OF, Palutikof JP, van der Linden PJ, Hanson CE (eds) Climate Change 2007: impacts, adaptation and vulnerability. Contribution of Working Group II to the fourth assessment report of the intergovernmental panel on climate change. Cambridge University Press, Cambridge, pp 717–743

Adger WN, Agrawal S, Mirza MMW, Conde C, O'Brien KL, Pulhin J, Pulwarty R, Smit B, Takahashi K (2007b) Assessment of adaptation practices, options, constraints and capacity. Climate Change 2007: impacts, adaptation and vulnerability. Contribution of Working Group II to the fourth assessment report of the intergovernmental panel on climate change. Cambridge University Press, Cambridge, pp 719–743

Adholeya A, Dadich PK (2008) Production and technology of bio-diesel: seeding a change. TERI Press, The Energy Resource Institute, New Delhi, p 256

African Development Bank (AfDB), Asian Development Bank, UK DfID, Netherlands DGIS, European Commission DG for Development, German BMZ, OECD, UNDP, UNEP, World Bank (2003) Poverty and climate change. Reducing the vulnerability of the poor through adaptation. Department for International Development

Arora M, Singh RD, Kumar R (2010) Impact of climate change on water resources. Yojana 54:36–40

Bandyopadhyay A, Bhadra A, Raghuwanshi NS, Singh R (2009) Temporal trends in estimates of reference evapotranspiration over India. J Hydrol Eng 14:508–515

Bhattacharya S (2007) Lessons learnt for vulnerability and adaptation assessment from India's First National Communication. Winrock International

Bhuiyan C, Singh RP, Kogan FN (2006) Monitoring drought dynamics in the Aravalli Region (India) using different indices based on ground and remote sensing data. Int J Appl Earth Obs Geoinf 8:289–302

Bhuiyan C, Flügel WA, Singh RP (2009) Erratic monsoon, growing water demand, and declining water table. J Spat Hydrol 9:1–19

Brenkert AL, Malone EL (2005) Modeling vulnerability and resilience to climate change: a case study of India and Indian States. Clim Change 72:57–102

Brown P, Reddy VR, Bandi M, Chiranjeevi T, Reddy DR, Roth C (2010) Adapting to climate variability in semi-arid regions: a study using sustainable rural livelihoods framework. LNRMI Working Paper No.1, Hyderabad.

Central Ground Water Board (2007) Ground Water Scenario: Western Region, Jaipur, Rajasthan. Ministry of Water Resources, Government of India

Central Ground Water Board (2010) Ground Water Scenario of India 2009–10. Ministry of Water Resources, Government of India

Central Ground Water Board (2011) Ground Water Scenario in Major Cities of India. Ministry of Water Resources, Government of India

Central Ground Water Board, Government of India, Western Region, Jaipur and Ground Water Department, Government of Rajasthan, Jodhpur (2011) Report on Dynamic Ground Water Resources of Rajasthan

Central Water Commission National Institute of Hydrology (2008) Preliminary Consolidated Report on Effect of Climate Change on Water Resources. Government of India

Commissionerate of Watershed Development & Soil Conservation (2010) Perspective & Strategic Plan for Development of Rainfed and Watershed areas in Rajasthan for 18 Years Period. Government of Rajasthan Rural Development and Panchayati Raj Department, Jaipur

Critchfield HJ (2012) Theories of climate change, general climatology. PHI Learning Private Limited, New Delhi, pp 238–245

Das S, Chatterjee A, Chatterjee K (2005) Community adaptation to drought in Rajasthan. Inst Dev Stud Bull 36:33–52

De US, Dube RK, Rao GSP (2005) Extreme weather events over India in the last 100 years. J Indian Geophys Union 9:173–187

Department of Environment (2010) Rajasthan State Environment Policy 2010, including Rajasthan Environment Mission. Government of Rajasthan

Department of Science and Technology (1994) Resource Atlas of Rajasthan. Government of Rajasthan, Jaipur, p 231

Dhaka BL, Chayal K, Poonia MK (2010) Analysis of farmers' perception and adaptation strategies to climate change. Libyan Agric Res Centre J Int 1:388–390
Dickson E, Jacobsen M, Danilenko A (2010) Climate change and urban water utilities: challenges and opportunities. Water Working Notes. Note No. 24
Disaster Management and Relief Department (2009) Drought management manual. Government of Rajasthan
Global Environmental Change 17, 12–24
Goyal RK (2009) Water Scenario of Rajasthan under the changing Climate. Workshop on Water Scenario, Efficient Use and Management in Rajasthan. Central Ground Water Board, Western Region, Jaipur, 13–14 Mar 2009
Guhathakurta P, Sreejith OP, Menon PA (2011) Impact of climate change on extreme rainfall events and flood risk in India. J Earth Syst Sci 120:359–373
Gupta V (2005) Climate change and domestic mitigation efforts. Econ Pol Wkly 40:981–987
Hahn MB, Riederer AM, Foster SO (2008) The livelihood vulnerability index: a pragmatic approach to assessing risks from climate variability and change—a case study in Mozambique, pp 74–76
Hahn M, Riederer M, Foster S (2009) The livelihood vulnerability index: a pragmatic approach to assessing risks from climate variability and change—a case study in Mozambique. Glob Environ Change 19(2009):74–88
Hamilton K, Fay M (2009) A changing climate for development. Finance Dev 46:10–12
Helm D, Hepburn C (2009) Economics and politics of climate change. Oxford University Press, New York, p 576
Hulme M (2003) Abrupt climate change: evidence, mechanisms and implications. Phil Trans: Math Phys Eng Sci 361:2001–2021
Hunt A, Watkiss P (2011) Climate change impacts and adaptation in cities: a review of the literature. Clim Change 104:13–49
Hurd BH (2008) Coping and adaptation to the changing climate: concepts, issues and challenges. Ann Arid Zone 47:221–231
Indian Meteorological Department (1988) Climate of Rajasthan State, Government of India, New Delhi
Indian Meteorological Department (2010) Climate of Rajasthan State, Government of India, New Delhi
IPCC (2001) Climate Change 2001: impacts, adaptation, and vulnerability. Contribution of Working Group II to the Third Assessment Report. Cambridge University Press, Cambridge, UK
IPCC (2007a) Climate Change 2007: impacts, adaptation, and vulnerability. Contribution of Working Group II to the Fourth Assessment Report (Chap. 9). Cambridge University Press, Cambridge, UK
IPCC (2007b) Climate Change 2007: the physical science basis. Contribution of Working Group I to the Fourth Assessment Report (Chap. 11). Cambridge University Press, Cambridge, UK
Jain JK (1993) Climatic conditions of Rajasthan. Natural and Human Resources of Rajasthan. Scientific Publishers, Jaipur, pp 37–50
Jain SK, Keshri R, Goswami A, Sarkar A, Chaudhry A (2009) Identification of drought-vulnerable areas using NOAA AVHRR data. Int J Remote Sens 30:2653–2668
Jaipur Singh RL (1995) India: A Regional Geography, National Geographical Society of India, Varanasi, pp 49–55
Kasperson RE, Kasperson JX (2001) Climate change, vulnerability, and social justice. Risk and Vulnerability Programme, Stockholm Environment Institute, Stockholm
Kasperson JX, Kasperson RE, Dow K (2001) Introduction: global environmental risk and society. In: Kasperson JX, Kasperson RE (eds) Global environmental risk. United Nations University Press, New York
Lindenberg M (2002) Measuring household livelihood security at the family and community level in the developing world. World Dev 30:301–318
Loaiciga H (2003) Climate change and ground water. Ann Assoc Am Geogr 93:30–41

Mahajan G (1989) Evaluation and development of groundwater. Ashish Publication House, New Delhi, p 317
Malhotra SP (1978) Socio-economic structure of population in Arid Rajasthan. CAZRl, Jodhpur, 51 p. Malhotra SP 1984. Traditional agroforestry practice in arid zone of Rajasthan. In: Shankarnarayan KA (ed) Agroforestry in Arid and Semi-Arid Zones, CAZRl, Jodhpur, pp 263–266
Mall RK, Singh R, Gupta A, Srinivasan G, Rathore LS (2006) Impact of climate change on Indian Agriculture: a review. Clim Change 78:445–478
Malone EL, Brenkert A (2008) Uncertainty in resilience to Climate Change in India and Indian States. Clim Change 91:451–476
Measham TG, Preston BL, Brook C, Smith TF, Morrison C, Withycombe G, Gorddard R (2010) Adaptation to climate change through local municipal planning: barriers and opportunities. Socio-Economics and the Environment in Discussion CSIRO. Working Paper Series 2010-05, 1–25
Ministry of Earth Sciences (2010) Climate Profile of India. India Meteorological Department. Government of India
Ministry of Environment and Forests (2006) National Environment Policy. Government of India
Ministry of Environment and Forests (2009a) Climate Change and India: towards preparation of a comprehensive climate change assessment. Government of India
Ministry of Environment and Forests (2009b) State of Environment Report: India, Government of India, New Delhi.
Ministry of Environment and Forests (2010a) India: Taking on Climate Change-24 Recent Initiatives Related to Climate Change, Government of India
Ministry of Environment and Forests (2010b) Climate Change And India: A 4×4 Assessment (A Sectoral and Regional Analysis for 2030s). Government of India
Mirza M (2003) Climate change and extreme weather events: can developing countries adapt? Clim Policy 3:233–248
Misra VC (1967) Geography of Rajasthan, National Book Trust, New Delhi, p 188
Mitchell T, Tanner T (2006) Adapting to climate change: challenges and opportunities for the development community. Institute of Development Studies
Moench M, Dixit A (2004) Adaptive capacity and livelihood resilience—adaptive strategies for responding to floods and drought in South-Asia. The Institute for Social and Environmental Transition, Boulder, Colorado, USA and the Institute for Social and Environmental Transition, Nepal
Moss RH, Watson RT, Zinyowera MC, Dokken DJ (1997) The regional impacts of climate change: an assessment of vulnerability. Intergovernmental Panel on Climate Change
Narain P, Khan MA, Singh G (2005) Potential for Water Conservation and Harvesting against Drought in Rajasthan, Drought Series: Paper 7, International Water Management Institute. Working Paper 104
Noss RF (2001) Beyond Kyoto: forest management in a time of rapid climate change. Conserv Biol 15:578–590
O'Brien K, Leichenko R, Kelkar U, Henry Venema G, Aandahl H, Tompkins A, Javed S, Bhadwal S, Barg LN, West J (2004) Mapping vulnerability to multiple stressors: climate change and globalization in India. Glob Environ Change 14:303–313
Olmos S (2001) Vulnerability and adaptation to climate change: concepts, issues, assessment methods. The Climate Change Knowledge Network
Organization for Economic Co-operation and Development (2002) Climate Change: India's perceptions, positions, policies and possibilities. Climate Change and Development
Pareek A, Trivedi PC (2011) Cultural values and indigenous knowledge of climate change and disaster prediction in Rajasthan, India. Indian J Tradit Knowl 10:183–189
Planning Commission (2011) Rajasthan Development Report. Academic Foundation, New Delhi, p 65
Rajasthan Pollution Control Board (RPCB) (2009) State Environmental Atlas, Rajasthan, Jaipur

Rajasthan State Pollution Control Board (1994) State Environment Atlas of Rajasthan, Government of Rajasthan, Jaipur

Rao AS (1992) Climate, climatic changes and Paleo-Climatic Aspects of Rajasthan. Geological Facets of Rajasthan, 38–44

Rao AS (1996) Climatic changes in the irrigated tracts of Indira Gandhi Canal Region of Arid Western Rajasthan, India. Ann Arid Zone 35:111–116

Rathore LS, Mall RK, Gupta A, Singh R, Singh RS (2006) Water resources and climate change: an Indian perspective. Curr Sci 90

Reddy VR (1996) Urban water crisis-rationale for pricing. Rawat Publications, Jaipur, p 180

Reddy BS, Assenza GB, Assenza D, Hasselmann F (2009) Energy efficiency and climate change: conserving power for a sustainable future. SAGE Publications India Pvt Ltd., New Delhi

Sahai S (2008) Climate change and agriculture: reinventing the crop cycle. Gene News 3:2–3

Sathaye J, Shukla PR, Ravindranath NH (2006) Climate change, sustainable development and India: global and national concerns. Curr Sci 90

Schipper ELF (2007) Climate change adaptation and development: exploring the linkages. Tyndall Centre for Climate Change Research. Working Paper No. 107, 1–28

Singh RB, Mal S (2009) Environmental Change and Biodiversity: Uttarakhand Experiences. Rawat Publications, New Delhi, p 246

Singh N, Singh S, Kar A, Joshi DC, Ram B, Kumar S, Vats PC, Raina P, Kolarkar AS, Dhir RP (1992) Desertification Mapping in Western Rajasthan. Ann Arid Zone 31:237–246

Singh VS, Pandey DN, Gupta AK, Ravindranath NH (2010) Climate change impacts, mitigation and adaptation: science for generating policy options in Rajasthan, India. Rajasthan State Pollution Control Board Occasional Paper No. 2/2010

Sreejith OP, Guhathakurta P, Menon P, Mazumdar AB (2010) Changes in extreme rainfall events and flood risk in India during the last century. National Climate Centre. India Meteorological Department. Pune

Srivastav R (2008) Need to buffer rainfed agriculture. Gene News 3:8

State Remote Sensing Application (1999) Centre Groundwater Atlas of Rajasthan, Department of Science and Technology, Government of Rajasthan, Jodhpur

State Remote Sensing Application Centre (2010) Soil Resource Atlas of Rajasthan. Department of Science and Technology, Government of Rajasthan, Jodhpur, p 270

Sullivan C (2006) Global change impacts: assessing human vulnerability at the sub-national scale. In: Presented at the International River Symposium, Brisbane

Sullivan C, Meigh JR, Fediw TS (2002) Derivation and testing of the water poverty index phase 1, Final Report. Department for International Development, UK, 2002

Sullivan C, Meigh J (2005) Targeting attention on local vulnerability using an integrated index approach: the example of the climate vulnerability index. Water Sci Technol 51:69–78

Tadross MA, Hewitson BC, Usman MT (2005) The interannual variability of the onset of the maize growing season over South Africa and Zimbabwe. J Clim 18:3356–3372

Tanner T, Yamin F, Mitchell T (2005) Linking climate adaptation: a research agenda. Inst Dev Stud Bull 36:126–130

The Energy and Resources Institute (2003) Coping with Global Change: Vulnerability and Adaptation in Indian Agriculture

The Energy and Resources Institute (2006) Adaptation to Climate Change in the context of Sustainable Development

Tideman E, Khatana V (2004) Drought coping Strategies in Semi-Arid India. Indo-German Bilateral Project

Todd C (1829) Annals and Antiquities of Rajasthan, GBD, Jaipur

Tompkins EL, Adger WN (2004) Does adaptive management of natural resources enhance resilience to climate change? Ecol Soc 9:10

UNEP (2004) Poverty-biodiversity mapping applications. In: Presented at: IUCN World Conservation Congress, Bangkok, Thailand, 17–25 Nov 2004

United Nations (2009) Drought risk reduction framework and practices: contributing to the implementation of the Hyogo framework for action. United Nations secretariat of the International Strategy for Disaster Reduction. Geneva, Switzerland

United Nations Environment Program (2011) A practical framework for planning pro-development climate policy

United Nations Framework Convention on Climate Change (2007) Climate change: impacts vulnerabilities and adaptation in developing countries

United Nations General Assembly, 1997. S/19-2. Programme for the Further Implementation Agenda 21. 19th Special Session, 23–27 June, New York

USAID (2007) Famine early warning systems network (FEWS-NET). http://www.fews.net/. Accessed 24 Dec 2007

Vincent K (2004) Creating an index of social vulnerability to climate change for Africa. Working Paper 56, Tyndall Centre for Climate Change Research and School of Environmental Sciences, University of East Anglia

Vincent K (2007) Uncertainty in adaptive capacity and the importance of scale

Vulnerability and Adaptation Program in Semi-Arid India (2009) Climate Change Vulnerability & Adaptation from Rajasthan and Andhra Pradesh. Swiss Agency for Development and Cooperation

White DA (2000) Drought Policy monitoring and management in arid lands. Ann Arid Zone 39:105–129

WHO (2005) Immunization coverage cluster survey—reference manual. WHO Document Production Services, Geneva, Switzerland

WHO/Roll Back Malaria (2003) Economic impact of malaria: household survey. Williamsburg Emergency Management, 2004. Household natural hazards preparedness questionnaire. Peninsula Hazard Mitigation Planning Committee, Williamsburg, VA

Wilhite DA (2000) Drought planning and risk assessment: status and future directions. Ann Arid Zone 39:211–230

World Bank (1997) Survey of living conditions: Uttar Pradesh and Bihar. Household Questionnaire, Dec 1997–Mar 1998

World Bank (2008) Climate change: impacts in drought & flood affected areas: case studies in India. Social, Environment and Water Resources Management Unit

World Bank (2010) Villages in Andhra Pradesh Innovate to Fight Drought and Climate Change

World Food Programme (2007) Vulnerability analysis and mapping (VAM). http://www.wfp.org/operations/vam/. Accessed 24 Dec 2007

Wu S, Bates B, Kundzewicz ZW, Palutikof J (2008) Climate change and water. Intergovernmental panel on climate change. IPCC Secretariat, Geneva

Yamin F, Rahman A, Huq S (2005) Developing the linking climate adaptation network: progress and prospects. Inst Dev Stud Bull 36:123–125

# Chapter 2
# Mining Related PCB in Wetland Sediments of the River Lippe (North Rhine-Westphalia, Germany)

**Sarah Ohlemacher, Claudia Post, and Klaus Baier**

**Abstract** This study aims to detect PCB (polychlorinated biphenyls) and their distribution in river sediments of the river Lippe near a discharge point of mine drainage water raised from the coalmine Haus Aden. Recommendations from former literature studies, European and German-Industry-Standards (DIN) standards were followed to conduct a reproducible field study. The sampling, processing, and analysis of the sediment and core samples were delineated in detail explored in order to open up a new discussion about the methodical execution of environmental examinations of PCB. The results show that the normally executed standardization processes to interpret PCB concentrations do not work as assumed. This refers especially to the normalization of the content of organic carbon in the sediment. The heterogeneity of observed concentrations dominates and the concentration fluctuates enormously instead of following an often-described correlation between PCB and organic carbon or the percentage of fine grain. Congener profiles, which show the amounts of the six indicator congeners, do not show an obvious trend or indicate one corresponding contamination source. The findings invoke a new debate and examinations concerning deposition and adsorption processes of PCB in sediments, especially against the background of flooding the derelict mines of the Ruhr district in the coming decades.

**Keywords** Polychlorinated biphenyls (PCB) · Uplifted mine water · Congeners · PCB distribution in river sediments · Standardization to organic carbon/ fine grain · Environmental quality standard

## 2.1 Introduction

Polychlorinated biphenyls (PCB) are still threatening human beings and the environment due to their high persistence in the environment and toxicity even though

S. Ohlemacher · C. Post (✉) · K. Baier
Department of Engineering Geology and Hydrogeology, RWTH Aachen University, Lochnerstraße 4-20, 52064 Aachen, Germany
e-mail: cpost@lih.rwth-aachen.de

 
P. Kumar et al. (eds.), *Water Resources Management and Sustainability*, Advances in Geographical and Environmental Sciences,
https://doi.org/10.1007/978-981-16-6573-8_2

their production and application were prohibited in Germany in the 1980s. Physical and chemical properties of PCB like heat resistance, chemical stability, good thermal and low electric conductivity (Lehnik-Habrink et al. 2006; IARC 2016; Hegnal 2010; Borja et al. 2005) led to around 1 to 1.5 million tons produced worldwide (Lehnik-Habrink et al. 2006; IARC 2016; Hegnal 2010; Liang et al. 2014; Demirtepe et al. 2015; Tehrani and Aken 2014) and intensive application. Large amounts of PCB were used in plasticizers of sealing joints, synthetic materials, colors, and varnishes, and as an insulating liquid in capacitors and current transformers (Lehnik-Habrink et al. 2006; Hegnal 2010; Borja et al. 2005; Erickson and Kaley 2011; Umweltbundesamt (2018).

PCB was used in the mining industry as a hydraulic fluid from 1960 until its prohibition (Schwarzbauer 2017; Engel and Kihl 1987), for example in coal mining in the Ruhr Area. When mining operations ceased in 2018 the question regarding the fate and behavior of PCBs in the environment remained. Studies were conducted to evaluate the remaining risk (e.g., LANUV, state office for nature, environment, and consumer protection) to surface water and aquatic life. A special issue of the present is the mine drainage water levels in the Ruhr district, which will rise in the coming decades and the following spread of its biphenyl content. This process evokes fear regarding groundwater quality.

Not only the mining industry employs PCB for technical purposes. Other industrial sectors such as mechanical engineering and the chemical industry benefit from the properties of PCB as well. One differentiation, in general, is the handling of the biphenyls. In practice, they ran either in open or closed systems, which refers to the possibility of the substances to get in contact with the environmental media (Hegnal 2010). There are more than 200 congeners differing in mass and chlorination, whose percentual distribution in the mixture gives an indication of their original industrial application (e.g., congeners 28 and 52 often appear in combination with coal mining).

The study presented here uses earlier studies as reference information and offers a deeper view into the distribution of PCB in river sediments nearby a discharge point of the uplifted water of a mining plant called Haus Aden near Dortmund. The results of the study originate from a cooperation between the Department of Engineering Geology and Hydrogeology of RWTH Aachen University and the Institute of Occupational Medicine of RWTH Aachen University.

The study focuses on the strategy of sampling and the detailed description of the processing and analyzing the samples in order to make the results replicable. Particularly the calculative standardization of PCB concentrations to the adsorptive parts of the sample will be discussed.

## 2.2 Methods

### *2.2.1 Analytical Reference System*

Sediments consist of heterogeneous matrices and PCB only adsorb at adsorptive materials like organic matter (IARC 2016; Rommel et al. 1998; Hölting and Coldewey 2005; Prinz and Strauÿ 2006). The pH-value (Rommel et al. 1998) and redox conditions also influence adsorption (Döring and Marschner 1998). Without standardization, the results of PCB measurements are comparable neither within the same study area nor with the results of other studies. The standardization has to refer to a reference system, which is the same for all samples, namely the adsorptive part of the samples. Without standardization, it is unsure, if the results are realistic. A sample with a low content of adsorptive material will have a low value for PCB. However, it is uncertain whether the value is low, because there is no more PCB or because the ion exchange capacity is low due to low content of adsorptive material in the sample. With carrying out the standardization procedure, the composition of the samples does not matter.

For this study, the parameters _ne grain and clay minerals were used synonymously. This simplification is based on the assumption that the _ne grain fraction (<0.063 mm) consists mostly of clay minerals and the non-adsorptive part is negligible (Schwarzbauer 2017). Busch et al. (Busch et al. 2007) found a significant correlation between the content of PCB and the fraction of 40–63 μm.

### *2.2.2 Sampling*

The samples were taken around about 20 km northeast of the German city Dortmund during wintertime. The starting point was the discharge point of mine drainage water from coalmine Haus Aden (Fig. 2.1).

Eleven saturated sediment samples were taken out of the riverbank. Additionally, five core samples were taken out of the embankment at 2 m distance from the waterline. The sediment sample named S1 and the core sample named BK1 were taken 30 m upstream of the discharge point. These two samples are taken with a 1 m long Pürkhauer drill stick for reference to estimate the background contamination with PCB in this area. The following samples were taken at the slip-off slopes and erosion banks of the meandering river Lippe.

The samples were filled into 250 ml wide mouth bottles made of brown glass. To prevent any contact between sample material and the plastic of the lid, which may lead to adsorption of PCB to the plastic, aluminum foil was put over the mouth. After that, the samples were stored at 5 °C in the dark until further analysis. The content of PCB was measured at the laboratory of the Institute of Occupational Medicine at the university hospital of Aachen. The contents of total organic carbon and fine grain were measured at the laboratory of the Department of Engineering Geology and Hydrogeology of the RWTH Aachen University.

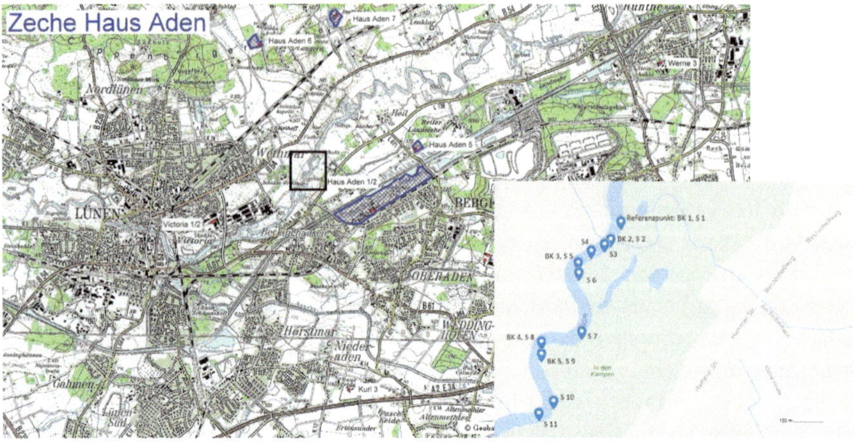


**Fig. 2.1** Sample area at the river Lippe near the city Lünen. Left: overview map, topographical map TK50. Right: Sample points at the river Lippe near the discharge point of mine drainage water from coalmine (German: Zeche) Haus Aden. Saturated sediment samples, BK = core samples (scale: 1 cm corresponds to about 80 m, map is north orientated)

## *2.2.3 Content of Organic Carbon*

The total organic carbon content of the samples was measured by estimating the ignition loss. The procedure followed the standard (DIN 18128 2002) (German Industry Norm, Soil—Investigation, and Testing—Determination of Ignition Loss). The samples dried in cups of porcelain in a heating cabinet at 105 °C until constant mass. Subsequently, the samples were put into a desiccator to cool down. The next step was homogenization with a mortar. Some samples contained organic material, that had not been decomposed yet. After homogenizing the samples were stored in a heating cabinet at 60 °C for 24 h and were afterward put into the desiccator again. The muffle furnace was heated up to 550 °C and the empty cups were calcined for 20 min. For fine grained and organic samples DIN 18128 requires a minimum weight of 15 g. The cups with the samples were ignited for 2 h until their mass stayed constant. The cups with the samples cooled down in the desiccator and were weighed again. The difference between the weights was the content of organic carbon.

## *2.2.4 Content of Fine Grain*

The sieving followed the standard (DIN 18123 2011) (German Industry Norm: Soil, Investigation, and Testing—Determination of Grain-size Distribution). The samples dried in a heating cabinet at 105 °C until constant mass. After that, they were weighed to 0.01 g, soaked with water, and stirred with the stirring device RW 20 RZM from the company IKA. A vibration-sieving machine (type AS 200 control, company

Retsch) executed the sieving. Mesh sizes ranged between 4 mm and 0.063 mm. Before weighing, the fractions dried in a heating cabinet at 105 °C and cooled down in a desiccator.

### 2.2.5 *Content of PCB*

The preparation of the samples for measuring the content of PCB followed the (DIN EN 15308 2016) (Euro Norm, Characterization of Waste—Determination of Selected Polychlorinated Biphenyls (PCB) in Solid Waste by Gas Chromatography With Electron Capture or Mass Spectrometric Detection).

Some congeners are volatile, so the samples were freeze dried with the tool Alpha 1–4 LOC-1 m from the company Christ. The first attempt to dry the samples within the glasses did not work, because this process proved excessively time-consuming. So bowls of aluminum were made to spread the material over a bigger area and reduce drying time.

The samples had to be homogenized after drying. The DIN 15308 allows this step with a ball mill (company Retsch, type PM 100 CM). 50 g of the sample was homogenized at 325 rpm (rounds per minute) for 15 min. The congeners were extracted with n-pentane in an ultrasonic device and measured with a gas-phase chromatograph (company Agilent Technologies, type 7890 A), which was linked to a time-of-flight mass spectrometer (company Leco, type Pegasus 4D GCcGX-TOFMS). The injection was executed by an autosampler from the company Gerstel Twister, type Multi Purpose Sampler MPS. The capillary column (company Restek, type RTX 5) had a length of 60 m and a diameter of 0.25 mm. Helium was used as carrier gas. PCB-54 (2,2′,6,6′-tetrachlorobiphenyl) was used as recovery standard. The oven program had the following steps: the initial temperature of 45 °C was held isothermally for 1 min. At 30 °C/min the oven heated until 190 °C, which was held for 0.01 min. Until 330 °C the oven was heated with 5 °C/min and this final temperature was held isothermally for 12.16 min.

## 2.3 Results

### 2.3.1 *Content or Organic Carbon and Fine Grain*

The results of the ignition loss and contents of _ne grain for the sediment and core samples are shown in Table 2.1. The results of the ignition loss for the sediment samples (S-samples) are very heterogeneous and are fluctuating between 1.26% (S11) and 16.06% (S3). The ignition losses for the core samples are more homogenous and spread between 3% (BK1) and 5.13% (BK5). Like for the content of organic carbon the results for the content of fine grain in the sediment samples show a wide range

**Table 2.1** Result for the ignition loss and content of ne grain for the sediment samples (a) and the core samples (b) %

| (a) | | |
|---|---|---|
| Sample | Content of organic carbon (%) | Content of ne grain (%) |
| S1 | 3.26 | 19.33 |
| S2 | 5.19 | 23.01 |
| S3 | 16.06 | 19.62 |
| S4 | 9.93 | – |
| S5 | 11.80 | 74.86 |
| S6 | 4.33 | 12.55 |
| S7 | 13.32 | 71.35 |
| S8 | 13.66 | 59.37 |
| S9 | 7.92 | 31.00 |
| S10 | 3.46 | 26.42 |
| S11 | 1.26 | 6.01 |
| (b) | | |
| Sample | Content of organic carbon (%) | Content of ne grain (%) |
| BK1 | 3.00 | 18.05 |
| BK2 | 4.00 | 22.80 |
| BK3 | 3.80 | 19.91 |
| BK4 | 3.86 | 21.67 |
| BK5 | 5.13 | 16.36 |

between 6.01% (S11) and 74.86% (S5). There was not enough material for sample S4 for sieving. And the percentage of the fine grain for the core samples ranges between 16.36% (BK5) and 22.80% (BK2).

## 2.3.2 *Content of PCB*

Table 2.2 contains the concentrations of the six indicator congeners in µg/kgDS (DS = dry substance) in the sediment samples.

Table 2.3 shows the concentrations of PCB in the whole samples and the concentrations, which are standardized to organic carbon and _ne grain. The results of all samples exceed the environmental quality standard of 20 µg/kg DS. The lowest levels are shown by samples S1 and S4 with 50 µg/kgDS and the highest level appears in S5 with 8610 µg/kgDS. Altogether four samples have concentrations above 2000 µg/kgDS.

After standardization to the content of organic carbon and fine grain, the PCB values are much higher. For the fine grain, the values range between 193 µg/kgDS

**Table 2.2** Concentrations of the six indicator congeners of the sediment samples (µg/kgDS)

| Sample<br>Congener | S1 | S2 | S3 | S4 | S5 | S6 | S7 | S8 | S9 | S10 | S11 |
|---|---|---|---|---|---|---|---|---|---|---|---|
| 25 | 17.0 | 163.9 | 47.4 | 19.5 | 1887.3 | 327.1 | 501.3 | 366.4 | 9.7 | 25.1 | 12.5 |
| 52 | 5.6 | 33.4 | 15.3 | 8.0 | 1161.3 | 379.5 | 320.3 | 293.3 | 4.9 | 7.7 | 7.6 |
| 101 | 7.7 | 27.5 | 18.8 | 3.3 | 2088.1 | 983.4 | 890.0 | 561.0 | 9.4 | 10.2 | 7.1 |
| 138 | 5.5 | 20.4 | 30.2 | 6.5 | 1930.8 | 935.7 | 800.8 | 820.9 | 16.5 | 15.0 | 12.0 |
| 153 | 6.6 | 14.5 | 38.5 | 12.0 | 1290.1 | 760.0 | 770.6 | 490.7 | 17.6 | 16.1 | 15.9 |
| 180 | 2.1 | 9.0 | 17.9 | 5.1 | 257.0 | 295.4 | 460.1 | 49.5 | 6.5 | 6.5 | 9.5 |

**Table 2.3** Concentrations of PCB in the sediment samples (S) (µg/kgDS)

| Sample | Concentrations PCB (µg/kgDS) | Concentrations PCB, standardized to organic carbon, rounded (µg/kgDS) | Concentrations PCB, standardized to ne grain, rounded (µg/kgDS) |
|---|---|---|---|
| S1 | 50 | 1533 | 258 |
| S2 | 270 | 5202 | 1173 |
| S3 | 170 | 1058 | 866 |
| S4 | 50 | 503 | – |
| S5 | 8610 | 72,966 | 11,501 |
| S6 | 3680 | 84,988 | 29,322 |
| S7 | 3740 | 28,078 | 5241 |
| S8 | 2580 | 18,887 | 4345 |
| S9 | 60 | 757 | 193 |
| S10 | 80 | 2312 | 302 |
| S11 | 60 | 4761 | 998 |

Left: Concentration in the whole sample. Middle: Concentrations standardized to organic carbon (µg/kgDS). Right: Concentrations standardized to ne grain (µg/kgDS)

(S9) and 29,322 µg/kgDS (S6). The lowest value for the concentration standardized to organic carbon is found in S4 with 503 µg/kgDS and the highest value for S6 with 84,988 µg/kgDS.

130 m downstream of the discharge point of mine drainage water the concentration sharply peaks at 8610 µg/kgDS (S5), while S8 [490 m downstream of the discharge point (following the waterline)] marks the abrupt end of the peak with 2580 µg/kgDS (Fig. 2.2).

The values for the indicator congeners in the core samples are shown in Table 2.4.

In Table 2.5 you find the total concentrations for the core sample. They are lower than those in the sediment samples. Without standardization to the adsorptive part of the samples, only BK3 and BK4 exceed the environmental quality standard

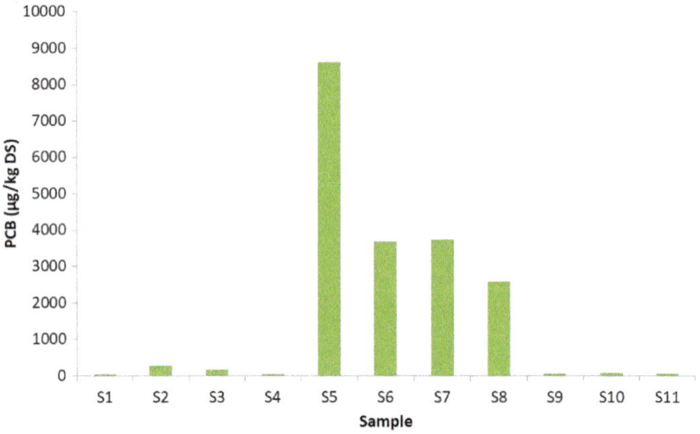


**Fig. 2.2** PCB concentration of the sediment samples (S) (µg/kgDS)

**Table 2.4** Concentrations of the six indicator congeners in the core samples (µg/kgDS)

| Samples congener | BK1 | BK2 | BK3 | BK4 | BK5 |
|---|---|---|---|---|---|
| 25 | 2.1 | 2.1 | <0.1 | 7.7 | <0.1 |
| 52 | 2 | 2.2 | 2.9 | 3.2 | 0.7 |
| 101 | 2 | 2.4 | 4.2 | 6.6 | 1.5 |
| 138 | 2.2 | 1.5 | 3.6 | 10.4 | 3 |
| 153 | 1.9 | 0.2 | 6.8 | 5.6 | 5.6 |
| 180 | 1.8 | <0.1 | <0.1 | 1.8 | 2 |

**Table 2.5** Concentrations of PCB in the core samples (BK) (µg/kgDS)

| Sample | Concentrations PCB (µg/kgDS) | Concentrations PCB, standardized to organic carbon, rounded (µg/kgDS) | Concentrations PCB, standardized to ne grain, rounded (µg/kgDS) |
|---|---|---|---|
| BK1 | 10 | 333 | 55 |
| BK2 | 10 | 250 | 43 |
| BK3 | 20 | 526 | 100 |
| BK4 | 40 | 1034 | 184 |
| BK5 | 10 | 194 | 61 |

Left: concentration in the whole sample. Right: concentrations scaled to the adsorptive part of the sample

(20 µg/kgDS BK3, 40 µg/kgDS BK4). However, with executing the standardization all BK samples exceed the limit value. The lowest level for the standardization of organic carbon is 194 µg/kgDS for BK5 and the highest value accounts

1036 μg/kgDS for BK4. After normalization to fine grain, the values range between 43 μg/kgDS (BK2) and 184 μg/kgDS (BK4).

## 2.4 Interpretation

The results show that all samples are contaminated with polychlorinated biphenyls, even 30 years after banning the production and application of PCB in Germany. Even the lowest level of PCB in the sediment samples (S1 and S4, 50 μg/kgDS) is higher than twice the environmental quality standard, set by EU legislation. Notable levels of PCB are also detectable in the core samples, which do not have direct contact with the river water containing mine drainage water. Contact between the river and the embankment, where the cores were taken, is imaginable only several times a year during floods.

The standardization of the content of organic carbon and fine grain rises the values to an even higher level (Tables 2.3 and 2.5).

A notable correlation between PCB concentrations and organic carbon and fine grain respectively was assumed, but could not be proved. The values for organic carbon and PCB as well as _ne grain and PCB are plotted against each other in Fig. 2.3a, b. The values for organic carbon content, fine grain content, and PCB show high heterogeneity within short distances and without any statistical trend. The distance between sample S5 and S6 amounts to only a few meters, but the content of organic matter in S5 represents 11.85% and in S6 only 4.33%. The values for the content of fine grain range between 74.86% (S5) and 12.55% (S6). The influence of sedimentation and degradation processes of organic matter on PCB adsorption is not yet well understood to explain the deposition and heterogeneity of PCB in river sediments.

Contents of organic carbon and _ne grain do not correlate to PCB concentration in the assumed way (Fig. 2.4) and so the standardized values for PCB show great differences. The data underlines the importance and difficulty to choose an appropriate reference system. A comparison of different studies about PCB in sediments by working groups and the scientific community in this field requires a standardized reference system.

The sampling area was selected downstream of a mine drainage water discharge point. One idea was to look at the congener profiles, which compare the concentrations of the six indicator congeners, to find out, if it is possible to name the source of the PCB contamination. Congeners 28 and 52 are considered typical for contamination from mining sources. The congener profiles for the sediment samples are shown in Fig. 2.5a, b.

Especially in S2, the value for PCB-28 is significantly raised compared to other congeners. S3 has high values for PCB-28, but also similar concentrations for PCB-153. PCB-138 and PCB-153 are typical for Clophen A60, which was used in open systems in Germany (Brusske et al. 2007). The usage of PCB in open systems was banned in Germany in 1978 (Hegnal 2010).

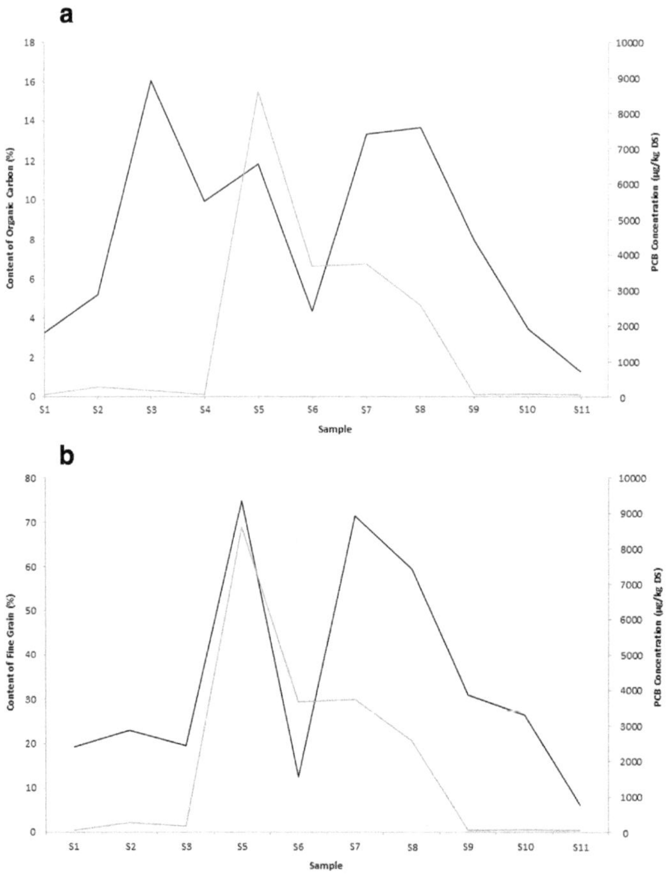


**Fig. 2.3** Missing/weak correlation between PCB and content of organic carbon and content of _ne grain. **a** PCB (μg/kgDS, light line) and content of organic carbon (%, black line). **b** PCB (μg/kgDS, light line) and content of _ne grain (%, black line)

There is no clear trend for the congener profile of the samples S5–S8. S5 has high values for PCB-28, but also for PCB-101 and PCB-138. For the other samples, the focus is on PCB-101 and PCB-138, both of which are not considered typical for mining sources (Rahm et al. 2015). Therefore, it was not possible to relate the contamination only to mining sources.

The heterogeneity and changes on short distances are also shown in Fig. 2.6a, b. S4 and S5 are around 30 m away from each other and the congener profile is significantly different. This shows again the complexity of sedimentation of particles, on which PCB molecules are adsorbed.

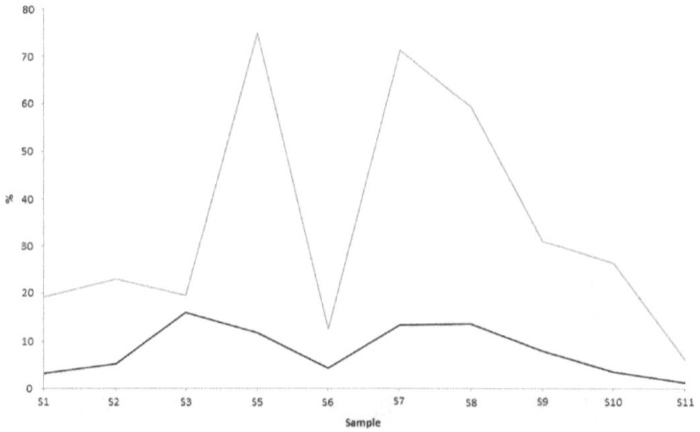


**Fig. 2.4** Correlation between content of organic carbon (black line) and content of fine grain (light line) in the sediment samples (%)

## 2.5 Conclusion

As the main outcome of this study, the statement is issued that the embankment sediments of the Lippe show significant concentrations of PCB downstream of the discharge point of mine drainage water. All values for the sediment samples exceed the environmental quality standard. Assumptions made in the literature about a strong correlation between PCB concentrations and the percentage of _ne grain (<0.063 mm) or organic carbon in the sediments could not be confirmed, so that an open question persists in the reference system. PCB deposition and processes of adsorption at river sediments should be focused on in further examinations. Neglecting factors such as deposition conditions or the aforementioned processes will put future studies at risk of misunderstanding or misinterpreting the measured PCB raw data. Another important result is the notable detection of PCB concentrations in slope sediments, which have no direct contact with the river water. This implies that these sediments show a kind of memory effect, even though direct contact exists only a few times a year during flooding. The results also show that PCB degradation after more than 30 years after the end of PCB application did not lead to extinction in the environment. Additionally, congener profiles in this study do not show an explicit source of contamination. The existence of mining-related congeners is in the same way proven as the existence of congeners which normally were used in open systems (banned 40 years ago). The difficulty of evaluating and interpreting PCB data due to their heterogeneity is clearly shown in this study, even though a relatively small study area with an intensive sampling procedure was chosen. The highly specific description of methodical processes and the usage of analytical apparatus shall increase transferability and reproducibility, which is essential for opening a scholarly debate. Responsibility in conducting such case studies to find objective observations is extremely relevant regarding the set-up of the examination. This is

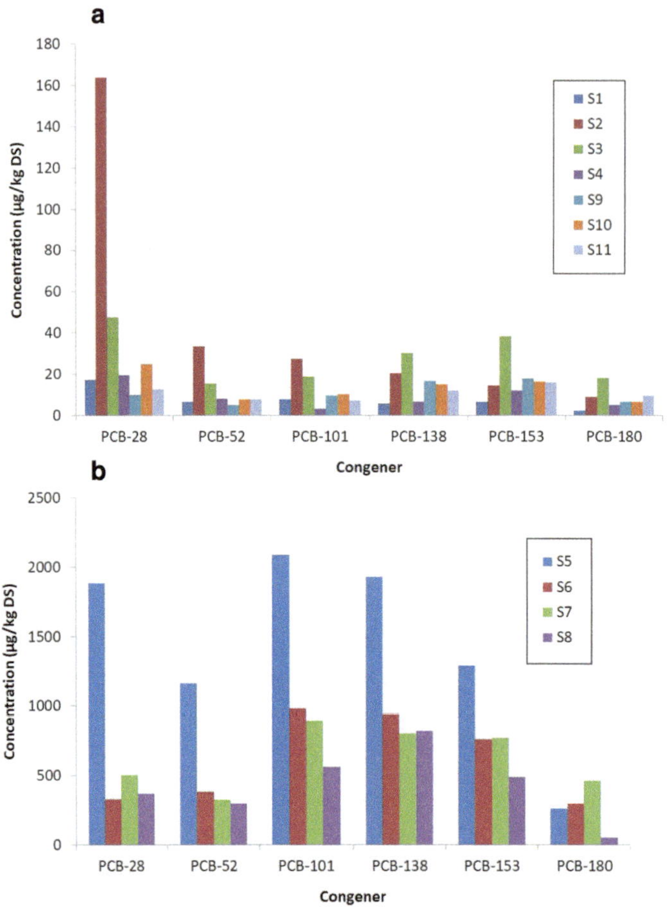


**Fig. 2.5** Congener profiles for the sediment samples (µg/kgDS). **a** S1–S4 and S9–S11. **b** S5–S8

particularly important in view of other areas with suspected high PCB contamination. In many Indian cities, there are areas with high levels of PCB contamination (Goswami 2017). For example, in the megacity of Chennai, very high PCB concentrations were found in an area where electronic waste was informally stored and recycled (Goswami 2017).

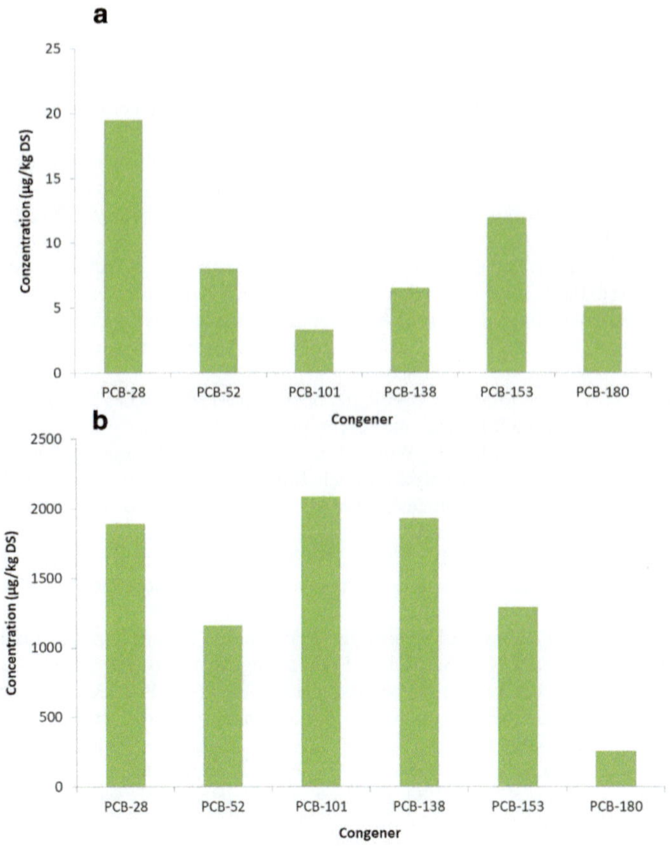


**Fig. 2.6** **a, b** Congener profiles for S4 (**a**) and S5 (**b**) (µg/kgDS)

# References


Borja J, Taleon DM, Auresenia J, Gallardo S (2005) Polychlorinated biphenyls and their biodegradation. Process Biochem 40(6):1999–2013

Brusske A, Eckartz-Nolden G, Fichthorn U, Grund I, Kalinka G, Müller W, Piegsa W, Reinders D, Schäfer K, Rosenbaum-Mertens J (2007) Gefährliche Stoffe _ Bericht zur Umsetzung der Gewässerschutzrichtlinie 76/464/EWG in NRW. Landesamt für Natur, Umwelt und Verbraucherschutz Nordrhein-Westfalen

Busch D, Furtmann K, Schneiderwind A, Zyuzina I, Reupert R, Sielex K (2007) Einfuss von Probenahme und Probenvorbereitung auf die Ergebnisse bei der Bestimmung ausgewählter prioritärer Stoffe nach der Wasserrahmenrichtlinie. In: Texte. Vol 32/07. Umweltbundeamt

Demirtepe H, Kjellerup B, Sowers KR, Imamoglu I (2015) Evaluation of PCB dechlorination pathways in anaerobic sediment microcosms using an anaerobic dechlorination model. J Hazard Mater 296:12–127

DIN 18123, April 2011. Soil, Investigation and Testing—Determination of Grain-size Distribution

DIN 18128, December 2002. Soil—Investigation and testing—Determination of ignition loss

DIN EN 15308, December 2016. Characterization of waste—Determination of selected polychlorinated biphenyls (PCB) in solid waste by gas chromatography with electron capture or mass spectrometric detection. DIN eV (ed)
Döring U, Marschner B (1998) Die Mobilisierbarkeit von ausgewählten organischen Schadstoffen in Rieselfeldböden. Bodenökologie und Bodengenese 26:42–51
Engel W, Kihl H (1987) Schwerentflammbare und gesundheitsunschädliche Hydraulikflüssigkeiten des Typs HFD für den Bergbau. Glückauf 123.2(22):1423–1426
Erickson MD, Kaley RG II (2011) Applications of polychlorinated biphenyls. Environ Sci Pollut Res 18(2):135–151
Goswami S (2017) Chennai's soil. Dehli's air most contaminated due to high PCB concentration: study. In: Down to Earth. https://www.downtoearth.org.in/news/environment/chennai-s-soil-and-delhi-s-air-most-contaminated-due-to-pcb-concentration-study-57217. Access July 2020
Hegnal A (2010) PCB, PCP, Dioxine und Furane. In: Dieckmann J, Hydvogel P, Stüdemann K, Wunderlich M (eds) Praxishandbuch Schadstoffe, pp 67–81
Hölting B, Coldewey WG (2005) Hydrogeologie. Spektrum Akademischer Verlag
IARC (2016) Polychlorinated Biphenyls and Polybrominated Biphenyls, vol 107. IARC Monogaphs
Lehnik-Habrink P, Hein S, Philipp R, Win T, Nehls I (2006) Erarbeitung und Validierung von Verfahren zur Bestimmung von polychlorierten Biphenylen und polychlorierten Terphenylen in organischen Materialien. Bundesanstalt für Materialforschung und -prüfung (BAM), ISBN 3-86509-481-3
Liang Y, Martinez A, Hornbuckle KC, Mattes TE (2014) Potential for polychlorinated biphenyl biodegradation in sediments from Indiana Harbor and Ship Canal. Int Biodeterior Biodegradation 89:50–57
Prinz H, Strauÿ R (2006) Abriss der Ingenieurgeologie, 4th edn. Spektrum Akademischer Verlag
Rahm H, Schweden U, Selent K, Düwel U (2015) Belastungen von Oberflächengewässern und von aktiven Grubenwassereinleitungen mit bergbaubürtigen PCB (und PCB-Ersatzstoffen)—Ergebnisse des LANUV-Sondermessprogramms 2015. Landesamt für Natur, Umwelt und Verbraucherschutz Nordrhein-Westfalen
Rommel R, Rommel J, Schneider J, Schneider V (1998) Literaturstudie zum Transfer von organischen Schadstoffen im System Boden/Pflanze und Boden/Sickerwasser. Landesanstalt für Umweltschutz Baden-Württemberg
Schwarzbauer J (2017) Detailbericht 6: Risikoanalyse PCB und weitere organische Stoffe. In: Gutachten zur Prüfung möglicher Umweltauswirkungen des Einsatzes von Abfall- und Reststoffen zur Bruch-Hohlraumverfüllung in Steinkohlenbergwerken in Nordrhein-Westfalen, Teil 1. Ministerium für Klimaschutz, Umwelt, Landwirtschaft, Natur- und Verbraucherschutz des Landes Nordrhein-Westfalen; Ministerium für Wirtschaft, Energie, Industrie, Mittelstand und Handwerk des Landes Nordrhein-Westfalen
Tehrani R, Van Aken B (2014) Hydroxylated polychlorinated biphenyls in the environment: sources, fate, and toxicities. Environ Sci Pollut Res 21(10):6334–6345
Umweltbundesamt (2018) Dioxine und dioxinähnliche PCB in Umwelt und Nahrungsketten. Bundesumweltamt, ISSN 2363-829X

# Chapter 3
# Wastewater Treatment Plants Advantage to Combat Climate Change and Help Sustainable Water Management

**Kapilkumar Nivrutti Ingle, Mulugeta Chanie Fenta, Koichi Harada, Akash Sopan Ingle, and Atsushi Ueda**

**Abstract** Wastewater treatment plants (WWTP) can provide water and nutrients for plant growth for small scale agricultural farming. Cultivation is possible if the parameters in wastewater treatment plants are available within the usable range. The waste treatment plants use wastewater mainly originating from underground, which is less affected by climate change than surface water and can be used for several purposes even during climate change. This chapter aims to assess the characteristics of treated wastewater at waste treatment plants and discuss its possibility to use the water, surrounding space and nutrients for small scale agricultural purposes. We assessed systematically the worldwide open sources works and use selected four wastewater treatment plants data of Kumamoto city of Japan. The previous results

---

**Supplementary Information** The online version contains supplementary material available at https://doi.org/10.1007/978-981-16-6573-8_3.

---

K. N. Ingle (✉)
Department of Ecology, University of Szeged, Középfasor 52, 6726 Szeged, Hungary

K. N. Ingle · A. Ueda
Department of Preventive and Environmental Medicine, Graduate School of Medical Sciences, Kumamoto University, Kumamoto, Japan

M. C. Fenta
Doctoral School of Geosciences, Department of Mineralogy, Geochemistry and Petrology, University of Szeged, Szeged, Hungary

School of Earth Sciences, Bahir Dar University, Bahir Dar, Ethiopia

K. Harada
Department of Microbiology and Environmental Chemistry, School of Health Sciences, Kumamoto University, Kumamoto, Japan

Department of Medical Technology, Faculty of Health Science, Kumamoto Health Science University, Kumamoto, Japan

A. S. Ingle
Department of Civil Engineering, Zeal College of Engineering and Research, Pune, India

A. Ueda
Asian Health Promotion Network Center (AHP-Net), Kumamoto, Japan

 
P. Kumar et al. (eds.), *Water Resources Management and Sustainability*,
Advances in Geographical and Environmental Sciences,
https://doi.org/10.1007/978-981-16-6573-8_3

and our data analysis of the pollutant load per capita to the plant and pollutant discharge per capita from the plant showed that all plants are efficiently reduced the pollutants loads. Still, there are variations in nitrogen and phosphorous concentration in the final effluent discharge. On that basis, we found that the treated water is useful for agricultural purposes, and moreover, the nitrates and phosphate values released by all plants are adequate to cultivate microalgae.

**Keywords** Wastewater treatment plant · Pollutant discharge · Nutrients · Small scale agriculture · Climate variability

## 3.1 Introduction

The recent global air and ocean temperature increase is considered indisputable evidence of climate change. It has already started to have adverse consequences for water resources and will have severe implications in the upcoming decades. Some of the factors that increase these values are population growth, urbanization, land-use changes, increasing demands for water and energy, improving living standards, shifting agricultural works, growing industrial works and change of economic activities. All these are profoundly likely to harm water resources. These adverse effects due to climate change and its impact on natural resources along with mitigation strategies must be studied for sustainability (Alcamo et al. 2007).

Climate predictions and current observations suggest that the hydrological system, including river flows and regional water resources, are anticipated to have the most substantial impact on climate change. As an example, the quantity of great inland flood devastations per decade that occurred between 1996 and 2005 is twice as large as between 1950 and 1980 that resulting in increased economic losses by a factor of 5 (Kron and Berz 2007). The increase in flood risk is observed due to increased climate variability as a result of an increase in temperature (Kundzewicz et al. 2007). Around the world, many major river systems are fed by snowpack and melting glaciers and global warming is likely to have an effect on snowmelt and related runoff. Glaciers are affected by several hydrological variables including wind speed, precipitation, humidity and typically the temperature that makes them a good indicator of global warming. Rivers found at higher latitude and altitude may experience an increase in discharge due to melting of glaciers though there is a decline of precipitation (Dyurgerov 2003). Global warming is having a noticeable influence on glacier retreat on every continent (Milner et al. 2017). Warmer temperatures can also affect water quality in many ways such as reducing dissolved oxygen level, and reduction in stream and river flow as well as an increase in contaminant load to water, algal blooms and the possibility of coastal region invasion by saltwater intrusion. The release of sulfuric and nitrogen compounds to the atmosphere results in acid rain which degrades the water quality and changes the chemical content variations in streamflow and increases chemical loads to rivers. Thus a proper water management technique, including its reuse is essential to provide a better quality of water.

Water, once an abundant natural resource, is now becoming a more valuable commodity due to its overconsumption and draughts. Water resources management encompassing the action of planning, developing, distributing, managing the optimum use and strategic reuse of water resources is required. Such a management plan will consider all competing water demands and seek to allocate water on an equitable basis to satisfy all of its uses and requirements. Therefore, for such water management plan to be successful, sustainable and to reduce the overexploitation of fresh groundwater, it is crucial to reuse wastewater after its treatment. Wastewater treatment and its reuse has been evolved and advanced throughout human history. The untreated municipal wastewater has been reused for many centuries to divert human waste outside the urban settlements. Also, for centuries, land application of domestic wastewater is a common practice that has seen many stages of development that led to a better understanding of treatment technology. Therefore, it leads to having processing techniques for the eventual outcome of water quality standards of the treated wastewater. The increase in population growth and urbanization with time imposes challenges for sustaining water resources and disposing of wastewater. Nowadays, wastewater is usually transported to a centralized wastewater treatment plant (WWTP) through collection sewers at the lowermost height of the gathering system generally close to the point of discarding to the environment. The centralized WWTPs are mostly arranged to route wastewater to these remote locations for treatment. Water reuse in urban areas is often inhibited due to lack of dual distribution systems (Metcalf & Eddy et al. 2014). These make it highly essential to provide proper wastewater treatment facility and water reuse strategy along with its reclamation in quality.

Treatment procedures in wastewater reclamation are implemented either separately or together with domestic water quality to achieve reclaimed water quality aims. Water reclamation might develop and use several stage treatment processes, flow diagrams and operations to meet the water quality requirements of a particular reuse application. The vital task is to point out significant factors affecting water reclamation technology. The choice of water reclamation technology might be affected by several factors. The critical aspect for such purpose are the kind of water reuse application, the objectives of reclaimed water quality, the characteristics of sources water to the wastewater, suitability with the present environments, process flexibility, energy and chemical requirements, operating and maintenance requirements, personnel and staffing requirements, residual disposal options and environmental constraints (Esposito et al. 2012). Water recycling is a crucial process in water treatment plants that applies the reuse of treated wastewater for valuable purposes including small scale agriculture and landscape irrigation, industrial applications, toilet flushing and replacing groundwater systems (groundwater recharge) (Exall 2004). Water reuse consents the public to be less reliant on groundwater and surface water resources and helps to decrease the alteration of water from sensitive ecosystems. Furthermore, water reuse may reduce the nutrient loads from wastewater emancipations into waterways, thereby reducing and avoiding pollution (Kim et al. 2008).

The purpose of this chapter is to assess the characteristics of treated wastewater at wastewater treatment plants at selected sites in Japan. The chapter aims to evaluate the nutrient content of treated water and its suitability for small scale agriculture and other purposes. The relevance of treated water from the treatment plants is planned to be assessed mainly based on the two main indicator parameters; pollutant loads per capita (PLC) to the wastewater treatment plant and pollutant discharge per capita (PDC) from the wastewater treatment plant (Tsuzuki 2006). Each of this parameters is calculated from the biochemical concentrations parameters of the treated and untreated water. Therefore the chapter objectives to calculate those indicators from the biochemical concentrations of input and output water of four selected treatment plants in Japan. The fresh groundwater is less affected by climate change and supplies water for the cities and then wastewater to the treatment plants. The chapter also aims to discuss the possibilities of using the treated water, surrounding space and nutrients for small scale agricultural purposes even during climate change.

### *3.1.1 Selected Case Study Area*

The case study area of our research is Kumamoto city, a capital city of Kumamoto Prefecture and a third populous city in Kyushu island of Japan, known as 'city of water' for its aquatic environment. The use of groundwater as one of the primary sources for domestic and societal purposes is a vital feature of Kumamoto city. The city public sewer started in 1948 as a part of post-war reconstruction of the city to serve a population of 48,000 residing in the city area of 278 hectares. The essential components of the wastewater treatment system are domestic wastewater sewerage, industrial non-hazardous wastewater sewerage, sanitary sewage pipes, pumping stations and treatment plants. Kumamoto city has four treatment plants which are managed by the city government and serve the city's major population. The four treatment plants of Kumamoto city namely; Chubu, Tobu, Nanbu and Seibu sewage treatment plants are situated and responsible for the disposal and treatment of sewerage from the central, eastern, southern and western parts of Kumamoto city respectively (Fig. 3.1).

The central and eastern part of the city is generally residential and commercial zones respectively, served by two treatment plants (Chubu and Tobu). These plants have two separate lines as Chubu line A and B and Tobu line A and B, which treats the main part of the city wastewater (Fig. 3.2). The western part is a new settlement and not much crowded.

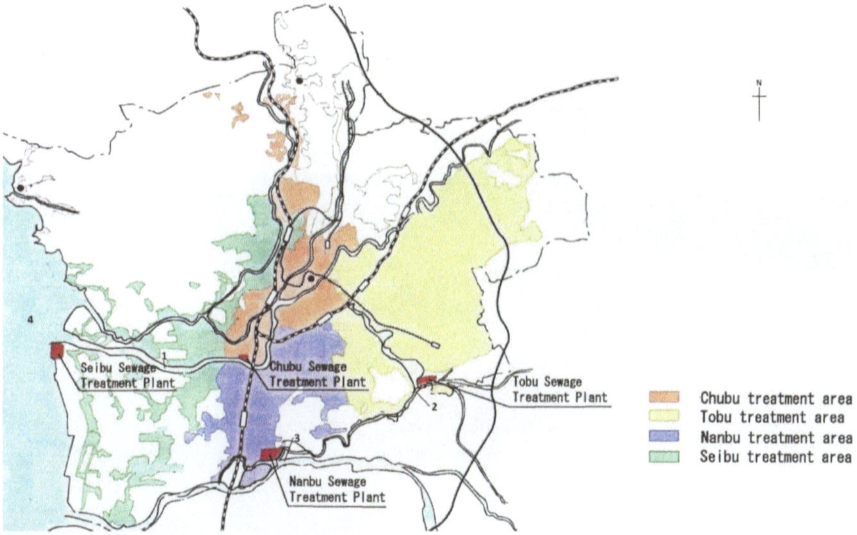


**Fig. 3.1** Map of Kumamoto city with sewerage treatment plants' service areas in 2008. *Source* Booklet of central wastewater treatment plant-2007, Kumamoto city

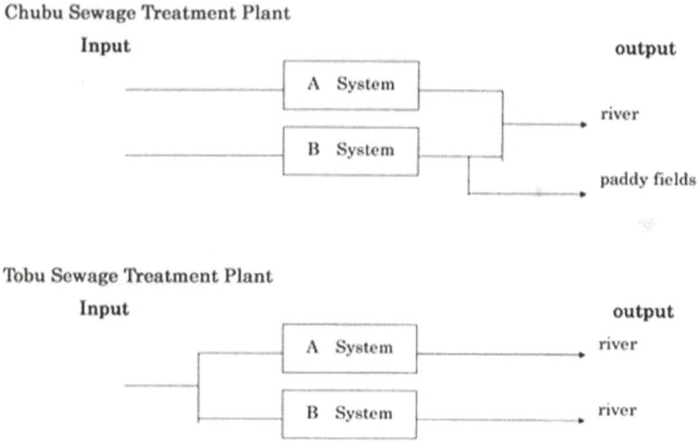


**Fig. 3.2** Arrangement of two parallel treatment systems of two treatment plants

## 3.2 Material and Methods

The necessary data and other information for this investigation were gathered from the four treatment plants and the wastewater management system of Kumamoto city. The data collection was focused on specific parameters that play an essential role for algal growth used to produce biofuel, vegetables and fruits (such as cabbage, salad, carrot, potatoes, tomatoes, etc.) that can be cultivated in a small farming area

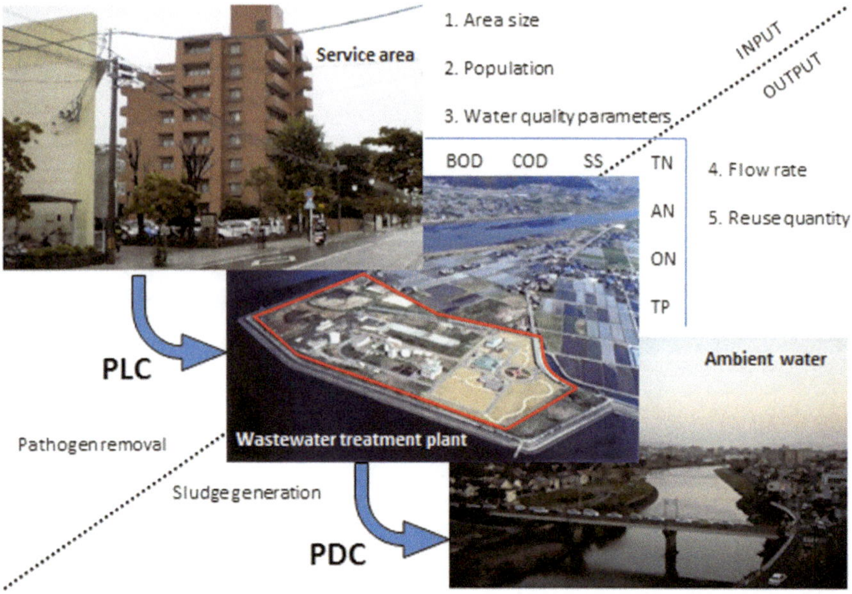


**Fig. 3.3** Indicators selected for this study are based on a few water quality parameters

close to the treatment plants. These data focus on the environmental parameters used to calculate the indicators, PLC and PDC, which are used to know the number of nutrients that come to the four WWTP and discharge to the ambient water bodies from them as shown in Fig. 3.3. They are also useful to estimate whether the treated water is suitable to cultivate the biofuel producing microalgae, vegetables and fruits in small farmland. These data were collected from the treatment plant test results by the plant authorities, governmental officials, websites, annual reports, visiting to plant and taking questionary survey results of the city and treatment plant personals.

These indicators are comprised of seven environmental parameters namely; biochemical oxygen demand ($BOD_5$), chemical oxygen demand (COD), suspended solids (SS), total nitrogen (TN), ammonium nitrogen (AN), organic nitrogen (ON) and total phosphorous (TP). The monthly data for the concentration of the seven environmental parameters were collected from April 2006 to March 2007 and the treated water agricultural sustainability parameters (PLC and PDC) were calculated using the monthly and yearly based formula (Tsuzuki 2006).

$$\mathrm{PLC} = \frac{\sum_{i} \mathrm{WQ}_{\mathrm{inf},i} \times \mathrm{FR}_i \times \mathrm{DAY}_i}{\mathrm{POP} \times 365} \tag{3.1}$$

and,

$$\mathrm{PDC} = \frac{\sum_{i} \mathrm{WQ}_{\mathrm{eff},i} \times \mathrm{FR}_i \times \mathrm{DAY}_i}{\mathrm{POP} \times 365} \tag{3.2}$$

where PLC is pollutant load per capita, $\mathrm{WQ}_{\mathrm{inf},i}$ is water quality in influent in the month $i$ (mg/l), PDC is pollutant discharge per capita, $\mathrm{WQ}_{\mathrm{eff},i}$ is water quality in the effluent in the month $i$ (mg/l), $\mathrm{FR}_i$ is flow rate or treated wastewater volume in the month $i$ ($m^3$/day), $\mathrm{DAY}_i$ is the number of days in a month $i$ and POP is population served by the treatment plant.

The data from two treatment plants (Chubu and Tobu) were taken as the average data of the two lines and considered as the data of the entire plant for this work. In Japan, WWTPs can treat nontoxic industrial wastewater if toxic elements are already removed or treated by industrial effluent treatment plant (Ingle et al. 2011). Here we are not considering the impacts of rainwater if any. The PLC (g/person/day) indicates the characteristics of input wastewater from the community served by WWTPs (Benedetti et al. 2008) and PDC (g/person/day) shows (Tsuzuki 2006) the discharge of pollutants to ambient water bodies like the coastal sea, rivers, etc. Moreover, the calculated PLC values enable us to infer the social life of the people and seasonal changes in the routine awareness of the environment.

## 3.3 Results

### *3.3.1 Pollutant Loads to Wastewater Treatment Plant Per Capita (PLC)*

A load of pollutants to the wastewater treatment plant is generally high if the plant is serving a large population of the city. The population size of Kumamoto city when this study was conducted was 556,806 who reside in a total area of 9313 hectors. Tobu WWTP plant served approximately 45% of people living in the area while Chubu plant served 14%, Nanbu plant 15% and Seibu plant 7% of the total population. The rest part of the city is served by two more plants managed by the prefectural government that we did not include in our case study. The large size of the site area of Seibu plant and Nanbu plant denotes future planning of urbanization in their regions. Chubu plant can use its total planned capacity to treat the wastewater amounting to 94,200 $m^3$/day, but average inflow recorded in 2006–2007 is 69,693 $m^3$/day. Nanbu plant has a planned ability to handle 73,500 $m^3$/day wastewater, with potential to treat 42,000 $m^3$/day; however, the recorded average inflow amount is 31,502 $m^3$/day. Seibu treatment plant is located near the coast and is developed for future prospects with a planned capacity of 54,900 $m^3$/day. However, the recorded average inflow of wastewater is 8355 $m^3$/day. Tobu plant has the potential to treat 141,000 $m^3$/day wastewater with a planned capacity of 204,000 $m^3$/day whereas its average inflow is 115,331 $m^3$/day.

The sources of sewage are generally domestic, included black water containing human faces with urine, greywater from kitchen sinks and bathrooms, outflows from washbasins and washing machines, etc. The average values of selected environmental parameters; $BOD_5$, COD, SS, TN, AN, ON and TP concentration to sewage treatment plants are given in Table 3.1.

The incoming loads to all pollutants present as influents, expressed in grams per capita per day, is higher for Chubu plant followed by Tobu, Nanbu and Seibu plant respectively, as shown in Fig. 3.4.

Although the treatment technologies of all WWTPs are similar, there are differences in the values of treated water quality parameters. These differences might be due to variations in the design of plants, the components of inflow water, operators skills, technical problems, local water consumption and the season of the year. The average daily inflow of wastewater per capita to Chubu plant is approximately

**Table 3.1** Environmental parameters and water characteristics to and from treatment plants

| Factors | Parameters | Chubu | Tobu | Nanbu | Seibu |
|---|---|---|---|---|---|
| Influents (water before treatment) | Water quantity treated (annual) (l) | 25,437,984 | 42,095,740 | 11,498,410 | 3,049,541 |
| | BOD (mg/l) | 236.5833 | 213.9167 | 193.1667 | 104.4167 |
| | COD (mg/l) | 82.2083 | 93.8750 | 83.5833 | 58.1667 |
| | SS (mg/l) | 159.5417 | 195.8750 | 133.6667 | 83.1667 |
| | TN (mg/l) | 34.7208 | 54.3792 | 48.4917 | 27.1833 |
| | A N (mg/l) | 21.4208 | 32.6542 | 36.3750 | 17.0917 |
| | ON (mg/l) | 12.9625 | 21.6208 | 12.0000 | 10.0917 |
| | TP (mg/l) | 4.0333 | 7.2417 | 5.0583 | 2.4083 |
| Effluents (Water after treatment) | Water quantity output (annual) (l) | 23,651,230 | 37,020,322 | 11,183,427 | 2,944,236 |
| | BOD (mg/l) | 5.0000 | 7.3333 | 5.6667 | 1.2833 |
| | COD (mg/l) | 7.3875 | 9.1083 | 10.6667 | 6.5417 |
| | SS (mg/l) | 2.2208 | 3.3750 | 3.8333 | 1.6000 |
| | TN (mg/l) | 14.9833 | 21.3792 | 34.2167 | 18.2083 |
| | A N (mg/l) | 5.1875 | 13.0792 | 27.4250 | 6.4750 |
| | ON (mg/l) | 2.8542 | 3.2333 | 2.8000 | 1.7500 |
| | TP (mg/l) | 1.5833 | 2.5708 | 1.7833 | 0.1500 |
| Population | | 79,646 | 249,953 | 85,727 | 41,442 |
| Area searve(ha) | | 1358 | 3653 | 1455 | 785 |
| Site area ($m^3$) | | 93,850 | 120,350 | 120,700 | 111,000 |
| Reuse water (l) | | 1,775,621 | 5,056,164 | 309,542 | 104,308 |
| Flow per day (l/day) | | 69,693 | 115,331 | 31,502 | 8355 |
| Sludge per year (kg/year) | | 1,113,242 | 1,925,360 | 544,065 | 99,661 |

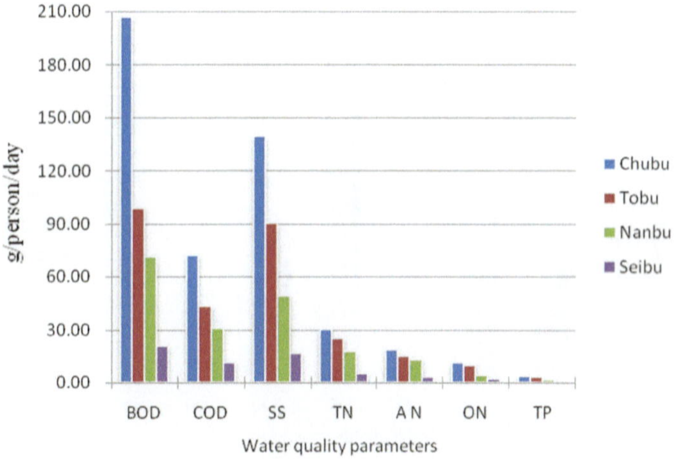


**Fig. 3.4** Average of annual loads of pollutants from the community to the treatment plant in g/person/day

twofold the sum amount of inflow to Tobu and Seibu plants and 5.6 times more than Seibu plant.

### *3.3.2 Pollutant Discharge from Wastewater Treatment Plant Per Capita (PDC)*

The amount of effluent discharged from a sewage system depends on the service area size, population and the amount of water used by the community. Figure 3.5 shows the pollutant discharges from all WWTPs. Chubu plants, with the exception of nitrogen, had a higher release of pollutants to the environment. Among all four WWTPs, Nanbu plant discharges a higher amount of nitrogen discharge per capita. BOD and SS are the most abundant constituents of WWTPs' outflow, which play an essential role to determine the environmental health of water bodies due to discharging treated wastewater. Chubu plant discharges effluents having BOD and SS concentration values of 5 and 2.2 milligrams per litre respectively and release a daily average of approximately 0.3 tonnes of BOD and 0.14 tonnes of SS into Shirakawa river. The discharge value of Tobu plant is less than half of Chubu plant, while more than twice the discharge value of Nanbu plant. Seibu plant discharges approximately 0.01 tonnes of BOD and SS per day. The National and City Government standards to release treated water into water bodies are less than 15 (mg/l) for BOD and 40 (mg/l) for SS, which are obeyed by all plants. Seibu plant, being the newest plant of all plants, releases the lowest PDC value compared to other plants. However, it discharges its effluent into ambient water bodies (the sea) results in a higher dilution capacity compared to other plants.

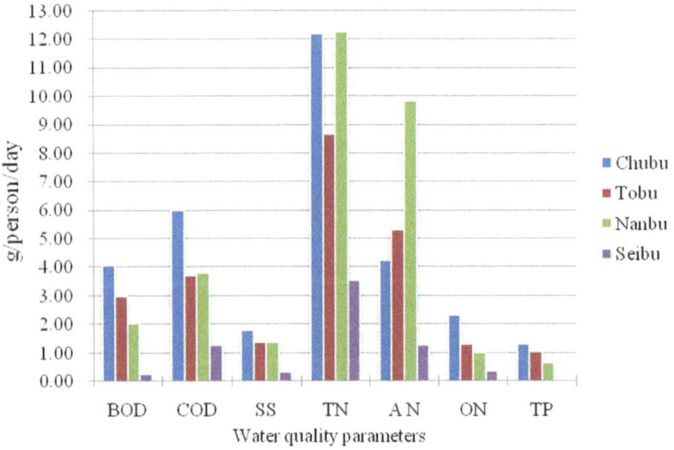


**Fig. 3.5** Average of annual pollutants discharge to ambient water bodies from a treatment plant in g/person/day

In some cases, few amounts of treated water have been already reused in a paddy field. However, a significant quantity of treated water is discharging to ambient water bodies of rivers and finally, the sea. The Chubu plant is located to the side of Shirakawa river, the main river in the prefecture and removes the treated water into it. Tobu plant is situated close to Kiyama river, a tributary river to Khase river and discharges treated water into it. Nanbu plant located near to Khase river and releases treated water into it. In contrast, Seibu plant is situated at the mouth of the Shirakawa river and removes treated water into the Ariake sea, the place where all rivers end and recharge the sea.

## 3.4 Discussion

Although all WWTP efficiently treated the wastewater, the nitrogen level in terms of PDC of treated water was higher compared to other pollutants which can be considered as nutrients for further use. In the process of urbanization, population growth in the rest of Kumamoto city can become an important reason to increase the number of pollutants discharged, through the increase in the use of water quantity, changes in lifestyle and land development. Generally, the urban population in any country would be more prosperous and affluent compare to the rural people. The growth of population leads to a change in the lifestyle and living standards due to the need for an improved quality of life. These interns will have a possibility of more consumption of natural resources, including water. As an example, changes in the toilet system from old-style Japanese toilets to flush toilets, or changes in the bath system leads to showers or tub baths with the use of different types of soaps. The changes in food style also play an essential role in the characteristics of greywater. This can result in a rising amount of wastewater generated per person and pollutant discharge to ambient water bodies. The impacts of improperly treated water on aquatic environments principally depend on, amount of effluent discharged, the quality of the effluent, receiving aquatic environment characteristics, assimilative capacity of the receiving ambient water bodies, climate and season.

The presence of chemicals in the effluent, particularly from industrial areas, may have long-term environmental effects. However, in the case of Kumamoto WWTPs, there is no possibility of such adverse impact because the industrial area was minimal and industries have their system to treat their effluents. Besides, there is a ban on the use of many chemicals, even in industrial areas. Therefore, there is no possibility of mixing industrial water into city wastewater. Kumamoto city treated wastewater has the potential to be reused for many purposes, including the issue of tackling global climate change.

### *3.4.1 Treated Water and Climate Change*

Most of the water supplies to urban cities come from groundwater and surface water. In comparison to surface water, groundwater is less affected by climate change, especially for the short term, seasonal factors. However, the associated structures and facilities are vulnerable to the adverse effects of climate change (Case 2008). The consummation of groundwater as freshwater is high and worldwide 3–5 billion people depend on groundwater as a drinking water source (Kundzewicz and Doll 2009). Population growth is pressure on water supply in addition to climate change and the overall factors lead to increase the need for reuse of water exceptionally diverse climate change (such as droughts) become more predominant (Major et al. 2011). In most places like Kumomanto city of Japan, where the sources of freshwater and the input to WWTP is groundwater, are expected to be less affected by climate change. It can form the basis of adaptation programs securing food through small scale agriculture. This is mainly because the groundwater systems create a buffer against more unpredictable rainfall (Kundzewicz and Doll 2009). Therefore, groundwater resources are more likely affected by increased demand for irrigation, domestic and industrials use than the changes to recharge due to climate change (Taylor et al. 2013). The global groundwater recharge is unlikely to be affected by more than 10% due to climate change (Kundzewicz and Doll 2009).

The treated water that is primarily used for household applications will have use in combating climate change for farmers who are mainly affected by extreme weather conditions including drought, severe heat, flooding and other shifting climatic trends. The small scale agriculture will help to combat climate change. If agricultural activities include cultivating crops like corn, wheat, rice in addition to vegetables and fruits, they will extract nitrogen from the air to use on their own and radically reduce the need for humanmade fertilizers. These kinds of agricultural works are an essential step towards a carbon zero future for agriculture. The large scale agricultural enlargements lead to having significant land that causes deforestation and diminishes draining of wait land. These agricultural works will later reduce the ability of the natural ecosystem to absorb and store carbon dioxide while small scale agriculture using the reuse of treated water can minimize the effect. Therefore, smalls scale agriculture using treated water will help to preserve the natural habitat and help to combat climate variation in turn.

The use of a large amount of fresh water for irrigation can broadly impact the world freshwater resources. As an example, the study by Döll et al. (2012) showed that in 2000, irrigation accounts for ~70% of global freshwater withdrawals and ~90% of consumptive water use. The same research indicated that globally, groundwater is the source of one-third of all freshwater withdrawals, supplying an estimated amount of 36%, 42% and 27% of the water used for domestic, agricultural and industrial purposes respectively. The use of groundwater for large scale agriculture will deplete the groundwater and it may fever future climate change and deficiency of food demand. The dependence on rainwater for agriculture similar to developing countries intern cannot combat climate change and even can trigger food demand shortage. The study and report by IPCC (2008) predicted that climate change over the next century would affect rainfall patterns, river flows and sea levels all over the world. According to the study by Jarvis et al. (2010), agriculture is one among which will be severely affected during the coming hundred years due to unprecedented rates of changes in the climate system. The use of recycled water (treated water) will, in turn, be intensified significantly for small scale agriculture and production of microalgae in a little farming land without affecting the ecosystems of the environment. The reuse of treated water can help to alleviate the shortage of water and food due to climate change.

Treated water is also used to circulate in the city as the climate change adaptation strategy. In various cities, rainwater on the surface circulates and flows in narrow channels that help to grow different green, attractive and ornamental plants (Fig. 3.6). The green plants help to increase the cities' beautification and combat climate change. Treated water can also help to reduce greenhouse gases (GHG) emissions from wastewater in the future due to the action of increasing wastewater collection and treatment. Moreover, as an increased scarcity of water resources, wastewater reuse will become more necessary as climate change accelerates. The value of treated water recuse is expected to increase in the coming decades due to intense climatic extremes associated with climate change which intern results in food scarcity. When the treated water sources recharge groundwater, the world's largest part of stored freshwater, it will play a significant role in sustaining ecosystems and enabling long-term human adaptation to climate variability and change.

### 3.4.2 *Treated Water and Agriculture*

The reuse of wastewater in agriculture involves further use in treated wastewater for crop irrigation (Claro 2008) which is an efficient means for water resources management. The reuse helps to minimize the requirement for planned water supply and compensate for water scarcities caused by seasonality and irregular availability of water resources for irrigation (Jaramillo and Restrepo 2017). Even though the reuse of wastewater is the earliest practice, it has not always been appropriately managed or met quality standards according to use. Consequently, the awareness about wastewater use has been progressed with the history of humankind. The Food and Agricultural Organization (FAO) of the United Nations has developed several guidelines relevant to the use of wastewater in agriculture. In 1987, the wastewater quality guidelines for agricultural use were issued that focused on the degree of restriction of water use to salinity, infiltration and toxicity parameters of specific ions (Ayers and Wescott 1985). The work by FAO in 1999 proposed guidelines for the agricultural reuse of treated waters and treatment requirements. The guidelines classified the types of agricultural water reuse depending on the kinds of irrigated crops (Pescod 1992).

The agricultural use of treated wastewater assist human wellbeing, the environment and the economy. It provides an alternative practice for diverse regions challenged with water deficiencies and increasing urban populations with growing water demands (Jaramillo and Restrepo 2017). It has crucial importance, particularly for areas affected by the decline of surface and groundwater resources instigated by climate variability and climate change. Wastewater sourced pollutions, which are not usually treated before reaching the surface channels and its associated aquifer pollution, affect the availability of freshwater resources (Jaramillo and Restrepo

**Fig. 3.6** Circulation of the rainwater from Kumamoto city to combat the climate change issue

2017). On the other hand, the agricultural use of treated wastewater has a tremendous advantage by reducing the pressure on freshwater resources serving as a vital alternative irrigation water source. The benefit is significant as agriculture is the highest global water user who consumes 70% of available water resources. Furthermore, wastewater reuse increases agricultural production in regions suffering from water shortages, thus contributing to food security (Roy et al. 2011).

### 3.4.3 Possible Utilization of Treated Water for Microalgae Cultivation

The sustainability of biofuel is uncertain due to the requirement of various resources such as land, water, nutrients, etc. and environmental issues possibilities and processing technological uncertainty (Gerbens-Leenes et al. 2009). For example, the use of soybean, palm, sunflower and rapeseed have no technical restrictions but require land and water (Sensoz and Kaynar 2006). This can have inquiries about food security and can have a dispute with food versus fuel and environmental problems. However, microalgae can serve as feedstock for fuel by converting the carbon dioxide in sunlight into potential biofuels, including biodiesel (Banerjee et al. 2002), biohydrogen and methane (Spolaore et al. 2006). The possibility of converting both types of microalgae, prokaryotic and eukaryotic types, into various biofuels and byproducts attracted many companies to cultivation at the mass level for their lipid accumulation properties (Chisti 2007). However, the mass level production of microalgal biomass is not cost-effective due to maintenance expense requirements.

The vast land requirement for large scale cultivation is more challenging. The use of transparent polymer films such as polyethylene or polyurethane (Kim et al. 2016) can be an expensive way of farming. This type of cultivation may have few environmental issues due to poor management. The open-pond outdoor microalgae cultivation system is the most popular and economical method which is implemented by many companies. Its low operations accost and low energy input requirement makes it be more attractive farming system while there are contamination possibility issues. Concerning the water footprint, microalgae bioenergy is beneficial compare to other bioenergy sources. A study by Yang et al. (2011) recorded a microalga, *Chlorella Vulgaris*; biodiesel has a water footprint of 3726 kg-water/kg-biodiesel. The reuse of treated water can reduce this and simplify the cultivation system.

Considering our case study WWTP of Kumamoto city of Japan, the nitrates and phosphate values are discharged by all plants in sufficient quantities to cultivate the microalgae. The phosphorous and nitrogen concentrations in city wastewater vary from 30 to 40 mg/l and 5 to 10 mg/l, respectively. The total nitrogen (TN) and total phosphorus (TP) requirement for algal growth vary from 15–90 mg/L and 5–20 mg/L values respectively and the nitrogen to phosphorus ratio (N/P) is approximately 3.3. On the other hand, the Chlorella and Scenedesmus species of microalgae can grow in a wide range of wastewaters and can decompose nitrogen and phosphate in

10 days. The microalgae that can be cultivated for the biofuel showed a harvesting cycle of 1–10 days, which makes it possible to harvest numerous times in a short duration (Chisti 2007). The open-pond system is a simple method based on ground level ponds in which intermixing of water may be possible by the use of paddle wheels. This system requires low operating, capital cost and power than closed-loop photobioreactor systems (Benemann et al. 2012), which make it accessible in commercial algae producers (Lundquist et al. 2010).

### *3.4.4 Can Treat Water be Used for the Household Appliance?*

Even after advanced treatment of the wastewater, there is no guarantee for continuous chemically and microbiologically indisputable drinking water quality. Thus, the substitution of drinking water with treated water may be possible for several purposes other than potable water, for instance, garden irrigation and toilet flushing. In addition to sludge arising after drinking water treatment, the addition of considerable amounts of chemicals can be minimized. The water reuse systems available from stormwater or greywater can be designed cost-effective with proper operation, presenting no hygienic risk or discomfort to the users (Nolde 1995). To use the treated water for household appliances, the treatment and the distribution of treated water should not demand more energy consumption and chemicals in comparison with conventional systems. The water from recycling systems should fulfil specific criteria such as hygiene, environmental tolerance, technical and economic feasibility (Nolde 2000) to be used for household appliances.

According to Delphi study, about 76% of surveyed international experts in "Water Technology, 2010" considered it as technically feasible to use treated wastewater in households by the year 2010 with no known risk. The "Guidelines for Water Reuse" published in 1992 by the US Environmental Protection Agency (EPA) describes the treatment stages, water quality requirements and monitoring tools (US EPA 1992). According to the report, reclaimed water used for toilet flushing must undergo final filtration and disinfection. The effluent should not have detectable faecal coliforms in 100 ml of the treated water, a $BOD_5$ of less than 10 mg/L and a residual $Cl_2$ of less than 1 mg/L and the resulting $Cl_2$ should be continuously monitored (US EPA 1992).

Treatment of wastewater for service use should follow a sedimentation stage, biological treatment, a clearing stage and final UV disinfection. According to the study, Funnel-shaped sedimentation tanks having automated sludge-removing devices proved to be most effective. Biological treatment can be followed in a plant itself with the provision of either a vertical-flow soil filter or a multiple-stage Rotary biological contactor (RBC) alternatively a trickling filter, coupled with a clearing tank to remove the existing biomass. The treated water should also be eventually disinfected by UV before storing in the service water tank. Distribution of service water is achieved with a booster pump for pumping to heights.

The quality requirements for non-potable, i.e. service water uses must be scientifically justified with a risk assessment analysis desirable in every case. As per sustainable water concepts, lower energy and chemical demand should be achieved in service water systems than conventional systems. Wastewater treatment for service use has proved to be technically feasible. There are enough positive examples to verify that the total water for toilet flushing (about 15 to 55 l/person/day) can be substituted with service water without a hygienic risk or discomfort. It should be possible soon to have a dual water system in households with two water qualities, the first one with high-quality drinking water originating primarily from natural freshwater resources and the second one with water quality for all other service uses. This should bring environmental and sustainable relief to both the water and energy sectors.

## 3.5 Conclusion

In this chapter, we reviewed the possibility of treated wastewater application in small scale agriculture, microalgae cultivations and household service water appliances which help to conserve water resources and combat climate change. A case study in Kumamoto city water treatment plant, Japan, water and treated water data has been taken and analyzed to find the BOD and COD values of the four treatment plants. We found that the city wastewater treatment plant is efficient to remove the solids in water, reducing the BOD and COD values and keeping the right amount of nutrients in the treated water which can be useful for small scale agriculture, microalgae cultivation. The clear treated water has the potential to penetrate the sunlight and create a better environment for microalgae. It is possible to make some arrangements in the treatment plant for use that treated water for cultivation of biofuel producing microalgae because a majority of requirements of such farming are already present in wastewater treatment plants. Moreover, it is identified that the treated water could also be applied in small scale farming land and possible also to use as a service household water for toilet flushing and gardening service water. In both cases, treated water can help to combat climate change by reducing vast farmlands that cause deforestation, minimizing the shortage of water for several applications during climate changes since most WWTPs are supplied water initially from groundwater.

**Acknowledgements** The authors are thankful to the research staff of the Department of Preventive and Health Medicine, Graduate School of Medical Sciences, Kumamoto University, Japan and authorities of wastewater treatment plants in Kumamoto city, Kumamoto city government and Kumamoto prefectural government authorities. KI is a former monbukagakusho scholar at Kumamoto University, Japan. KI and MF are Stipendium Hungaricum scholars.

## References


Alcamo J, Flörke M, Märker M (2007) Future long-term changes in global water resources driven by socio-economic and climatic changes. Hydrol Sci J 52:247–275. https://doi.org/10.1623/hysj.52.2.247

Ayers R, Wescott D (1985) Water quality for agriculture. FAO, Rome, Italy

Banerjee A, Sharma R, Chisti Y, Banerjee UC, Banerjee A, Sharma R, Chisti Y, Banerjee UC (2002) Botryococcus braunii: a renewable source of hydrocarbons and other chemicals. Crit Rev Biotechnol 22:245–279. https://doi.org/10.1080/07388550290789513

Benedetti L, Dirckx G, Bixio D, Thoeye C, Vanrolleghem PA (2008) Environmental and economic performance assessment of the integrated urban wastewater system. J Environ Manage 88:1262–1272. https://doi.org/10.1016/j.jenvman.2007.06.020

Benemann J, Woertz I, Lundquist T (2012) Life cycle assessment for microalgae oil production. Disruptive Sci Technol 1:68–78. https://doi.org/10.1089/dst.2012.0013

Case T (2008) Climate change and infrastructure issues. AWWA Res Found Drink Water Res Chang Specail Issue 1:15–17

Chisti Y (2007) Biodiesel from microalgae. Biotechnol Adv 25:294–306. https://doi.org/10.1016/j.biotechadv.2007.02.001

Claro R (2008) Reúso de água no contexto da gestão de recursos hídricos: impacto, tecnologias e desafios. Rodrigo Braga Moruzzi [1]. 8:271–294

Döll P, Hoffmann-Dobrev H, Portmann FT, Siebert S, Eicker A, Rodell M, Strassberg G, Scanlon BR (2012) Impact of water withdrawals from groundwater and surface water on continental water storage variations. J Geodyn 59–60:143–156. https://doi.org/10.1016/j.jog.2011.05.001

Dyurgerov M (2003) Mountain and subpolar glaciers show an increase in sensitivity to climate warming and intensification of the water cycle. J Hydrol 282:164–176. https://doi.org/10.1016/S0022-1694(03)00254-3

Esposito K, Tsuchihashi R, Anderson J, Selstrom J (2012) the Role of water reclamation in water resources management in the 21st century. Proc Water Environ Fed 2005:8621–8633. https://doi.org/10.2175/193864705783812792

Essa AM, El-Gayar KE, Awad MN, Al-Shaeri M, Salam MA, Alnashiri HM (2019) Bacterial purification of sewage wastewater effluents and its influence on germination indices of Vigna radiata and Hordeum vulgare. Polish J Environ Stud 28:3145–3153. https://doi.org/10.15244/pjoes/92523

Exall K (2004) A review of water reuse and recycling, with reference to Canadian practice and potential: 2. Applications. Water Qual Res J Canada 39:13–28. https://doi.org/10.2166/wqrj.2004.004

Gerbens-Leenes W, Hoekstra AY, van der Meer TH (2009) The water footprint of bioenergy. Proc Natl Acad Sci U S A 106:10219–10223. https://doi.org/10.1073/pnas.0812619106

Ingle KN, Harada K, Wei CNCN, Minamoto K, Ueda A (2011) Policy framework for formulating environmental management strategy for sustainable development of tanneries in India. Environ Health Prev Med 16:123–128. https://doi.org/10.1007/s12199-010-0168-8

IPCC: Climate change and water, Intergovernmental panel on climate change technical report IV (2008)

Jaramillo MF, Restrepo I (2017) Wastewater reuse in agriculture: a review about its limitations and benefits. Sustainability 9. https://doi.org/10.3390/su9101734

Jarvis A, Ramirez J, Anderson B, Leibing C, Aggarwal P (2010) Scenarios of climate change within the context of agriculture. In: Climate change and crop production. CAB International. In: Climate Change and Crop Production. CAB International

Kim SM, Im SJ, Park SW, Lee JJ, Benham BL, Jang T (2008) II: assessment of wastewater reuse effects on nutrient loads from paddy field using field-scale water quality model. Environ Model Assess 13:305–313. https://doi.org/10.1007/s10666-007-9093-7

Kim Z-H, Park H, Hong S, Lim S-M, Lee C (2016) Development of a floating photobioreactor with internal partitions for efficient utilization of ocean wave into improved mass transfer and algal culture mixing. Bioprocess Biosyst Eng 39:713–723. https://doi.org/10.1007/s00449-016-1552-6

Kron W, Berz G (2007) Flood disasters and climate change: trends and options—a (re-) insurer's view. Glob Chang Enough water all? 268–273

Kundzewicz Z, Doll P (2009) Will groundwater ease freshwater stress under climate change? Hydrogol Sci 54:65–75

Kundzewicz ZW, Mata LJ, Arnell NW, Döll P, Kabat B, Jimenez B, Miller KA, Oki T, Sen Z, Shiklomanov IA (2007) Freshwater resources and their management. Clim Chang 2007 Impacts, Adapt. Vulnerability. Contrib. of Working Gr. II to Fourth Assess. Rep. Intergov. Panel Clim. Chang, pp 173–210

Lundquist TJ, Woertz IC, Quinn NWT, Benemann JR (2010) A realistic technology and engineering assessment of algae biofuel production. https://doi.org/10.1556/1848.2015.6.1.6

Major D, Omojola A, Dettinger M, Hanson R, Sanchez-Rodriguez R (2011) Climate change, water, and wastewater in cities. In: Rosenzweig C, Solecki W, Hammer S, Mehrotra S (eds) Climate change and cities: first assessment report of the urban climate change research network, Cambridge, UK, pp 113–143

Metcalf & Eddy I, G.T. and H.S. and R.T. and F.B. (2014) Wastewater engineering: treatment and resource recovery

Milner AM, Khamis K, Battin TJ, Brittain JE, Barrand NE, Füreder L, Cauvy-Fraunié S, Gíslason GM, Jacobsen D, Hannah DM, Hodson AJ, Hood E, Lencioni V, Ólafsson JS, Robinson CT, Tranter M, Brown LE (2017) Glacier shrinkage driving global changes in downstream systems. Proc Natl Acad Sci U S A 114:9770–9778. https://doi.org/10.1073/pnas.1619807114

Nolde E (1995) Greywater recycling systems in Germany—results, experiences and guidelines. 203–210

Nolde E (2000) Erratum to "Greywater reuse systems for toilet flushing in multi-storey buildings—over ten years experience in Berlin" [Urban Water 1 (1999) 275–284]. Urban Water 2:260. https://doi.org/10.1016/s1462-0758(01)00002-4

Pescod MB (1992) Wastewater treatment and use in agriculture—FAO irrigation and drainage. M-56

Roy A, Das BK, Bhattacharya J (2011) Development and validation of a spectrophotometric method to measure sulfate concentrations in mine water without interference. https://doi.org/10.1007/s10230-011-0140-x

Sensoz S, Kaynar I (2006) Bio-oil production from soybean (Glycine max L.); fuel properties of Bio-oil. Ind Crops Prod 23:99–105. https://doi.org/10.1016/j.indcrop.2005.04.005

Spolaore P, Joannis-cassan C, Duran E, Isambert A, Génie LD, Paris EC (2006) Commercial applications of microalgae. J Biosci Bioeng 101:87–96. https://doi.org/10.1263/jbb.101.87

Taylor RG, Scanlon B, Döll P, Rodell M, Beek R van Wada Y, Laurent L, Leblanc M, Famiglietti JS, Edmunds M, Konikow L, Green TR, Chen J, Taniguchi M, Bierkens MFP, MacDonald A, Fan Y, Maxwell RM, Yechieli Y, Gurdak JJ, Allen DM, Shamsudduha M, Hiscock K, Yeh PJ-F, Holman I, Treidel H (2013) Ground water and climate change. Nat Clim Change 3. https://doi.org/10.1038/nclimate1744

Tsuzuki Y (2006) An index directly indicates land-based pollutant load contributions of domestic wastewater to the water pollution and its application. Sci Total Environ 370:425–440. https://doi.org/10.1016/j.scitotenv.2006.07.014

US EPA (1992) Manual on guidelines for water reuse

Yang J, Xu M, Zhang X, Hu Q, Sommerfeld M, Chen Y (2011) Life-cycle analysis on biodiesel production from microalgae: water footprint and nutrients balance. Bioresour Technol 102:159–165. https://doi.org/10.1016/j.biortech.2010.07.017

# Chapter 4
# Assessment of Community Participation in Water Governance in the Wassa Amenfi East District of the Western Region, Ghana

**Kofi Adu-Boahen, Ishmael Yaw Dadson, Faustina Ankomah, and Sender Kyeremeh**

**Abstract** This paper investigates the effectiveness of water governance at the local level using selected communities in the Wassa Amenfi East District of Ghana as a case study. Descriptive design was employed and eighty (80) respondents were selected by purposive and simple random sampling techniques. Interview guide and questionnaire were used in collecting data. Results indicate that chiefs and the traditional set up (informal actors) are the main architects involved in water governance in the selected communities and traditional management practices such as taboos and cultural practices were employed and considered effective and sustainable. Local decisions on water utilization were shaped by traditional or bottom-up water management principles with little or no assistance from the Government. The study, therefore, recommends awareness campaigns on the policy and the legal framework regulating access and control of water resources. This will ensure sustainable utilization and management of water resources. Traditional water management approaches should be evaluated, documented and used to complement scientific water resources management regimes.

**Keywords** Customs · Livelihoods · Management · Sustainable water resource management · Water governance · Traditional approach

## 4.1 Introduction

Water governance refers to the scope of the political, social, economic and administrative structures put up to control the development of water services at different levels of society (Atampugre et al. 2016). Efficient water governance helps to avoid

K. Adu-Boahen (✉) · I. Y. Dadson · S. Kyeremeh
Department of Geography Education, University of Education, Winneba, Ghana
e-mail: kadu-boahen@uew.edu.gh

F. Ankomah
Department of Geography, Adu Gyamfi Senior High School, Agona Jamasi, Ghana

 
P. Kumar et al. (eds.), *Water Resources Management and Sustainability*,
Advances in Geographical and Environmental Sciences,
https://doi.org/10.1007/978-981-16-6573-8_4

the recurrent mismatch between central government policies and goals, on the one hand, and the interest and expectations of the people on the other (Nompumelelo 2001). Water play essential roles in sustainable development by reducing poverty and promoting economic growth. It adds quality to human life and its absence or inadequacy makes life unbearable for the individual at the household level (Ghana Government 1996).

Nompumelelo 2001) has agreed that water crisis in the developing world and rural African communities, in particular, is one resulting from governance but not necessarily a crisis from water scarcity. Portable water coverage is very low in Ghana with about 45% as of 2009 for rural areas. If we desire to achieve the Sustainable Development Goals (SDGs), then issues of water governance have to be taken into consideration with all seriousness. This is so because water plays a key role in the various aspects of the MDGs (Eugene 2010).

The introduction of governance into water sectors is a growing concept, especially in the developing world. This has therefore raised a myriad of questions whose underpinnings are of great importance to this study. Some of these questions addressed are; who are the key actors in water governance at the local level; and how effective are they in playing their respective role? The paper will provide more awareness of the policy-making model, decision-making among the informal or the non-state actors, researchers, and development practitioners will be well informed about the existing local conditions.

### 4.1.1 Objectives of the Study

The study was tailored according to the following objectives.

1. To examine the key actors in water governance in the Wassa Amenfi East District (W.A.E.D).
2. To assess the extent of decision-making among the various stakeholders at the local level
3. To evaluate the effectiveness of the institutions responsible for water governance at the local level.

### 4.1.2 Institutions' Role for Water Governance

Over the past decade, a significant focus has emerged in literature on water governance. Bakker (2003) defines water governance as the range of political, organizational and administrative mechanisms by which communities communicate their concerns, absorb their inputs, make and execute decisions, and keep decision-makers responsible for the production and management of water resources. Karikari (2000) outlined the importance of maintaining governments' role in water governance, noting that when governments are committed to educating and empowering people, building trust and helping to develop common awareness, shared water governance works best. They also noted the importance of sustained capacity; noting that the benefit of shared water governance will only be realized in situations where financial and other resources are guaranteed and sustained over the long-term.

Institutions that are in charge of water governance play a vital role in ensuring the sustainable use of water. According to Nowlan and Bakker (2010), water institutions are agencies that put up rules that together describes action situations, delicate action sets, provide incentives and influence outcomes both in individual and collective decisions related to water development, allocation use and management. These institutions can be group into two main types; formal and informal institution. The formal is accepted and recognized by law and are documented, whilst the latter is comprised of the behavior and conventions (Leroy 2002).

### 4.1.3 Local Level Decision-Making on Water Governance

In 1994, the government launched the National Community Water and Sanitation Program (NCWSP) and in 1998, the NCWSP was transformed into Community Water and Sanitation (CWSA) by Act 564 and is responsible for coordinating and facilitating the implementation of the NCWSP in the District Assemblies. Political administration, planning and decision-making processes overburden the central government, and devolution of political power to the local level and has been borne out of one basic fact: that community participation in governance, planning and decision-making will make the process of development self-sustaining. The assumption is that bringing policy decisions that reflect local conditions closer to the people will induce total participation in programs which will bring about improvement in the well-being of the people. This has been the principal idea behind Ghana's decentralization policy and the concept of District Assembly. Thus the concept has led to the creation of basic units of government called the District Assemblies which receives the devolved powers.

Decentralization refers to the transfer of responsibility for planning management and resource allocation to the local level agencies. This transfer helps to overcome many of the problems of decentralized service provision. Decentralization is aimed at enhancing efficiency, equity, and sustainable resource use; mainly by limiting the distance between the decision-makers and beneficiaries (Karr 1991). The district assemblies are responsible for ensuring that adequate and wholesome water is provided in the district. This is by the help of decentralized policy where power, resource, and means are transferred to the local level to enable them to make policy and decisions on water. The Community Water and Sanitation Agency supports the District Assemblies in promoting sustainable safe water service in rural communities and encourage the participation of communities in the management of water services.

### *4.1.4 Water Management and Local Participation*

Water is essential in all aspects of our lives and an optimum quantity and quality is required to sustain life. With the increasing human population globally with the growth of technology, human societies must devote strict attention to the protection and adequate supplies of water. Degradation of water resource has long been a worry for human societies (Gumisai 2004). The focus of water management has shifted from where all attention was on the means of getting water for domestic consumption, hydropower generation and irrigation in agriculture to the setting where water management does not only consider the delivering of water and related services but by doing so in the way that balances and or compensates the competing interests of individuals, industries, agriculture and wildlife. It is also to maintain a good relationship between all the users who shares water resource and develop systems that will favor future generations (Global Water Partnership 2000).

Customary right and formal right (North 1993) govern the use of water in most sub-Saharan African countries. Public education and participation are identified as the basis for commitment, and coherence in the implementation of effective water governance (Global Water Partnership 2000). Rural communities in the Volta basin include the use of taboos and other cultural practices to protect natural resource including water over the past years (Meinzen-Dick and Knox 1999). Studies in other African countries according to Leroy and Tatenhove (2000) have shown that traditional practices are still in use and proven to be efficient in the management of water resource. Evidence of successful self-governance of natural resources by the local users themselves has engendered considerable optimism that delegating responsibility to organized local communities will enhance the efficacy, equity and sustainability of the resource base, and decrease the financial load on the state and its institutions (Hooghe and Marks 2003).

### *4.1.5 Conceptual Perspective on Water Governance*

In the scientific literature, there are a plethora of approaches to conceptualize water governance. Governance factors the increasing important mode of governing, where non- state and private corporate actors and networks participate, in the concoction and implementation of public policy or development of policy instruments that co-exist with existing government policy processes (Merriam-Webster 2008). Governance includes coordination and steering processes to determine the behavior of formal and informal institutions (Marks 2005). Self-organization, emergence and diverse leadership thus characterize governance. Major characteristics of governance regimes are the varied roles of non-state actors in the provision of water resources. Participatory approaches have become a foundational pillar in environmental resources management. Among them are discussed below.

#### 4.1.5.1 Political Modernization

Political modernization is the shifting of the relationship between the traditional politics of left and right towards novel, where hybrid arrangements are formed among states, market players and civil society, which leads to the production and redistribution of resources and the formation of new rules aimed at shaping society (Arts et al. 2006). The concept, in the context of this paper, refers to structural processes of institutional change and their impact on the water governance among the communities in the Wassa Amenfi East District. Due to all kinds of socio-economic and political processes and ideas such as transnationalism, reflexive modernization, ecological modernization, individualization, commercialization and globalization, new relationships are coming into being with different ideas and policy formulations and practices (GSS 2010).

#### 4.1.5.2 Multilevel Governance

An early explanation saw multilevel governance as a system of continuous negotiation among nested governments at several territorial tiers and described how supranational, national, regional, and local governments are enmeshed in territorially policy networks (Bache 2005). In recent times, multilevel governance has also explored the varying relationships between actors at various territorial levels, both in the public and private sectors (Atampugre et al. 2016). The multilevel governance theory intersects the traditionally separate domains of local and international politics and highlights the increasingly languishing distinctions between these domains. The theory illustrates both the increasingly frequent and dynamic interactions between government actors and the increasingly critical dimension of non-state actors involved in the development of cohesion policy. According to Ekow and Adu-Boahen (2016), these interactions could be viewed in two dimensions, the vertical and the horizontal dimensions. The vertical dimension refers to the linkages, including their structural dimensions, between higher and lower levels of government. Local capacity building and incentives for the effectiveness of sub-national levels of government are key issues for enhancing public policy efficiency and coherence. The horizontal dimension refers to cooperation agreements typically between states and non-state actors.

## 4.2 Materials and Methods

### *4.2.1 Profile of Study Area*

The Wassa Amenfi East is one of the districts in the Western Region of Ghana. It lies between latitudes 5°, 30′ N and 6°, 15′ N and longitudes 1°, 45′ W and 2°,

11′ W. It is bounded in the west by Amenfi West District, to the east by Mpohor Wassa East District, to the south by Prestea, Huni Valley and the north by Upper Denkyira West and East District. The district covers a total land area of 1558 $km^2$ which is about 7.5% of the total size of the Western Region. Wassa Akropong is the capital town of the district, which is about 180 km away from the regional capital Sekondi-Takoradi. The District has an average annual rainfall ranging between 1400 and 1,730 mm. There are two main rainfall regimes, the first starts from March and ends in July, second from September to the early part of November. Temperature is generally high in most of the year ranging from 24 to 29 °C (75–83 °F). The maximum temperature is experienced in March and the coolest month is August. The population of the district as of 2010 was projected to be 83,478 representing 3.5% of the regions total population (Wassa Amenfi East District Assembly 2014). The district is noted for illegal mining rendering the water resources in the area polluted and poorly managed. A good network of rivers and streams are found in the study area. Notable among them are the Ankobra, Ashire and Manse rivers (Wassa Amenfi East District Assembly 2014). During the dry seasons, the flow of these rivers and streams decreases dramatically. In the dry seasons, most of the streams also dry out entirely when they are mainly used for farming purposes. Coupled with the heavy rainfall trend, the network of rivers and streams make road building complicated and costly. The constant demand for expansion and provision of basic infrastructure is the critical fallout of the huge population phenomenon in the area. Available safe water sources in the District are boreholes, hand-dug wells and Small Town Water Supply (STWS), serving 80.1% of the total population (Wassa Amenfi East District Assembly 2014). Other non-portable sources, such as lakes, reservoirs, springs, rivers and rainwater, accompany these portable sources. Streams, streams, reservoirs and rivers serve as the most important sources of drinking water in rural areas. Boreholes and hand-dug wells are often shallow and dry up in the dry season, which are the main sources of safe water supply. Water and Sanitation Management Teams (WSMTs) (Wassa Amenfi East District Assembly 2014) manage these networks.

### 4.2.2 *Research Design*

The study employed a mixture of methods, which were descriptive in nature and emphasized on sequential exploratory approach. This is because no research method is free from limitation and biases and one method could potentially be checked by the other method (UNEP 2008). The case study research design was used with emphasized on the descriptive design. UNEP (2008) describe the design of a case study as an empirical inquiry that explores a current phenomenon in its real-life context where there is no clear boundary between the phenomenon and the context, and where several sources of evidence are used. The strategy offered an opportunity to gather as much information as possible to support the observation and debates, to draw concrete conclusions for the report. The approach provided an opportunity to collect as many as possible information needed to support the analyses and discussions to draw meaningful conclusions for the study. The paper employed primary and secondary data sources. Primary data were obtained from the residents such as household heads, assembly members, unit committee members, traditional council, institutions involved in water governance, district water and sanitation team and other stakeholders in the community. Secondary data were also obtained from already processed information available within the institutions involved in water governance, traditional leaders and also the Internet was heavily relied on for another relevant source of secondary data like magazines, news items and publications related to the topic under discussion.

### 4.2.3 Sample Frame

All the communities of Wassa Amenfi East District form the respondents for the study but the selected communities within the district namely Ankonsia, Asuadi and Bawdie, (Fig. 4.1). Form the focus of the research and were purposively selected. The selection of these communities was because of the important role they attach to water governance and the recent issue of water conflict in these areas. Subsequently, 'galamssey' or illegal mining is predominant in these communities and hence most water bodies are subjected to a series of environmental injustices (Ekow and Adu-Boahen 2016). Institutions and stakeholders responsible for water governance at the local level were also consulted.

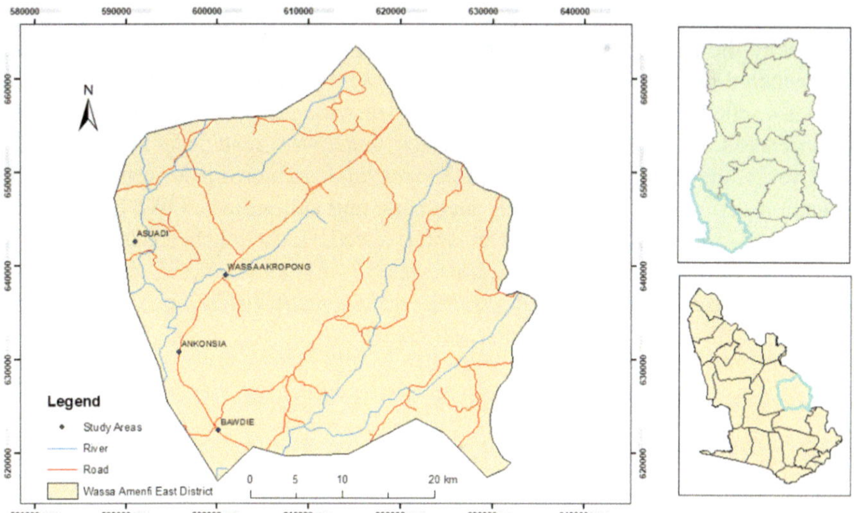


**Fig. 4.1** Map of Wassa Amenfi East District. *Source* Authors' construct (2019)

### 4.2.4 Sample Size

A total of eighty (80) respondents were sampled for the study from the three communities. Two representatives from each of the following institutions and stakeholders involved in water governance were interviewed. They included; the assembly members, unit committee members, DWST, traditional leaders, and environmental NGOs who are engaged in water resources management. The remaining 70 respondents included household heads or their surrogates.

### 4.2.5 Sampling Technique

The study used purposive and random sampling technique to select the respondents. Purposive sampling was used to select the institutions and the key actors involved in water governance as well as the three communities. Probability sampling that is simple random sampling technique was used to identify household heads or their surrogates in the selected communities namely Ankonsia, Asuadi and Bawdie. The reasons for using both sampling techniques were to ensure that quality information is obtained for the study. Household heads were given equal chances to be selected for the study.

### 4.2.6 Methods and Instruments for Data Collection

To explore the issue of water governance in the Wassa Amenfi East District, data was gathered using an interview schedule and semi structured-questionnaires. The interview schedule required the participants to provide extensive details on the topics under discussion. A well-structured interview guide was created and a 20-min interview was arranged for the interviewees. The respondents were made aware of the data collection period, thus the assembly members, unit committees, DWST, traditional leaders, and environmental NGOs were interviewed. Household heads or their surrogates from the three selected communities were given semi structured-questionnaires to respond. To detail the condition of the water sources visited on issues of water governance, an observation checklist was created.

### 4.2.7 Data Presentation and Analysis

The data obtained was evaluated in qualitative and quantitative terms. The qualitative data obtained from the in-depth and key informants interviews were classified into themes and the appropriate responses that have related to the objectives of the study were analyzed and where necessary, some of them were directly quoted to support the quantitative data collected from the field. The qualitative data obtained from the in-depth and key informant interviews were categorized into themes, the related responses relating to the study's objectives were analyzed, and some of them were specifically cited in support of the quantitative data collected from the field. By drawing parallels to the literature and norms available, each of the themes was presented in details. To provide a graphical representation of the answers or responses, frequency and percentages were used. The questionnaire responses were coded into SPSS version 21 and analyzed using descriptive statistics. Tables and a chart were used to present the results of the study.

## 4.3 Results and Discussion

### *4.3.1 Source of Water in Study Communities*

Wassa Amenfi East District is a unique district in Ghana when it comes to issues of water and its governance. Water, as they say, is life is manifested in myriads of ways in the district because of the way issues on water are handled. However, it was very noteworthy that the communities visited within the district had differing opinions regarding their sources of potable water. The sources of potable water in the three communities visited included; boreholes, rivers and streams with few households having pipe-borne water as a supplement to the other sources.

### *4.3.2 The Key Actors Responsible for Water Governance in the Wassa Amenfi East District (W.A.E.D.)*

The conservation of resource whether natural or artificial is a collective responsibility of every member in a given locality (Held 1995). Water is a natural resource, which should be conserved and properly taken care of to ensure its sustainability. From the study, it was revealed that aside the community members who through their leaders make laws to promote water governance, other institutions like government/assembly members and non-governmental organizations are part of water resources management. These institutions enact laws, which ensure frequent supply and sustainable usage of water. Table 4.1 and Fig. 4.2 show the key actors responsible for water governance as responded by the key informants during the study.

**Table 4.1** Role and functions of institutions

| Institutions | Roles and functions |
|---|---|
| Government | • Responsible for the ownership and control of the water resource<br>• The government resource the district assembly through the district assembly common fund to ensure the supply of water source |
| District Assembly | • Provision of water sources to communities that lack access to potable water |
| District water and sanitation team | • Works in collaboration with the WATSAN committee to ensure the maintenance of water source |
| Non-governmental organization | • Help in the maintenance and construction of water source |
| *Community leaders* | |
| • Chief | • Resolve water-related issues<br>• Protection and conservation of the water source<br>• Ensure the progress of the community |
| • Unit committee | • Mobilization of members for community labor |
| • Community Water and Sanitation Team | • Responsible for the supply of safe water in the communities<br>• Collection of water bills |
| • Water and sanitation (WATSAN) committee | • Responsible for the maintenance and constructions of the water source<br>• Fixes the prices of water sources |

*Source* Fieldwork (2019)

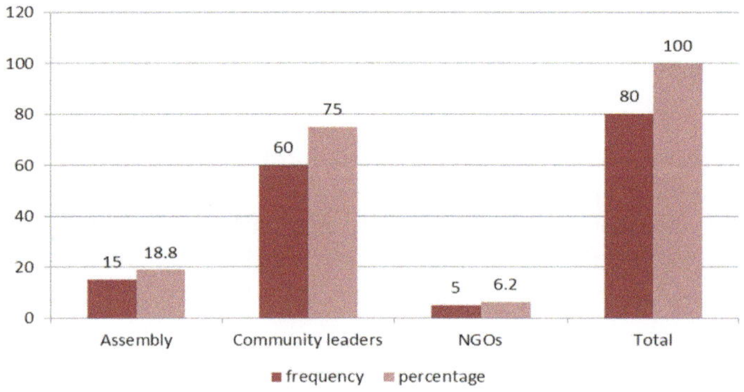


**Fig. 4.2** Key actors responsible for water governance in W.A.E.D. *Source* Field data (2019)

#### 4.3.2.1 Government and the District Assembly

The water resources commission (WRC) is mandated by the Act of parliament (Act 522 of 1966) to inflect and manage Ghana's water resources and to coordinate the relevant government policies. The Act specifies that on behalf of the people, the president is given the ownership and control of all water resources and explicitly defines the Water Resources Commission (WRC) as the ultimate body responsible for the management of water resources in Ghana. From Fig. 4.2, 15 out of 80 informants (both institutions and the household respondent) who represent 18.8% said an issue of water governance is the responsibility of the government (assembly). In their view, the government is a key stakeholder who should spearhead the management and conservation of water (water governance). This confirms the local government Act 1997 (Act 462) which mandates the District Assemblies as the exclusive segment answerable to the preparation, execution, processing and preservation of water and the legal owner of public amenities among rural populations and villages.

#### 4.3.2.2 Community Leaders

In the community, both the members and the leaders such as the unit committee, the community water and sanitation team, and other leaders work together to ensure water sustainability. From the study, 60 out of the 80 respondents which constitute (75%) of the respondents opined that the community leaders, as well as the members, are responsible for water governance. This is so because some of the informants saw water as a common property which no individual or institution can deny them from having access to and since they benefit directly, they should be responsible for it maintenance. This assertion aligns with (Held 1995) view on communal ownership of resources and their governance. For instance, during the interview, an informant revealed that,

> *We are drinking from the borehole, the government or the assemblyman does not live here. So we should protect and maintain it ourselves, we do not need anybody to tell us what to do or how to do it.*

#### 4.3.2.3 Non-Governmental Organizations

Parliamentary Act 522 of 1996 requires NGOs to participate in the WRC and defines the WRC as the body in Ghana that is charged with the administration of water resources. From the study, 5 out of the 80 respondent representing 6.2% were of the view that the Non-Governmental Organizations (NGOs) are participants in water resource governance and control. In their view, most of the wells in the communities were possessed by the NGOs and as such, they should be responsible for their maintenance and usage. From the results above, it can be deduced that institutions responsible for water governance can be formal or informal as opine by Saleth and Dinar

(2005) who saw water management as networks that are principally administered by formal and informal organizations, as well as state and non-state actors.

The in-depth interviews with the key stakeholders outline the following functions as performed by the key actor to ensure sustainable usage of water.

### *4.3.3 How Stakeholders Make Decisions on Water in Wassa Amenfi East District*

Water plays an integral role in sustainable development through the elimination of malnutrition, promoting economic growth and development. It adds quality to human life and its absence or inadequacy makes life unbearable for the individual at the household level (Global Water Partnership 2000). Maintaining the availability of water in Wassa Amenfi East District, the communities visited during the data collection revealed that there are numerous policies and governance approaches, which regulate right to use to and ownership of water in the district. During the in-depth interview, it was revealed that before decisions are taken on water governance, certain criteria are taken into considerations to shape decision-making in the study area. These are as follows;

- The total number of population in the community.
- Water source available in the community.
- Water source needed to supplement the existing ones.
- Source of funds for the provision of an additional water source.
- Quality, quantity and demand for the water.
- Amount to be paid by the individual before having access.
- Which age group will have the privilege to access or fetch the water.

From the issues raised above, the Ghana Water and Sewage Corporation (GWSC) explained that the population living in a specific community determines the available water source(s) to be provided. Communities with huge populations according to the information gathered will need more water sources as compared to those with less population. Sometimes the existing water sources in the area may not be able to serve the needs of the inhabitants in the community hence the need to provide new or alternative sources to assist. The results also revealed that children below ten years old are prevented from accessing the water especially wells, boreholes since fetching water from these sources demand some skills and strength that children in most cases do not have, and this can cause the problem to the individual or the facility as a whole.

However, there were diverse views about the issues raised above. During the interview with the key stakeholders responsible for water governance, 80% of them said they usually consider the quality and demand for the water before any decision can be made on water governance while 20% said they normally consider the quantity of the water. These views support the claim that, regardless of their degree of advancement and their economic and social circumstances, all persons have the right to have access to drinking water in amounts and with a consistency equal to their basic needs (Doe 2007). From the three communities studied, it was revealed that water decision is done every year unless there are important issues that call for action to be taken with immediate effect. From all the communities visited, it was noted that decisions on water governance were a priority. They stated that communal labor is organized every first week of the month to keep the surroundings of the water sources clean. They also pay money for repairing the water facilities and this amount varies from community to community. Some pay it daily or monthly whiles others pay it yearly.

### *4.3.4 Effectiveness of Institutions Responsible for Water Governance at Wassa Amenfi East District*

Institution responsible for water governance in the district can be categorized as formal and informal in nature. These institutions play a pivotal role in ensuring effective utilization and sustainability of water in the district. They consider water as an important element in people's lives. Given this, they have developed steps to ensure that sufficient drinking water is available in the district and that citizens have convenient access to water sources. As described in the Dublin Water and Sustainable Development Principles (Global Water Partnership 2002), in all competitive uses, water has an economic value and should be accepted as an economic resource. As stated in the Dublin principles on water and sustainable development by Global Water Partnership (2002), water has an economic value in all its competing uses. Water should therefore be recognized as an economic good. Given this, the following criteria were used to evaluate the effectiveness of the institutions responsible for water governance in the district.

#### 4.3.4.1 The Use of Money Earned from the Sale of Water

This criterion was included to know what the institutions used the money obtained from the sale of water for whether it is used for the provision of additional water facility; maintenance; or for any other purposes. This criterion as used here, confirms the meaning of efficiency as expressed by many scholars including (Wassa Amenfi East District Assembly 2014) who said efficiency is the measure of the extent to which goals are achieved with minimum expenditure of resource. One of the principles for achieving effective water governance is efficiency (Opoku-Ankomah et al. 2006). The study revealed that the proceeds earned from the sale of the water are used for the provision of additional water facilities and or maintenance of outmoded or ill-functioning ones. Others were of the view that no money is earned from the sale of water. Table 4.2 depicts the views expressed by household respondents.

From the table above, 15 (21%) of the respondents indicated that the money earned is used for the provision of additional water facilities. 35 (50%) said the money is

**Table 4.2** Uses of money earned from the sale of water

| Money earned uses | Frequency | Percentage% |
|---|---|---|
| Provision of additional water facility | 15 | 21 |
| Maintenance | 35 | 50 |
| No money is earned | 20 | 29 |
| Total | 70 | 100 |

$N = (70)$
*Source* Field data (2019)

used for maintenance, and the remaining which constitute 20 (29%) said no money is earned because they see no reason for paying for a natural resource which is a free gift from nature. With them, their major source of water is from the river that does not demand payment. The views expressed above support the notion that water resources are a gift of nature and so users are at liberty to take and make use of them.

#### 4.3.4.2 Satisfaction Level of Household Respondents

The satisfactory level of the respondent was used as a proxy to evaluate the effectiveness of the institution. The respondents were to provide whether they are satisfied or not satisfied with the water supply in the community. The satisfactory level as used as an assessment criterion confirms equity as an assessment criterion for achieving effective water governance as expressed by institutions including (United Nation Development Programme 2003). The respondent confirmed the sustainability of the management procedures been used by the various stakeholders as efficient and effective. Table 4.3 shows the responses given by the respondents.

Fifty respondents representing (71%) of the respondent stated that they were satisfied with services whiles, 20 representing (29%) were not satisfied. Those who said they were satisfied provided the following reasons as the water is safe to drink; they have easy access to the water sources; the water sources are in proximity to their residence. Those who said they were not satisfied also provided the following reasons; like the water, sources were not limited and most of the time have to rely on the rivers and streams. During dry seasons these rivers dry up with an almost impossibility for them to have contact to water and the ones found in the communities are located far away from their homes, hence long distance has to be covered before having access

**Table 4.3** Satisfactory level of the household respondent

| Satisfactory level | Frequency | Percentage% |
|---|---|---|
| Satisfied | 50 | 71 |
| Not satisfied | 20 | 29 |
| Total | 70 | 100 |

*Source* Field data (2019)

to the water source. This means that the theory of distance decay is influencing their decision on satisfaction.

#### 4.3.4.3 Household Views on Accountability

Meinzen-Dick and Knox 1999) defined accountability as an obligation or willingness to accept responsibility for one's actions. The study revealed that the leaders responsible the water governance in the three communities render their accounts annually except one community where no money is earned from the water sources since their source of water is mainly from the rivers. About 60% of the respondents said the leaders are making transparent account concerning monies accrued from the sale of water. They recounted accountability and transparency as an effective water governance tenet in the district.

The respondents (both the household and the institutional) that apart from the government policies enacted to ensure water governance, the traditional institution also have their management system known as the traditional management approach also revealed it. This is in line with (Adu-Boahen et al. 2018) assertion on traditional management approaches to water resources management in Ghana, where the chiefs are given the power to enact laws and convention to regulate the utilization and conservation of water. About 70% of the respondents commented on the effectiveness of these beliefs and taboos as the best management practice of water governance. As confirmed by Nompumelelo (2001), the use of taboo and other cultural practices for the conservation of natural resources, including water, is still in use in rural communities. Some of the beliefs and taboos outlined by the respondents included;

- On Thursday and Saturdays, it is a taboo to fetch water from the riverside.
- It is a taboo to wear sandals into the river
- Women in their menstrual period cannot fetch the water of the rivers.

It was however noted by one respondent that because of migration, urbanization, education and emergence of a new religion, these taboos and beliefs are being disregarded. Nevertheless, through public education on the need to manage and conserve water effectively and to ensure sustainable water supply, policies regarding the conservation of water are gradually being addressed in the study communities.

## 4.4 Conclusions

Water has an economic advantage in all its competitive applications and should be regarded as an economic asset. A participatory strategy involving users, planners and decision-makers at all levels should be focussed on the management and development of water resources. Research has shown that the various community leaders in each of the communities within the district are the main actors in charge of water governance. There are also various actors which in one way or the other help in governance in the district. These actors included the Government, Non-Governmental Organizations, the District Assembly, and the like. Good governance and good water governance in this regard need all policies and decisions to be transparent so that the communities can readily adopt the measures taken in policy formulation.

The study revealed that decisions in the district are taken in relation to certain criteria as discussed above and the decisions are made on yearly basis. Apart from the institutional regulations on access to and regulations of water. Aside the institutional laws on the access to and control of water; one important policy on access to the water source in the communities under study is shrouded in the traditional management approach where inhabitants are not to fetch from the riverside on Thursdays and Saturdays. However, there are many traditional management approaches but the above is considered the major one among all the findings on the traditional management approach on the access to a water source and these have helped to protect the water sources in the district.

Effective water governance ensures water sustainability; the study revealed that the institutions in the district are however performing their duties especially the community leaders and these were revealed based on the criteria used to access their effectiveness. The most effective of criteria were accountability, inhabitant's satisfactory level and the use of money earned from the sale of the water. These criteria used in relation to the respondents' responses helped the research to identify that these criteria are most needed when assessing institutions' effectiveness on water governance. However, the other criteria also help in arriving at a relevant conclusion but thc above are considered as more effective criteria.

## 4.5 Policy Recommendations

- There is a need to improve awareness campaigns on the policy and legal framework for access and management of water sources and facilities. Effective traditional practices related to access and control of water need to be analyzed and promoted.
- Communities in rural areas can form water user associations aside the WATSAN committee to help in addressing their water needs. Such an organization will help to ensure the security of water supplies by mobilizing the workforce and other tools required to enhance the management of water through the development and implementation of access rules and user responsibilities.
- The Government, NGO's, CBOs and donor organizations should help the communities in executing their functions effectively and when this is taken into consideration, effective water governance can be made.
- If the long-term course of the WATSAN committee was established, the governance of the water system will be improved. To function as a legal body, the WATSAN Committee must be supported in establishing its laws, processes and procedures, as well as its own rules and regulations. This will ensure successful

water governance: and the particular roles of other stakeholders in small communities should be identified as how they will help to conserve and restore their water supplies.

## References

Adu- Boahen K, Dadson IY, Atugbiga JA (2018) Customary practices and wetland management in Ghana: a case of Muni lagoon Ramsar site in the central region. KNUST J Geogr Dev 2:1

Arts B, Leroy P, van Tatenhove J (2006) Political modernisation and policy arrangements: a framework for understanding environmental policy change, vol 6. Springer Science + Business Media, pp 93–106

Atampugre G, Adu-Boahen K, Eshun J (2016) Trans-boundary-boundary river governance in West Africa: a case of the Volta River. African J Interdisc Stud 9:25–40

Bache I (2005) Europeanization and Britain: towards multi-level governance? Paper prepared for the EUSA 9th Biennial conference in Austin, Texas

Bakker K (2003) Good governance in restructuring water supply: a handbook. Federation of Canadian municipalities, Ottawa

Contribution and lessons of decentralized management of rural water to decentralization in Ghana. Institute of Local Government Studies conference on decentralized management in the context of GPRS and MDBS. Kumasi, Ghana. http://www.discap.org/publication/Ghana_Decentration_Mngmnt_Water_Discap

Doe HW (2007) Assessing the challenges of water supply in urban Ghana. Stockholm, Sweden. http://www.wr.kth.se/publikationer/PDFFiles/LWREX0706

Ekow FJ, Adu-Boahen K (2016) Coastal environmental injustice in Ghana: the activities of coastal sediment miners in the Elmina, Cape Coast and Moree area. GeoJournal 79:6. ISSN 0343-2521. https://doi.org/10.1007/s10708-014-9612-4

Eugene OS (2010) Assessment of water governance in Ghana. Available at http://www.gwpao.org/GHANA-Governance.PDF

Ghana Government (1996) Water Resource Commission Act. Act 522 of parliament of Ghana

GSS (2010) National analytical report: 2010 population and housing census

Global Water Partnership (2002) Introducing effective water governance. Global Water Partnership. Stockholm

Global Water Partnership (2000) Integrated water resource management technical advisory committee. Background paper No. 4, Stockholm. Available online at http://www.cepis.ops-oms.org/bvsargli/fulltext/tac4/tact4.pdf

Gumisai M (2004) Rough road to sustainable development. Africa Renewal 18

Held D (1995) Democracy and the global order—from the modern state to cosmopolitan governance. Cambridge

Hooghe L, Marks G (2003) Unraveling the central state, but how? Types of multi-level governance. Am Polit Sci Rev 97:233–243

Karr JR (1991) Biological integrity: a long-neglected aspect of water resource management. Ecol Appl 1:66–84

Karikari K (2000) Water supply and management in rural Ghana: overview and case studies. In: Rached E, Rathgeber E, Books DB (eds) Water management in Africa and the Middle East: challenges and opportunities. IDRC, Ottawa Canada, pp 25–31

Leroy P (2002) Environmental politics, participation and political inequality. In: Draetta L, Lai F (eds) Naturalia. Sciences Socialeset Environment: entre cadres theoriques et approaches empiriques. EUROPAEA 8:1–2, 153–174

Leroy P, van Tatenhove J (2000) Political modernization theory and environmental politics. In: Spaargaren G, Buttel FH (eds) Environment and global modernity. SAGE, London
Meinzen-Dick R, Knox A (1999) Collective actions. Property rights and devolution of natural resource management: a conceptual framework. Draft workshop paper, 15 July 1999, Mimeo
Merriam-Webster (2008) Online definition and thesaurus. http://www.merriam-webster.com
Marks G (2005) Structural policy and multi-level governance in the EC. In: Cafruny A, Rosenthal G (eds) The state of the European community: the maastricht debate and beyond, pp 391–411
Nowlan L, Bakker K (2010) Practicing shared water governance in Canada: a primer Vancouver: UBC program on water governance office of research ethics the University of Waterloo. Ethics review process. Available athttp://www.Research.uwaterloo.ca/ethics/human/ethicsReview/ReviewProcess.utm
North DC (1993) Economic performance through time: Les Prix Nobel: The noble prices 1993 Stockholm. The noble foundation and the Royal Swedish Academy of Science
Nompumelelo TB (2001) Water resource management: the challenge of integration in the implementation of Zimbabwe's new water policy: a case study of the catchment level institution surrounding the Pungwe-Mature supply project. University of the Western Cape P/Bayx17, Belleville 7535, South Africa
Opoku-Ankomah Y, Dembele Y, Ampomah BY, Some L (2006) Hydro-political assessment of water governance from the top-down and review of literature on local level institution and practices in the Volta Basin. Colombo, Sri Lanka. International Water Management Institute, 36, IWMI working paper 111
Saleth RM, Dinar A (2005) Water institutional reforms: theory and practice. Water Policy 7:1–19
United Nation Development Programme (2003) Clean water and sanitation for the poor. www.undp.org
UNEP (2008) Annual report. http://hdl.handle.net/20.500.11822/7742
Wassa Amenfi East District Assembly (2014) Draft medium-term development plan 2014–2017 district planning coordinating unit, Ghana

# Chapter 5
# Impact of Evaporation Losses from On-Farm Reservoirs on the Economics of the Farming System

**Aniket Deo, Amit Arora, and Subhankar Karmakar**

**Abstract** This study explored the change in farm economics when evaporation losses (EL) from on-farm reservoir (OFR) were saved. A generic framework was developed wherein water balance simulations of soil and OFR was integrated with multi-stage mathematical programming. Five scenarios of EL in different sizes of OFR were created to access the change in irrigation potential. For a study system, the simulation results suggested that the conservation of EL would enable more options for production plans, scope of crop diversification and an improvement in farm income. Moreover, if 100% EL were saved, the volume of water available for one season of Rabi cultivation increased by 280 $m^3$, 468 $m^3$ and 704 $m^3$ while income increased by INR 15,200, INR 25,900 and INR 32,600 for OFR size 400 $m^2$, 625 $m^2$ and 900 $m^2$ respectively. The framework and results are useful for farmers and government agencies to plan water resources at farm level.

**Keywords** Cropping plans · Evaporation losses · Farm economics · On-farm reservoirs · Water management

## 5.1 Introduction

The Indian farming systems have been struggling with lack of availability and accessibility to reliable irrigation (Singh et al. 2020). This is limiting their growth and raising serious concerns for the food security of the country. Therefore, development of knowledge and strategies for sustainable on-farm water management and self-sustained farming systems have been the core of water researchers (Singh 2010).

Among a few strategies to improve irrigation accessibility to farmers, on-farm reservoirs (OFR) systems have proven to be reliable and profitable (Banerjee 2019).

A. Deo (✉) · A. Arora
Center for Technology Alternatives for Rural Areas, Indian Institute of Technology, Bombay, Mumbai 400076, India

S. Karmakar
Environment Sciences and Engineering, Indian Institute of Technology, Bombay, Mumbai 400076, India



P. Kumar et al. (eds.), *Water Resources Management and Sustainability*,
Advances in Geographical and Environmental Sciences,
https://doi.org/10.1007/978-981-16-6573-8_5

However, due to its limited storage capacity, the water use efficiency of OFR systems are often questioned (Faulkner et al. 2008). The availability of water in OFR is a function of water inflows and outflows. Water resources such as rainwater, canal water and groundwater are the inflows in OFR while irrigation discharge and evaporation and seepage losses are the outflows. It is an established knowledge that water resources worldwide are either reducing, degrading or uncertain. While the water inflows are often beyond the control of the farmer, the best efforts she could do to maximize the water use efficiency of OFR is to minimize the evaporative and seepage losses.

Evaporation losses account to a substantial fraction of water outflow from the reservoir (Gupta et al. 2002; Mugabe et al. 2003). A study conducted by Craig et al. (2005) in Queensland, Australia suggested that the evaporation losses OFRs accounted for 40% of the total storage capacity. They further estimated that the lost water could irrigate around 125,000 hectares and generate agricultural production worth US$ 375 million annually. In a study done by Gökbulak and Özhan (2006) in Turkey, the evaporation losses in lakes and dams were estimated to be more than the groundwater pumped. In another study done in Segura basin of Spain, (Alvarez et al. 2008)concluded that the evaporation losses were 8% more that the water used for agriculture in this basin. Furthermore, the climate change forecasts are worrisome because the projections indicate a rise in temperature by 1.5 °C globally (Masson-Delmotte et al. 2018). This rise in temperature could increase the evaporation losses significantly (Trenberth et al. 2005). A few studies have estimated the effect of temperature rise on evaporation. For example Jurak (1989) estimated a rise of 11–24% in evaporation losses for a temperature rise of 2.8 °C in Europe. Schindler and Donahue (2006) reported that for a 2 °C rise in air temperature, the temperature of lake in Kenora (Canada) rose by 1.5 °C, wherein a significant increase in evaporation losses was expected. Since, it is well established in literature that evaporation losses contribute to a significant fraction of water lost for irrigation, it is of paramount importance that the evaporation losses be assessed and its economic implications be studied (Stanhill 2002). Moreover, such assessment could contribute for better water resources surveying, design and management (Morton 1994).

There are very few studies in literature which address the economic impact of the evaporation losses (Maestre-Valero et al. 2013). Like, work done by Craig et al. (2005) has quantified the forgone production opportunity on a region scale due to the cumulative evaporation losses from all OFRs in the region of study. Similarly, (Martínez-Granados et al. 2011) assessed the economic impact of evaporation losses at basin level. Maestre-Valero et al. (2013) studied the socio-economic impact of evaporation losses in Segura basin by creating different water availability scenarios which represented climate change conditions. Bou-Fakhreddine et al. (2019) assessed the economic impact of the evaporation losses in Qaraoun dam, Lebanon on the hydropower generation and irrigation sector. However, we did not find any study in the literature which addressed the impact of evaporation losses on the economics of the farming system.

Agro-economic mathematical models have proven to be appropriate for economic evaluation of water resources in the farming systems (Singh 2012; Maestre-Valero et al. 2013). These models have the capabilities to capture the agro-ecological-economic as well as social aspects of the farming system (Itoh et al. 2003; Benli and Kodal 2003; Roy et al. 2009; Galán-Martín et al. 2015; Jaiswal et al. 2020; Zhai et al. 2020). These models are widely used to design alternative strategies for farm management and can operate at multi-spatial and temporal scales. From the literature survey of mathematical models in the context of evaporation losses, we could derive that these models could be used to (a) estimate the quantum of evaporation losses (b) generate potential production plans under different evaporation scenarios (c) assess the variation in economics of the farming system under different evaporation scenarios. Amongst the few studies who have assessed the economic impacts of evaporation losses, all have used mathematical models but they have considered a fixed cropping plan. These models which are designed for regional/basin scale analysis are simulated by not considering farmers as the primary stakeholders. In that case, the model results are merely the summation of individual farmer results which are considered to be homogeneous across the entire region. This is imprecise, specifically in case of India where each farmer possess dissimilar profiles/resource capacities hence considering farm outputs to be uniform across a region in India is impractical. Therefore farm level analysis is essential to get the actual scenario even in the context of economic evaluation of evaporation losses.

Understanding the limitations in the current literature, this study presented a framework for assessing the impact of evaporation losses from OFR on the economics of the farming system. This framework was demonstrated for a study system. We used hydrological modelling to estimate the inflows and outflows from OFR. Scenarios of evaporation losses saved (ELS) in different sizes of OFRs were developed and using mathematical programming, we studied the potential for change in cropping pattern and farm income under given system constraints. Hydrological modelling included water balancing of soil and OFR with historical meteorological data wherein we used percent of exceedance (PE) to forecast irrigation demand, water availability in OFR, and evaporation losses. A multi-stage analytical procedure was used to generate all feasible cropping pattern under different ELS scenarios. The analytical procedure included combination algorithm, linear programming, k-means clustering, TOPSIS, and monte-carlo simulation.

## 5.2 Material and Methods

### *5.2.1 Study System*

Ganeshpuri is a farming village located around 80 km from Mumbai. It lies in the north Konkan region of western part of Maharashtra, India. It receives an average rainfall of 2600 which is significantly higher than the state average of 1450. Farming is generally practiced in two seasons, i.e. Kharif (monsoon) and Rabi (winter). The Kharif season prolongs from June to October and Rabi from November to March. The major crop cultivated during Kharif season (Monsoon) is rainfed rice which is grown for subsistence. Besides, the farmer prefer four crop mix commercial vegetable cultivation for Rabi season. Rabi cultivation is practiced by very few farmers because of lack of irrigation facility. The farmers who practice Rabi cultivation have constructed OFR systems which is used for supplement irrigation during dry spells of Kharif season and the remaining water post-Kharif is used for Rabi cultivation. The crops and cropping areas for Rabi cultivation are determined by the amount of water remaining post-Kharif. The OFR design is standard in this region. The OFR shape is square trapezoidal with a 3 m depth, 1:1 slope, 0.5 m Bern and low density polyethylene (LDPE) lining. However the OFR top area varies wherein the OFR sizing is generally based on the land holding, financial capacity of the farmer and availability of subsidy schemes. Common OFR sizes are 20 × 20 × 3 m, 25 × 25 × 3 m and 30 × 30 × 3 m. Since this village is near the coastal region, the groundwater is often saline at many sites. Therefore, rainwater harvesting is the sole source for recharge of the OFR. This limits the irrigation potential of OFR and constraints the farmers for limited options of cropping patterns.

Figure 5.1 represents the rainfall pattern over the past few years in Ganeshpuri. It is evident that the amount of rainfall has been irregular in the past and hence is expected to remain irregular in the following years as well (the chances being more due to climate change). This variability in the rainfall compel the farmers in Ganeshpuri to modify their cropping pattern every year since rainwater is the only source in their OFRs. Moreover, the rainfall intensity during Kharif season is erratic, creating many dry spells during the monsoon. Figure 5.2 represents the daily rainfall figures for the year 2000. The figure highlights some days of high intensity rainfall along with certain dry spells.

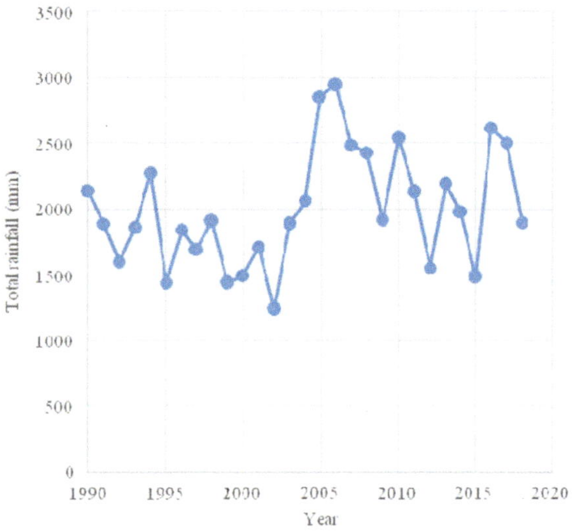


**Fig. 5.1** Total rainfall in Ganeshpuri for year 1990–2020

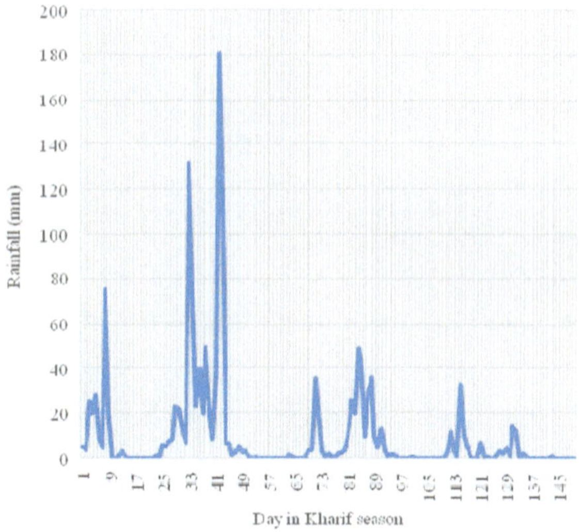


**Fig. 5.2** Daily rainfall in the Kharif season for a sample year 2000

Systems like OFR could prevent crop failures during these dry spells, however this depends on the amount of water available in OFR during the dry spells. An excessive rainfall could overflow the OFR, losing out a portion of rainwater while prolonged dry spells in Kharif season can exhaust the water in OFR. Hence, a misbalance between water inflow and water outflow is not desirable for an OFR system. To satisfy the irrigation demands for Kharif and Rabi crops, the OFR capacity should be well designed. The irrigation potential of OFR can be enhanced by increasing the OFR area or reducing the losses. Increasing the OFR area is difficult as compared to reducing the losses because firstly, the cost inquired for construction is very high and secondly the valuable cultivable land is lost. Therefore, in this study we explored the potential gains in the irrigation potential, cropped area and the net returns from cultivation when the evaporation losses from the OFR system were reduced/saved.

### *5.2.2 Simulation Modelling Approach*

The analysis of the study system is carried out using simulation modelling. Figure 5.3 presents the simulation modelling approach in the form of flow chart. Inputs concerning the agro-ecology and socio-economics of the study system were considered to ensure a holistic approach. Farm properties which included cropping pattern and physical soil properties such as soil texture, bulk density, and available water content were considered. Crop properties such as crop duration, root depth and local irrigation practices were captured through field level investigations. Historic meteorological data ranging for years 1990–2018 was acquired for accurate predictions of climatic parameters. Using the inputs mentioned above, daily soil moisture balance (DSMB) was simulated for Kharif season of all years. For this model, the Kharif

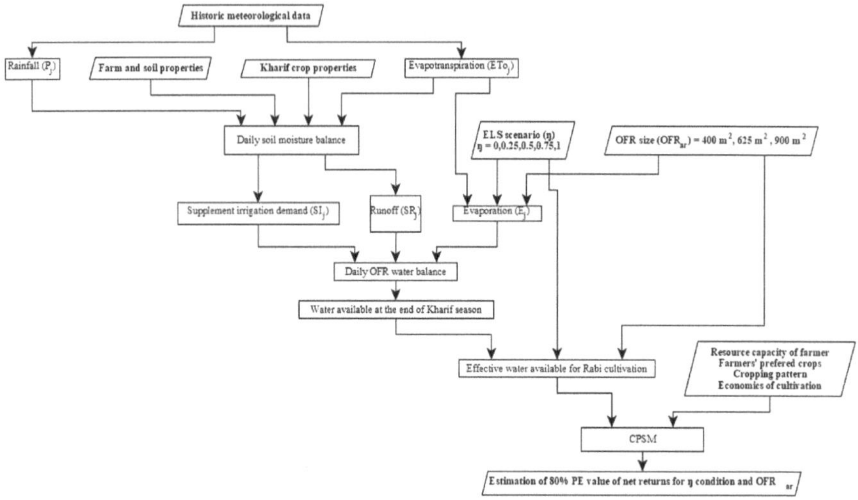


**Fig. 5.3** Flowchart to represent simulation approach

season was considered from 1st June till 31st October for all years such that, $j = 1{:}153$. The DSMB involved hydrological parameters such as rainfall, evapotranspiration, runoff and infiltration. The rainfall values was directly taken from the meteorological database while daily evapotranspiration, runoff and infiltration were estimated using methods described in next sub-sections. DSMB simulation resulted in daily supplement irrigation demand. Percent of exceedance (PE) was calculated to predict the value of a parameter in terms of chances of occurrence. An 80% PE value was estimated [using methodology described by Helsel and Hirsch (2002)] for runoff and supplement irrigation demand. For the study system analysis, three different sizes of OFRs are considered to understand the relation between evaporation losses, size of OFR and economics of farming system. The sizes ($OFR_{ar}$) in the form of length × breadth × depth were 20 × 20 × 3 m, 25 × 25 × 3 m and 30 × 30 × 3 m (the length × breadth determined the top area of the OFR). These are the most common sizes of OFR in the selected study region.

As a next step, a daily OFR water balance was simulated for Kharif season of all years. Daily water inflows and outflows were considered for this simulation. The water inflows comprised of daily rainfall and runoff while water outflows was supplement irrigation demand and evaporation and seepage losses. The values of daily runoff and supplement irrigation demand were obtained from DSMB. The estimation method of evaporation loses is described in next sub-section. For this study, five scenarios of evaporation losses saved (ELS) were created to understand the impact of evaporation losses on the water availability in OFR. The evaporation losses could be saved by various methods such as covering the top area of OFR with plastic or any opaque material while another method is shading using plantation around the OFR. All these methods conserve/save evaporation losses with different efficiencies. Therefore, the scenarios consisted of 0%, 25%, 50%, 75% and 100%

ELS where in the reducing factor ($\eta$) was 0, 0.25, 0.5, 0.75 and 1 respectively. Further, the OFR water balance simulation was used to predict the amount of water available at the end of Kharif season. An 80% PE value of water availability was calculated which was used for further steps in planning Rabi cultivation.

The inputs for planning Rabi cultivation consisted of farmer's preferred crops, resource capacity of the farming system, economics of cultivation and preferences in cultivation. This data was acquired through field level investigation and secondary sources (AGMARKNET 2020; CDAP 2019). Since, Rabi cultivation involved selecting crops and allotting correct acreages to them, it involved designing appropriate cropping plans. This was a typical case of "crop planning problem". A model developed in the previous study called cropping plan selection model (CPSM), was used for designing cropping plans for Rabi cultivation, considering the most practical objective of profit maximization and resources constraints such as limited land, water, budget and labour. The CPSM is briefly described in following sub-section.

Considering all the scenarios of ELS and different sizes of OFR, a total of 15 cases were simulated using the above mentioned approach. The output of CPSM for each case was analysed to generate conclusive patterns and implications of evaporation losses on the economics of the farming system. Following subsections describe the analytical tools, scientific relations, coefficients and constants considered in this simulation modelling.

**Daily Soil Water Balance Simulation For Kharif Season**: DSWB is a prevalent tool to evaluate the hydrological parameter in effective root zone of the crop which was rice in this case. DSWB was simulated using Eq. (5.1) for soil in an unsaturated state.

$$\mathrm{SW}_j = \mathrm{SW}_{j-1} + P_j + \mathrm{SI}_j - \mathrm{AET}_J - \mathrm{SP}_j - \mathrm{SR}_J \tag{5.1}$$

where, $\mathrm{SW_j}$ and $\mathrm{SW_{j-1}}$ was the soil water/moisture (mm) of *j*th and *j* – 1th day respectively. $P_j$, $\mathrm{SI}_j$, $\mathrm{AET}_j$, $\mathrm{SP}_j$, and $\mathrm{SR_j}$ is the rainfall (mm), supplement irrigation (mm), actual evapotranspiration (mm), seepage and percolation losses (mm) and surface runoff (mm) respectively on the *j*th day. The *j* here represented a day in Kharif season such that j = 1 for 1st June. The groundwater table in the region of study was much below the root zone of rice crop therefore the capillary rise was ignored in this simulation. The soil water content on j = 1 was considered at the wilting point of the soil. The measured soil volumetric moisture content was 27.7, 39.7 and 48.6% at permanent wilting point, field capacity and saturation respectively for a sample soil. The $\mathrm{SP_j}$ losses was considered to be 4 mm/day for unsaturated state of soil (Brouwer and Heibloem 1985).

Irrigation of Rice in the study region was through flood irrigation method. The rice was cultivated by preparing saplings and later transplanting the saplings in the mail field. The saplings were prepared in a nursery which was developed in about 8% of the total land. The ideal irrigation practices for the local variety of rice were documented and used in this simulation. Usually, after sowing the seeds in nursery, the soil was kept at saturation for one week from sowing. After this, ponding of depth 2.5 cm was maintained in the nursery until transplanting. The main fields were puddled by the time the saplings were ready for transplantation. After the transplantation of saplings in the main field, a ponding depth of 5 cm is maintained in the field. This ponding is d in the last two weeks from harvest after wherein the soil and crop is allowed to dry so as to facilitate the harvest.

For the saturated and ponded state of the soil, the DSWB was simulated using Eq. (5.2).

$$D_j = D_{j-1} + P_j + \mathrm{SI}_j - \mathrm{AET}_J - \mathrm{SP}_j - \mathrm{SR}_J \tag{5.2}$$

where $D_j$ and $D_{j-1}$ are the ponding depth (mm) of $j$th and $j-1$th day respectively. The excess water above the ponding depth is drained off which contributes to runoff. The $SP_j$ losses was considered to be 1.5 mm/day for unsaturated state of soil (Razavipour and Farrokh 2014).

*Actual Evapotranspiration And Supplement Irrigation*: The actual evapotranspiration was calculated using methodology described by Idike et al. (1982). The relationship used for its estimation is mentioned in Eq. (5.3).

$$\mathrm{AET}_j = K_c \times K_s \times \mathrm{ET}_{oj} \tag{5.3}$$

Where $AET_j$ and $ET_{oj}$ was the actual evapotranspiration (mm) and reference evapotranspiration (mm) respectively for $j$th day. $K_c$ and $K_s$ were the crop coefficient and soil moisture stress factor respectively. A FAO Penman–Monteith method described by Allen et al. (1998) was used for estimation of the reference evapotranspiration. The $K_c$ of rice were taken from Tyagi et al. (2000) which were 1.15, 1.23, 1.14 and 1.02 for initial, crop development, reproductive and maturity stage respectively. $K_s$ ranges from 0 to 1 and for soil at saturated state, the $K_s$ is 1, while in unsaturated state of soil, it is linear with soil moisture (Allen et al. 1998) which is represented by Eq. (5.4) (Panigrahi 2001).

$$K_s = \frac{\mathrm{SW}_j}{\mathrm{SAT}} \tag{5.4}$$

where SAT was the moisture content at saturation. The effective root zone of rice was 0.5 m and the calculated value of SAT was 243 mm.

*Surface Runoff*: Due to the generic framework and its wide applications, the Soil–Water Assessment Tool (SWAT) is a popular tool used in hydrology (Jaiswal et al. 2020). SWAT was used in this study to estimate the $SR_j$. The SWAT offers two methods for estimation of runoff, one is SCS curve number method while the other is Green and Ampt infiltration method. In this study we used SCS curve number method because of its moderate input data complexity and direct formulations. In SCS runoff procedure, the soil in the study region fits to the hydrological soil type D and the soil runoff curve number (CN2) is 89 (Neitsch et al. 2009).

*Bare Soil Evaporation*: When the field is ideal, i.e. during pre-germination period, evaporation from bare soil takes places. This reduces the soil moisture content. $ET_{oj}$ was used for estimation of the Bare soil evaporation ($ES_j$) in mm which was subjected to $P_j$ condition of the *j*th day using the following relations (Jensen et al. 1993):

$$ES_j = 0.1\,ET_{oj} \quad \text{if } P_j = 0 \tag{5.5}$$

$$ES_j = ET_{oj} \quad \text{if } P_j > ET_{oj} \tag{5.6}$$

$$ES_j = P_j \quad \text{if } 0 < P_j < ET_{oj} \tag{5.7}$$

Equations (5.5), (5.6) and (5.7) were used to compute the DSWB in ideal state of field by replacing $AET_j$ with $ES_j$ in the Eq. (5.1).

**Water Balance of OFR during Kharif Cultivation**: A square sized trapezoidal OFR with low density polyethylene (LDPE) lining and a depth of 3 m, side slope of 1:1 and bern of 0.5 m around the OFR was considered for this study. The simulations were done for all the three OFR sizes mentioned in Sect. 5.2.2. The daily inflows in OFR were rainfall and runoff while daily outflows were evaporation losses and supplement irrigation. Generally, due to the LDPE lining its hot sealing, the seepage losses in OFR are trivial therefore they were ignored for simulation in this study. Equation (5.8) represents the daily water balance simulation in OFR.

$$\text{Vol}_j = \text{Vol}_{j-1} + P_j \times \text{OFR}_{\text{ar}} + \text{SR}_j \times \text{Land} - E_J \times \text{ar}_J \times (1-\eta) - \text{SI}_j \tag{5.8}$$

where $\text{Vol}_j$ and $\text{Vol}_{j-1}$ are the OFR volume ($m^3$) on *j*th and $j - 1$th day respectively.$\text{ar}_J$, Land, $\text{OFR}_{\text{ar}}$ and $E_J$ are the top water surface area in OFR ($m^2$) on the *j*th day, cultivated land ($m^2$), area occupied on field by the OFR ($m^2$) and evaporation losses (m) respectively. Since, OFR was square trapezoidal spaced, $\text{ar}_J$ was a variable parameter and depended on the water level in OFR. Figure 5.4 shows the $\text{ar}_J$ in OFR with its formulation. Land was the difference between total land and $\text{OFR}_{\text{ar}}$. $E_J$ was estimated by using Eq. (5.3) where, the $AET_j$ can be replaced with $E_J$ while $K_c$ and $K_s$ were equal to 1 for open water bodies like OFR (Kohli and Frenken 2015). $E_J$ was multiplied by a factor $(1 - \eta)$ which denoted the ELS scenario that are described in Sect. 0. For ELS 0%, 25%, 50%, 75% and 100%, the value of $\eta$ was 0, 0.25, 0.5, 0.75 and 1 respectively. These ELS scenarios were simulated for each OFR size.

During high intensity rainfall, the water inflows are usually large which might overflow the OFR. In such a scenario, the excess water is drained out of the farm. The overflow condition of the OFR is represented by Eq. (5.9).

$$\text{Vol}_j = \text{Vol}_{\max}; \quad \text{if } \text{Vol}_{\max} - \text{Vol}_{j-1} < P_j \times \text{OFR}_{\text{ar}} + \text{SR}_j \times \text{Land} \tag{5.9}$$

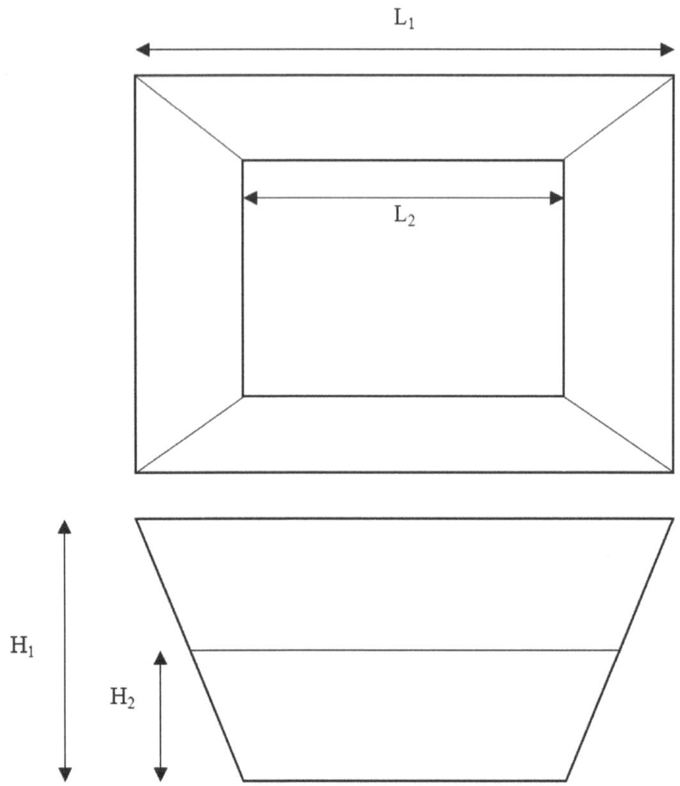


$$V=\frac{H_1}{3}\left(L_1^{\,2}+L_2^{\,2}+\left(\sqrt{L_1^{\,2}\times L_2^{\,2}}\right)\right)$$
$$ar_j=(L_2+2H_2)^2$$

**Fig. 5.4** Schematic diagram of square trapezoidal OFR. "*V*" is the total volume of the OFR at its full depth $H_1$ and "$ar_j$" is the top surface area at depth $H_2$

where $Vol_{max}$ is the maximum volume of the OFR. The $Vol_{153}$ is calculated for all years and an 80% PE value is considered for Rabi cultivation. The index j = 153 represent the last day of Kharif season, i.e. 31st October.

**Planning Rabi Cultivation**: The cropping plans for Rabi cultivation was developed using a multi-stage mathematical model called cropping plan selection model. This model was created in previous study (Deo et al. 2020). This model has the capability to generate full factorial cropping plan options and filter a list of most appropriate plans. The cropping plans were generated considering the objective function of profit maximization of farmer under resource constraints such as limited land, water, budget and labour. Along with this optimization based approach in the model, it evaluated sustainability indices for all cropping plans which were generated. These indices were net returns, irrigation requirement, human work hours, diversity index and relative time dispersion. These indices are described in detail in Deo et al. (2020). For the process of determining the list of most appropriate plans, it considers farmer preferences among the sustainability indices to segregate and rank the cropping plans. As per the weightage given by the farmer to each sustainability index, the cropping plans are filtered such that these cropping plans have higher values of indices which are given higher weightages. Finally, to capture the variability in the farming system such as fluctuating market rates, variable production and cost of cultivation, an uncertainty analysis was performed which provided the mean net returns and coefficient of variance (CV) of net returns value.

Following were the analytical tools used in this five-stage model. First, combination algorithm was applied to generate crop combinations. The farmers in this region generally practice a four crop mix pattern, however a number of vegetables are cultivable on their land. Therefore, they choose four crops among a list of crop options. Combination algorithm enabled the possibility to explore full factorial crop combinations. Second, linear programing was used as an optimization technique to allocate optimal acreage to crops considering an objective function and resource constraints. Optimization was useful to utilize the available resources for maximum gains. Third, K-means clustering algorithm was applied to group/cluster similar cropping plans based on their sustainability indices values. Clustering algorithm reduced the sample space for further analysis. Forth, a multi-criteria decision making tool known as Technique of Order Preference Similarity to the Ideal Solution (TOPSIS) was used to rank the clusters formed in stage-3 using the weightages of sustainability indices given by the farmer. The cropping plans in rank-1 cluster were considered as more appropriate than cropping plans in other clusters. This was first level filtering of cropping plans. Fifth, a monte-carlo simulation was used for uncertainty analysis. The cropping plans which had CV less than 25th percentile and mean net returns higher than 75th percentile were screened as the most appropriate plans for the given context and field conditions. These screened cropping plans are most profitable and least risky plans.

*Stakeholder's Inputs for Planning Rabi Cultivation*: Farmer was the primary stakeholder for this study, hence a comprehensive field level study was conducted in Ganeshpuri to capture the farmer's decision making, resource capacities, local agricultural practices and economics of cultivation. A farmer was selected for this study whose profile exhibited a common case in this region. The farmer's profile or portfolio was obtained through semi-structured interviews which consisted of questionnaires related his farming system. The interviews were directed towards estimating his resource capacity in terms of land, budget, labour hours and transportability. These capacities were quantified and were used as constraints in the optimization stage mentioned in Sect. 0. The quantified resource capacities are mentioned in Table 5.1. Moreover, the questionnaires included an inquiry about the preference of criteria while planning Rabi cultivation. These criteria were basically sustainability indices which were weighted and its order of preference was used in MCMD mentioned in Sect. 0. The weights of these indices are mentioned in Table 5.2.

**Table 5.1** Resource profile of the farmer

| Resources | Capacity |
|---|---|
| Land holding | 10,000 $m^2$ |
| Budget | INR 25,000 |
| Human work hours | 400 hours |
| Minimum production target (transportability constraint) | 400 kg |

**Table 5.2** Weightages of sustainability indices

| Sustainability indices | Weightage (%) |
|---|---|
| Net returns | 26.2 |
| Diversity index | 16.5 |
| Relative time dispersion | 14.5 |
| Water requirement | 30 |
| Human work hours | 12.6 |

The farmer was inquired about the preferred list of crops for Rabi cultivation. The crops in this list were cultivable and marketable in this region. Local agricultural practices along with wholesale market rates (AGMARKNET 2020) were documented to estimate the economics of cultivation. Table 5.3 mentions the list of preferred crops and their economics of cultivation. The cost of cultivation in Table 5.3 included material inputs cost such as seeds, fertilizers, pesticides and labour costs such as land preparation, sowing, weeding, harvesting and transporting. The crop water requirement was calculated using FAO Penman–Monteith method (Allen et al. 1998). For all crops, the water requirement was calculated as per date of sowing as 15th November. In this region, the Rabi irrigation is done using drip irrigation systems. An irrigation efficiency of 90% (Brouwer et al. 1989) for drip irrigation systems was considered for calculating irrigation requirements. The irrigation requirement values were calculated for all years and an 80% PE value was mentioned in Table 5.3. As usually used in agricultural studies, the relationship between productivity, cost of cultivation, irrigation requirement, and labour hours were considered linear with the cropped area (Harrington 1992).

## 5.3 Results and Discussion

The proposed framework was used to compare ELS scenarios for different sizes of the OFR and evaluate its economic implications on the Rabi cultivation. The formulations and simulations were done in MATLAB 2016b software package. Hydrological parameters such as rainfall, evapotranspiration and surface runoff were estimated to conduct DSWB simulations. DSWB resulted in daily supplement irrigation demand for Kharif crop. The hydrological parameters and supplement irrigation demand were

**Table 5.3** Characteristics for the preferred crops

| Crop index | Preferred crop | Crop duration (days) | Market rates (INR/kg) | Coefficient of variance for market rates | Productivity (kg/ha) | Cost of cultivation (INR/acre) | Profit (INR/acre) | Water requirement (mm/season) | Human work hours (h/acre) |
|---|---|---|---|---|---|---|---|---|---|
| 1 | Cluster beans | 90 | 32.94 | 0.34 | 5000 | 17,584 | 49,100 | 264 | 193 |
| 2 | Carrot | 100 | 11.50 | 0.24 | 25,000 | 18,367 | 98,029 | 308 | 203 |
| 3 | Cucumber | 105 | 16.18 | 0.35 | 8000 | 18,987 | 33,418 | 297 | 204 |
| 4 | Eggplant | 130 | 15.82 | 0.35 | 25,000 | 19,624 | 140,511 | 447 | 216 |
| 5 | Onion/dry | 150 | 12.57 | 0.36 | 25,000 | 20,674 | 106,593 | 602 | 223 |
| 6 | Capsicum | 165 | 23.67 | 0.49 | 15,000 | 21,162 | 122,59 | 668 | 237 |
| 7 | Radish | 35 | 11.58 | 0.27 | 15,000 | 16,975 | 53,348 | 107 | 167 |
| 8 | Spinach | 60 | 6.15 | 0.24 | 8000 | 18,654 | 1264 | 151 | 176 |
| 9 | Tomato | 135 | 11.31 | 0.26 | 30,000 | 21,049 | 116,284 | 470 | 217 |
| 10 | Okra | 100 | 26.07 | 0.19 | 5000 | 18,147 | 34,622 | 302 | 199 |
| 11 | Bitter gourd | 105 | 29.03 | 0.36 | 6000 | 17,584 | 52,925 | 310 | 202 |
| 12 | Chilly | 125 | 25.73 | 0.42 | 10,000 | 17,584 | 86,565 | 424 | 212 |
| 13 | Pumpkin | 105 | 13.53 | 0.29 | 18,000 | 17,584 | 80,978 | 312 | 202 |
| 14 | Cow pea | 100 | 31.12 | 0.20 | 4000 | 17,584 | 32,817 | 274 | 197 |
| 15 | Coriander leaves | 110 | 12.56 | 0.28 | 6000 | 17,454 | 13,061 | 417 | 201 |

used for daily water balance of OFR. Figure 5.5 represents the supplement irrigation demand, runoff potential in farmland generated due to rainfall, evaporation losses from OFR and water level in OFR on daily basis for the Kharif season of year 1993, for a case when there is no conservation of evaporation losses (ELS-0). The water level in OFR was calculated by considering the daily rainfall captured by the OFR, water outflow for supplement irrigation, runoff inflow in OFR and evaporation losses from OFR. Similarly, these parameters were estimated for all the year (1990–2018) and an 80% PE value for water availability in OFR at the end of Kharif season (i.e. for j = 153) was computed. This procedure was followed for all the 15 cases as mentioned in Sect. 0.

Since the evaporation losses in Kharif season was relatively low as compared to Rabi season, the availability of water in OFR at the end of Kharif season is almost same for all ELS scenarios. The water level of OFR at the end of Kharif season are 2.67 m, 2.70 m and 2.73 m for $OFR_{ar}$ 400 $m^2$, 625 $m^2$ and 900 $m^2$ respectively. This suggest that an average 90% of the total OFR capacity is available at the end of Kharif season. However, it would be incorrect to consider this amount of water for planning Rabi cultivation because the water availability during Rabi season would reduce significantly due to evaporation losses in OFR. We computed the evaporation losses in OFR during Rabi season and the effective water available for Rabi cultivation is represented in Fig. 5.6. The values of effective water availabilities are presented as the fraction of total capacity of OFR. For $OFR_{ar}$ 400 $m^2$, 625 $m^2$ and 900 $m^2$, the total capacities are 876 $m^3$, 1461 $m^3$ and 2196 $m^3$ respectively. Figure 5.6 suggests that with every 25% saving of evaporation losses, the average effective water availability

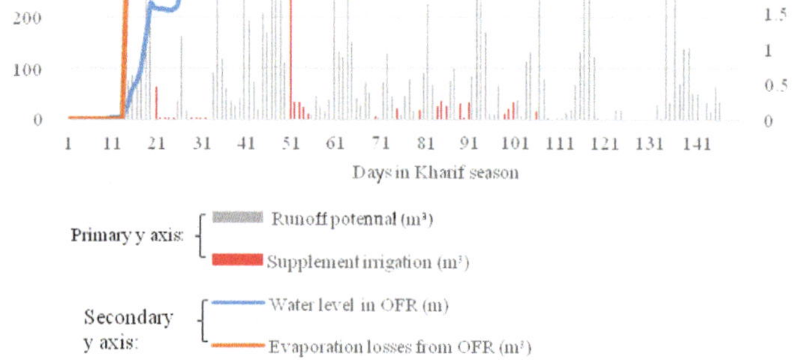


**Fig. 5.5** Daily supplement irrigation demand, runoff potential, evaporation losses from OFR and water level in OFR

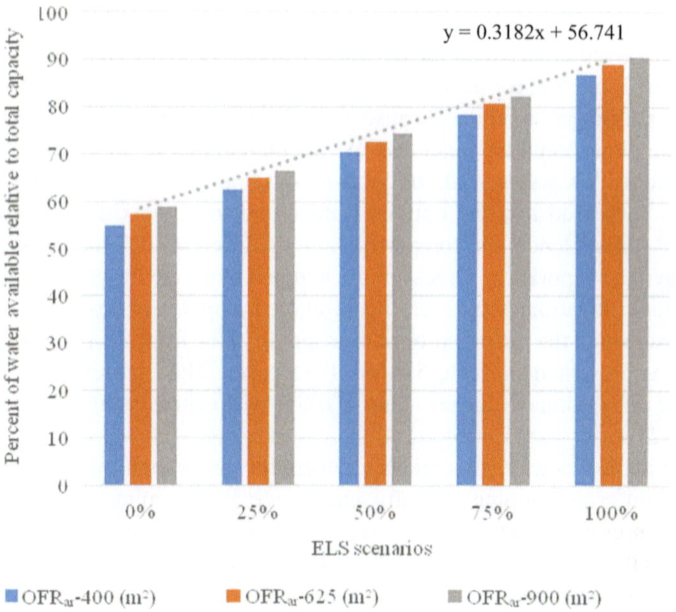


**Fig. 5.6** Effective water available for Rabi cultivation as a fraction of total OFR capacity

for all $OFR_{ar}$ increased by 8%, i.e. an increase of 70 $m^3$, 117 $m^3$ and 176 $m^3$ for $OFR_{ar}$ 400 $m^2$, 625 $m^2$ and 900 $m^2$ respectively.

Further, we analysed the potential gains in Rabi cultivation if the water lost through evaporation losses were saved. For this we used CPSM, as mentioned in Sect. 0. CPSM was simulated for all the 15 cases wherein it generated appropriate cropping plans for different water availability and $OFR_{ar}$ scenarios. Table 5.4 shows the number of potential cropping plan options for different cases. A four crop mix plans were generated wherein the model made full factorial crop combinations by selecting 4 crops from a list of 15 crops. In such a scenario, the maximum number of crop combinations that could be formed were 1365 ($^{15}C_4$). In the second stage of the CPSM wherein linear programming algorithm was used, only those crop combinations could form cropping plans which satisfied the resource constraints (mentioned in Table

**Table 5.4** Number of potential cropping plans for different ELS and $OFR_{ar}$ scenarios

| ELS (%) | $OFR_{ar}$ | | |
|---|---|---|---|
| | 400 $m^2$ | 625 $m^2$ | 900 $m^2$ |
| 0 | 22 | 328 | 953 |
| 25 | 94 | 801 | 1331 |
| 50 | 294 | 1245 | 1365 |
| 75 | 646 | 1362 | 1365 |
| 100 | 1038 | 1365 | 1365 |

5.1). Therefore, many crop combinations were rejected as their combined resource requirements were higher than the resource availability in the study system. This is evident from Table 5.4, for example in case of $OFR_{ar}$ 400 $m^2$and ELS-0, maximum crop combinations got rejected because of low water availability condition and only 22 cropping plans could be generated. While, the number of cropping plan options were limited for low water availability conditions, the options increased as the water availability increased and the options were found to be maximum in high water availability cases. Hence, the comparison among different ELS scenarios suggested that by saving evaporation losses, more number of cropping plans options were available for cultivation. Moreover, this promoted crop diversification.

CPSM screened the most appropriate cropping plans from the potential cropping plan options presented in Table 5.4 for all 15 cases. The first level screening was done based on ranking of clusters using TOPSIS with criteria weightages mentioned in Table 5.2.

Further, in the second level screening, uncertainty analysis eliminated the high risk and low profit cropping plans. Finally, an average 5% cropping plans were derived to be most appropriate meaning, they were less risky and more profitable relative to the others in the lot for an individual case. The average net returns of these screened cropping plans for all the 15 cases is represented in Fig. 5.7.

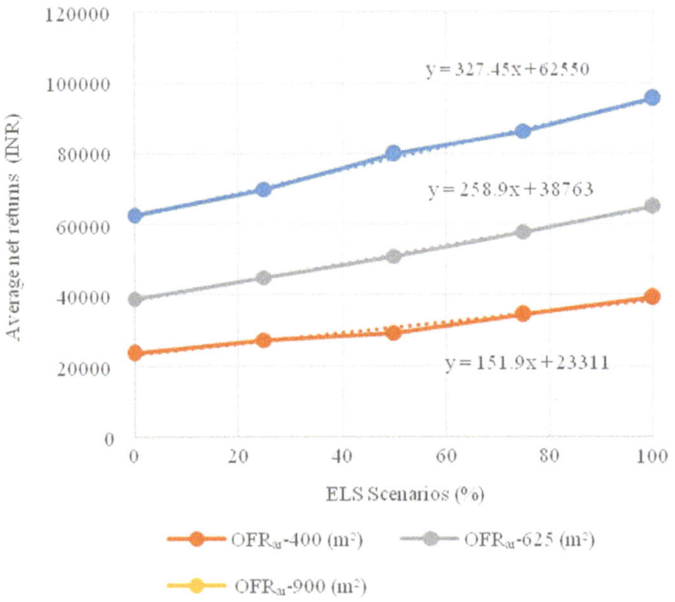


**Fig. 5.7** Average net returns from most appropriate cropping plans for different ELS and $ORF_{ar}$ scenarios

The equations in Fig. 5.7 represented the relationship between the percentage of ELS ($x$) and the net returns ($y$) for the corresponding $OFR_{ar}$ scenarios. The equations suggested the effect of ELS on the net returns. As depicted in Fig. 5.7, the slope of equation corresponding to $OFR_{ar}$ 900 $m^2$ was largest among the three which meant the potential gains of ELS were more significant in $OFR_{ar}$ 900 $m^2$. Since the amount of evaporation losses is proportional to the area of OFR therefore as ELS increased ($x$-axis in Fig. 5.7) the water available for cultivation increased with a higher rate in $OFR_{ar}$ 900 $m^2$, hence indicating the highest rate of increase in net returns. It can be elicited that for all the $OFR_{ar}$ the increase in net returns was significant when evaporation losses were saved while the rate of increase in net returns was found to be proportional to the OFR size.

In the most appropriate cropping plans for all ELS scenarios, the average area under Rabi cultivation was found to be almost similar ($5508 \pm 206$ $m^2$) in case of $OFR_{ar}$ 625 and 900 $m^2$. The cropped area was limited to this value because of system constraints. While, for $OFR_{ar}$ 400 $m^2$, the cropped area was observed to increase from 0 to 100% ELS scenario wherein, with every 10% saving of evaporation losses, the cropped area was increased by 190 $m^2$. The increase in net returns with ELS scenario in case of $OFR_{ar}$ 400 $m^2$ (as represented in Fig. 5.7) is justified with an increase in cropped area.

However even though the cropped area remained similar for all ELS scenario in $OFR_{ar}$ 625 and 900 $m^2$, the net returns increased. This happened because the selection of crops in the cropping plans changed. The cropping plans which contained high profit but high irrigation requirement crops were eliminated in low water availability scenarios due to lack of water. Figure 5.8 represent the case of $OFR_{ar}$ 900 $m^2$ wherein different crop choices are available at various crop orders. Here, the crop order means the order of crop in the cropping plan, i.e. primary, secondary, tertiary and quaternary order. These orders are based on the acreage distribution among the crops in the four crop system where primary crop occupy the maximum acreage among the four. For the OFRcase presented in Fig. 5.8, the average land use distribution for all ELS scenarios was 55%, 32%, 9% and 4% for primary, secondary, tertiary and quaternary crop order respectively.

According to the land use distribution among crop orders, it could be established that the primary crop produced major portion of the net returns since it occupied the maximum fraction of land. As depicted in Fig. 5.8, the choice of crop at primary order were different in the ELS-0 and 100% scenario. In the latter case, eggplant was recommend as the primary crop while in former, carrot and radish were recommended. Since, the profit/unit land for eggplant is higher than carrot and radish (as suggested in Table 5.3), it is evident that the net returns in the latter case was higher as compared to the former even though the total cropped area and land use distribution remained same. As, the water requirement for eggplant is high as relative to carrot and radish, for ELS-0%, the CPSM did not recommend eggplant at the primary order due of lack of water availability. Therefore, it can be elicited that saving evaporation losses could enable selection of profitable crops and allotment of larger acreages to them.

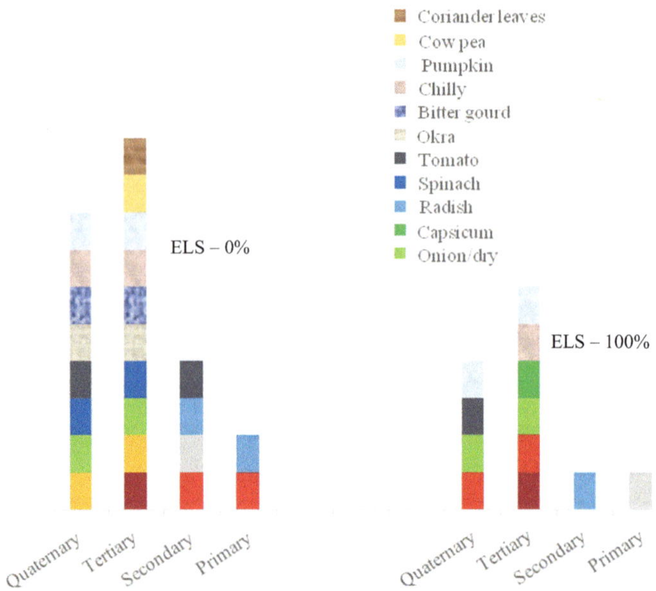


**Fig. 5.8** Crop choices at various crop orders for ELS-0% and ELS-100% in $OFR_{ar}$ 900 $m^2$

## 5.4 Conclusion

The evaporation losses from water reservoirs constitute a significant amount of water. In Indian farming systems where irrigation accessibility and availability is a major concern, conservation of evaporation losses from on-farm water storage facility (such as OFR) becomes a necessity. In this study, we developed a framework to analyse the impact of evaporation losses on the economics of the farming system. We created ELS scenarios for different sizes of OFR and designed potential production plans for each case. The framework is a novel integration of hydrological modelling (water balance simulations of soil and OFR) and multi-stage mathematical programming (optimization procedure).

The framework was demonstrated for a study system. The results suggested that the conservation of evaporation losses enabled more options for production plans, scope of crop diversification and an improvement in farm income. Moreover, the choices for crop selection increased enabling the farmers to cultivation high profit crops in more acreages. For 100% saving of evaporation losses, the volume of water available for one season of Rabi cultivation increased by 280 $m^3$, 468 $m^3$ and 704 $m^3$while income increased by INR 15,200, INR 25,900 and INR 32,600 for $OFR_{ar}$ 400 $m^2$, 625 $m^2$ and 900 $m^2$ respectively. These figures were 80% PE values which were computed by simulating this framework with past 28 years of data. The irrigation potential of the OFR was found to increase with the OFR size however, the evaporation losses also increased proportionally. Therefore it implies that optimal sizing of OFR and appropriate shading are crucial for sustainable economics of the farming system. The shading techniques and their efficiencies are not covered in this study because it is already available in literature (Craig et al. 2005; Alvarez et al. 2006; Hipsey and Sivapalan 2003).

The proposed framework is generic and applicable at multiple scales and geographies. The applications of this framework extends to policy makers and researchers for planning water resources at farm level. Also, the agencies who provide subsidy schemes for OFR systems could get benefits from this study. This framework could be seen as the foundation for developing advanced models, that would include complex hydrological process for water balancing, biophysical (biotic and abiotic) components for crop planning, economic relationships and uncertainties in the farming system. Moreover, a techno-economic analysis of OFR sizes with appropriate shading techniques could be an extension of this work.

## References

AGMARKNET (2020) Directorate of Marketing & Inspection (DMI), Ministry of Agriculture, Government of India. http://agmarkweb.dacnet.nic.in/SA_Pri_Month.aspx. Last accessed 2020/06/23

Allen RG, Pereira LS, Raes D, Smith M (1998) Crop evapotranspiration-guidelines for computing crop water requirements-FAO irrigation and drainage paper 56. Fao, Rome, 300, D05109

Alvarez VM, Baille A, Martínez JMM, González-Real MM (2006) Efficiency of shading materials in reducing evaporation from free water surfaces. Agric Water Manage 84:229–239

Alvarez VM, González-Real MM, Baille A, Valero JFM, Elvira BG (2008) Regional assessment of evaporation from agricultural irrigation reservoirs in a semiarid climate. Agric Water Manage 95:1056–1066

Banerjee PS (2019) Impact study of Hapa and its multiple uses in Bankura District. Gates Open Res 3

Benli B, Kodal S (2003) A non-linear model for farm optimization with adequate and limited water supplies: application to the South-east Anatolian project (GAP) Region. Agric Water Manage 62:187–203

Bou-Fakhreddine B, Mougharbel I, Faye A, Pollet Y (2019) Estimating daily evaporation from poorly-monitored lakes using limited meteorological data: a case study within Qaraoun dam-Lebanon. J Environ Manage 241:502–513

Brouwer C, Heibloem M (1985) Irrigation water management: training manual no. 3: irrigation water needs. Irrig Water Manage Train Man 3

Brouwer C, Prins K, Heibloem M (1989) Irrigation water management: irrigation scheduling. Train Man 4

CDAP Report-Palghar district, Department of Agriculture, Government of Maharashtra. http://krishi.maharashtra.gov.in/1240/CDAP. Last accessed 2019/12/25

Craig I, Green A, Scobie M, Schmidt E (2005) Controlling evaporation loss from water storages

Deo A, Arora A, Karmakar S (2020) Model to select appropriate production plans from factorial crop combinations: a case of vegetable farmers in India

Faulkner JW, Steenhuis T, van de Giesen N, Andreini M, Liebe JR (2008) Water use and productivity of two small reservoir irrigation schemes in Ghana's Upper East Region. Irrig Drain J Int Comm Irrig Drain 57:151–163

Galán-Martín Á, Pozo C, Guillén-Gosálbez G, Vallejo AA, Esteller LJ (2015) Multi-stage linear programming model for optimizing cropping plan decisions under the new common agricultural policy. Land Use Policy 48:515–524

Gökbulak F, Özhan S (2006) Water loss through evaporation from water surfaces of lakes and reservoirs in Turkey. Official Publication of European Water Association EWA
Gupta V, Tadros M, Bradshaw RW, Showalter SK, Miller JE, Evans L, Sasaki DY (2002) Progress report: advanced concepts group Fy01–1362-LDRD, solutions to national and global security issues based on limited freshwater resources. SAND2002-0781. Sandia National Laboratories, Albuquerque NM
Harrington LW (1992) Wheat and Rice in Karnal and Kurukshetra Districts, Haryana, India: farmers' practices, problems, and an agenda for action. CIMMYT
Helsel DR, Hirsch RM (2002) Statistical methods in water resources. US Geological Survey Reston, VA
Hipsey MR, Sivapalan M (2003) Parameterizing the effect of a wind shelter on evaporation from small water bodies. Water Resour Res 39
Idike FI, Larson CL, Slack DC (1982) Modeling soil moisture and effects of basin tillage. Trans ASAE 25:1262–1267
Itoh T, Ishii H, Nanseki T (2003) A model of crop planning under uncertainty in agricultural management. Int J Prod Econ 81:555–558
Jaiswal RK, Yadav RN, Lohani AK, Tiwari HL, Yadav S (2020) Water balance modeling of Tandula (India) reservoir catchment using SWAT. Arab J Geosci 13:148
Jensen JR, Mannan SMA, Uddin SMN (1993) Irrigation requirement of transplanted monsoon rice in Bangladesh. Agric Water Manage 23:199–212
Jurak D (1989) Effect of climate change on evaporation and water temperature
Kohli A, Frenken K (2015) Evaporation from artificial lakes and reservoirs. FAO-AQUASTAT, Rome, p 10
Maestre-Valero JF, Martínez-Granados D, Martínez-Alvarez V, Calatrava J (2013) Socio-economic impact of evaporation losses from reservoirs under past, current and future water availability scenarios in the semi-arid Segura Basin. Water Resour Manage 27:1411–1426
Martínez-Granados D, Maestre-Valero JF, Calatrava J, Martínez-Alvarez V (2011) The economic impact of water evaporation losses from water reservoirs in the Segura basin, SE Spain. Water Resour Manage 25:3153
Masson-Delmotte V, Zhai P, Pörtner H-O, Roberts D, Skea J, Shukla PR, Pirani A, Moufouma-Okia W, Péan C, Pidcock R (2018) Global warming of 1.5 OC: an IPCC special report on the impacts of global warming of 1.5 °C above pre-industrial levels and related global greenhouse gas emission pathways, in the context of strengthening the global response to the threat of climate Chang. World Meteorological Organization Geneva, Switzerland
Morton FI (1994) Evaporation research—a critical review and its lessons for the environmental sciences. Crit Rev Environ Sci Technol 24:237–280
Mugabe FT, Hodnett MG, Senzanje A (2003) Opportunities for increasing productive water use from dam water: a case study from semi-arid Zimbabwe. Agric Water Manage 62:149–163
Neitsch SL, Arnold JG, Kiniry JR, Williams JR (2011) Soil and water assessment tool theoretical documentation version 2009. Texas Water Resour Inst
Panigrahi B (2001) Simulation of ponding and soil moisture status through water balance model for rainfed upland rice. Agric Eng J 10:39–56
Razavipour T, Farrokh AR (2014) Measurement of vertical water percolation through different soil textures of paddy field during rice growth season
Roy D, Panda SN, Panigrahi B (2009) Water balance simulation model for optimal sizing of on-farm reservoir in rainfed farming system. Comput Electron Agric 65:114–124
Schindler DW, Donahue WF (2006) An impending water crisis in Canada's western prairie provinces. Proc Natl Acad Sci 103:7210–7216
Singh A (2010) Decision support for on-farm water management and long-term agricultural sustainability in a semi-arid region of India. J Hydrol 391:63–76
Singh A (2012) An overview of the optimization modelling applications. J Hydrol 466:167–182
Singh PR, Sharma S, Bihari B (2020) Learnings from community-based small-scale irrigation in tribal areas of Jharkhand. The World Bank, India

Stanhill G (2002) Is the class A evaporation pan still the most practical and accurate meteorological method for determining irrigation water requirements? Agric Meteorol 112:233–236
Trenberth KE, Fasullo J, Smith L (2005) Trends and variability in column-integrated atmospheric water vapor. Clim Dyn 24:741–758
Tyagi NK, Sharma DK, Luthra SK (2000) Determination of evapotranspiration and crop coefficients of rice and sunflower with lysimeter. Agric Water Manage 45:41–54
Zhai Z, Martínez JF, Beltran V, Martínez NL (2020) Decision support systems for agriculture 4.0: survey and challenges. Comput Electron Agric 170:105256

# Chapter 6
# Estimating Sediment Rate Through Stage-Discharge Rating Curve for Two Mountain Streams in Sikkim, India

**Sonu Kumar, Santosh Rangrao Yadav, and Triambak Baghel**

**Abstract** Sediment transport in streams is associated with a wide variety of environmental and engineering issues. Rates of sediment discharge are related to the sources, transport, and storage of sediment and erosion hazards within a watershed, which are related to the tectonic regime, climate, land cover, land use, and river setting. In the present study, an attempt has been made to estimate the rates of sedimentation in two tributaries Ranikhola and Busuk-khola of the Teesta River in Sikkim, India. The river water sampling was done on weekly basis for monsoon months, and suspended sediment concentration was estimated. To obtain rates of sedimentation, river discharge is required. For this purpose, a stage-discharge rating curve was developed by measuring flow discharge ($Q$) using standard current meter method, whereas stage ($h$) obtained by automatic water level recorder. The developed stage-discharge rating curve equations $h = 1.8196$ Q0.168 (Ranikhola) and $h = 1.9184$ Q0.156 (Busuk-khola) are useful for computing flow discharge from the river stages that will aid in the estimation of sediment discharge in the rivers. The result shows that the total sediment load in the Ranikhola and Busuk-khola rivers was ranged between 18.00–4071.51 and 1.92–603.73 tonnes per day, respectively.

**Keywords** Suspended sediment · Bed load · Stage · Discharge · Rating curve · Ranikhola · Busuk-khola

## 6.1 Introduction

The prediction of river sediment load constitutes an important issue in hydraulic and sanitary engineering. Sediment is responsible for transporting a significant proportion of many nutrients and contaminants including their uptake, storage, release, and

S. Kumar · T. Baghel (✉)
Water Engineering and Management, Department of Civil and Infrastructure Engineering, School of Engineering and Technology, Asian Institute of Technology, Pathum Thani 12120, Thailand

S. R. Yadav
Soil and Water Conservation Engineering, College of Agricultural Engineering and Post-Harvest Technology (CAU, Imphal), Ranipool, Sikkim 737135, India

 
P. Kumar et al. (eds.), *Water Resources Management and Sustainability*, Advances in Geographical and Environmental Sciences,
https://doi.org/10.1007/978-981-16-6573-8_6

transfer between environmental compartments. Most sediment in surface waters is derived from surface erosion and comprises a mineral component, arising from the erosion of bedrock, and an organic component arising during soil-forming processes including biological and microbiological production and decomposition. An additional organic component may be added by biological activity within the water body.

Sedimentation in rivers has long been an issue of serious concern worldwide, which has broad effects upon both terrestrial and aquatic aspects of life within a river basin. The study of river suspended sediments is becoming more important, nationally and internationally, as the need to assess fluxes of nutrients and contaminants to lakes and oceans, or across international boundaries, increases. River suspended sediment concentrations are most important with respect to pollution, channel navigability, reservoir filling, hydroelectric-equipment longevity, fish habitat, river aesthetics, and scientific interests and provide insights to the erosion and transport of materials from a landscape, and changes in concentrations with time that result from landscape processes or human disturbance. Moreover, it is also used to evaluate the erosion hazard, management of water resources, water quality, hydrology project management (dams, reservoirs, and irrigation), and to determine the extent of the damage that occurred in the catchment. Sediment carried in a stream is classified as either suspended load (fine-grained soils, e.g. clay and silt) or bed load (coarser fractions like sand and gravel). Coarser sediments will be deposited first and suspended sediment load moves at the approximately same velocity as that of the flowing water (Martin and Mcybcck 1979).

The stage-discharge rating curve at a river cross-section is a fundamental technique in hydrology employed for determining discharge from catchments. It is a common practice to measure the discharge of streams at suitable times, usually by a current meter or other methods and the corresponding stage followed by plotting and fitting stage against discharge data with a power or polynomial curve. The traditional and simple way to gather information on flow discharge is then to measure the stage with gauges and to use the stage-discharge curve to estimate the flow discharge (Rantz 1982; Braca 2008; Chen 2013).

Recently, many researchers worldwide have estimated rating curves for rivers by measuring velocities either with a cup type current meter (Alfa et al. 2018) or an ultrasonic current meter (Adegbola and Olaniyan 2017). Alexandrov et al. (Alexandrov et al. 2003) analysed relations between suspended sediment concentration and water discharge during flash floods in an ephemeral stream.

In Sikkim, India, the two perennial rivers, viz. Ranikhola and Busuk-khola, are tributaries of Teesta River on which two large hydro-power projects are operational. Annually, these projects need to be closed due to high inflow of sediment with large flow discharge during four monsoon months from June to September. Therefore, it is utmost important to estimate the sediment rates of those tributaries which has considerable contribution to sediment rates in the Teesta river. To estimate these sediment rates, river discharge was required to be measured, and it is mostly difficult to measure flow discharge directly during monsoon months because of high river stages coupled with chaotic turbulence. Therefore, an attempt was made to arrive at a stage-discharge rating curves for channel cross-sections below two bridges on the selected rivers. For this purpose, two sites were selected, viz. Ranipool Bridge (27° 17′ 37″ N latitude, 88° 35′ 19″ E longitude and altitude of 867.77 m above mean sea level (AMSL)) on Ranikhola River and Jalipool Bridge (27° 17′ 26″ N latitude, 88° 35′ 44″ E, longitude and altitude of 829 m AMSL) on Busuk-khola River.

## 6.2 Materials and Methods

### 6.2.1 *Stage-Discharge Rating Curve*

To measure flow velocity in the selected rivers, the standard current meter method was used which is a velocity area method that involves measuring flow velocity in the flow cross-section with the help of a vertical axis cup type current meter and flow cross-section area by using automatic water level recorder. At the selected gauging sites, the bridge spans of 23 m and 11 m were divided into 23 and 11 segments, respectively, each being 1 m wide. To measure depth from bridge span, measuring tape attached with a galvanized iron (G.I.) wire was lowered down after attaching a sounding weight to its one end. This assembly of G.I. wire, measuring tape, and sounding weight was lowered down to the river water by using a pulley arrangement and reading was taken when the sounding weight touched to the river bed. This was necessitated by inaccessibility of the selected gauging sites during high flood conditions. Whereas, during low flow conditions, measurements were done directly by using a staff gauge. To measure flow depth in the channel, water surface depth from the bridge span was measured randomly at six different points and average of six readings was taken as the water surface depth from the bridge span. To convert depth measured from the bridge span into water surface elevation, the datum was taken at 15 m depth from the bridge span because the maximum depth of river bed observed from the bridge span was 13.9 m. The river stage was computed by deducting the depth of water surface at each segment from the datum.

As per criteria given by Subramanya (2013), the flow velocity at each segment was measured at the depth of 0.6 times the depth of flow because the depth of water is less than 3 m. The discharge in each segment estimated by measuring velocities was found to be about 8% of total discharge in the river, which was within 10% of the total discharge. After measuring flow velocity in each segment with the current meter, total flow discharge in the river was computed by using mean section method as illustrated in Fig. 6.1 in which following formulae were used.

$$d_{\mathrm{m}} = \frac{d_2 + d_3}{2} \tag{6.1}$$

$$V_{\mathrm{m}} = \frac{V_2 + V_3}{2} \tag{6.2}$$

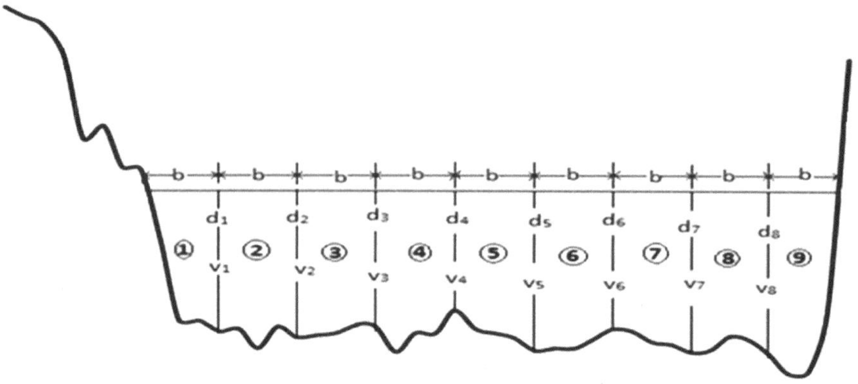

**Fig. 6.1** Measuring flow discharge by mean section method

$$\Delta Q = b \times d_{\mathrm{m}} \times V_{\mathrm{m}} \quad (6.3)$$

$$Q = \sum \Delta Q \quad (6.4)$$

where $b$ = width of a segment (m), $d_2$ = depth of water flow at point 2 (m), $d_3$ = depth of water flow at point 3 (m), $V_2$ = flow velocity at point 2 (m/s), $V_3$ = flow velocity at point 3 (m/s), $d_{\mathrm{m}}$ = mean depth of a segment (m), $V_{\mathrm{m}}$ = mean flow velocity in a segment (m/s), $\Delta Q$ = discharge in a segment ($m^3/s$), and $Q$ = total discharge in river ($m^3/s$).

The stage and corresponding discharge values were plotted along $Y$-axis and $X$-axis, respectively, on arithmetic and logarithmic scales to get the stage-discharge rating curves.

### *6.2.2 River Sedimentation Rate*

In the present study, water samples were collected on weekly to fortnightly basis from the Ranikhola and Busuk-khola rivers by using two litre plastic sampling bottles from 28 July 2018 to 25 November 2018 to measure suspended sediment load. While collecting samples, it was assumed that the suspended sediment was distributed uniformly across the entire cross-section of the river and samples were collected at a depth of six-tenth of the depth of river flow. After thoroughly stirring, the samples were taken into a pre-weighed aluminium container (weight of empty aluminium container = $W_1$) that was kept into a hot air oven which was set at 110 °C until all water evaporated. The container was allowed to cool till it reached room temperature, and its weight was measured again by using an electronic balance (weight of container and suspended sediment = $W_2$). The difference in the above-mentioned two weights divided by two gave suspended sediment concentration ($C$ in g/L) in the sample. The rate of suspended sediment (also known as suspended sediment load, SSL in tonnes per day) was estimated by Eq. (6.5).

$$\mathrm{SSL} = Q \times C \times 86.4 \tag{6.5}$$

where $Q$ = flow discharge ($m^3/s$), $C$ = suspended sediment concentration (g/L).

As for mountainous rivers, the bed load in a stream lies in the range of 20–40% of the SSL (Schroder and Theune 1984; Dadson et al. 2003; Turowski et al. 2008), in the present study, bed load is assumed to be 20% of the SSL; considering the lowest value of the above-mentioned range. Total sediment rates in the Ranikhola and Busuk-khola rivers during monsoon months were estimated by adding SSL and bed load.

## 6.3 Results and Discussion

### *6.3.1 Measurement of Channel Dimensions*

To compute flow discharge in the Ranikhola and Busuk-khola rivers, the cross-sectional details of the two selected rivers were measured before taking each reading. Sample readings taken on 28 July 2018 and 28 July 2018 at Ranipool and Jalipool bridges are given in Tables 6.1 and 6.2, respectively.

The measured channel cross-sections of the Ranikhola and Busuk-khola rivers at Ranipool and Jalipool bridges were plotted (Figs. 6.2 and 6.3). In Figs. 6.2 and 6.3,

**Table 6.1** Depth of river bed measured from the bridge span at Ranipool Bridge (28 July 2018)

| Distance (m) | Depth of river bed from bridge span (m) | Distance (m) | Depth of river bed from bridge span (m) |
|---|---|---|---|
| 3 | 15.84 | 14 | 17.5 |
| 4 | 15.9 | 15 | 17.6 |
| 5 | 16.42 | 16 | 17.1 |
| 6 | 16.15 | 17 | 16.8 |
| 7 | 16.34 | 18 | 15.45 |
| 8 | 16.9 | 19 | 15.4 |
| 9 | 16.8 | 20 | 16.41 |
| 10 | 17 | 21 | 16.8 |
| 11 | 17.11 | 22 | 16.5 |
| 12 | 17.35 | 23 | 16 |
| 13 | 17.4 | – | – |

**Table 6.2** Depth of river bed measured from the bridge span at Jalipool Bridge (29 July 2018)

| Distance (m) | Depth of river bed from bridge span (m) | Distance (m) | Depth of river bed from bridge span (m) |
|---|---|---|---|
| 0 | 7.8a>9.95 | 6 | 15.7 |
| 1 | 3.39a>15.65 | 7 | 15.1 |
| 2 | 15.15 | 8 | 15.1 |
| 3 | 15.65 | 9 | 14.95 |
| 4 | 15.8 | 10 | 14.95 |
| 5 | 15.6 | 11 | 14.95 |

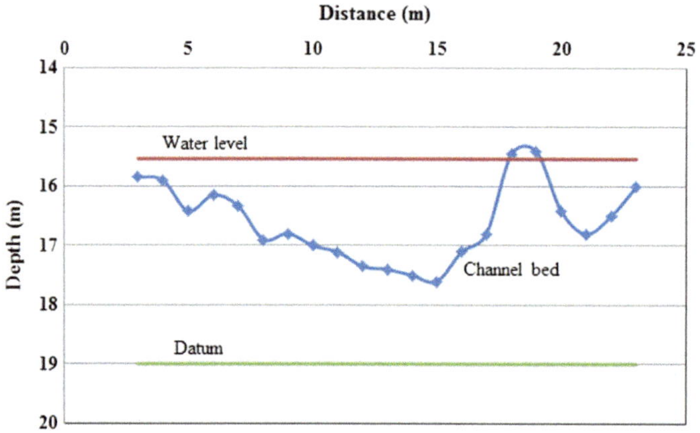


**Fig. 6.2** The Ranikhola River cross-section at Ranipool Bridge

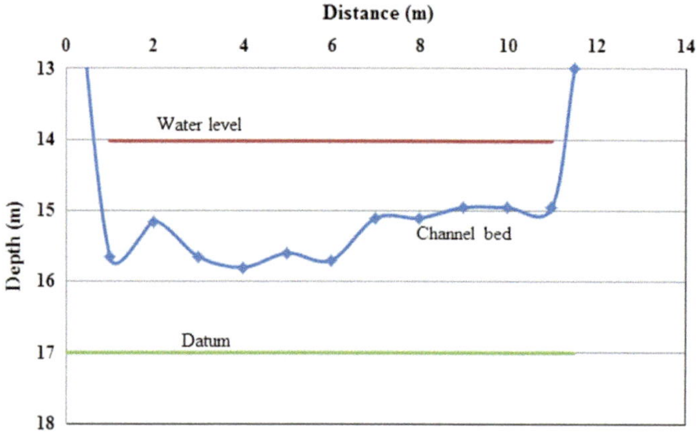


**Fig. 6.3** The Busuk-khola River cross-section at Jalipool Bridge

the distance from the left-hand side while facing upstream is plotted on the $X$-axis, whereas depths of the river bed and that of water surface (both measured from the bridge span) are plotted on $Y$-axis.

### 6.3.2 Flow Velocities and Discharge Measurement

To estimate the flow discharge in the Ranikhola and Busuk-khola rivers, the flow velocity measurements were made by using current meter at the frequency of about a week and daily basis based on discharge variation in the rivers. One of the sample reading and corresponding computational procedure is shown in Table 6.3, in which the flow velocity (V) was estimated by using calibration equations ($V = 0.719$ Ns + 0.009 and $V = 0.8649$ Ns, where Ns = revolution per second of the current meter) of the two current meters used in the present study.

### 6.3.3 Stage-Discharge Relationship

The river stage and corresponding flow discharge measurements were taken for about four months duration, i.e. from 28 July 2018 to 27 November 2018. The stage-discharge data generated through the present study are given in Tables 6.4 and 6.5. The measured river stages were plotted against the corresponding estimated flow discharge values in arithmetic and logarithmic plots with stage as ordinate and discharge as abscissa (Figs. 6.4, 6.5, 6.6 and 6.7). After plotting the stage versus discharge to the arithmetic scale, a smooth curve through the plotted points was drawn, whereas a straight line was drawn for logarithmic plots and power form equations were chosen. The coefficients of determination ($R^2$) values for the above-mentioned plots were observed to be about 0.968 and 0.989 for Ranikhola and Busuk-khola rivers, respectively. The corresponding power form equations for the two rivers are given in Eqs. (6.6) and (6.7).

$$h = 1.8196\,Q^{0.168} \tag{6.6}$$

$$h = 1.9184\,Q^{0.156} \tag{6.7}$$

where $h$ = river stage (m) and $Q$ = flow discharge ($m^3/s$).

**Table 6.3** Discharge computation by mean section method (28 July 2018)

| Distance (m) | Depth of river bed from bridge span (m) | Depth of water level from bridge span (m) | Depth of water (m) | Revolutions of current meter | Time (s) | Velocity (m/s) | Avg. velocity (m/s) | Avg. depth (m) | Elemental discharge (m3/s) |
|---|---|---|---|---|---|---|---|---|---|
| 3 | 15.84 | 15.5 | 0.31 | 40 | 30 | 1.15 | 1.59 | 0.34 | 0.54 |
| 4 | 15.9 | 15.5 | 0.37 | 70 | 30 | 2.02 | 2.36 | 0.63 | 1.49 |
| 5 | 16.42 | 15.5 | 0.89 | 94 | 30 | 2.71 | 2.36 | 0.76 | 1.78 |
| 6 | 16.15 | 15.5 | 0.62 | 70 | 30 | 2.02 | 1.85 | 0.72 | 1.32 |
| 7 | 16.34 | 15.5 | 0.81 | 58 | 30 | 1.67 | 2.19 | 1.09 | 2.39 |
| 8 | 16.9 | 15.5 | 1.37 | 94 | 30 | 2.71 | 2.65 | 1.32 | 3.5 |
| 9 | 16.8 | 15.5 | 1.27 | 90 | 30 | 2.59 | 2.67 | 1.37 | 3.65 |
| 10 | 17 | 15.5 | 1.47 | 95 | 30 | 2.74 | 2.75 | 1.53 | 4.2 |
| 11 | 17.11 | 15.5 | 1.58 | 96 | 30 | 2.77 | 2.93 | 1.7 | 4.97 |
| 12 | 17.35 | 15.5 | 1.82 | 107 | 30 | 3.08 | 2.84 | 1.85 | 5.24 |
| 13 | 17.4 | 15.5 | 1.87 | 90 | 30 | 2.59 | 2.36 | 1.92 | 4.54 |
| 14 | 17.5 | 15.5 | 1.97 | 74 | 30 | 2.13 | 2.1 | 2.02 | 4.25 |
| 15 | 17.6 | 15.5 | 2.07 | 72 | 30 | 2.08 | 1.86 | 1.82 | 3.38 |
| 16 | 17.1 | 15.5 | 1.57 | 57 | 30 | 1.64 | 1.66 | 1.42 | 2.35 |
| 17 | 16.8 | 15.5 | 1.27 | 58 | 30 | 1.67 | 0.84 | 0.64 | 0.53 |
| 18 | 15.45 | 15.5 | – | – | 30 | – | – | – | – |
| 19 | 15.4 | 15.5 | – | – | 30 | – | 0.58 | 0.44 | 0.25 |
| 20 | 16.41 | 15.5 | 0.88 | 40 | 30 | 1.15 | 0.92 | 1.08 | 0.99 |
| 21 | 16.8 | 15.5 | 1.27 | 24 | 30 | 0.69 | 0.63 | 1.12 | 0.71 |

(continued)

**Table 6.3** (continued)

| Distance (m) | Depth of river bed from bridge span (m) | Depth of water level from bridge span (m) | Depth of water (m) | Revolutions of current meter | Time (s) | Velocity (m/s) | Avg. velocity (m/s) | Avg. depth (m) | Elemental discharge (m3/s) |
|---|---|---|---|---|---|---|---|---|---|
| 22 | 16.5 | 15.5 | 0.97 | 20 | 30 | 0.58 | 0.72 | 0.72 | 0.52 |
| 23 | 16 | 15.5 | 0.47 | 30 | 30 | 0.86 | 0.43 | 0.235 | 0.1 |
| Total discharge | | | | | | | | | 46.72 |

**Table 6.4** Stage-discharge data for the Ranikhola River at Ranipool Bridge

| Date | Discharge | Stage | Date | Discharge | Stage |
|---|---|---|---|---|---|
| 7/29/2018 | 46.72 | 3.47 | 11/7/2018 | 4.2 | 2.31 |
| 8/11/2018 | 38.76 | 3.23 | 11/8/2018 | 4.22 | 2.3 |
| 8/13/2018 | 42.11 | 3.42 | 11/9/2018 | 3.67 | 2.29 |
| 8/19/2018 | 32.61 | 3.3 | 11/10/2018 | 3.65 | 2.27 |
| 8/21/2018 | 25.31 | 3.14 | 11/11/2018 | 3.386 | 2.28 |
| 8/23/2018 | 22.51 | 3.08 | 11/12/2018 | 3.66 | 2.33 |
| 8/25/2018 | 37.55 | 3.4 | 11/13/2018 | 4.07 | 2.29 |
| 8/27/2018 | 27.88 | 3.2 | 11/14/2018 | 3.41 | 2.25 |
| 8/29/2018 | 20.1 | 3.04 | 11/18/2018 | 3.57 | 2.27 |
| 8/31/2018 | 13.09 | 2.9 | 11/20/2018 | 3.31 | 2.21 |
| 9/3/2018 | 11.23 | 2.71 | 11/22/2018 | 2.74 | 2.2 |
| 9/13/2018 | 7.80 | 2.5 | 11/24/2018 | 2.48 | 2 |
| 9/17/2018 | 17.61 | 2.88 | 11/25/2018 | 2.5 | 2.1 |
| 11/5/2018 | 3.44 | 2.26 | 11/27/2018 | 2.4 | 2.11 |

**Table 6.5** Stage-discharge data for the Busuk-khola River at Jalipool Bridge

| Date | Discharge | Stage | Date | Discharge | Stage |
|---|---|---|---|---|---|
| 8/11/2018 | 17.9 | 3.01 | 11/9/2018 | 4.29 | 2.41 |
| 8/13/2018 | 19.41 | 3.1 | 11/10/2018 | 4.19 | 2.43 |
| 8/22/2018 | 18.34 | 2.98 | 11/11/2018 | 3.71 | 2.35 |
| 8/26/2018 | 13.88 | 2.9 | 11/12/2018 | 3.63 | 2.34 |
| 8/31/2018 | 21.61 | 3.2 | 11/14/2018 | 3.25 | 2.33 |
| 9/10/2018 | 10.2 | 2.72 | 11/18/2018 | 3.29 | 2.32 |
| 9/20/2018 | 7.86 | 2.59 | 11/20/2018 | 3.23 | 2.31 |
| 9/18/2018 | 12.14 | 2.75 | 11/22/2018 | 3.07 | 2.3 |
| 10/8/2018 | 6.05 | 2.55 | 11/23/2018 | 3.16 | 2.31 |
| 10/15/2018 | 15.85 | 2.96 | 11/25/2018 | 2.97 | 2.27 |
| 11/5/2018 | 5.46 | 2.5 | 11/28/2018 | 2.51 | 2.2 |
| 11/7/2018 | 4.3 | 2.42 | 12/2/2018 | 1.85 | 2.11 |
| 11/8/2018 | 4.13 | 2.41 | – | – | – |

### 6.3.4 *Sediment Discharge in Two Rivers*

The suspended sediment concentration was estimated through laboratory analysis, and the resultant data is given in Table 6.6. The suspended sediment load in the Ranikhola River was computed by using Eq. (6.5).

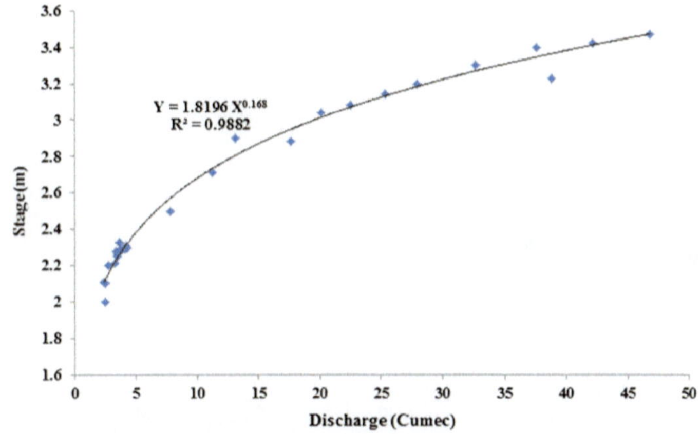


**Fig. 6.4** Stage-discharge rating curve for the Ranikhola River at Ranipool Bridge (arithmetic)

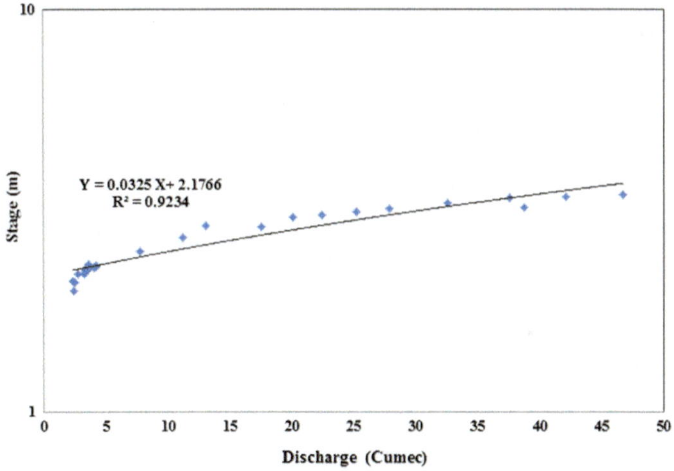


**Fig. 6.5** Stage-discharge rating curve for the Ranikhola River at Ranipool Bridge (logarithmic)

From Tables 6.6 and 6.7, it was observed that during the study period of four months, the total sediment loads in the Ranikhola and Busuk-khola rivers were ranged between 18–4071.51 and 1.92–603.73 tonnes per day, respectively.

## 6.4 Conclusions

The developed stage-discharge rating curve equations are: $h = 1.8196\ Q^{0.168}$ and $1.9184\ Q^{0.156}$ with coefficients of determination ($R^2$ values) of 0.968 and 0.989 which

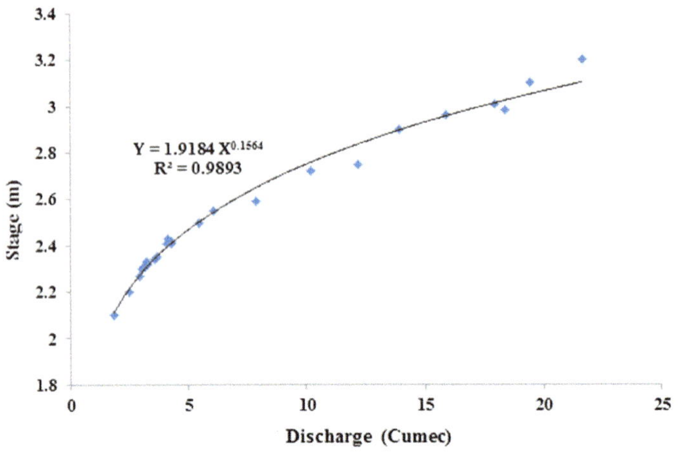


**Fig. 6.6** Stage-discharge rating curve for the Busuk-khola River at Jalipool Bridge (arithmetic)

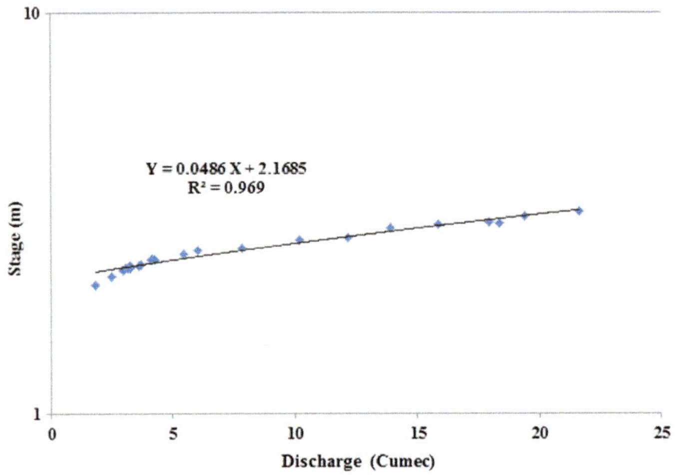


**Fig. 6.7** Stage-discharge rating curve for the Busuk-khola River at Jalipool Bridge (logarithmic)

may be useful for computing flow discharge in the Ranikhola and Busuk-khola rivers for the measured stages at Ranipool and Jalipool bridge cross-sections, respectively. It was observed that at smaller stages, increase in discharge is comparatively lesser than the same at higher stage values, which can be attributed to the more increase in flow cross-sectional area at higher stage values as compared to the lower increase at lesser stage values. The range of sediment rates in the two rivers was observed to be 18–4071.51 tonnes per day in Ranikhola River and 1.92–603.73 tonnes per day in Busuk-khola River. These large sediment inflow rates into the Teesta River necessitates the large-scale catchment area treatment programmes including both

**Table 6.6** Sediment load in the Ranikhola River

| Date | Flow discharge ($m^3/s$) | Sediment concentration (g/L) | Suspended load (tonnes/day) | Bed load (tonnes/day) | Total load (tonnes/day) |
|---|---|---|---|---|---|
| 7/29/2018 | 46.72 | 0.53 | 2139.59 | 427.92 | 2567.5 |
| 8/11/2018 | 38.76 | 0.41 | 1373.23 | 274.65 | 1647.87 |
| 8/13/2018 | 42.11 | 0.88 | 3201.71 | 640.34 | 3842.05 |
| 8/19/2018 | 32.61 | 0.62 | 1746.85 | 349.37 | 2096.22 |
| 8/21/2018 | 25.31 | 0.33 | 721.64 | 144.33 | 865.97 |
| 8/23/2018 | 22.51 | 0.31 | 602.91 | 120.58 | 723.49 |
| 8/25/2018 | 37.55 | 0.34 | 1103.07 | 220.61 | 1323.68 |
| 8/27/2018 | 27.88 | 0.2 | 481.77 | 96.35 | 578.12 |
| 8/29/2018 | 20.1 | 0.21 | 364.69 | 72.94 | 437.63 |
| 8/31/2018 | 13.09 | 3 | 3392.93 | 678.59 | 4071.51 |
| 9/3/2018 | 11.23 | 0.2 | 194.05 | 38.81 | 232.87 |
| 9/13/2018 | 7.8 | 0.3 | 202.18 | 40.44 | 242.61 |
| 9/17/2018 | 17.61 | 0.19 | 289.09 | 57.82 | 346.9 |
| 11/5/2018 | 3.44 | 0.4 | 118.89 | 23.78 | 142.66 |
| 11/7/2018 | 4.2 | 0.47 | 170.55 | 34.11 | 204.66 |
| 11/8/2018 | 4.22 | 0.8 | 291.55 | 58.31 | 349.86 |
| 11/9/2018 | 3.7 | 2.6 | 830.94 | 166.19 | 997.13 |
| 11/10/2018 | 3.65 | 0.3 | 94.61 | 18.92 | 113.53 |
| 11/11/2018 | 3.39 | 0.45 | 131.67 | 26.33 | 158 |
| 11/12/2018 | 3.66 | 0.1 | 31.65 | 6.33 | 37.98 |
| 11/13/2018 | 4.07 | 0.25 | 87.91 | 17.58 | 105.49 |
| 11/14/2018 | 3.41 | 0.45 | 132.62 | 26.52 | 159.14 |
| 11/18/2018 | 3.57 | 0.24 | 74.03 | 14.81 | 88.83 |
| 11/20/2018 | 3.31 | 0.3 | 85.87 | 17.17 | 103.05 |
| 11/22/2018 | 2.74 | 0.2 | 47.35 | 9.47 | 56.82 |
| 11/24/2018 | 2.48 | 0.07 | 15 | 3 | 18 |
| 11/25/2018 | 2.5 | 0.08 | 17.28 | 3.46 | 20.74 |
| 11/27/2018 | 2.4 | 0.20 | 41.47 | 8.29 | 49.77 |

structural and non-structural control measures in the watershed areas of the two selected rivers so that the sediment rates may be reduced. Further, it is necessary to control the landslides during monsoon season that are prime source of sediments apart from the erosion of terraced agricultural fields.

**Table 6.7** Sediment load in the Busuk-khola River

| Date | Flow discharge ($m^3$/s) | Sediment concentration (g/L) | Suspended load (tonnes/day) | Bed load (tonnes/day) | Total load (tonnes/day) |
|---|---|---|---|---|---|
| 8/11/2018 | 17.9 | 0.23 | 355.71 | 71.14 | 426.85 |
| 8/13/2018 | 19.41 | 0.3 | 503.11 | 100.62 | 603.73 |
| 8/22/2018 | 18.34 | 0.21 | 332.74 | 66.55 | 399.29 |
| 8/26/2018 | 13.88 | 0.3 | 359.77 | 71.95 | 431.72 |
| 8/31/2018 | 21.61 | 0.1 | 186.73 | 37.35 | 224.07 |
| 9/10/2018 | 10.2 | 0.11 | 96.98 | 19.4 | 116.37 |
| 9/20/2018 | 7.86 | 0.1 | 67.94 | 13.59 | 81.53 |
| 9/18/2018 | 12.14 | 0.15 | 157.33 | 31.47 | 188.8 |
| 10/8/2018 | 6.05 | 0.2 | 104.49 | 20.9 | 125.39 |
| 10/15/2018 | 15.85 | 0.12 | 164.33 | 32.87 | 197.2 |
| 11/5/2018 | 5.46 | 1 | 471.74 | 94.35 | 566.09 |
| 11/7/2018 | 4.3 | 0.45 | 167.11 | 33.42 | 200.53 |
| 11/8/2018 | 4.14 | 0.3 | 107.18 | 21.44 | 128.62 |
| 11/9/2018 | 4.29 | 0.12 | 44.5 | 8.9 | 53.4 |
| 11/10/2018 | 4.19 | 0.2 | 72.37 | 14.47 | 86.84 |
| 11/11/2018 | 3.71 | 0.8 | 256.64 | 51.33 | 307.97 |
| 11/12/2018 | 3.63 | 0.4 | 125.38 | 25.08 | 150.46 |
| 11/14/2018 | 3.25 | 0.31 | 86.99 | 17.4 | 104.39 |
| 11/18/2018 | 3.29 | 0.13 | 36.95 | 7.39 | 44.34 |
| 11/20/2018 | 3.23 | 0.1 | 27.92 | 5.58 | 33.5 |
| 11/22/2018 | 3.07 | 0.5 | 132.67 | 26.53 | 159.2 |
| 11/23/2018 | 3.16 | 0.32 | 87.28 | 17.46 | 104.74 |
| 11/25/2018 | 2.97 | 0.11 | 28.2 | 5.64 | 33.84 |
| 11/28/2018 | 2.51 | 0.09 | 19.53 | 3.91 | 23.44 |
| 12/2/2018 | 1.85 | 0.01 | 1.6 | 0.32 | 1.92 |

# References

Adegbola AA, Olaniyan OS (2017) Computation of River discharge from formulated rating equation in hydrological catchments with inadequate data: River Omi, South Western Nigeria, as case study. Aquademia Water Environ Technol 1:1, 02. ISSN: 2468-1946

Alexandrov Y, Laronne JB, Reid I (2003) Suspended sediment concentration and its variation with water discharge in a dryland ephemeral channel, Northern Negev, Israel. J Arid Environ 53(1):73–84

Alfa MI, Adie DB, Ajibike MA, Mudiare OJ (2018) Development of rating curve for Ofu River at Oforachi hydrometric station. Nigerian J Technol Dev 15(1):14–19

Braca G (2008) Stage-discharge relationships in open channels: practices and problems. Univ. Degli Studi Di Trento, Dipartimento di Ingegneria Civile e Ambientale

Chen YC (2013) Flood discharge measurement of a mountain river–Nanshih river in Taiwan. Hydrol Earth Syst Sci 17(5):1951–1962

Dadson SJ, Hovius N, Chen H, Dade WB, Hsieh ML, Willett SD, Lague D (2003) Links between erosion, runoff variability and seismicity in the Taiwan Orogen. Nature 426(6967):648

Martin JM, Meybeck M (1979) Elemental mass-balance of material carried by major world rivers. Mar Chem 7(3):173–206

Rantz SE (1982) Measurement and computation of streamflow. US Geol Survey Water Supply 1:2175–2190

Schroder W, Theune C (1984) Denudation and Reservoir Sedimentation in Middle Europe (Feststoffabtrag Und Stauraumverlandung In Mitteleuropa). Wasserwirtschaft 74(7/8)

Subramanya K (2013) Engineering hydrology. McGraw-Hill Education (India) Private Limited, New Delhi, p 534

Turowski JM, Hovius N, Wilson A, Horng MJ (2008) Hydraulic geometry, river sediment and the definition of bedrock channels. Geomorphology 99(1–4):26–38

# Chapter 7
# Relative Contribution of Climate Variables on Long-Term Runoff Using Budyko Framework

**Uttam Puri Goswami and Manish Kumar Goyal**

**Abstract** Climate change and anthropogenic activities and their intensification have unprecedented alterations to hydrological cycle. The quantification of relative contribution of climate variable and vegetation changes to streamflow alteration is essential for water resources sustainability and management, but also challenging. The purpose of this study is to assess the relative contribution of climate variables and their effects on long-term annual runoff using Budyko framework induced by climate change. The result exhibits that the snow ratio has decreased, but this does not influence the annual runoff significantly as compared to precipitation, but region may receive less precipitation as snowfall in future. Precipitation is the major factor that affects the runoff changes with the ranges of 2–19% over the region. The study highlights that the change in precipitation can affect the temporal variation of intra-annual runoff, which can cause the increase in chance of flood disasters in spring season and water crises in summer.

**Keywords** Climate variables · Budyko · Relative contribution · Snow ratio · Himalaya

## 7.1 Introduction

Climate change and global warming are becoming an immense important issue in the field of hydrology over the recent decades (Immerzeel and Bierkens 2012). The intensification of global warming and climate extremes has led to increase in precipitation, temperature and evaporation. Kaab et al. (2012). In addition, the global water

U. P. Goswami (✉)
Department of Civil Engineering, Indian Institute of Technology Guwahati, Guwahati 781039, India

Department of Environmental and Water Resources Engineering Engineering, Chhattisgarh Swami Vivekanand Technical University, Bhilai 491107, India

M. K. Goyal
Discipline of Civil Engineering, Indian Institute of Technology Indore, Indore 453552, India

 
P. Kumar et al. (eds.), *Water Resources Management and Sustainability*,
Advances in Geographical and Environmental Sciences,
https://doi.org/10.1007/978-981-16-6573-8_7

cycle such as use of ground water, surface water and changing land cover is disturbed by human interactions along with the climate change (Zhang et al. 2004; Findell et al. 2007). Therefore, the consideration of water availability is important for its applicability and sustainability which can be assessed by the distribution of precipitation in terms of streamflow and evaporation (Immerzeel and Bierkens 2012; Greve et al. 2015). The hydrological responses of a catchment which are affected by vegetation processes is vital for assessment; on the other hand, due to the lack of understanding of interaction of data availability and processes, it is difficult to develop the model for prediction of hydrological responses at the catchment scales (Zhang et al. 2004). The diminishing water supply and variability in the dynamics of hydrological cycle prove the sensitiveness of river basins in India to climate and change in LULC. Therefore, it is important to evaluate the contribution of climate change and human activities to runoff change (Jiang et al. 2015). The global temperature is likely to exceed 1.5 °C at the end of the twenty-first century (Pachauri and Meyer 2014). Due to the increase in global air temperature, this may lead to less occurrence of precipitation as snowfall over the snow dominant regions (Berghuijs et al. 2014). Thus, the change in precipitation may affect the available water resources, especially in the snowmelt-dominated river catchments such as Hindu Kush Himalayas, which alone supply water to about 20% of world's population (Barnett et al. 2005). The temperature rise can alter the precipitation form that can change the available water resources as change in runoff (earlier peak in spring and water shortage in summer and autumn) (Zhang et al. 2015). This implies that the change in precipitation due to global warming can change the temporal intra-annual variability of runoff, resulting increase in chance of flood disasters during and water crises in summer (Allamano et al. 2009). The shift in precipitation state (form of precipitation snowfall to rainfall) and their effects on long-term mean annual runoff is significant. This factor can play an important role to control the available water resources (Zhang et al. 2015). This study is focused to understand the relationship between climate variables (snow ratio, precipitation and evapotranspiration) and mean annual runoff variation induces by climate change. To assess the water–energy balance, there are number of models (global and regional land surface model) such as hydrological modelling, statistical methods and Budyko framework. However, the Budyko framework is the robust approach to evaluate the water–energy balance (Williams et al. 2012; Xu et al. 2013). The Budyko framework is a simple but effective tool for assess the linkages between land surface and climatic factors to characterize the water and energy cycles at catchment scales (Xu et al. 2013). This framework has been applied globally because of their simple calculation and inclusion of physical mechanisms (Yang et al. 2007; Williams et al. 2012; Xu et al. 2013; Wang and Tang 2014; Greve et al. 2015; Jiang et al. 2015; Zhang et al. 2015; Gao et al. 2016). Budyko (1974), developed a framework which suggests that the importance of quantification of effect of climate variables of change in long-term annual runoff.

The main focus of this study is to understand the consequence of change in climate variables (snow ratio, precipitation and evapotranspiration) on annual runoff using Budyko framework under changing climate. In order to accomplish this work, historical and projected climate data (precipitation, temperature and evapotranspiration) used for Sikkim Himalayas. The projected precipitation and temperature are extracted from climate models [Coupled Model Intercomparison Project 5 (CMIP5)].

## 7.2 Study Area and Data

### *7.2.1 Study Area*

The recent trends of global warming and climate change, Himalayas and its climate responses, wind circulation and complex topography; make it centre of attention for global discussion topic. In India, major rivers such as Ganges, Brahmaputra, Indus, Sutlej, Beas, Yamuna, Chenab and Teesta originate from Himalayan glaciers (Goyal and Goswami 2018). The Teesta River part of Eastern Himalaya is considered as study area. The geographical extent of the study area lies from 25° N to 29° N latitude to 86.25° E to 91.25° E longitude as shown in Fig. 7.1. The covered area is about 2587.4 km$^2$. The river starts from the Tso Lhamo Lake, at 5033 m altitude (Singh et al. 2016). The study area up to Chugthang, the upper part of the study area is mostly occupied by glaciers and snow, whereas the downstream area nearby

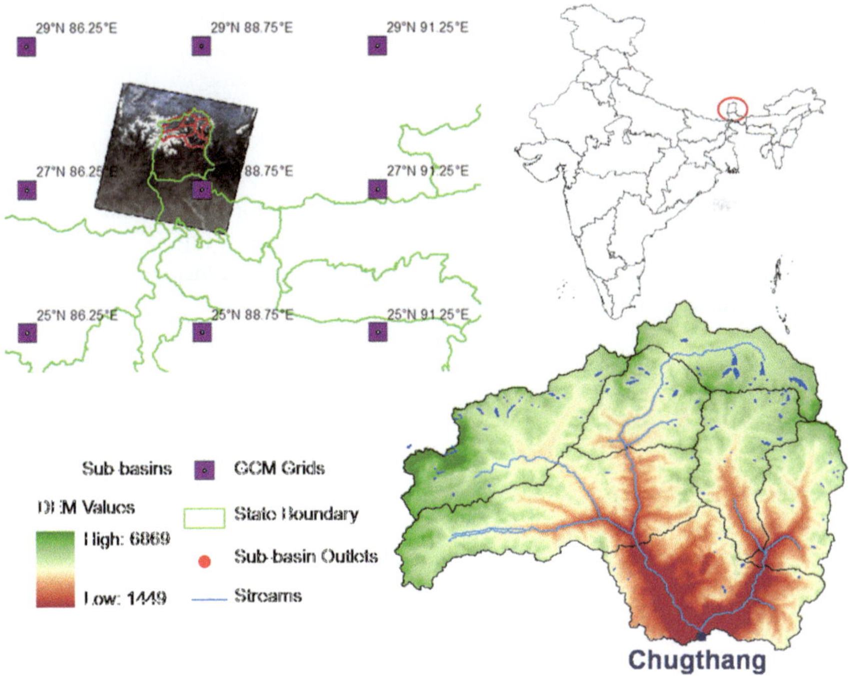


**Fig. 7.1** Location map of study area (Teesta River, Eastern Himalaya, India) and GCM grid-points

Chugthang includes hard rocks and dense forest (Meetei et al. 2007; Goswami et al. 2018a, b).

### 7.2.2 *Data Used*

The meteorological dataset (precipitation and temperature) for the historical duration (1982–2005) is collected at $0.5° \times 0.5°$ spatial resolution from IMD (India Meteorological Department). Further, the discharge data is obtained from CWC (central water commission) for the same duration (1982–2005). For climate change analysis, global climate model (GCM) outputs are taken namely GFDL-ESM2G, GFDL-ESM2M and GFDL-CM3. In addition, three Representative Concentration Pathway (RCP) scenarios (low-RCP2.6, moderate-RCP4.5 and high-RCP8.5) are considered (http://cmip-pcmdi.llnl.gov/CMIP5/) (Taylor et al. 2012). The GCM outputs are used for temperature and precipitation. Where, RCPs are stands for Representative Concentration Pathway for different emission scenarios; estimates the radiative forcing (RF). The downscaled projected precipitation (Goswami et al. 2018a, b) and temperature (Goswami et al. 2018a, b) data is taken from our previous research articles. Physical, hydrological and climatologically significances are considered to construct the GCM combinations; total seven combinations are constructed. For more details, see (Goswami et al. 2018a, b).

## 7.3 Methodology

The overall methodology framework of relative contribution of climate variables over the long-term annual runoff changes has been shown in Fig. 7.2. The overall methodology includes the calculation of snow ratio, evapotranspiration (past and future) and change in runoff due to climate variables.

### 7.3.1 *Snow Ratio*

Due to unavailability of precipitation type at any meteorological stations, the relationship based on the threshold temperature ($T = 0$) used to categorize the precipitation type, e.g. if the average temperature is below zero and precipitation occurs on the same day, so the precipitation will occur as snow, else rainfall.

$$\text{Precipitation type} = \begin{cases} \text{Snowfall } T_d < 0 \\ \text{Rainfall } \ T_d > 0 \end{cases} \tag{7.1}$$

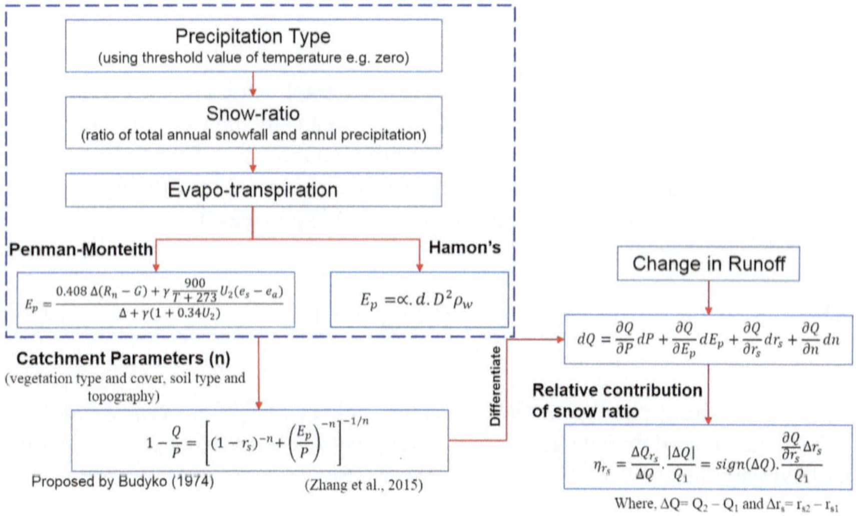


**Fig. 7.2** Overall methodology framework of the relative contribution of climate variables to annual runoff changes

The ratio of total annual snowfall and total annual precipitation is known as snow ratio ($r_s$). The concept of snow ratio eliminates the effect of phase difference generating form the accumulation of snow and melting of snow in the different years (Zhang et al. 2015).

$$\text{Snow - ratio}(r_s) = \frac{\text{Total annual snowfall}}{\text{Total annual precipitation}} \tag{7.2}$$

## 7.3.2 Evapotranspiration

The daily evapotranspiration was estimated using Penman–Monteith FAO equation (Allen et al. 1998) and Hamon's equation (Hamon 1960) for historical time period (1982–2005). The daily evapotranspiration compared to obtain the adjustment factor for Hamon's equation. The adjusted factor is calibrated by minimizing the differences of mean annual evapotranspiration values (1982–2005) estimated using both Penman–Monteith and Hamon's methods. Further, future monthly precipitation, temperature and snowfall outputs are extracted from downscaled CMIP5 GCM scenarios (Goswami et al. 2018a, b). The Penman–Monteith is the most suitable method for $E_p$ calculation, but it requires large dataset. Due to lack of data availability in future time period, $E_p$ values are calculated using Hamon's equation, which is a faction of temperature [$f(t)$] with adjusted parameter (adjustment factor).

### 7.3.3 Estimation of Runoff Change

The assessments of catchment characteristics (topography, soil type, vegetation type and cover) are essential for estimation of annual mean streamflow ($Q$). To estimate the $Q$, water–energy balance mathematical expressions are used. The catchment characteristic's information is estimated as synthesis parameter ($n$) by utilizing empirical equation (Choudhury's Equation) (Zhang et al. 2015) given as below:

$$1 - \frac{Q}{P} = \left[1 + \left(\frac{E_P}{P}\right)^{-n}\right]^{-1/n} \tag{7.3}$$

where $n$ is synthesis parameter, $Q/P$ is wetness index and $E_p/P$ is dryness index.

Over the snow-dominant regions, Eq. 7.3 (water–energy balance relationship) modified with the consideration of snow, shown as:

$$1 - \frac{Q}{P} = \left[(1 - r_s)^{-n} + \left(\frac{E_p}{P}\right)^{-n}\right]^{-1/n} \tag{7.4}$$

where $r_s$ = snow ratio.

The synthesis parameter ($n$) is calculated using long-term precipitation ($P$), discharge ($Q$), snow ratio ($r_s$) and potential evapotranspiration during 1982–2005. For the validation of $n$ value, the data divided in two parts, 1982–1993 and 1994–2005, to check the variation in synthesis parameter ($n$) values.

The Choudhury's equation (Eq. 7.4) again used to assess the long-term water balance over the catchment, where snowfall included for assessment of hydrological processes. Moreover, Eq. 7.4 is differentiating to calculate the change in runoff change climate factors and ($P$, $E_p$, $r_s$) and catchment characteristics ($n$). The first-order equation of Eq. 7.4 (Zhang et al. 2015):

$$dQ = \frac{\partial Q}{\partial P}dP + \frac{\partial Q}{\partial E_p}dE_p + \frac{\partial Q}{\partial r_s}dr_s + \frac{\partial Q}{\partial n}dn \tag{7.5}$$

Total change in annual runoff is the sum of small changes due to precipitation ($P$), evapotranspiration ($E$), snow ratio and catchment characteristics ($n$). The small changes due to ($P$, $E_p$, $r_s$) are calculated using empirical relationships given as below (Zhang et al. 2015).

$$\frac{\partial Q}{\partial P} = 1 - \frac{P - Q}{P}\frac{E_p^n}{[P(1 - r_s]^n + E_p^n} \tag{7.6}$$

$$\frac{\partial Q}{\partial E_p} = -\frac{P - Q}{E_p}\frac{[P(1 - r_s]^n}{[P(1 - r_s]^n + E_p^n} \tag{7.7}$$

$$\frac{\partial Q}{\partial r_s} = \frac{P-Q}{1-r_s}\frac{E_p^n}{[P(1-r_s]^n + E_p^n} \tag{7.8}$$

$$\frac{\partial Q}{\partial n} = -\frac{P-Q}{n} \left( \frac{\ln\left[P^n(1-r_s)^n + E_p^n\right]}{n} - \frac{[P(1-r_s)]^n \ln[P(1-r_s)] + E_p^n \ln(E_p)}{\left[P^n(1-r_s)^n + E_p^n\right]} \right) \tag{7.9}$$

The relative contributions of climate variables $(P, E_p, r_s)$ variation to annual runoff change are:

$$\eta_P = \frac{\Delta Q_p}{\Delta Q}.\frac{|\Delta Q|}{Q_1} = \text{sign}(\Delta Q).\frac{\frac{\partial Q}{\partial P}\Delta P}{Q_1} \tag{7.10}$$

Similarly, the elative contribution of evapotranspiration and snow ratio variation to annual changes can be estimated as:

$$\eta_{E_p} = \frac{\Delta Q_{E_p}}{\Delta Q}.\frac{|\Delta Q|}{Q_1} = \text{sign}(\Delta Q).\frac{\frac{\partial Q}{\partial E_p}\Delta E_p}{Q_1} \tag{7.11}$$

$$\eta_{r_s} = \frac{\Delta Q_{r_s}}{\Delta Q}.\frac{|\Delta Q|}{Q_1} = \text{sign}(\Delta Q).\frac{\frac{\partial Q}{\partial r_s}\Delta r_s}{Q_1} \tag{7.12}$$

where $\Delta Q = Q_2 - Q_1$; $\Delta P = P_2 - P_1$; $\Delta E_p = E_{p2} - E_{p1}$ and $\Delta r_s = r_{s2} - r_{s1}$.

## 7.4 Results and Analysiss

### *7.4.1 Estimation of Snow Ratio*

From the precipitation and temperature data, the precipitation type distinct as snowfall or rainfall based on the temperature threshold ($T = 0$). Further, the snow ratio is calculated for historical period (1982–2005) as well for the projected scenarios of twenty-first century. Figure 7.3 shows the annual snow ratio trends of total 63 ensembles (includes three GCMs, three RCPs and seven GCM grid-point combinations) for the projected scenarios and GCM grid-point combinations (see Goswami et al. 2018a, b). All the ensembles are showing falling trend in the projected snow ratio; however, the projected total precipitation trend is increasing. This implies that the precipitation is less likely to occur as snowfall in future over the region. The less occurrence of snow may lead to less deposition of snow in glaciers, causing the rivers

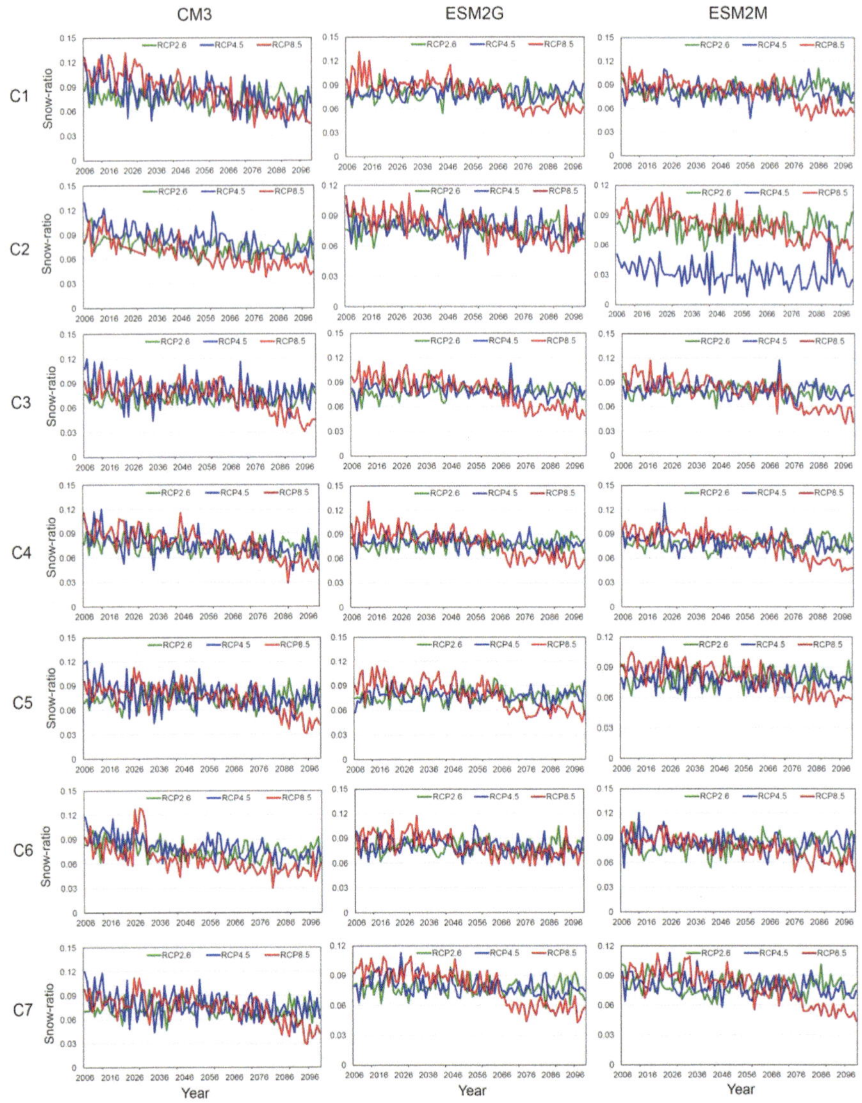


**Fig. 7.3** Annual projected snow ratio trends over the North Sikkim Himalaya for three GCMs and their RCP scenarios for all combinations, where *C*'s are combination

to dry up and drought to occur in long-term future. In last thirty years, 2070–2100, the RCP 8.5 (extreme scenario) shows the significant fallings trend, this clearly indicates that the snow occurrence would be vary less in end of the twenty-first century.

**Table 7.1** Mean annual changes in precipitation, runoff and snow ratio during historical period

| Parameters | 1982–1993 | 1994–2005 | Changes |
|---|---|---|---|
| Correction factor ($\alpha$) for ET | 3.78 | | NA |
| Specific catchment parameter (n) | 0.898 | | Negligible |
| Annual average precipitation (mm) | 2407.83 | 2401.39 | – 6.5 mm |
| Annual average discharge ($m^3/s$) | 139.17 | 126.72 | – 12.51 $m^3/s$ |
| Snow ratio | 0.10 | 0.15 | + 0.05 |

### 7.4.2 *Estimation of Change in Annual Runoff*

Firstly, the specific catchment parameter ($n$) is calculated using Eq. 7.5, which is about 0.898 for the Teesta River catchment for baseline period (1982–2005). Further, the historical data is divided in two part (1982–1993 and 1994–2005) and change in annual precipitation, runoff and snow ratio estimated for the historical period shown in Table 7.1. The change in annual mean runoff and mean precipitation is estimated, which is decreased about 12.51 $m^3/s$ and 6.5 mm, respectively, during 1994–2005 with respect to 1982–1993; while, the snow ratio is increased by 0.05 during 1994–2005 with respect to 1982–1993.

Further, snow ratio is calculated for the duration of 2020–2060 and compared with baseline period (1982–2005). The snow ratio is decreased by 0.038–0.096 during the 2020–2060 w.r.t. 1982–2005. In addition, the runoff is increase by 2–18% during 2020–2060 w.r.t. base periods (1982–2005). The estimated specific catchment parameter $n$ is 0.898 for the baseline period. As we have observed that the North Sikkim Himalaya is quite untouched and found less human activities during the field visits (by our research team); therefore, we are assuming that the catchment characteristics (topography, soil type, vegetation type and cover) are same and historical period. Therefore, the specific catchment parameter is considered as same as historical period ($n = 0.898$) for future duration. Further, the percentage relative contribution is calculated using Eq. 7.5 with the consideration of $n = 0.898$. Furthermore, the percentage relative contribution of climate variable to change in annual runoff is assessed for projected scenarios (total 63 ensembles, includes, three GCMs, three RCPs and seven GCM grid-point combinations). Table 7.2 exhibits the detailed percentage relative contribution of climate variable for all the ensembles. The result exhibits that the annual runoff is constantly decreasing up to 2% due to snow ratio for all the ensembles. Whereas, precipitation and evapotranspiration showing the increasing percentage change in annual runoff ranging from 2 to 19% and 0 to 0.4%, respectively. This signifies that the snow ratio and evapotranspiration are not the major factor for affecting the annual runoff; however, precipitation is the major factor to runoff changes. As the possible warming climate quantifies that the change in annual runoff resulting from snow ratio is negative, as the snow ratio decreased during 2020–2060 relative to 1982–2005. This implies that there may change of less occurrence of precipitation as snowfall. A shift from snowfall towards rainfall over the region is considered not to influence the annual runoff significantly.

**Table 7.2** Percentage relative contribution of snow ratio, precipitation and evapotranspiration variation to annual mean runoff change, where *C*'s represent the GCM grid-point combinations

| Scenarios | Relative change (%) in annual runoff due to climate factors | | | |
|---|---|---|---|---|
| | | *Snow ratio* ($r_s$) | *Precipitation* ($P$) | *Evapo-transpiration* ($E_p$) |
| Observed | | – 0.96 | 0.33 | 1.36 |
| *C1* | | | | |
| ESM2G | RCP2.6 | – 1.00 | 10.24 | 0.21 |
| | RCP4.5 | – 0.96 | 9.56 | 0.24 |
| | RCP8.5 | – 0.77 | 6.94 | 0.39 |
| ESM2M | RCP2.6 | – 0.98 | 11.34 | 0.18 |
| | RCP4.5 | – 0.98 | 8.62 | 0.24 |
| | RCP8.5 | – 0.83 | 6.39 | 0.37 |
| CM3 | RCP2.6 | – 1.08 | 8.41 | 0.08 |
| | RCP4.5 | – 0.92 | 9.10 | 0.20 |
| | RCP8.5 | – 0.71 | 4.75 | 0.31 |
| *C2* | | | | |
| ESM2G | RCP2.6 | – 1.04 | 17.78 | – 0.02 |
| | RCP4.5 | – 1.02 | 16.23 | 0.06 |
| | RCP8.5 | – 0.91 | 12.35 | 0.23 |
| ESM2M | RCP2.6 | – 1.05 | 19.33 | 0.00 |
| | RCP4.5 | – 1.91 | 17.03 | 0.13 |
| | RCP8.5 | – 0.89 | 15.24 | 0.24 |
| CM3 | RCP2.6 | – 1.08 | 9.21 | 0.16 |
| | RCP4.5 | – 0.81 | 11.65 | 0.25 |
| | RCP8.5 | – 0.98 | 2.33 | 0.30 |
| *C3* | | | | |
| ESM2G | RCP2.6 | – 0.98 | 8.76 | 0.12 |
| | RCP4.5 | – 0.96 | 8.35 | 0.15 |
| | RCP8.5 | – 0.82 | 6.31 | 0.30 |
| ESM2M | RCP2.6 | – 1.00 | 8.94 | 0.14 |
| | RCP4.5 | – 0.98 | 8.89 | 0.21 |
| | RCP8.5 | – 0.88 | 5.48 | 0.31 |
| CM3 | RCP2.6 | – 1.16 | 8.94 | 0.13 |
| | RCP4.5 | – 1.02 | 8.63 | 0.23 |
| | RCP8.5 | – 0.98 | 2.76 | 0.38 |
| *C4* | | | | |
| ESM2G | RCP2.6 | – 1.04 | 9.95 | 0.14 |
| | RCP4.5 | – 0.96 | 9.35 | 0.16 |

(continued)

**Table 7.2** (continued)

| Scenarios | Relative change (%) in annual runoff due to climate factors | | | |
|---|---|---|---|---|
| | | *Snow ratio* ($r_s$) | *Precipitation* ($P$) | *Evapo-transpiration* ($E_p$) |
| | RCP8.5 | – 0.84 | 6.96 | 0.32 |
| ESM2M | RCP2.6 | – 1.08 | 8.74 | 0.17 |
| | RCP4.5 | – 1.00 | 8.68 | 0.22 |
| | RCP8.5 | – 0.84 | 5.12 | 0.31 |
| CM3 | RCP2.6 | – 1.10 | 8.86 | 0.14 |
| | RCP4.5 | – 1.04 | 8.85 | 0.24 |
| | RCP8.5 | – 0.90 | 2.85 | 0.35 |
| *C5* | | | | |
| ESM2G | RCP2.6 | – 1.06 | 7.57 | 0.13 |
| | RCP4.5 | – 1.00 | 8.57 | 0.16 |
| | RCP8.5 | – 0.84 | 6.28 | 0.30 |
| ESM2M | RCP2.6 | – 1.04 | 8.34 | 0.12 |
| | RCP4.5 | – 1.02 | 7.48 | 0.20 |
| | RCP8.5 | – 0.86 | 4.85 | 0.28 |
| CM3 | RCP2.6 | – 1.08 | 8.86 | 0.13 |
| | RCP4.5 | – 1.00 | 8.70 | 0.24 |
| | RCP8.5 | – 0.94 | 2.79 | 0.40 |
| *C6* | | | | |
| ESM2G | RCP2.6 | – 1.00 | 16.40 | 0.02 |
| | RCP4.5 | – 0.94 | 14.41 | 0.11 |
| | RCP8.5 | – 0.86 | 8.60 | 0.28 |
| ESM2M | RCP2.6 | – 1.04 | 16.03 | 0.03 |
| | RCP4.5 | – 0.88 | 13.07 | 0.11 |
| | RCP8.5 | – 0.92 | 9.13 | 0.30 |
| CM3 | RCP2.6 | – 1.08 | 7.88 | 0.15 |
| | RCP4.5 | – 0.96 | 8.37 | 0.25 |
| | RCP8.5 | – 1.15 | 9.23 | 0.28 |
| *C7* | | | | |
| ESM2G | RCP2.6 | – 1.02 | 7.41 | 0.12 |
| | RCP4.5 | – 1.00 | 7.34 | 0.19 |
| | RCP8.5 | – 0.82 | 5.23 | 0.31 |
| ESM2M | RCP2.6 | – 1.06 | 9.71 | 0.19 |
| | RCP4.5 | – 0.98 | 8.19 | 0.20 |

(continued)

**Table 7.2** (continued)

| Scenarios | Relative change (%) in annual runoff due to climate factors | | | |
|---|---|---|---|---|
| | | *Snow ratio* ($r_s$) | *Precipitation* ($P$) | *Evapo-transpiration* ($E_p$) |
| | RCP8.5 | – 0.84 | 3.65 | 0.31 |
| CM3 | RCP2.6 | – 1.10 | 9.21 | 0.16 |
| | RCP4.5 | – 1.04 | 9.11 | 0.24 |
| | RCP8.5 | – 0.96 | 3.05 | 0.38 |

Similar results were found in previous studies (Barnett et al. 2005; Berghuijs et al. 2014); those are also snow-dominated regions.

## 7.5 Conclusions

The assessment of relative contribution of climate variables to annual runoff is important to sustainable water resources, especially for the snow-dominant regions under the climate change scenarios. The global temperature rise having the significant impact all around the world, but the most affected area are the snow/glacier regions (e.g. Himalaya). The relative contribution of climate variables (precipitation, evapotranspiration and snow ratio) to long-term annual runoff is assessed using Budyko framework and total 63 future ensembles (includes 3-GCMs, 3-RCPs, and 7-GCM grid-point combinations). As the possible warming climate quantifies that the change in annual runoff resulting from snow ratio is negative, as the snow ratio decreased during 2020–2060 relative to 1982–2005 for different combinations and ensembles. Moreover, the projected snow ratio scenarios are showing decreasing trends in twenty-first century. The precipitation is the predominant factor which is affecting the annual runoff with the range of 2–19%. In addition, it has been also observed that the less change of occurrence of precipitation as snowfall over the region; this signifies in future, there may be chances of spring flood disasters and water crises in summer. In long term, the less occurrence of snow may also lead to the disappearing to the glaciers, river dry up and cause the drought conditions.

## References

Allamano P, Claps P, Laio F (2009) Global warming increases flood risk in mountainous areas. Geophys Res Lett 36:L24404

Allen RG, Pereira LS, Raes D, Smith M (1998) Crop evapotranspiration: guidelines for computing crop requirements. FAO Irrigation and Drainage Paper No. 56. FAO, Rome, Italy

Barnett TP, Adam JC, Lettenmaier DP (2005) Potential impacts of a warming climate on water availability in snow-dominated regions. Nature 438(17):303–309

Berghuijs WR, Woods RA, Hrachowitz M (2014) A precipitation shift from snow towards rain leads to a decrease in streamflow-supplement. Nat Climate Change 4(7):583–586

Budyko MI (1974) Climate and life. Academic Press, New York, p 508

Findell HL, Shevliakova E, Milly P, Stouffer RJ (2007) Modeled impact of anthropogenic land cover change on climate. Am Meteorol Soc 20:3621–3634

Gao G, Fu B, Wang S, Liang W, Jiang X (2016) Determining the hydrological responses to climate variability and land use/cover change in the Loess Plateau with the Budyko framework. Sci Total Environ 557–558:331–342

Goswami UP, Hazra B, Goyal MK (2018a) Copula-based probabilistic characterization of precipitation extremes over North Sikkim Himalaya. Atmos Res 212:273–284

Goswami UP, Bhargav K, Hazra B, Goyal MK (2018b) Spatiotemporal and joint probability behavior of temperature extremes over the Himalayan region under changing climate. Theoret Appl Climatol 134(1–2):477–498

Goyal MK, Goswami UP (2018) Teesta river and its ecosystem. In: The Indian Rivers. Springer, Singapore, pp 537–551

Greve P, Gudmundsson L, Orlowsky B, Seneviratne SI (2015) Introducing a probabilistic Budyko framework. Geophys Res Lett 42:2261–2269

Hamon WR (1960) Estimating potential evapotranspiration (Doctoral dissertation, Massachusetts Institute of Technology)

Immerzeel WW, Bierkens MFP (2012) Asia's water balance. Nat Geosci 5(12):841–842

Jiang C, Xiong L, Wang D, Liu P, Guo S, Xu C-Y (2015) Separating the impacts of climate change and human activities on runoff using the Budyko-type equations with time-varying parameters. J Hydrol 522:326–338

Kaab A, Berthier E, Nuth C, Gardelle J, Arnaud Y (2012) Contrasting patterns of early twenty-first-century glacier mass change in the Himalayas. Nature 488(7412):495–498

Meetei LI, Pattanayak SK, Bhaskar A, Pandit MK, Tandon SK (2007) Climatic imprints in Quaternary valley fill deposits of the middle Teesta valley, Sikkim Himalaya. Quatern Int 159:32–49

Pachauri RK, Meyer LA (2014) Climate change 2014: synthesis report. Contribution of working groups I, II and III to the fifth assessment report of the intergovernmental panel on climate change. Core Writing Team. IPCC. https://doi.org/10.1017/CBO9781107415324.004

Singh V, Goyal MK (2016) Analysis and trends of precipitation lapse rate and extreme indices over north Sikkim eastern Himalayas under CMIP5ESM-2M RCPs experiments. Atmos Res 167:34–60

Taylor KE, Stouffer RJ, Meehl GA (2012) An overview of CMIP5 and the experiment design. Bull Am Meteor Soc 93(4):485–498

Wang D, Tang T (2014) A one-parameter Budyko model for water balance captures emergent behavior in darwinian hydrologic models. Geophys Res Lett 41:4569–4577

Williams CA, Reichstein M, Buchmann N, Baldocchi D, Beer C, Schwalm C, Wohlfahrt G, Hasler N, Foken T, Papale D, Schymanski S, Schaefer K (2012) Climate and vegetation controls on the surface water balance: synthesis of evapotranspiration measured across a global network of flux towers. Water Resour Res 48:1–13

Xu X, Liu W, Scanlon BR, Zhang L, Pan M (2013) Local and global factors controlling water-energy balances within the Budyko framework. Geophys Res Lett 40:6123–6129

Zhang D, Cong Z, Ni G, Yang D, Hu S (2015) Effects of snow ratio on annual runoff within the Budyko framework. Hydrol Earth Syst Sci 19(4):1977–1992

Zhang L, Hickel K, Dawes WR, Chiew F, Western A, Briggs P (2004) A rational function approach for estimating mean annual evapotranspiration. Water Resour Res 40:1–14

# Chapter 8
# Irrigation Planning with Fuzzy Parametric Programming Approach

**J. B. Gurav, R. U. Kamodkar, and D. G. Regulwar**

**Abstract** The present book chapter deals with tackling of uncertainty and vagueness involved in the irrigation planning problem with the concept of sustainability. Multi Objective Fuzzy Linear Programming (MOFLP) models have been developed using Fuzzy Parametric Programming (FPP) approach and applied it to the case study of Jayakwadi Project Stage-I, Maharashtra State, India. The various objectives such as NB, CP, EG, and MU, which are maximized. The involvement of Decision Maker (DM) is permitted in the proposed model to handle the fuzziness in the resources, technological coefficients, and coefficients of objective function. The results of the study show that the level of satisfaction ($\lambda$) is maximum when the precision level ($\mu$) is minimum and vice versa. It is also seen that for each level of precision which is varying from 0 to 1 with an interval of 0.1, the aggregated level of satisfaction ($\lambda$) is above 68%. The obtained results help to provide insight into various alternative optimal cropping patterns with different degree of precision and allow the DM to take judicious decision(s) in the context of the socio-economic development of a particular region.

**Keywords** Sustainability · Irrigation planning · Uncertainty · Membership functions: Monotonously increasing · Linear concave · Linear convex

---

J. B. Gurav (✉)
Department of Civil Engineering, Amrutvahini College of Engineering, Sangamner, Maharashtra 422608, India

Savitribai Phule Pune University, Pune, India

R. U. Kamodkar
Department of Civil Engineering, Government Polytechnic Jalgaon, Jalgaon 425002, India

D. G. Regulwar
Department of Civil Engineering, Government College of Engineering, Aurangabad, Aurangabad, Maharashtra 431005, India

 
P. Kumar et al. (eds.), *Water Resources Management and Sustainability*,
Advances in Geographical and Environmental Sciences,
https://doi.org/10.1007/978-981-16-6573-8_8

# 8.1 Introduction

## *8.1.1 General*

The irrigation planning problem can be formulated with reference to the availability of various resources such as land availability, availability of water, cropping pattern at the time of inception of a particular project, cropping seasons, climatic conditions requirement of the food for a particular region, and socio-economic development, etc.

## *8.1.2 Sustainable Irrigation Planning*

The formulation of irrigation planning problem and its execution with reference to proper use of the natural resources available for the betterment of mankind and other living creatures, which may be recognized as sustainable irrigation planning. In the mathematical sense, this refers to the adoption of irrigation practice of sustainable objectives and constraints. The sustainable objectives can be clubbed together as maximization of manure utilization, employment generation, crop production, net benefits, green energy, etc. These objectives can be subjected to various constraints/stipulations such as cultivable land, availability and quality of water, labor availability, manure availability, fixed water demand for industrial and drinking purpose, cropping pattern, etc. The active involvement of decision (policy) maker plays a vital role in all phases of the irrigation planning and its execution.

## *8.1.3 Uncertainty and Vagueness in Irrigation Planning*

In the real-world scenario, various parameters are related to irrigation planning such as hydrology, hydrogeology, soil and its characteristics, climatic variables, hydrometeorological parameters, management practice data, crop data, and basin data. Uncertainty and vagueness are associated with these parameters, which needs proper attention to work out adequate irrigation planning model and its solution. In the case of handling the number of objectives in case of irrigation planning, focusing only on a single objective does not result in optimal cropping pattern planning because various purposes are conflicting with each other. Similarly, dealing with the aim of net benefits maximization, the irrigation policymaker needs to think from different aspects, such as the generation of employment, sufficient food availability, etc.

### *8.1.4 Necessity of Fuzzy Logic for Uncertainty and Vagueness*

The fuzzy set theory/concept can be used to express mathematically the uncertainty and vagueness related to a variety of parameters of sustainable irrigation planning. Zadeh (1965), the founder of the term fuzzy, developed the fuzzy set theory which has given the substitute to classical set theory. This theory involves the infinite range of alternatives between zero and one. It also involves the parameters and its grade of membership in the form of a fuzzy set, which can be used to deal with uncertainty and vagueness.

### *8.1.5 Literature Review*

The methodology has developed and suggested for the solution of fuzzy linear programming (Gasimov and Yenilmez 2002). Jimenez et al. (2007) have developed a number of LP problems with a variety of parameters as fuzzy numbers however whose decision variables are crisp in nature and proposed an interactive approach for the solution of LP with fuzzy numbers. A two-phase method has introduced for solving MOFLP problems with the involvement of DM (Arikan and Gungor 2007). The fuzzy dynamic programming model has developed with the use of fuzzy inference system, to include experience and professional judgments of DM and farmers, which helps to work out optimal cropping pattern with conjunctive use of surface water and groundwater to a wide range of climatic conditions (Safavi and Alijanian 2010).

A number of LP models have been developed with the use of fuzzy logic. These models tackle uncertainty and vagueness associated with irrigation planning with the concept of sustainability (Regulwar and Gurav 2011, 2012; Gurav and Regulwar 2012). The LP model with robust optimization method has been proposed and utilized it for agricultural water resource management under uncertainty to the case study of the irrigation network in the province of Isfahan, Iran (Sabouni and Mardani 2013). An inexact fuzzy chance-constrained nonlinear programming method has proposed for the management of agriculture water resources under various uncertainties, which is an improvement over the existing stochastic programming methods (Guo et al. 2013).

A bi-level programming model involving the fuzzy inputs, which has developed and utilized it for the solution to regional water resources allocation problem (Xu et al. 2013). Kumari and Mujumdar (2015) have developed fuzzy a stochastic dynamic programming model and applied to the case study of Bhadra reservoir Karnataka, India.

The biobjective programming model with fuzzy inputs has been developed and applied successfully with its capability to increase agricultural water productivity and to reduce the shortage of irrigation water, along with proper justification to the various concerns of DM and farmers (Li et al. 2016). The particle swarm optimization techniques along with the fuzzy approach have been proposed to deal with uncertainty for irrigation planning problem with the use of hyperbolic and exponential membership functions (Morankar et al. 2016). Fuzzy optimization model along with fuzzy inference system, which has proposed for conjunctive use of surface and groundwater resources and applied to the case study in the Astaneh-Kouchesfahan Plain in the north of Iran (Sami et al. 2018).

The fuzzy multi objective heuristic model has been proposed along with the uncertainties involved in water resources and economic parameters in a basin. The results of the study show that optimal operation policies provide better results than the deterministic model (Banihabib et al. 2019). Yue et al. (2020) have developed a full fuzzy-interval credibility-constrained nonlinear programming (FFICNP) model to deal with uncertainties in planning irrigation water allocation and found that, higher net system benefit and system efficiency for lower credibility level.

From the above literature review, it is found that various models of LP and MOFLP have been used to tackle the uncertainty and vagueness associated with parameters of irrigation planning.

## 8.2 Methodology and Model Development

### 8.2.1 *Objectives*

The various objectives are considered in the formulation of irrigation planning model depending upon the requirement of the particular region and national importance. In the present study, four objectives of maximization type are considered from the analysis point of view.

**Net Benefits (NB)**

Most of the time, the farmers have the tendency to get the maximum net benefits with the cultivation of particular crops for the economic-prosperity, which insists the DM to incorporate this as part of irrigation planning policy.

In the present study, discussion with farmers, agricultural and field experts is the basis for deciding the input cost. The input cost is taken as of twenty percent of gross benefit for each crop in the particular crop season. On the basis of average yield of a particular crop per ha and the current market price of a crop, the gross benefits are worked out for each crop.

$$\text{Maximize NB} = \left[\left(\sum_{i=1}^{2} J_i^K F_i^K + \sum_{i=1}^{3} J_i^R F_i^R + \sum_{i=1}^{2} J_i^{\text{TS}} F_i^{\text{TS}} + \sum_{i=1}^{2} J_i^P F_i^P + \sum_{i=1}^{1} J_i^{\text{HW}} F_i^{\text{HW}}\right)\right.$$

$$-\left(\sum_{i=1}^{2} J_i^K G_i^K + \sum_{i=1}^{3} J_i^R G_i^R + \sum_{i=1}^{2} J_i^{TS} G_i^{TS} + \sum_{i=1}^{2} J_i^P G_i^P + \sum_{i=1}^{1} J_i^{HW} G_i^{HW}\right)\Bigg] \tag{8.1}$$

| | |
|---|---|
| $i$ | crop index, 1 ha = 10,000 m$^2$. |
| $J_i^K$ | Area (ha) for various crops in Kharif Season; |
| $J_i^R$ | Area (ha) for various crops in Rabi Season; |
| $J_i^{HW}$ | Area (ha) for various crops in Hot Weather Season; |
| $J_i^P$ | Area (ha) for Perennial crops; |
| $J_i^{TS}$ | Area (ha) for Two Seasonal crops; |
| $F_i$ | Coefficient of benefit for $i$th crop in particular season; |
| $G_i$ | Input cost for $i$th crop in particular season. |

**Crop Production (CP)**

To expect maximum production of a particular crop is the natural tendency of every farmer, which is essential for the DM to consider it for the irrigation planning with the objective of maximization of crop production. The average yield for crop per ha is considered as crop production coefficient.

$$\text{Maximize (CP)} = \left[\left(\sum_{i=1}^{2} J_i^K U_i^K + \sum_{i=1}^{3} J_i^R U_i^R + \sum_{i=1}^{2} J_i^{TS} U_i^{TS} + \sum_{i=1}^{2} J_i^P U_i^P + \sum_{i=1}^{1} J_i^{HW} U_i^{HW}\right)\right] \tag{8.2}$$

$U_i$ Average yield of $i$th crop (Tons per ha).

The food sufficiency for a particular region can be thought of with the help of maximization of crop production, which mainly focus on the survival of people with satisfaction food demand of the particular region, with of view of this the second objective can be thought of sustainability-related.

**Employment Generation (EG)**

Socio-economic development of a region cannot be possible without availing employment generation to people. This suggests the DM focus on the need for maximization of employment generation related to irrigation planning.

$$\text{Maximize(EG)} = \left[\left(\sum_{i=1}^{2} J_i^K M_i^K + \sum_{i=1}^{3} J_i^R M_i^R + \sum_{i=1}^{2} J_i^{TS} M_i^{TS} + \sum_{i=1}^{2} J_i^P M_i^P + \sum_{i=1}^{1} J_i^{HW} M_i^{HW}\right)\right] \tag{8.3}$$

$M_i$ Requirement of Man Days for $i$th crop per ha.

The required labor/Man Days for a particular crop per ha have been considered based on the basis of discussions with farmers and experts from the agricultural field. This objective related to sustainability with the view of socio-economic development for developing countries like India, where there is uneven distribution of agricultural land.

**Manure Utilization (MU)**

$$\text{Maximize (MU)} = \left[\left(\sum_{i=1}^{2} J_i^K M_i^K + \sum_{i=1}^{3} J_i^R M_i^R + \sum_{i=1}^{2} J_i^{TS} M_i^{TS} + \sum_{i=1}^{2} J_i^P M_i^P + \sum_{i=1}^{1} J_i^{HW} M_i^{HW}\right)\right] \tag{8.4}$$

$M_i$ Manure Utilization Requirement (Tons/ha for $i$th crop).

The discussion with farmers and agricultural field experts is used to fix up the manure utilization requirement per ha for the various crop in respective seasons. The concept of utilization of green manure has introduced in the present study with a view to promote the natural ability of soil to sustain micronutrient sufficiency for various crops in the respective crop season. The waste generated out of various activities of farming and livestock, which can be used as a green manure after decomposition. Such prepared manure is free from harmful chemicals, fertilizers, and pesticides. This attracts the DM to include it in the form of sustainability for irrigation planning.

### 8.2.2 Constraints

**Total Sowing Area**

It can be represented as below:

$$\left(\sum_{i=1}^{2} J_i^K + \sum_{i=1}^{3} J_i^R + \sum_{i=1}^{2} J_i^{TS} + \sum_{i=1}^{2} J_i^P + \sum_{i=1}^{1} J_i^{HW}\right) \leq \text{TJ} \tag{8.5}$$

TJ Total command area.

**Maximum Sowing Area**

It can be represented as below:

Kharif

$$\left(\sum_{i=1}^{2} J_i^P + \sum_{i=1}^{2} K_i^{TS} + \sum_{i=1}^{2} J_i^K\right) \leq \text{TJ}_i^P + \text{TJ}_i^{TS} + \text{TJ}_i^K \tag{8.6}$$

Rabi

$$\left(\sum_{i=1}^{2} J_i^P + \sum_{i=1}^{2} K_i^{TS} + \sum_{i=1}^{3} K_i^R\right) \leq \text{TJ}_i^P + \text{TJ}_i^{TS} + \text{TJ}_i^R \tag{8.7}$$

Hot Weather and Perennial

$$\left(\sum_{i=1}^{2} J_i^P + \sum_{i=1}^{1} J_i^{\text{HW}}\right) \le \text{TJ}_i^P + \text{TJ}_i^{\text{HW}} \tag{8.8}$$

$\text{TJ}_i^K, \text{TJ}_i^R, \text{TJ}_i^{\text{HW}}, \text{TJ}_i^P, \text{TJ}_i^{TS}$ Command area for *i*th crop (ha) in different seasons.

**Affinity**

Perennial

$$J_1^P \le \text{TJ}_i^P \tag{8.9a}$$

$J_1^P$ Area under Sugarcane (Perennial)

$$J_2^P \le \text{TJ}_i^P \tag{8.9b}$$

$J_2^P$ Area under Banana (Perennial).

Two Seasonal

$$J_3^{\text{TS}} \le \text{TJ}_i^{\text{TS}} \tag{8.9c}$$

$J_3^{\text{TS}}$ Area under Chillies (Two seasonal)

$$J_4^{\text{TS}} \le \text{TJ}_i^{\text{TS}} \tag{8.9d}$$

$J_4^{\text{TS}}$ Area under LS Cotton (Two seasonal).

Kharif

$$J_5^K \le \text{TJ}_i^K \tag{8.9e}$$

$J_5^K$ Area under Sorghum (Kharif)

$$J_6^K \le \text{TJ}_i^K \tag{8.9f}$$

$J_6^K$ Area under Paddy (Kharif).

Rabi

$$J_7^R \leq \mathrm{TJ}_i^R \tag{8.9g}$$

$J_7^R$ Area under Sorghum (Rabi)

$$J_8^R \leq \mathrm{TJ}_i^R \tag{8.9h}$$

$J_8^R$ Area under Wheat (Rabi)

$$J_9^R \leq \mathrm{TJ}_i^R \tag{8.9i}$$

$J_9^R$ Area under Gram (Rabi).

Hot Weather

$$J_{10}^{\mathrm{HW}} \leq \mathrm{TJ}_i^{\mathrm{HW}} \tag{8.9j}$$

$J_{10}^{\mathrm{HW}}$ Area under Groundnut (Hot Weather).

**Labor Availability Constraint**

To deal with issue of unavailability of labor for the duration of cultivation period of different crops in different season, it is proposed that to address the issue of uncertainty concerned with the labor availability; the labor requirement should not be more than the whole labor availability for the duration of that specific crop season.

Kharif

$$\left(\sum_{i=1}^{2} J_i^P M_i^P + \sum_{i=1}^{2} J_i^{\mathrm{TS}} M_i^{\mathrm{TS}} + \sum_{i=1}^{2} J_i^K M_i^K\right) \leq \sum_{i=1}^{2} \mathrm{AM}_i^P + \sum_{i=1}^{2} \mathrm{AM}_i^{TS} + \sum_{i=1}^{2} \mathrm{AM}_i^K \tag{8.10}$$

Rabi

$$\left(\sum_{i=1}^{2} J_i^P M_i^P + \sum_{i=1}^{2} J_i^{\mathrm{TS}} M_i^{\mathrm{TS}} + \sum_{i=1}^{3} J_i^R M_i^R\right) \leq \sum_{i=1}^{2} \mathrm{AM}_i^P + \sum_{i=1}^{2} \mathrm{AM}_i^{\mathrm{TS}} + \sum_{i=1}^{3} \mathrm{AM}_i^R \tag{8.11}$$

Perennial and Hot Weather

$$\left(\sum_{i=1}^{2} J_i^P M_i^P + \sum_{i=1}^{1} J_i^{\mathrm{HW}} M_i^{\mathrm{HW}}\right) \leq \sum_{i=1}^{2} \mathrm{AM}_i^P + \sum_{i=1}^{1} \mathrm{AM}_i^{\mathrm{HW}} \tag{8.12}$$

$AM_i$ Availability of labor for particular crop;
$M_i$ Man Days Requirement for particular crop per ha.

**Manure Availability**

It can be represented as below:
Kharif

$$\left(\sum_{i=1}^{2} J_i^P M_i^P + \sum_{i=1}^{2} J_i^{TS} M_i^{TS} + \sum_{i=1}^{2} J_i^K M_i^K\right) \le \sum_{i=1}^{2} AM_i^P + \sum_{i=1}^{2} AM_i^{TS} + \sum_{i=1}^{2} AM_i^K \tag{8.13}$$

Rabi

$$\left(\sum_{i=1}^{2} J_i^P M_i^P + \sum_{i=1}^{2} J_i^{TS} M_i^{TS} + \sum_{i=1}^{3} J_i^R M_i^R\right) \le \sum_{i=1}^{2} AM_i^P + \sum_{i=1}^{2} AM_i^{TS} + \sum_{i=1}^{3} AM_i^R \tag{8.14}$$

Perennial and Hot Weather

$$\left(\sum_{i=1}^{2} J_i^P M_i^P + \sum_{i=1}^{1} J_i^{HW} M_i^{HW}\right) \le \sum_{i=1}^{2} AM_i^P + \sum_{i=1}^{1} AM_i^{HW} \tag{8.15}$$

$AM_i$ Availability of Manure for particular crop;
$M_i$ Manure Utilization Requirement for particular crop per ha.

**Water Availability**

It can be represented as below:

$$\left(\sum_{i=1}^{2} J_i^K W_i^K + \sum_{i=1}^{3} J_i^R W_i^R + \sum_{i=1}^{2} J_i^{TS} W_i^{TS} + \sum_{i=1}^{2} J_i^P W_i^P + \sum_{i=1}^{1} J_i^{HW} W_i^{HW}\right) \le AW_i^j$$

$$j = 1, 2, 3, 4, 5 \quad \text{(No of crop Seasons)} \tag{8.16}$$

$W_i$ Irrigation water requirement for $i$th crop;
$AW_i^j$ Total water availability for $i$th crop (all crops) for $j$th interval (all seasons).

**Non-negativity Constraint**

$$\begin{aligned} &J_i^K, J_i^R, J_i^{TS}, J_i^P, J_i^{HW}, U_i, M_i, TJ, TJ_i^K, TJ_i^R, TJ_i^P,\\ &TJ_i^{TS}, TJ_i^{HW}, AM_i^P, AM_i^{TS}, AM_i^K, AM_i^R,\\ &AM_i^{HW}, M_i^P, M_i^{TS}, M_i^K, M_i^R, M_i^{HW},\\ &W_i^P, W_i^{TS}, W_i^K, W_i^R, W_i^{HW}, AW_i^j \ge 0 \quad \forall i, j \end{aligned} \tag{8.17}$$

### 8.2.3 Fuzzy Linear Programming Problem with Fuzzy Parameters

The proposed model, for fuzziness associated with various parameters to deal with uncertainty and vagueness, which can be expressed as below in the form of Eq. (8.18).

$$\begin{aligned} \max = \sum_{j=1}^{n} \tilde{c}_j x_j \\ \text{s.t.} \quad \sum_{j=1}^{n} \tilde{a}_{ij} x_j \leq \tilde{b}_i, (i \in \mathbb{N}_m) \\ x_j \geq 0 (j \in \mathbb{N}_n) \end{aligned} \tag{8.18}$$

where $\tilde{c}_j$,$\tilde{a}_{ij}$, $\tilde{b}_i$ = fuzzy numbers can be represented by membership functions(linear); $x_j$ = variables whose states are fuzzy numbers ($i \in \mathbb{N}_m$, $j \in \mathbb{N}_n$); the operations of addition and multiplication are treated as the fuzzy arithmetic operations, and fuzzy numbers ordering denoted by $\leq$.

The FLP approach can be used to solve the above problem (Eq. 8.18). While using such approach, the best solutions are used to formulate the intervals of membership functions (linear). Because of this, it doesn't provide the scope to include the fuzziness acquired from the DM. However, such difficulty can be tackled by the utilization of Fuzzy Parametric Programming (FPP) method. In the present study, this method is proposed and used to develop the multiple objective decision problems for various grades of precision suggested by the DM based on his/her expertise in the relevant field, which helps to reflect the fuzziness in every model. Such a designed model is used to find out the optimal solutions for all competing objectives.

### 8.2.4 Algorithm for MOFLP Model Using FPP Approach

The algorithm for the proposed MOFLP model using the approach of Fuzzy Parametric Programming, which has represented as below:

i. Design of mathematical programming model, which allows the numeral of contending objectives and various types of constraints.
ii. Include the input from DM based on his/her experience and rational view regarding fuzziness allowed in the various values of the parameters (i.e., *a*, *b*, and *c*), and different possible combination(s) of parameters.
iii. It is using the FPP concept, the imprecision/fuzziness included in the various values of the parameters (i.e., *a*, *b*, and *c*) as discussed in the above step (ii), which is used to fix up the type of membership function along with its construction. The trade-off between the membership function is formulated

in any one of the forms of different types of membership functions such as linear/nonlinear (convex/concave) and/or piecewise linear. It is considered that, $[c^0, c^1), [a^0, a^1), [b^0, b^1)$ are the possible interval values of fuzzy parameters designated by the DM. $c^0$ represent "practically acceptable value (PAV)" and $c^1$ as "practically unacceptable value (PUAV)" value and similarly for $b^0$ and $b^1$. $a^0$ represent "practically unacceptable value (PUAV)" value and $a^1$ as "practically acceptable value (PAV)." The lower bounds represent "practically acceptable value (PAV)," which indicate that the solution is implementable. In contrast, the upper bound shows that the obtained solution cannot be implemented practically on the field, as these bounds show that its values are in the form of unrealistic and impossible. The formulation of multi objective decision problem concerning various levels of precision (μ) for multiple types of membership functions. The FPP model is developed for a different level of precision based on formulated membership functions for fully trade-off membership ($\mu = (\mu_{ck} or \mu_{zk}) = \mu_A = \mu_b$), and which are considered to be linear in nature. Consequently, the developed models using FPP approach can be represented in the form of Eq. (8.19), which is shown below.

$$\begin{aligned}&\text{Max } Z_k = [C^1_{kj} - \mu c_{kj}(C^1_{kj} - C^0_{kj})]x_j\\&\text{s.t.} \left[A^1_{ij} + \mu A_{ij}\left(A^0_{ij} - A^1_{ij}\right)\right]x_j \le b^1_i - \mu b_i\left(b^1_i - b^0_i\right), \forall i, j \text{ and } \mu \text{ level}\\&x_j \ge 0.\end{aligned} \tag{8.19}$$

iv. The solution of the above Eq. (8.19), which leads to the various values for each objective functions, based these values a pay-off table is prepared, and with the help of which upper ($Z_u$) and lower ($Z_l$) values for each objective are decided.

v. The membership function, which is considered as linear, is constructed with the help of values represented in the form of pay-table as obtained in the above step (v).

vi. The developed MOFLP model, which is represented in the form of Eq. (8.20), it is having the new objective function of maximization type and represented with a new dummy variable (λ) for a different level of precisions (i.e., $\mu = 0, 0.1, 0.2, \ldots, 1.0$). This proposed model is subjected to newly added constraints because of fuzziness involved in the objective function values along with original fuzzy constraints.

$$\begin{aligned}&\text{Max } \lambda\\&\text{s.t.}(Z_k - Z_{lk})/(Z_{uk} - Z_{lk}) \ge \lambda\\&\left[A^1_{ij} + \mu A_{ij}\left(A^0_{ij} - A^1_{ij}\right)\right]x_j \le b^1_i - \mu b_i\left(b^1_i - b^0_i\right),\\&0 \le \lambda \le 1,\\&x_j \ge 0.\end{aligned} \tag{8.20}$$

where $\lambda = \mu_k(x) = (Z_k - Z_{lk})/(Z_{uk} - Z_{lk})$, $\mu = \mu_s = \mu_A = \mu_b = \mu_c$ and $Z_k = [C_{kj}^1 - \mu c_{kj}\left(C_{kj}^1 - C_{kj}^0\right)]x_j$.

vii. Obtain the solution of the above problem (Eq. 8.20) for various precision levels.

viii. The DM can choose the most appropriate solution from the solution obtained in the above step (vii), and if not satisfied, then return to step (ii) or (iii) and reiterate the entire process.

Figure 8.1 shows the algorithm for the solution using the above procedure.

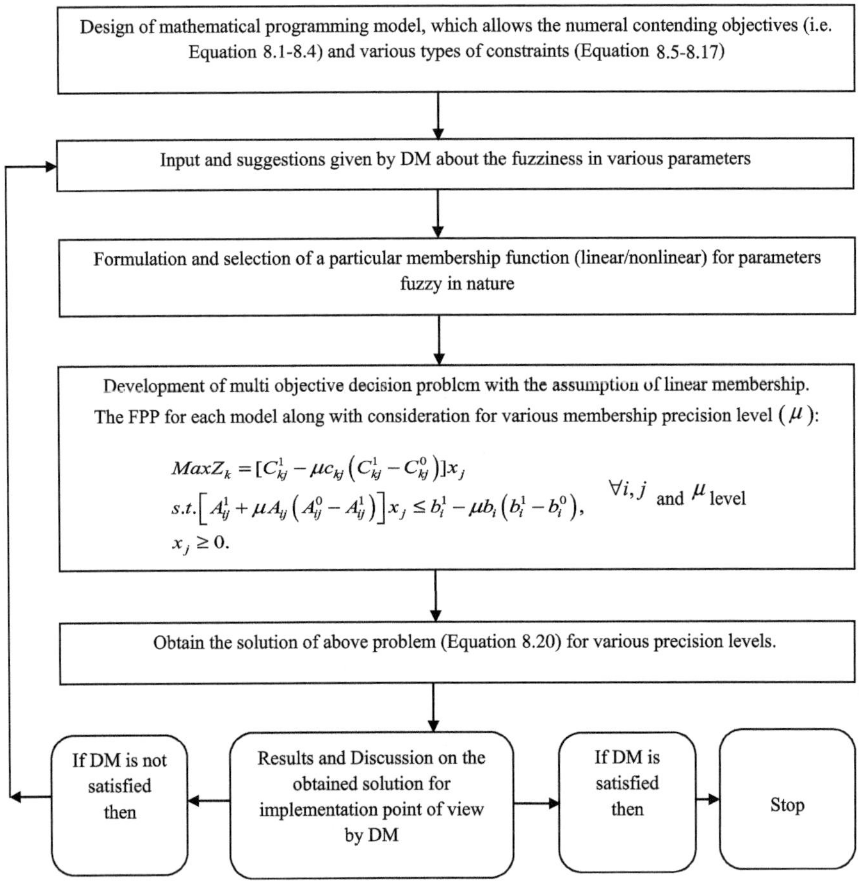


**Fig. 8.1** The FPP approach-based algorithm for MOFLP with fuzziness for various parameters (i.e., *a*, *b*, and *c*)

## 8.3 Results and Discussion

The fuzziness involved in various parameters (a, b, and c) has taken into account in the proposed MOFLP model. The solution has been obtained concerning the algorithm and procedure mentioned in the above section. Figures 8.2, 8.3 and 8.4 shows the construction of the membership function for various fuzzy parameters (*a*, *b*, and *c*). The membership functions (Figs. 8.2, 8.3 and 8.4) associated with $\mathbf{a}_{ij}$, $\mathbf{b}_i$, and $\mathbf{c}_{kj}$ parameters, which are assumed as linear monotonously increasing and linear concave/convex, where $\mathbf{A} = [a_{ij}]$, $\mathbf{b} = [b_i]$, $\mathbf{c}_k = [c_{kj}]$, $i = 1 \ldots 17$, and $j = 1 \ldots 10$, $k = 1 \ldots 4$.

$$\mu a_{ij} = (a_{ij} - a_{ij}^1)/(a_{ij}^0 - a_{ij}^1),$$

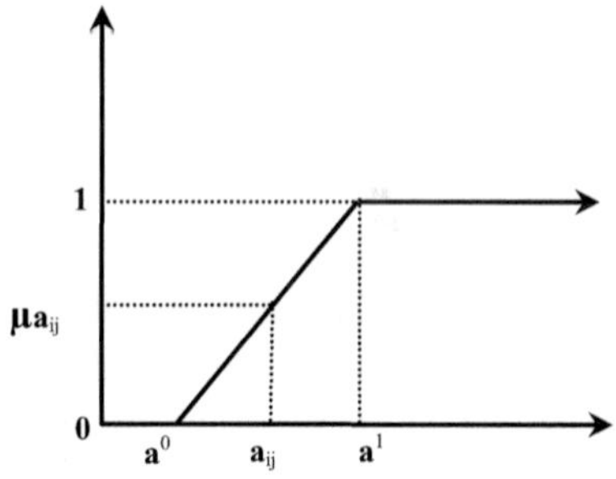


**Fig. 8.2** Membership function-monotonously increasing for technological coefficients ($\mathbf{a}^1$—practically acceptable value and $\mathbf{a}^0$—practically unacceptable value

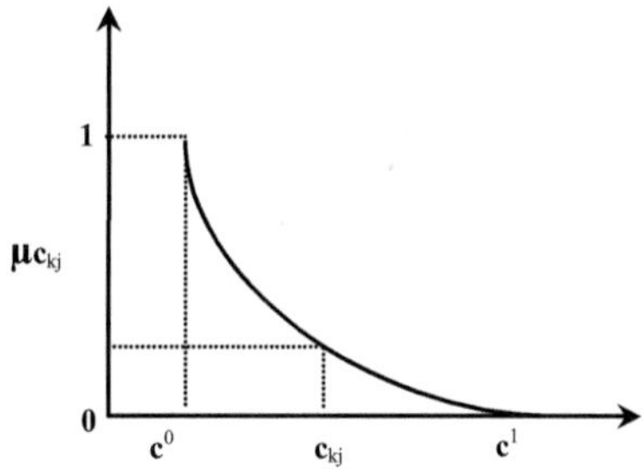


**Fig. 8.3** Membership function-linear concave for cost coefficients $\mathbf{c}^1$—practically unacceptable value and $\mathbf{c}^0$—practically acceptable value

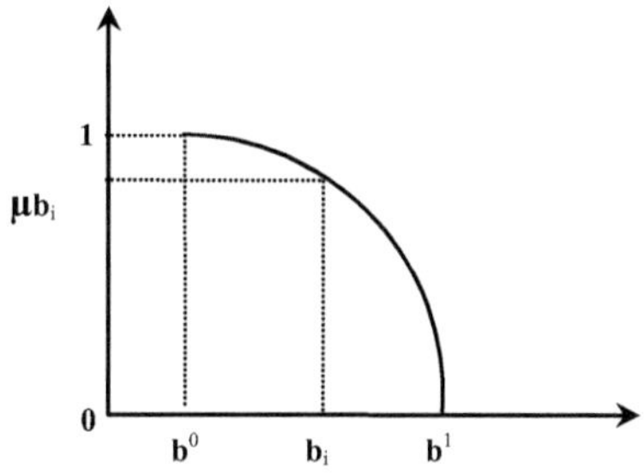


**Fig. 8.4** Membership function-linear convex for stipulations/resources $\mathbf{b}^0$—practically acceptable value and $\mathbf{b}^1$—practically unacceptable value

$$\mu b_i = 1/(1 - \exp(-0.8))[1 - \exp(-0.8(b_i - b_i^1)/(b_i^0 - b_i^1))],$$
$$\mu c_{kj} = 1/(1 - \exp(3))[1 - \exp(3(c_{kj} - c_{kj}^1)/(c_{kj}^0 - c_{kj}^1))],$$

The membership functions are used to determine the various model parameters; these parameters are represented as follows:

$$a_{ij} = a_{ij}^1 + \mu a_{ij}(a_{ij}^0 - a_{ij}^1),$$
$$b_i = b_i^1 - (1/0.8)\ln[1 - \mu b_i(1 - \exp(-0.8))](b_i^0 - b_i^1),$$
$$c_{kj} = c_{kj}^1 + (1/3)\ln[1 - \mu c_{kj}(1 - \exp(3))](c_{kj}^0 - c_{kj}^1),$$

After this, all the coefficients can be parameterized with reference to its membership functions. The parametric multiple objective linear programming problem can be formulated and expressed as below:

$$\text{Max } Z_k = [c_{kj}^1 + (1/3)\ln[1 - \mu c_{kj}(1 - \exp(3))](C_{kj}^1 - C_{kj}^0)]x_j,$$
$$\left[A_{ij}^1 + \mu A_{ij}\left(A_{ij}^0 - A_{ij}^1\right)\right]x_j \le b_i^1 - (1/0.8)\ln[1 - \mu b_i(1 - \exp(-0.8))](b_i^0 - b_i^1),\ \forall i, j \text{ and } \mu \text{ level}$$
$$x_j \ge 0.$$

The best value for the objective function is found out when $\mu s = \mu a = \mu b = \mu c$ for all $i, j, k$, because every membership function replicates the degree of precision and the precision evaluated from the optimal solution equals the precision of the most risky of the parameters ($\mu s = \min(\mu a, \mu b, \mu c)$). In other words, at a particular level of precision, the best value of the objective function can be determined using values of parameters at the same specific level of precision.

It is recommended that one should not use the developed MOLP model as it is to find out the optimal solutions due to the different nature of membership functions; therefore, it is essential to carry out sequences of the model run for different values of membership with precision level (i.e., $\mu s = \mu a = \mu b = \mu c$) varying in between zero to one with an equal interval of 0.1. Such a developed Multi Objective Linear Programming model has analyzed for each precision level with priority to one objective among the four objectives, based on these ideal solutions obtained, the pay-off table (Table 8.1) has prepared. With reference to the comparisons of values presented in Table 8.1, values (bold figures) for upper bound ( $Z_u$) and lower bound ($Z_l$) have been decided for each level of precision ($\mu$) and every objective function, with the help of these values and using FLP approach, various membership functions are constructed which are linear. The Eqs. 8.21–8.24 shows the formulated membership functions when the precision level is zero (i.e., $\mu = 0$). In case of irrigation planning and management, the DM decides the level of precision for all decision parameters which are fuzzy, with the help of his/her expertise and field experience associated with the implementation of irrigation policy.

**Table 8.1** Pay-off table

| Precision level ($\mu$) | NB (Million Rupees) | CP (Tons) | EG (Million Man Days) | MU (Tons) | Precision level ($\mu$) | NB (Million Rupees) | CP (Tons) | EG (Million Man Days) | MU (Tons) |
|---|---|---|---|---|---|---|---|---|---|
| 0 | **1976.02** | **517,825.50** | **37.43** | **119,208.40** | 0.6 | **1231.27** | **337,904.40** | **22.22** | **74,212.58** |
| | 1950.05 | 516,103.30 | 34.10 | 99,319.41 | | 1206.91 | 336,721.90 | 19.31 | 58,983.19 |
| | **1606.95** | **177,867.20** | **31.47** | **82,930.79** | | **977.19** | **86,736.79** | **17.80** | **51,732.58** |
| | 1875.09 | 506,347.90 | 32.02 | 101,499.90 | | 1160.73 | 329,662.90 | 17.96 | 60,422.87 |
| 0.1 | **1736.53** | **458,701.10** | **32.61** | **104,684.70** | 0.7 | **1157.18** | **319,811.30** | **20.74** | **69,544.24** |
| | 1709.47 | 457,046.10 | 29.57 | 87,243.55 | | 1135.35 | 318,770.60 | 17.82 | 54,655.98 |
| | **1398.39** | **139,973.00** | **27.13** | **72,896.98** | | **917.60** | **81,399.37** | **16.52** | **48,593.92** |
| | 1644.72 | 448,300.80 | 27.61 | 88,306.19 | | 1085.79 | 310,017.10 | 16.68 | 56,455.39 |
| 0.2 | **1596.59** | **424,969.40** | **29.74** | **96,181.25** | 0.8 | **1085.50** | **302,036.70** | **19.31** | **64,982.39** |
| | 1567.35 | 423,318.60 | 26.84 | 80,178.75 | | 1066.00 | 301,124.80 | 16.42 | 50,562.09 |
| | **1273.56** | **114,388.70** | **24.49** | **67,041.67** | | **860.01** | **76,245.92** | **15.31** | **45,558.86** |
| | 1509.45 | 415,127.60 | 24.92 | 80,405.94 | | 1012.61 | 290,396.20 | 15.49 | 52,655.20 |
| 0.3 | **1487.84** | **399,023.10** | **27.47** | **89,586.07** | 0.9 | **1014.86** | **284,224.70** | **17.93** | **60,517.18** |
| | 1457.06 | 397,389.80 | 24.67 | 74,469.34 | | 997.50 | 283,430.50 | 15.09 | 46,628.38 |
| | **1183.91** | **105,265.80** | **22.42** | **62,506.09** | | **803.50** | **71,179.32** | **14.14** | **42,569.73** |
| | 1404.53 | 389,597.50 | 22.81 | 74,367.40 | | 941.29 | 271,093.20 | 14.33 | 48,951.97 |
| 0.4 | **1393.76** | **376,840.80** | **25.53** | **83,954.67** | 1.0 | **943.91** | **266,021.40** | **16.56** | **56,071.37** |
| | 1363.48 | 375,323.90 | 22.68 | 68,730.80 | | 928.56 | 265,335.50 | 13.79 | 42,787.96 |
| | **1108.08** | **98,467.73** | **20.66** | **58,619.26** | | **746.55** | **66,106.50** | **13.00** | **39,571.06** |
| | 1315.04 | 367,785.10 | 21.01 | 69,229.96 | | 896.74 | 260,858.60 | 13.25 | 45,279.42 |
| 0.5 | **1309.39** | **356,729.20** | **23.81** | **78,899.24** | | | | | |
| | 1282.23 | 355,389.20 | 20.91 | 63,633.50 | | | | | |
| | **1040.10** | **92,373.79** | **19.17** | **55,043.33** | | | | | |
| | 1234.84 | 348,090.60 | 19.42 | 64,642.42 | | | | | |

Regulwar and Gurav (2012)

$$\mu z_1(x) = \begin{cases} 1 & z_1 \geq 1976.02 \\ (Z_1 - 1606.95)/(1976.02 - 1606.95) & 1606.95 < z_1 < 1976.02 \\ 0 & z_1 \leq 1606.95 \end{cases} \tag{8.21}$$

$$\mu z_2(x) = \begin{cases} 1 & z_2 \geq 517{,}825.50 \\ (Z_2 - 177{,}867.20)/(517{,}825.50 - 177867.20) & 177{,}867.20 < z_2 < 517{,}825.50 \\ 0 & z_2 \leq 177{,}867.20 \end{cases} \tag{8.22}$$

$$\mu z_3(x) = \begin{cases} 1 & z_3 \geq 37.43 \\ (Z_3 - 31.47)/(37.43 - 31.47) & 31.47 < z_3 < 37.43 \\ 0 & z_3 \leq 31.47 \end{cases} \tag{8.23}$$

$$\mu z_4(x) = \begin{cases} 1 & z_4 \geq 119208.40 \\ (Z_4 - 82{,}930.79)/(119{,}208.40 - 82{,}930.79) & 82{,}930.79 < z_4 < 119208.40 \\ 0 & z_4 \leq 82{,}930.79 \end{cases} \tag{8.24}$$

A dummy variable $\lambda$ is newly added as the fuzzy achievement function ($\lambda = \min[\mu Z_1(x), \mu Z_2(x), \mu Z_3(x), \mu Z_4(x)]$), a final modified form of the optimization problem as a MOFLP model has represented as:

$$\begin{aligned} &\text{Maximize } \lambda \\ &\text{s.t.,} \\ &(Z_1 - 1606.95 \times 10^6)/(1976.02 \times 10^6 - 1606.95 \times 10^6) \geq \lambda \\ &(Z_2 - 17{,}7867.20)/(517{,}825.50 - 17{,}7867.20) \geq \lambda \\ &(Z_3 - 31.47 \times 10^6)/(37.43 \times 10^6 - 31.47 \times 10^6) \geq \lambda \\ &(Z_4 - 82{,}930.79)/(119{,}208.40 - 82{,}930.79) \geq \lambda \end{aligned}$$

Along with other constraints involving fuzziness (i.e., Eqs. (8.5)–(8.17)) in the model; $1 \geq \lambda \geq 0$.

According to each precision level ($\mu$), the various solutions of the above model is shown in Table 8.2. In the present study, the application of the above MOFLP model using the FPP approach has been shown with the help of a case study of Jayakwadi Project Stage-I, Maharashtra State, India.

The results presented in Table 8.2, which suggests various decision plans in the form of optimal cropping pattern planning. Also, it is seen that for each level of precision which is varying from 0 to 1, the aggregated level of satisfaction ($\lambda$) is above 68%. These values of the level of satisfaction ($\lambda$) represent that all four objectives under consideration are analyzed at a time and out of which one objective is satisfied at $\lambda$ level, and other objectives are satisfied at least at $\lambda$ level. The highest value of the level of satisfaction is 0.727 for the lowest value of $\mu$ (i.e., zero), and the lowest value of the level of satisfaction is 0.682 for the highest value of $\mu$ (i.e., one), such a

**Table 8.2** The MOFLP optimization model solutions for various level of precision ($\mu = 0.0.1, 0.2\ldots 1.0$)

| Precision level ($\mu$) | Level of satisfaction ($\lambda$) | NB (Million Rupees) | CP (Tons) | EG (Million Man Days) | MU (Tons) | Acreages for different crops (ha) | Total area for Irrigation (ha) and irrigation intensity (%) |
|---|---|---|---|---|---|---|---|
| 0 | 0.727 | 1875.27 | 381,720.70 | 35.80 | 109,306.00 | $X_1 =$ 2536.84<br>$X_2 =$ 2048.17<br>$X_3 =$ 4472.84<br>$X_4 =$ 5665.60<br>$X_5 =$ 18,885.33<br>$X_6 =$ 15,737.78<br>$X_7 =$ 12,878.92<br>$X_8 =$ 47,213.33<br>$X_9 =$ 7454.73<br>$X_{10} = 0$ | 116,893.09 (82.52) |
| 0.1 | 0.718 | 1641.40 | 369,034.80 | 31.06 | 95,991.77 | $X_1 =$ 3202.12<br>$X_2 =$ 1597.70<br>$X_3 =$ 4370.66<br>$X_4 =$ 5385.79<br>$X_5 =$ 18,347.20<br>$X_6 =$ 15,289.10<br>$X_7 =$ 9105.37<br>$X_8 =$ 44,881.58<br>$X_9 =$ 7246.50<br>$X_{10} = 0$ | 109,426.02 (77.25) |

(continued)

**Table 8.2** (continued)

| Precision level ($\mu$) | Level of satisfaction ($\lambda$) | NB (Million Rupees) | CP (Tons) | EG (Million Man Days) | MU (Tons) | Acreages for different crops (ha) | Total area for Irrigation (ha) and irrigation intensity (%) |
|---|---|---|---|---|---|---|---|
| 0.2 | 0.715 | 1504.54 | 340,704.60 | 28.44 | 87,878.35 | $X_1 =$ 3140.23 $X_2 =$ 1506.88 $X_3 =$ 4264.85 $X_4 =$ 5117.82 $X_5 =$ 17,801.12 $X_6 =$ 14,833.79 $X_7 =$ 7457.26 $X_8 =$ 42,648.51 $X_9 =$ 7034.80 $X_{10} = 0$ | 103,625.26 (73.16) |
| 0.3 | 0.710 | 1399.96 | 329,196.70 | 26.00 | 81,756.59 | $X_1 =$ 3298.90 $X_2 =$ 1339.78 $X_3 =$ 4154.85 $X_4 =$ 4859.91 $X_5 =$ 17,244.87 $X_6 =$ 14,370.00 $X_7 =$ 5153.98 $X_8 =$ 40,499.32 $X_9 =$ 6818.76 $X_{10} = 0$ | 97,740.37 (69.00) |

(continued)

**Table 8.2** (continued)

| Precision level ($\mu$) | Level of satisfaction ($\lambda$) | NB (Million Rupees) | CP (Tons) | EG (Million Man Days) | MU (Tons) | Acreages for different crops (ha) | Total area for Irrigation (ha) and irrigation intensity (%) |
|---|---|---|---|---|---|---|---|
| 0.4 | 0.707 | 1310.23 | 321,893.20 | 24.10 | 76,546.99 | $X_1 =$ 3494.90 $X_2 =$ 1146.85 $X_3 =$ 4040.01 $X_4 =$ 4610.37 $X_5 =$ 16,675.82 $X_6 =$ 13,895.52 $X_7 =$ 2933.66 $X_8 =$ 38,419.78 $X_9 =$ 6597.33 $X_{10} = 0$ | 91,814.24 (64.82) |
| 0.5 | 0.701 | 1228.98 | 309,244.90 | 22.42 | 71,776.64 | $X_1 =$ 3529.21 $X_2 =$ 1015.79 $X_3 =$ 3919.55 $X_4 =$ 4367.50 $X_5 =$ 16,090.79 $X_6 =$ 13,407.72 $X_7 =$ 1386.23 $X_8 =$ 36,395.83 $X_9 =$ 6369.27 $X_{10} = 0$ | 86,481.89 (61.05) |

(continued)

**Table 8.2** (continued)

| Precision level ($\mu$) | Level of satisfaction ($\lambda$) | NB (Million Rupees) | CP (Tons) | EG (Million Man Days) | MU (Tons) | Acreages for different crops (ha) | Total area for Irrigation (ha) and irrigation intensity (%) |
|---|---|---|---|---|---|---|---|
| 0.6 | 0.696 | 1154.13 | 297,506.30 | 20.87 | 67,387.58 | $X_1$ = 3550.35 $X_2$ = 895.45 $X_3$ = 3792.46 $X_4$ = 4129.57 $X_5$ = 15,485.91 $X_6$ = 12,903.36 $X_7 = 0$ $X_8$ = 34,413.13 $X_9$ = 6079.66 $X_{10} = 0$ | 81,249.89 (57.36) |
| 0.7 | 0.691 | 1083.26 | 271,381.50 | 19.43 | 63,080.88 | $X_1$ = 3183.67 $X_2$ = 959.80 $X_3$ = 3657.92 $X_4$ = 3894.76 $X_5$ = 14,856.34 $X_6$ = 12,378.40 $X_7 = 0$ $X_8$ = 32,456.41 $X_9$ = 5560.14 $X_{10} = 0$ | 76,947.04 (54.32) |

(continued)

**Table 8.2** (continued)

| Precision level ($\mu$) | Level of satisfaction ($\lambda$) | NB (Million Rupees) | CP (Tons) | EG (Million Man Days) | MU (Tons) | Acreages for different crops (ha) | Total area for Irrigation (ha) and irrigation intensity (%) |
|---|---|---|---|---|---|---|---|
| 0.8 | 0.689 | 1015.38 | 246,261.20 | 18.06 | 58,942.82 | $X_1 =$ 2813.51 $X_2 =$ 1028.54 $X_3 =$ 3513.13 $X_4 =$ 3661.05 $X_5 =$ 14,195.92 $X_6 =$ 11,827.71 $X_7 = 0$ $X_8 =$ 30,508.77 $X_9 =$ 5100.05 $X_{10} = 0$ | 72,648.68 (51.29) |
| 0.9 | 0.686 | 948.53 | 222,457.00 | 16.74 | 54,885.51 | $X_1 =$ 2461.54 $X_2 =$ 1079.80 $X_3 =$ 3357.19 $X_4 =$ 3426.06 $X_5 =$ 13,496.61 $X_6 =$ 11,244.58 $X_7 = 0$ $X_8 =$ 28,550.52 $X_9 =$ 4687.88 $X_{10} = 0$ | 68,304.18 (48.22) |

(continued)

**Table 8.2** (continued)

| Precision level ($\mu$) | Level of satisfaction ($\lambda$) | NB (Million Rupees) | CP (Tons) | EG (Million Man Days) | MU (Tons) | Acreages for different crops (ha) | Total area for Irrigation (ha) and irrigation intensity (%) |
|---|---|---|---|---|---|---|---|
| 1.0 | 0.682 | 881.26 | 202,565.30 | 15.43 | 50,838.54 | $X_1 =$ 2211.14 $X_2 =$ 1064.16 $X_3 =$ 3186.90 $X_4 =$ 3186.90 $X_5 =$ 12,747.60 $X_6 =$ 10,620.00 $X_7 = 0$ $X_8 =$ 26,090.44 $X_9 =$ 5058.57 $X_{10} = 0$ | 64,165.71 (45.30) |

Regulwar and Gurav (2012)

scenario helps the DM to take judicious decision with reference to particular crops in the specific crop season.

It is seen that for a minimum level of precision (i.e., zero), values of various objectives under consideration are maximum such as NB = 1875.27 Million Rupees, CP = 381,720.70 Tons, EG = 35.80 Million Man Days and MU = 109,306 Tons. These values are minimum for the maximum level of precision (i.e., one) such as NB = 881.26 Million Rupees, CP = 202,565.30 Tons, EG = 15.43 Million Man Days and MU = 50,838.54 Tons, respectively. Similarly, for a minimum level of precision (i.e., zero) the net cropped area = 116,893.09 ha and irrigation intensity = 82.52%, as well as these values, are minimum for the maximum level of precision (i.e., one) such as net cropped area = 64,165.71 ha and irrigation intensity = 45.30%, respectively. It is also seen that when the level of precision is minimum (i.e., zero), all the crops under existing cropping pattern suggests the maximum area to be irrigated and vice versa for the maximum value of the level of precision. Apart from these, Groundnut (HW) is having zero acreage irrespective of the any value of the level of precision (varying from zero to one). Also, when $\mu \geq 0.6$, there is crop such as Sorghum ($R$) having zero acreage.

## 8.4 Conclusion

The Multi Objective Fuzzy Linear Programming Model using Fuzzy Parametric Programming Approach has been developed, and its applicability has been demonstrated through its application for the case study of Jayakwadi Project Stage-I in Godavari River sub-basin of Maharashtra State, India.

The proposed methodology and models focus on how to tackle the uncertainty and vagueness involved in the form of fuzziness in the different decision parameters of the irrigation planning problem (i.e., technological coefficients, stipulations/resources, and coefficients of the objective function). The solution for the same has also presented.

The proposed model provides a series of the solution along with the level of precision required to the field situation with the view of irrigation planning based in sustainability concept, which allows the DM to take a judicious decision.

## References

Arikan F, Gungor Z (2007) A two phase approach for multi-objective programming problems with fuzzy coefficients. Inf Sci 177:5191–5202

Banihabib ME, Tabari MMR, Tabari MMR (2019) Development of a fuzzy multi-objective heuristic model for optimum water allocation. Water Resour Manage 33:3673–3689

Gasimov RN, Yenilmez K (2002) Solving fuzzy linear programming problems with linear membership functions. Turk J Math 26:375–396

Guo P, Wang X, Zhu H, Li M (2013) Inexact fuzzy chance-constrained nonlinear programming approach for crop water allocation under precipitation variation and sustainable development. J Water Resour Plann Manage 140(9):05014003

Gurav JB, Regulwar DG (2012) Multi objective sustainable irrigation planning with decision parameters and decision variables fuzzy in nature. Water Resour Manage 26(10):3005–3021

Jimenez M, Arenas M, Bilbao A, Rodriguez VM (2007) Linear programming with fuzzy parameters: an interactive method resolution. Euro J Oper Res 177(3):1599–1609

Kumari S, Mujumdar PP (2015) Reservoir operation with fuzzy state variables for irrigation of multiple crops. J Irrig Drain Eng 141(11):04015015

Li M, Guo P. Singh VP (2016) Biobjective optimization for efficient irrigation under fuzzy uncertainty. J Irrig Drain Eng 05016003

Morankar DV, Raju KS, Vasan A, Vardhan LA (2016) Fuzzy Multiobjective irrigation planning using particle swarm optimization. J Water Resour Plann Manage 05016004

Regulwar DG, Gurav JB (2011) Irrigation planning under uncertainty—a multi objective fuzzy linear programming approach. Water Resour Manage 25(5):1387–1416

Regulwar DG, Gurav JB (2012) Sustainable irrigation planning with imprecise parameters under fuzzy environment. Water Resour Manage 26(13):3871–3892

Sabouni MS, Mardani M (2013) Application of robust optimization approach for agricultural water resource management under uncertainty. J Irrig Drain Eng 139(7):571–581

Safavi HR, Alijanian MA (2010) Optimal crop planning and conjunctive use of surface water and groundwater resources using fuzzy dynamic programming. J Irrig Drain Eng 137(6):383–397

Sami GM, Roozbahani A, Banihabib ME (2018) Fuzzy optimization model and fuzzy inference system for conjunctive use of surface and groundwater resources. J Hydrology 566:421–434

Xu J, Tu Y, Zeng Z (2013) Bilevel optimization of regional water resources allocation problem under fuzzy random environment. J Water Resour Plann Manage 139(3):246–264

Yue Q, Zhang F, Zhang C, Zhu H, Tang Y, Guo P (2020) A full fuzzy-interval credibility-constrained nonlinear programming approach for irrigation water allocation under uncertainty agric. Water Manage 230(1):105961

Zadeh LA (1965) Fuzzy sets. Inf Control 8:338–353

# Chapter 9
# Application of High Resolution Hydrological and Hydraulic Models for Sustainable Water Resources Management

**Sangameswaran Shyamprasad**

**Abstract** Rapidly declining water quality in India has become a major cause of concern, with several sources citing the country's water quality among the poorest across the globe. A framework is needed to reverse the situation. The approach presented here would involve collecting very fine resolution data using state of the art technologies such as remote sensing from satellites and Light Detection and Ranging (LiDAR) technology supplemented by locally placed sensors or from sensors mounted and flown on drones. Hydrological and hydraulic models need to be developed to generate flow paths and pollutant flows. High resolution Digital Elevation Models (DEM) need to be developed using input from remote sensing data. Flow monitoring plan and rainfall measurement methodologies will be developed as a part of this approach. The primary goal is to develop framework that allow urban water managers to reinstate water quality of rivers as they leave an urban centre to the same (or better) state as compared to the water quality with which they entered the area.

**Keywords** Hydrological and hydraulic modelling · High resolution data · Geographical information systems (GIS) · Pollutants · Groundwater · Stormwater management · Integrated urban water resources management plan · Sewer systems · Urban river management · Sustainable water resource management · USEPA SWMM5 · Environmental flows · Combined sewer overflows · Baseflow · Flow monitoring plan · Point and non-point source pollution

## 9.1 Introduction

There are several factors that have contributed towards the present levels of water quality deterioration: lack of access to improved sanitation, untreated industrial water being discharged either into groundwater or surface water sources, and significant

S. Shyamprasad (✉)
Clear Water Dynamics Pvt. Ltd., #3009/2, 19th Cross, 2nd Main, BSK II Stage, Bengaluru 560070, India
e-mail: shyam@clearwaterdynamics.com



P. Kumar et al. (eds.), *Water Resources Management and Sustainability*, Advances in Geographical and Environmental Sciences,
https://doi.org/10.1007/978-981-16-6573-8_9

fertilizer runoff towards water bodies. In addition, there are regional issues that further exacerbate the situation, ranging from presence of arsenic in groundwater, to natural radioactive substances finding their way into surface or groundwater systems in a region (Bach et al. 2018). As India is expected to host more people in its urban centres than in its rural villages over next decade, a robust water quality management plan for its urban centres will go a long way in alleviating water quality issues across the country (USEPA 1979, 1999).

Furthermore, to understand the spatial variability of water quality within a surface water body such as a river, urban centres across India emerge as water quality hotspots where a significant portion of pollutants enter the water body in a very short spatial scale. This challenges the capacity of the water body to recover from the pollution (Tuomela et al. 2017). Typically, a fraction of pollution from upstream urban centres persists in the river and enters downstream urban centres. The downstream urban centres then add to the pollutants already existing, resulting in the scenario that many times rivers are more akin to flowing sewers than fresh water bodies (Tuomela 2017).

While desirable, the aforementioned objective has two major challenges. First, there is only a limited understanding of the source and fate of pollutants in urban centres. To address this challenge, we will begin with identification of major pollutants, and develop methods to quantify them via areal mapping or by utilizing existing data with water managers in these areas. Second, to our knowledge there are limited (or no) attempts at modelling the flow of pollutants through an urban centre where flow paths are a combination of manmade and natural storm sewers (Burns et al. 2012). Using high resolution data on flow paths, the transport of pollutants through the system can be mapped. This will in turn indicate potential areas where polluted water can be treated before getting diluted by freshwater (Bach et al. 2018; Tuomela 2017).

There is a clear lack of high-resolution data (spatial and temporal) required to manage water flows and water quality in these urban rivers. Through this approach, it is proposed to carryout intensive data collection by installation of new flow monitoring metres and automatic rain gauge stations through pre-defined guidelines for the same. This would be used for development of hydrological and hydraulic models for the region. In turn this would be input into water quality models. Various domestic, industrial or agricultural sources of point and non-point sources of pollution entering urban river streams will be identified (Smith and Vidmar 1994). This will help the policy makers to understand where exactly the problem area is and which sector needs more attention. Wastewater management plan would be implemented with the use of low cost green infrastructural options which can be contextualized to address the local situation. Hence, new wastewater management policies can be used by the policy makers for efficient decision making (WaPUG 2002).

A logical conclusion from this approach is to develop an effective integrated urban water resources management plan that would address all aspects of water related issues faced by each City or region. The plan will provide blue print tailored for each City or region to improve the water quality issues by addressing the causes identified by the previously described modelling exercise (Castonguay et al. 2018; Fletcher et al. 2007; Schubert et al. 2017).

## 9.2 Methodology

The schematic of the discussed approach is presented in the figure below.

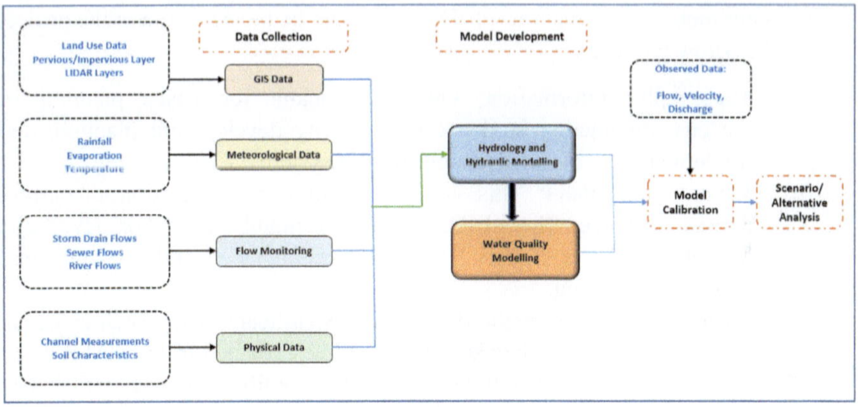


This methodology covers high-definition LIDAR survey to hydrological modelling and results. This methodological flow chart is an idea derived from (Fletcher et al. 2015; Rauch et al. 2017; Ellisa et al. 2004; Palla et al. 2010; Shoemaker et al. 2009).

## 9.3 Data Collection

Data review and collection is a critical first step in understanding the extent of data availability and developing hydrological and hydraulic (H/H) and water quality model. As part of this approach, it is required to collect, review and assess available data for the urban cities to develop a complete understanding of the various aspects of urban environment for a comprehensive urban water quality modelling. GIS data, flow monitoring plan, field measurements and EPA Storm Water Management Model (SWMM) have to be combined to develop a model that fulfils these objectives.

### *9.3.1 GIS Data*

An important aspect of the model development approach will be applying the available GIS data and supplementing the missing information with surveyed or data collected from other means. This will provide the backbone necessary for a successful modelling effort and typically includes the following categories of data:

1. LiDAR (Light Detection And Ranging) data: 3D GIS, based on LiDAR technology will be used for collecting data as this is a powerful tool to assist the city in gathering details of the entire city at a very high resolution with up to a few centimetres level accuracy. LiDAR data can be incorporated in the project for the below purposes:

   (a) Topographic information: This is applicable for master planning of the city, conducting land use studies, city development planning and implementation of any decision support system.
   (b) Water management: This is the field of core application in current approach. Under this, LiDAR data is mainly used for Hydraulic Modelling, Water Supply Planning, Maintenance and Management, Sewerage Planning.
   (c) Hydrological & hydraulic data layers: Significant layers such as canal, drains, lakes, ponds, dam structures will be extracted from LiDAR data.
   (d) Sewage structures: Manholes, pipe outlets, gully traps and similar data will be obtained from LiDAR. It will also be used for assessment of open storm water drainage system conditions, open sewage conditions, present construction quality, requirement assessment for repairs and upgrading, sanitation conditions of open systems, debris removal; and asset management related to the area.

2. Land Use Land Cover (LULC) data: Using topographic maps and LiDAR imagery, land use and land cover information will be extracted for the cities. Documentation of different land cover types such as forest, agriculture, built up, water bodies for the study area depending on the level of classification. The main objective of studying LULC in this project is to find the influence of different land use types on the quantity and quality of runoff that might eventually reach the river stream. This is mainly because these will in turn influence the spatial and temporal water availability and quality. In order to contribute to a better understanding of the LULC effects on hydrological processes, this approach will incorporate LULC related parameters within the hydrologic models.
3. Pervious and Impervious layers information: Percentage of area covered by pervious surfaces such as vegetation, open spaces and impervious information such as parking lots, pavements, sidewalks for the city and its environment will be extracted. This is taken into consideration as it majorly influences the surface runoff estimation. Hence, this will be mainly incorporated in H/H modelling.

### 9.3.2 Meteorological Data

1. Rainfall data: Rainfall information serves as the most important input into a hydrological model. Methodology for a comprehensive, qualitative review process to facilitate analysis of large rainfall datasets have to be developed as a part of the study. Where there are insufficient numbers of rain gauges in the study area, supplementary gauges will be installed with the aim of modelling the system with a good precision depending on the terrain and area. As pollutants are likely to be flushed out towards the streams during high intensity short duration storms, it is required to collect the rainfall data at 15 min interval. The attributes that will be a part of rainfall data are (a) Number of rain gauges in the location and its location details in $X$, $Y$ coordinates. (b) Frequency of data collection (i.e. whether it has been collected hourly or sub hourly). (c) Intensity and volume of the rainfall in mm/h.
2. Evaporation data: Evaporation is an important part of the water balance of a catchment, and estimates of potential evaporation (PE) are an important input to hydrological models. Pan evaporation coefficients will be collected and used as an input for hydrological models.
3. Temperature: This is spatially and temporally highly variable parameter. To gain insight into the spatial and temporal variability, temperature data needs be collected for the study area at a reasonable resolution.

### 9.3.3 Flow Monitoring Plan

Quality flow monitoring data is critical for development and calibration of H/H models. A comprehensive flow monitoring plan has to developed for identifying the preliminary locations for obtaining the flows for the entire study area. Finalizing the site selection, maintenance and QA/QC procedures should be part of the plan. Flow estimation will be given utmost importance as it directly influences the dilution of pollutant levels in the streams.

The flow measurements will be done for the following categories:

1. Storm water drains: Flow in important or major storm water drains during monsoon or rainy days will be measured. Flow monitoring data collected during monsoon season from these drains will be used for model calibration and development of flood mitigation measures.
2. Sewerage systems: Flow measurements in both sanitary and combined sewers will be carefully performed and made use of in the model calibration.
3. Rivers and tributaries: Flow measurements will be carried out by installation of stream gauges at suitable locations in rivers. This will be done by adopting suitable measures for determining the stream velocity at suitable cross sections of the river systems.

### 9.3.4 Physical Data

1. Channel cross sections measurements: River channel measurements such as measurement of channel width, depth and cross sectional area to aid in flow measurement calculations of the river channel will be conducted.
2. Soil characteristics: The soil characteristics play an important role in the hydrology and hence a robust plan for soil data collection and analysis is essential for both developing and calibration of a hydrological model. Understanding soil parameters such as effective porosity, wetting front capillary pressure, and hydraulic conductivity are key for calculating infiltration which is an important component of hydrological cycle are required generate.

### 9.3.5 Additional Data

1. Wastewater treatment plant discharges: In order to investigate the impact of discharge of wastewater treatment plants into the receiving streams, discharge as well as water quality samples will be collected from upstream and downstream of major municipal sewage treatment plants.
2. Barrages and reservoirs: Water storage in reservoirs and its contribution to the water quality changes in the river system needs to be investigated. For this, physical, chemical and thermal changes the flowing river water undergoes when it is stilled by a reservoir will be examined. This is important because the water flowing downstream may be affected due to these changes.

## 9.4 Model Development

Based on information obtained from data review, primary data from field and other data collected from secondary sources as explained earlier, a hydrologic and hydraulic model will be developed. The proposed hydrologic modelling approach using Storm Water Management Model (SWMM) is of high resolution and physically-based. Critical to completion of the model is the development of a detailed modelling plan. This section outlines model development, calibration and quality assurance procedures required for successful implementation of this approach.

1. Hydrological Modelling: To accurately generate the runoff response in the model, catchment delineation can be performed in two steps:

   Step 1: Delineation of runoff catchments to each storm inlet.
   Step 2: Delineation of sewersheds.

2. Hydraulic Modelling: In addition to storm, sewer, open drains and rivers, capturing storm inlets and representing the streets explicitly in the hydraulic model where necessary will further increase the resolution of the model.

One of main advantages of the high resolution, physically-based modelling approach is that it provides detailed information as to exactly 'where' the flows are coming from leading to much lower model uncertainty and offers a better platform for robust alternatives analysis.

3. Water Quality Modelling: The above H/H model from SWMM will be used as a basis for development of a water quality model. Numerical form of the advection diffusion equation can be incorporated within SWMM to model the concentration of chosen pollutants at different locations. Both point and non-point sources of pollution can be considered as inputs, that are already mapped and quantified at data preparation stage. Output from the models will be in the form of pollutographs at desired locations.

## 9.5 Model Calibration

Model calibration and validation is the repetitive process during which select parameters are adjusted until the model can reasonably simulate the response to a series of wet weather and dry weather events. Successfully calibrating a model is the key to reliability of the subsequent model results and recommendations. By selecting the entire monitoring period, rather than discrete storms, the model's ability to replicate hydrologic and hydraulic conditions continuously is improved. Thus, stream flow and pollutant loadings are further refined as part of the calibration.

Models needs to be calibrated using industry-standard guidelines applied on modelling projects elsewhere, including quantitative measures for assessing model calibration. Goals suggested in the WaPUG Code of Practice guidelines can be applied, but these will not be the only calibration metric. Graphical 'goodness of fit' comparison of model and monitored data to supplement the assessed model calibration will also be applied. A critical final step will be an independent verification of the model. This will be accomplished through:

1. Comparing the model predicted flows against monitoring or observed data not used during the model calibration and
2. Comparing against maintenance records and historical observations.

## 9.6 Evaluating Existing Condition

Once the calibrated model is available, a number of baseline scenarios can be developed. A few of these baseline scenarios are listed below:

1. Baseline scenarios:

    (a) Develop detailed maps identifying critical sources of specific pollutants, i.e. sewage sources, fertilizers, and pesticides.

(b) A series of mapping activities will take place to allow evaluation of how these sources have changed as the cities have developed and also how they change throughout the year, principally between the monsoon and non-monsoon seasons.
(c) Spatial and temporal trends are to be analysed to assess how pollutant sources change along the course of a river system, i.e. at what point does the trend shift—what is causing this? And also analyse the pathway pollutants are taking through the surface water.
(d) Using these maps as a base, an export coefficient approach will be adopted to estimate the loadings associated with each source, thus identifying critical source areas within systems which can

- be modelled in further detail, and
- be the focus of any mitigation strategies.

(e) Detailed site specific analysis/modelling.
(f) Catchment data collected will be used to validate the mapping activities and provide more specific details on the industries/activities responsible for the pollution.
(g) The more detailed site specific data will then be analysed to consider variations across a range of temporal scales, i.e. seasonal to daily.
(h) Model future scenarios of change.
(i) Study of the impact of sustainable management options.
(j) Opportunity mapping of mitigation options: To target key sources of urban pollution and model response on river system; this should include stakeholder assessment of priority services and expert opinion derived from analysis of data sets.

The export coefficient approach will provide a means to develop 'standard' loadings associated with different pollution sources, and the scenarios of future change will provide an indication of the anticipated direction and scale of change in these loadings. Together with the mapping methodology, this will provide a toolbox for regulators in different cities.

2. Urban waste loading: Methodologies must be set out for the collection, treatment and discharge of urban wastewater, with the objective of protecting the natural river systems from the adverse effects of wastewater discharges. Effluent discharge standards will be redefined for the study areas.
3. Industrial loading: Industries play a major role in deterioration of the urban rivers. Database of the number of industries in the banks on the rivers will be developed and for each of which strict effluent discharge standards will be reviewed. Drains that are discharging the industrial treated water to the river systems will be monitored.

Finally, building on model simulation results, feasible alternatives will be evaluated to mitigate water quality issues. After the completion of all the above steps, an project implementation plan should be prepared to summarize all activities performed under the study including data collection, hydrologic and hydraulic model building, focus on the selected alternatives and recommendations. In addition, the implementation plan should provide conceptual figures, implementation schedule and final cost estimates. QA/QC system has to be provided and implemented through the full duration of the project.

## 9.7 Development of an Integrated Urban Water Resources Management Plan for Each City

A key output of this approach would be to develop an effective integrated urban water resources management plan that would address all aspects of water related issues faced by every City. The plan will provide blue print tailored for each City to improve the water quality issues by addressing the causes identified by the previously described modelling exercise.

A well calibrated hydrodynamic and water quality model will serve as the most important tool for developing a decision matrix to formulate an informed, economical and effective plan. The plan would address:

(a) All key uses of water: urban, rural, agricultural, ecological
(b) Water source availability: water abstraction by cities, rural areas, agriculture, etc.
(c) Ground water: quality and quantity of ground water flow, interactions between surface and groundwater in the selected stretch
(d) All drainage flows: natural drains, manmade drains and mapping flow of runoff from urban areas
(e) All water bodies: large and small scale water bodies
(f) Major ecological indicators: indicators for water quality, especially ecological indicators like flora, fauna, biodiversity would be in built into the model to indicate status of water quality in the selected stretch
(g) All wastewater flow: flow of wastewater from the city, not only in terms of quantity, but also in terms of water quality. The impact of wastewater discharge from urban areas (whether treated or untreated) on the main stream of the river would also be studied
(h) Waste from cities which leads to pollution of water resources.

## 9.8 Conclusion

A combination of high resolution hydrological and hydraulic model, water quality model, robust primary and secondary datasets to support the modelling, calibration of the models to existing conditions will provide a framework to identify pollution pathways to the receiving streams. This framework can be leveraged to develop an economical and effective integrated water resource management plan. Further, these solutions at finer scales would be enriched with primary data and willingness

**Table 9.1** Framework for integrated modelling execution

| Phase of survey and analysis | Data collection method and parameters | Analysis and outputs/remarks |
|---|---|---|
| Phase I | Secondary data collections<br>Institutional<br>Socio-economic<br>Water quality<br>Land use | (a) Create social accounting matrix<br>(b) Calibrate applied general equilibrium model<br>(c) Based on the availability of secondary data at municipal corporation level<br>(d) Coarse/meso level data |
| Phase II | Primary data collection<br>Institutional (community and individual)<br>Socio-economic<br>Water quality<br>Land use/geospatial | Estimate production functions<br>With water, including industrial, domestic, commercial, other sectors |
| Phase III | Scenario generation<br>With effective allocation<br>i. Direct impacts of water quality improvement<br>ii. Indirect impacts of water quality improvement<br>iii. Willingness to pay exercise | Valuation of water quality improvement<br>Policy options through scenarios:<br>i. Budgetary allocation<br>ii. Community level options<br>iii. Public–private partnership options |

to pay data, which can be collected at sub-urban levels. Direct and indirect impacts of water quality improvements for each sector would be based on general equilibrium models and case studies focusing on high priority regions within the overall theme as given in the framework in Table 9.1.

## References

Bach PM, Deletic A, Urich C, McCarthy DT (2018) Modelling characteristics of the urban form to support water systems planning. Environ Model Softw 104:249–269

Burns MJ, Fletcher TD, Walsh CJ, Ladson AR, Hatt BE (2012) Hydrologic shortcomings of conventional urban stormwater management and opportunities for reform. Landscape Urban Plan 105:230–240

Castonguay AC, Iftekhar MS, Urich C, Bach PM, Deletic A (2018) Integrated modelling of stormwater treatment systems uptake. Water Res 142:301–312

Ellisa JB, Deutschb JC, Mouchelb JM, Scholesa L, Revitta MD (2004) Multicriteria decision approaches to support sustainable drainage options for the treatment of highway and urban runoff. Sci Total Environ 334–335:251–260

Fletcher T, Mitchell V, Deletic A, Ladson T, Seven A (2007) Is stormwater harvesting beneficial to urban waterway environmental flows? Water Sci Technol 55(4):265–272

Fletcher TD, Shuster W, Hunt WF, Ashley R, Butler D, Arthur S, Trowsdale S, Barraud S, Semadeni-Davies A, Bertrand-Krajewski J-L, Mikkelsen PS, Rivard G, Uhl M, Dagenais D, Viklander M (2015) SUDS, LID, BMPs, WSUD and more—the evolution and application of terminology surrounding urban drainage. Urban Water J 12(7):525–542

Palla A, Gnecco I, Lanza LG (2010) Hydrologic restoration in the urban environment using green roofs. Water J 2:140–154

Rauch W, Urich C, Bach PM, Rogers BC, de Haan FJ, Brown RR, Mair M, McCarthy DT, Kleidorfer M, Sitzenfrei R, Deletic A (2017) Modelling transitions in urban water systems. Water Res 126:501–514

Schubert JE, Burns MJ, Fletcher TD, Sanders BF (2017) A framework for the case-specific assessment of Green Infrastructure in mitigating urban flood hazards. Adv Water Resour 108(Supplement C):55–68

Shoemaker L, Riverson J, Alvi K, Zhen JX, Paul S, Rafi T (2009) SUSTAIN—a framework for placement of best management practices in urban watersheds to protect water quality. U.S. Environmental Protection Agency, Cincinnati, Ohio

Smith M, Vidmar A (1994) Data set derivation for GIS-based urban hydrological modeling. Photogramm Eng Remote Sens 60(1):67–76

Tuomela C (2017) Modelling source area contributions of stormwater pollutants for stormwater quality management. M.Sc. thesis, Department of Built Environment, School of Engineering, Aalto University, Finland

Tuomela C, Sillanpää N, Koivusalo H (2017) Modelling source area contributions of stormwater pollutants with Storm Water Management Model (SWMM). In: Proceedings of 14th IWA/IAHR conference on urban drainage, Prague, Czech Republic

USEPA (1979) Proceedings stormwater management model (SWMM) users group meeting, 24–25 May 1979. Report No 600/9-79-026. United States Environmental Protection Agency, Washington, DC

USEPA (1999) Combined sewer overflows guidance for monitoring and modeling. Report No 832-B-99-002. United States Environmental Protection Agency, Washington, DC

WaPUG (2002) Code of practice for hydraulic modeling of sewer systems version 3.001, Nov 2002, amended Dec 2002. Wastewater Planning Users Group, London. http://www.ciwem.org/wp-content/uploads/2016/05/Code-of-Practice-for-the-Hydraulic-Modelling-of-Sewer-Systems.pdf

# Chapter 10
# Development of a Three-Dimensional Mathematical Groundwater Flow Model in Raipur City Area, Chhattisgarh, India

**Suvendu Kumar Sahu and D. C. Jhariya**

**Abstract** Groundwater is one of the most important natural resources. In the current situation, it is important to consider the condition of groundwater to suggest a proper exploration pattern and management strategy to fulfill the need of society without changing its natural characteristics. Among the methods suggested for the overall assessment of aquifer characteristics, groundwater modeling is considered as is an effective management tool to study the aquifer response with different hydrological stress conditions. This present study is concentrating on the study of aquifer conditions in Raipur city with the suggestion of a fruitful management strategy. In order to understand the causes of water table declination, a two-layered finite-difference flow model was formulated for the region by using Visual MODFLOW software. Water level data of 21 well have been collected from all over the study area. Data such as, hydrological data, empirical values, and equations were used for the development of the groundwater transient flow model, water budget estimation, and to know the impact of over-pumping on the aquifer system.

**Keywords** Groundwater model · MODFLOW · Groundwater · Raipur city Chhattisgarh

## 10.1 Introduction

The high demand of groundwater for the swiftly growing population has increased the rate of requirement and trigger the effective management of available groundwater resources. The management of water resources needs a proper assessment and visualization of its overall structure and which is effectively visualize with the help of groundwater modeling software. The groundwater modeling has consid-

S. K. Sahu · D. C. Jhariya (✉)
Department of Applied Geology, National Institute of Technology Raipur, G.E. Road, Raipur, Chhattisgarh 492010, India
e-mail: dcjhariya.geo@nitrr.ac.in

 
P. Kumar et al. (eds.), *Water Resources Management and Sustainability*, Advances in Geographical and Environmental Sciences,
https://doi.org/10.1007/978-981-16-6573-8_10

ered as a multidisciplinary management tool which can carry out multiple functions like furnishing a framework for arranging hydrological data, assessing the behavior and properties of the aquifer system, and allowing both quantitative and qualitative prediction of responses of the system on applied stress condition (Rojas and Dassargues 2007; Senthilkumar and Elango 2004). Researchers all over the world have attempted the development of groundwater modeling for the effective management of groundwater resources in their respective areas (Akbariyeh et al. 2018; Lee et al. 2005; Yehia et al. 2013).

As the capital city, Raipur is experiencing rapid growth. The consumption of groundwater in the city has been noticeable increases to satisfy the growing demand for groundwater for domestic and agricultural purposes. Due to the over-extraction of groundwater, the water table in the area is showing a trend that is gradually falling with time to time (Central Ground Water Board 2012–13; Khan and Jhariya 2018). This study has been designed by considering this scenario by investigating the aquifer condition with present stress conditions and as a result, it suggests a possible management strategy to overcome the withstanding groundwater-related issues. The objectives of this study that has considered to solve the problem are (a) development of a flow model to assess the flow pattern and budget, (b) development of a transient flow model to assess the water table and to forecast for 14,600 days, and (c) fixing a suitable pumping rate and to decide recharge site as a management strategy for groundwater extraction without harming the natural aquifer condition.

## 10.2 Materials and Method

### *10.2.1 Study Area*

The area selected for the current study is falling under the Raipur district and is bounded by two water bodies, namely Kharun River and Chhokra Nala (Fig. 10.1). The study area falls under Survey of India toposheet no. 64G/11 and 12 in a scale of 1:50,000. The area is located in between latitude 21°12′ N to 21°25′ N and longitude 81°31′ E to 81°42′ E. The total area is around 192 km$^2$ with an almost gentle sloped landmark. It is situated about 300 m above mean sea level. The climate of the area is tropical, warm, and semi-arid with temperatures varying from 10 to 40 °C. The average humidity is more or less 55%. The southwest monsoon is prominent in the area. It receives an average rainfall of 1230 mm from June to September (Central Ground Water Board 2016–17).

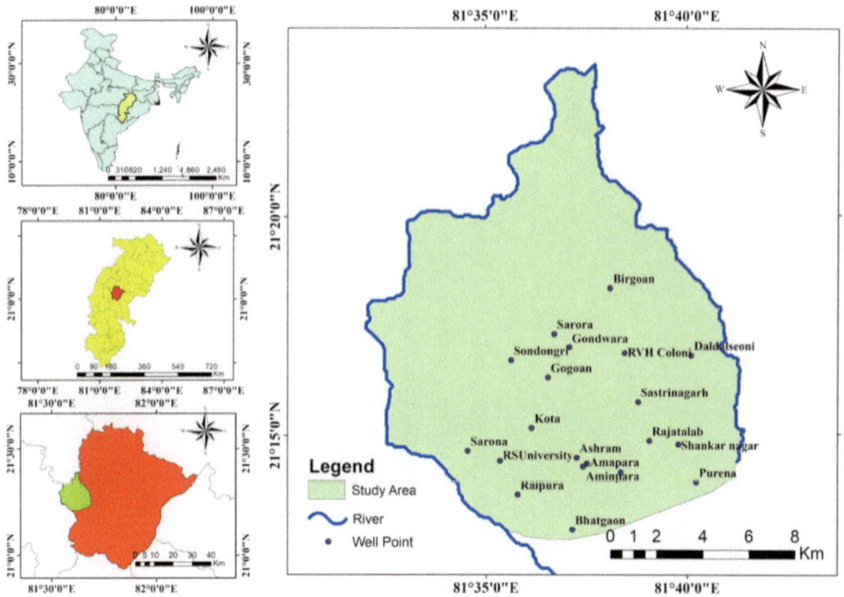


**Fig. 10.1** Location of study area

### *10.2.2 Local Geology*

The study area mainly consists of sedimentary rocks of the Precambrian age, coming under the Raipur group of the Chhattisgarh supergroup. Raipur group of rocks is consisting of Bejepur, Charmuriya, Gunderdehi, Chandi, Tarenga, Hirri, and Maniyari formations. The Raipur group is underlined by the Chandrapur group and overlain by the Mesozoic Gondwana supergroup (Table 10.1). Rocks of the Raipur group are mainly argillite-carbonate sequence and consist of limestone, shale, dolomite, and sandstone (Vaidyanadhan and Ramakrishnan 2010).

The Chandi formation is the major geological unit of the Raipur group which is exposed in the Raipur city. Its thickness varies from 103 to 136 m (Sinha et al. 2002). Chandi formation comprises Deodonger shale, limestone, and sandstone. The nature of the limestone exposed in the city is cavernous and jointed. Chandi formation overlies the Gundardehi formation with a sharp contact. Gundardehi formation occurs in the Raipur city comprises mainly shale and limestone.

### *10.2.3 Hydrogeology*

The aquifer of the area is defined by the Chandi formation and is categorized under unconfined aquifer type. Groundwater in the aquifer is mainly occupied in fractures

**Table 10.1** Stratigraphic succession of Raipur group of Chhattisgarh supergroup after Das et al. (1992) and Deb (2004)

| Group | Formation (thickness in m) | Lithology |
|---|---|---|
| Raipur group (1900 m) | Maniari shale (70) | Purple shale, dolomite, gypsum |
| | Hirri dolomite (70) | Gray dolomite |
| | Tarenga shale (180) | Dolomitic shale, shale-chert beds, purple shale, limestone |
| | Chandi limestone (670) | Stromatolitic dolomite, limestone, glauconitic sandstone, shale |
| | Gunderdehi shale (430) | Shale with limestone interbeds, arenite-shale, ignimbrite |
| | Charmuria limestone (490) | Phosphatic limestone with shale interbeds, cherty limestone and phosphatic dolomite, chert-shale interbeds |
| | Bijepur shale (100) | Green and brown calcareous shale with sandy interbeds |
| ……………………Unconformity…………………… | | |
| Chandrapur group (400 m) | | |

and solution cavities developed in the formation. The length of casing installed in the formations varies from 8.00 to 40.00 m with respect to the variable thickness of this formation. The average numbers of fractures encountered in the formations are of 2–12 numbers, within an average depth of 13–137 m. Determined yields are varying from 0.1 to 40 $m^3$/day in the study area with a maximum drawdown of 35.47 m. The static water level of the area is varying between 1.99 and 18.07 mbgl (Central Ground Water Board 2011).

## 10.2.4 *Model Conceptualization*

The hydrogeological system of the study area is conceptualized according to the overall picture that developed from the detailed study of geology, geomorphology, borehole lithology, well location, and data of water level fluctuation. Groundwater in the area is distributed in both Chandi and Gundardehi formations. Based on the collected information, the model is conceptualized as a single-layered unconfined aquifer having a variable thickness from 80 to 190 m.

### 10.2.4.1 Boundary Condition

The study area is surrounded by Kharun River on the north and west side and Chhokra Nala is on the eastern side. Both rivers are considered as constant head boundaries. The third boundary is the southern one, which covers 1150 km is considered as a specified flux boundary. The flux across the boundary has been assigned from the amount of flow calculated from the existing datasets (Fig. 10.2).

### 10.2.4.2 Grid Design

The model grid of the study area was discretized into 646 cells including 17 columns, 19 rows, and 2 layers. The cross-sectional length of the layers is 19 km toward north–south and 17 km toward east–west, respectively (Fig. 10.3).

### 10.2.4.3 Initial Groundwater Head

Initial groundwater head is one of the important parameters uses in the groundwater modeling has been collected during seasons of pre-monsoon and post-monsoon during the years of 1998–2000. The water level data of twenty numbers wells from different locations of the study area have been processed and converted into water table data for assigning the initial head value (Table 10.2).

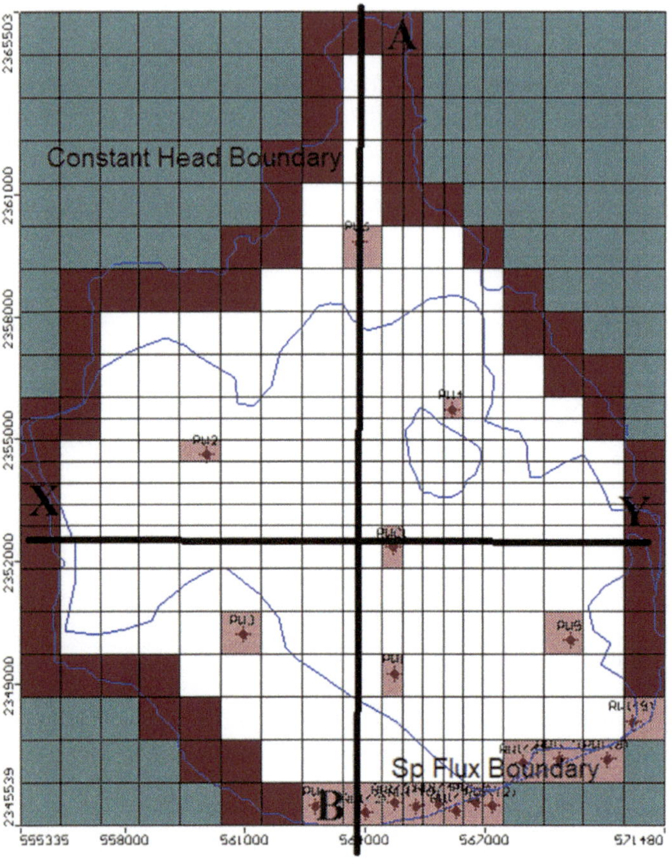


**Fig. 10.2** Discretization of study area

#### 10.2.4.4 Aquifer Characteristics

Major aquifer parameters like hydraulic conductivity ($K$), transmissivity ($T$), storativity ($S$), and specific yield ($S_y$) were gathered from the groundwater exploration report of Chhattisgarh state (CGWB 2011). Hydraulic conductivity ($K$) is varying between 10 and 21 m/day. Specific yield ranges from 0.01 to 0.025. The thickness of the aquifer varies from around 80 to 190 m. The reported value of storativity is between $3.0 \times 10^{-5}$ and $3.7 \times 10^{-4}$, and transmissivity is 69–1500 $m^2$/day (Central Ground Water Board 2016).

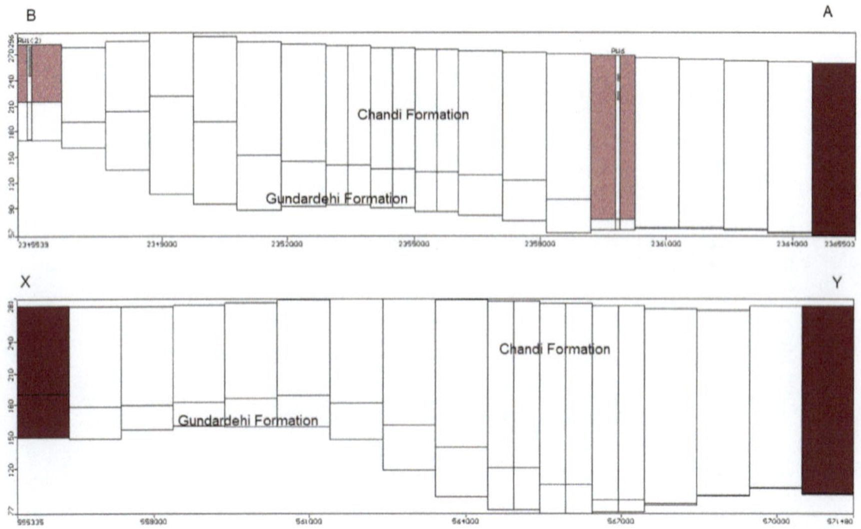


**Fig. 10.3** Cross-sectional view along AB and XY

#### 10.2.4.5 Groundwater Abstraction and Recharge

The major utilization groundwater withdrawal in Raipur city is for agriculture and domestic use. Throughout the year, majority of the localities in the Raipur city depends on groundwater source for different activities related to agriculture. The rate of discharge of groundwater on average is determined as 13,400 $m^3$/day. It shows an increasing trend with 5% in each successive year. Groundwater extraction was calculated based on the population report of census 2001 and 2011, which is about 22 l per capita per day (lpcd).

Rainfall is the main source of groundwater recharge in the area. Rainfall data are collected from annual reports of the Central Ground Water Board (Central Ground Water Board 2016–17) and as per the Groundwater Resources Estimation Committee report of the Government of India, infiltration capacity ranges from 15 to 20% (GREC 1997).

### *10.2.5 Model Description*

The developed model is an anisotropic and heterogeneous three-dimensional groundwater flow model. The model is developed by considering equivalent porous media (EPM) approach. It is considered to possess constant density, described by partial differential equation (Brown et al. 1998; Ding et al. 2014; McDonald and Harbaugh 1988; Wei et al. 2018).

**Table 10.2** Pre- and post-monsoon water level data of different well

| Area | Latitude | Longitude | Elevation (m) | 1998 Jan | 1998 May | 1999 Jan | 1999 June | 2000 March | 2000 June |
|---|---|---|---|---|---|---|---|---|---|
| | | | | DTW (m) | DTW (m) | DTW (m) | DTW (m) | DTW (m) | DTW (m) |
| Bhatgaon | 21.21389 | 81.61889 | 289.99 | 6.26 | 5.1 | 10.7 | 11.78 | 11.48 | 12.03 |
| Raipura | 21.22729 | 81.59645 | 285.83 | 2.27 | 6.3 | 5.25 | 6.75 | 5.94 | 7.69 |
| Sarona | 21.24394 | 81.57584 | 276.16 | 2.32 | 3.22 | 3.42 | 3.47 | 3.22 | 3.52 |
| Kota | 21.25255 | 81.6023 | 297.23 | 7.44 | 7.89 | 8.76 | 9.04 | 8.94 | 9.33 |
| Ashram | 21.24123 | 81.62076 | 302.58 | 1.46 | 1.54 | 1.58 | 1.72 | 1.62 | 1.78 |
| Amapara | 21.23898 | 81.62498 | 297.67 | 2.4 | 6.63 | 3.01 | 3.68 | 3.76 | 4.83 |
| Burapara | 21.23567 | 81.63891 | 302 | 8.7 | 9.7 | 10.34 | 11.87 | 11.68 | 12.95 |
| Aminpara | 21.2379 | 81.62337 | 298.45 | 1.27 | 1.13 | 2.63 | 2.39 | 3.22 | 2.67 |
| Sondongri | 21.27844 | 81.59385 | 281.99 | 2.22 | 1.74 | 2 | 2.22 | 2.05 | 2.23 |
| Sarora | 21.28833 | 81.61167 | 281.84 | 2.52 | 8.35 | 4.2 | 8.76 | 5.92 | 8.82 |
| Gondwara | 21.28337 | 81.61779 | 276.49 | 1.41 | 5.7 | 6.12 | 7.75 | 7.68 | 8.6 |
| Gogoan | 21.27188 | 81.60907 | 276.18 | 2.99 | 8.35 | 3.1 | 8.67 | 3.97 | 9.05 |
| Birgoan | 21.30574 | 81.63464 | 294 | 8.6 | 9.37 | 9.78 | 10.35 | 10.86 | 11.34 |
| RVH Coloni | 21.28109 | 81.64063 | 288 | 5.66 | 6.96 | 4.51 | 7.86 | 7.25 | 7.96 |
| Rajatalab | 21.24757 | 81.65086 | 273.83 | 2.05 | 1.18 | 1.42 | 1.77 | 2.71 | 3.07 |
| Daldalseoni | 21.28008 | 81.66837 | 286.6 | 1.07 | 2.36 | 1.67 | 2.02 | 2.87 | 6.53 |
| RSUniversity | 21.24012 | 81.58918 | 261.66 | 4.74 | 6.82 | 5.76 | 7.12 | 7.43 | 8.09 |
| Sastrinagarh | 21.26237 | 81.6462 | 293.99 | 3.77 | 4.05 | 4.07 | 4.16 | 3.13 | 4.52 |
| Shankar nagar | 21.24611 | 81.66306 | 295.39 | 7.46 | 8.12 | 8.4 | 9.36 | 9.74 | 10.03 |
| Purena | 21.23177 | 81.67038 | 293.5 | 2.38 | 4.15 | 2.3 | 5.74 | 6.94 | 7.58 |

$$\frac{\mathrm{d}}{\mathrm{d}x}\left[K_{xx}\frac{\mathrm{d}h}{\mathrm{d}x}\right]+\frac{\mathrm{d}}{\mathrm{d}y}\left[K_{yy}\frac{\mathrm{d}h}{\mathrm{d}y}\right]+\frac{\mathrm{d}}{\mathrm{d}z}\left[K_{zz}\frac{\mathrm{d}h}{\mathrm{d}z}\right]-Q=S_{\mathrm{S}}\frac{\mathrm{d}h}{\mathrm{d}t}$$

where

| | |
|---|---|
| $K_{xx}$, $K_{yy}$, $K_{zz}$ | Hydraulic conductivity along $x$, $y$, $z$ axes |
| $h$ | Head |
| $Q$ | Volumetric flux per unit volume |
| $S_{\mathrm{S}}$ | Specific storage coefficient. |

The simulation of the model has been attempted with the help of three-dimensional finite-difference MODFLOW and Visual MODFLOW.

## 10.3 Model Calibration

Model calibration can simply achieve by minimization of the error in the final result. The model has been calibrated to match or reduce the difference between computed value and the observed/field value by changing the influencing factors such as aquifer parameters and stress value (Akbariyeh et al. 2018; Vetrimurugan et al. 2017). Here, two steps of calibration have been used. The first one is a trial and error method, and the second best method is the trial and error followed by the PEST (parameter estimation) method (Doherty et al. 1994). The steady-state calibration was adopted for the year 1998. The primary special distribution of hydraulic conductivity was 20 m/day obtained from the available data, and the corresponding value was distributed in three zones based on lithological properties to compare the historical water table values. During trial and error, calibration different hydraulic conductivity values were adopted. After trial and error calibration, the PEST was used for improvisation of the steady-state calibration.

Transient state calibration was adopted for the year 1998–2001. Similarly, to the parameter hydraulic conductivity here, storage coefficient has been used with trial and error calibration. Finally, to get the improved calibration minor change of hydraulic conductivity value of steady state was made by keeping all other parameters constant. Calibration targets are arbitrary defined as for steady state, normalized root mean squared (NRMS) is 3.63%, absolute residual mean (ARM) is 1.3 m, root mean squared (RMS) is 1.996 m and for transient state NRMS is 4.87%, ARM is 1.67 m, and RMS is 1.999 m (Fig. 10.4).

The model simulation was operated out under a transient state for a duration of 14,600 days from the year 2000 to 2040. PEST and trial and error were used for the calibration of the transient model. After numerous trial runs a fair match between observed and computed head was obtained. There is very little difference between the calibrated value of hydraulic conductivity of steady and transient states. Observed

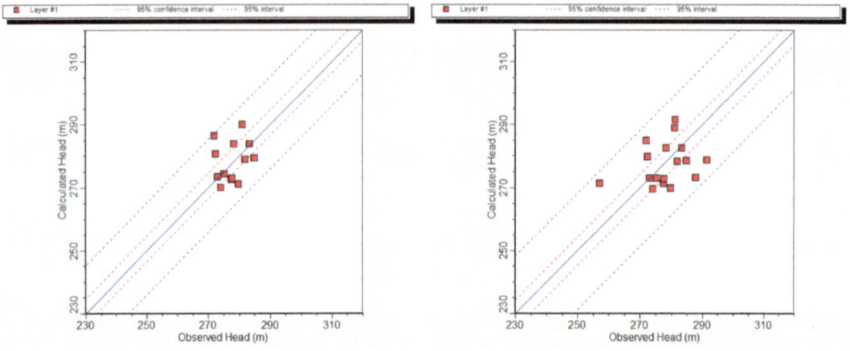


(a) (b)

**Fig. 10.4** Calibration: **a** comparison of observed and computed groundwater head in steady state. **b** Comparison of observed and computed groundwater head in the transient state

and simulated head values are comparable and having an alike trend for the duration from 1998 to 2000 and from 2000 to 2040.

This study reveals that the groundwater head is maximum at the southeast side and minimum at the northwest side which follows the general topographic tend. The flow of groundwater is from the southeast to the northwest side in this study area. The transient groundwater table of this area with the present average pumping rate from 2000 to 2040 is given in Fig. 10.5.

## 10.4 Sensitivity Analysis

Sensitivity analysis has been performed during model calibration by changing parameters such as hydraulic conductivity, recharge, and specific yield one by one to match the computed data with observed data. Sensitivity analysis is to determine the input parameters that have much more influence on the output result. Sensitivity analysis helps to understand the response of the aquifer system to different parameter conditions (Sathe and Mahanta 2019; Senthilkumar and Elango 2004). Here, hydraulic conductivity is the most sensitive parameter for this aquifer system by changing the hydraulic conductivity value 3–8 m/day for different spatial distribution, it matches the computed value with real-time aquifer condition.

## 10.5 Results and Discussion

The developed steady and transient flow model have concluded an idea about the present and transient groundwater flow direction, water table changing pattern with

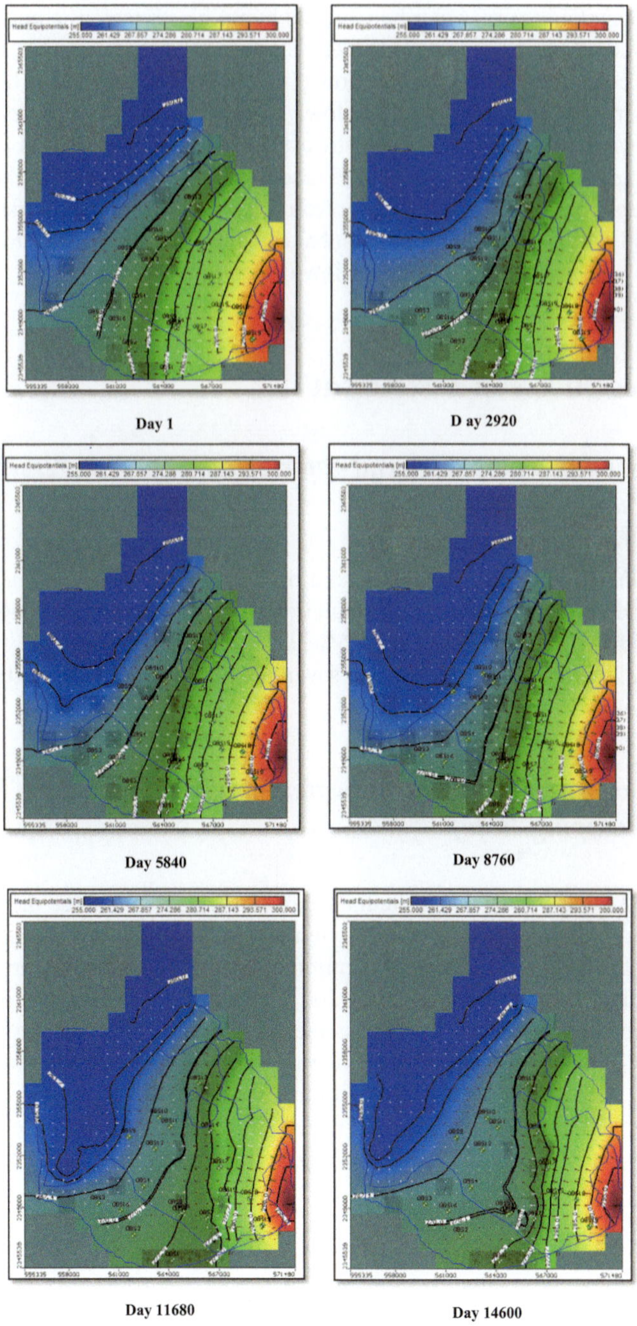


**Fig. 10.5** Transient water table with persisting pumping rate

reference to applied stress, water budget, and pumping strategy as a mitigation plan for future water table decline to protect the aquifer of the study area for a long duration. While considering groundwater flow direction, the water table is higher in the southeast part and sequentially lower down toward the northwest side so groundwater is flowing from southeast to northwest by following the general topographic trend and from the steady-transient flow model has shown an output like the water table will deplete near about seven meter from 2001 to 2040 with the persisting stress condition (Fig. 10.5). Flow budget is not balanced by total inflow with total outflow. Total outflow is higher than the total inflow in this aquifer system.

## 10.6 Prediction and Assessment

### *10.6.1 Scenario One (Groundwater Condition with Persisting Pumping Rate)*

In and around, Raipur city people depend on agriculture so they extract water for domestic and cultivation purposes. Simulation with the persisting average pumping rate of 13,400 $m^3$/day with increasing 5% each year resulting groundwater table in the Raipur city area will decrease up to seven meters up to 2040 (Fig. 10.5).

### *10.6.2 Scenario Two (Groundwater Condition with 22% Less Pumping Rate)*

The second scenario is all about the management strategy to reduce the water table depletion by minimize the water extraction and establishing a recharge well. The promising step for the management is minimizing pumping from the area (Abu-El-Sha'r and Hatamleh 2007; Rejani et al. 2007; Yehia et al. 2013). By numerous trial and run with keeping the all other parameters constant, if the pumping rate will be reduced 22% than the persisting pumping rate and three recharge well will establish in Ashram, Hatband, and Urla area with an average recharge rate of 110 $m^3$/day to 25 m screen depth, then there will be very little change in the water table. The transient flow model with the prescribed suggestion is shown with different duration (Fig. 10.6).

## 10.7 Conclusion

The transient flow model of the Raipur city area has provided an idea about the present scenario of the aquifer and its behavior to persisting stress conditions. The

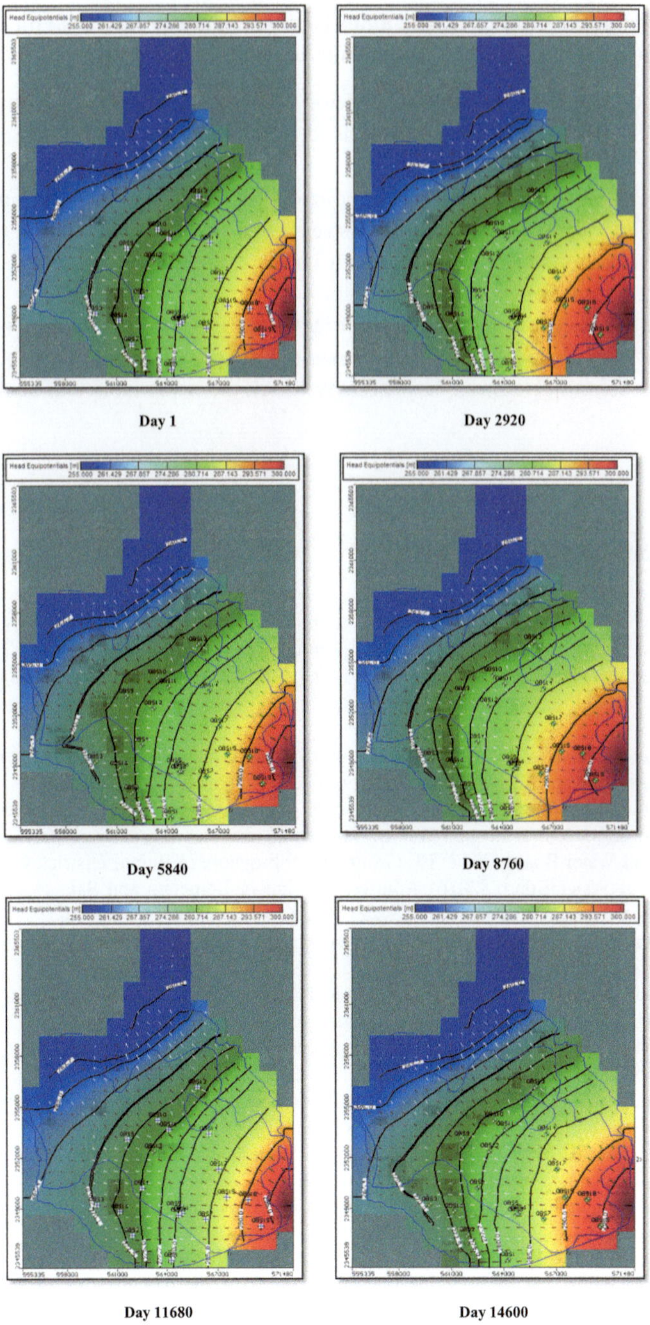


**Fig. 10.6** Transient water table with a suggested strategy

developed model is simulated with a fair level of accuracy. Based on the simulation result, the followings can be reckoned and directed for the study area:

a. This study represents the flow direction, i.e., from southeast to northwest direction, and water budget is not balanced with higher outflow in static condition with present stress condition.
b. The transient model has given a conclusion like with an average pumping rate of 13,400 $m^3$/day which is persisting stress condition, the water table will deplete nearly seven meters by 2040, which indicates that the study area is under the threat of depletion of the water table.
c. Lastly, the most promising management strategy for the study area is to reduce the 22% pumping rate than the persisting pumping rate and establish a recharge well at Ashram, Hatband, Urla area by which the water table will be maintained same.

## References

Abu-El-Sha'r WY, Hatamleh RI (2007) Using modflow and MT3D groundwater flow and transport models as a management tool for the Azraq groundwater system. Jordan J CivEng 1:153–172

Akbariyeh S, Hunt SB, Snow D, Li X, Tang Z, Li Y (2018) Three-dimensional modeling of nitrate-N transport in Vadose zone: roles of soil heterogeneity and groundwater flux. J Contam Hydrol 15–25

Brown PL, Guerin M, Hankin SI, Lowson RT (1998) Uranium and other contaminant migration in groundwater at a tropical Australian uranium mine. J Contam Hydrol 295–303

Central Ground Water Board (2011) Ground water exploration in Chhattisgarh State Ministry of Water Resources. Central Ground Water Board, pp 67–68

Central Ground Water Board (2012–13) Groundwater brochure of Raipur District, Chhattisgarh

Central Ground Water Board (2016) Aquifer mapping in Bemetara and Saja blocks, Bemetara district, Chhattisgarh. Central Ground Water Board, pp 22–25, 37–39, 62

Central Ground Water Board (2016–17) Groundwater year book Chhattisgarh. Central Ground Water Board, pp 7, 82

Das DP, Kundu A, Das N, Dutta DR, Kumaran K, Ramamurthy S, Thanavelu C, Rajaiya V (1992) Lithostratigraphy and sedimentation of Chhattisgarh basin: Indian minerals. 46:271–288

Deb PS (2004) Lithostratigraphy of the neoproterozoic Chhattisgarh sequence: its bearing on the tectonics and Palaeogeography. Gondwana Res 7:323–337

Ding F, Yamashita T, Lee HS, Pan J (2014) A modelling study of seawater intrusion in the Liao Dong bay coastal plain, China. J Mar Sci Technol 22(2):103–115

Doherty J, Brebber L, Whyte P (1994) PEST: model-independent parameter estimation. User's manual. Watermark Computing, Corinda, Australia

GREC (1997) Groundwater Resources Estimation Committee report of the groundwater resources estimation methodology. Government of India, New Delhi, p 107

Khan R, Jhariya DC (2018) Hydrogeochemistry and groundwater quality assessment for drinking and irrigation purpose of Raipur City, Chhattisgarh. J Geol Soc India 91:475–482

Lee MS, Lee KK, Hyun Y, Clement TP, Hamilton D (2005) Nitrogen transformation and transport modeling in groundwater aquifers. J Ecol Model 143–159

McDonald MG, Harbaugh AW (1988) A modular three-dimensional finite-difference ground-water flow model. U.S. Geol Surv 6A1:89

Rejani R, Jha MK, Panda SN, Mull R (2007) Simulation modelling for efficient groundwater management in Balasore Coastal Basin, India. Water Resour Manage 22:23–50
Rojas R, Dassargues A (2007) Groundwater flow modelling of the regional aquifer of the Pampa del Tamarugal, northern Chile. Hydrogeol J 15:537–551
Sathe SS, Mahanta C (2019) Groundwater flow and arsenic contamination transport modeling for a multi aquifer terrain: assessment and mitigation strategies. J Environ Manage 213:166–181
Senthilkumar M, Elango L (2004) Three-dimensional mathematical model to simulate groundwater flow in the lower Palar River basin, southern India. Hydrogeol J 12(4):197–208
Sinha UK, Kulkarni KM, Sharma S, Ray A, Bodhankar N (2002) Assessment of aquifer system using isotope techniques in urban centers Raipur, Calcutta and Jodhpur, India. No. IAEA-TECDOC-1298
Vaidyanadhan R, Ramakrishnan M (2010) Geology of India. Geological Society of India, p 1475
Vetrimurugan E, Senthilkumar M, Elango L (2017) Solute transport modelling for assessing the duration of river flow to improve the groundwater quality in an intensively irrigated deltaic region. Int J Environ Sci Technol
Wei X, Bailey RT, Records RM, Wible TC, Arabi M (2018) Comprehensive simulation of nitrate transport in coupled surface subsurface hydrologic systems using the linked SWAT-MODFLOW-RT3D model. Environ Model Softw 1–33
Yehia MM, Monem M, Mansour NM, El-Fakharany MA (2013) Using modflow and MT3D groundwater flow and transport models as a management tool for the quaternary aquifer, East Nile Delta, Egypt. Int J Sci Commer Humanit 1:219–233

# Chapter 11
# Flood Modelling Using HEC-RAS for Purna River, Navsari District, Gujarat, India

**Darshan J. Mehta and Yennam Varun Kumar**

**Abstract** The small and medium river banks are seriously flooded due to high-intensity of rainfall in monsoon seasons. The floods threatened human safety, life and property. This paper presents a process of 1D steady flow analysis used Hydrologic Engineering Center River Analysis System software. The one-dimensional modelling is applied on research Purna River, this river is one of the non-perennial rivers in Gujarat. This work contains flood model in which station, elevation of each cross-section was assessed at a particular section of the study reach. Steady flow analysis and hydraulic design analysis were carried out and after providing slope and discharge (flood event) at particular cross section software will compute the water surface elevation, depth of water and velocity of the water. The result from the research analysis could be used by flood management authorities to mensurate the flood at various cross-section of the study region.

**Keywords** HEC-RAS · Purna River · Steady flow analysis

## 11.1 Introduction

Flooding may occur due to the overflow of water from the river and lake (Patel and Sanjay 2019). This event occurrence due to high-intensity of rainfall in the river catchment area (Khattak et al. 2016). Flash floods are caused by a steep slope and highly erodible mountains, especially in mountain ranges (Mehta et al. 2013a, 2021). Floods damage human lives and property. Most of the floods occurred suddenly those are natural and artificially, most of flood disasters naturally occurred, but some disasters caused due to dam and poor reservoir management (Kumara and Mehtab 2020). In India, flood affected area is about 98.8422 million acres or nearly one by eight of the countries geographical area is flood-prone (Mehta et al. 2013b, 2017). Flooding is a significant aggravation that impacts sea-going aquatic ecosystems and the ecosystem services that they provide. Anticipated expansions in worldwide

D. J. Mehta (✉) · Y. V. Kumar
Dr. S & S. S. Ghandhy Government Engineering College, Surat, Gujarat, India



P. Kumar et al. (eds.), *Water Resources Management and Sustainability*,
Advances in Geographical and Environmental Sciences,
https://doi.org/10.1007/978-981-16-6573-8_11

flood hazards because of land use change and water cycle intensification will likely only increase the frequency and severity of these impacts. Sometimes, cyclones and tsunamis also cause floods (Mehta and Yadav 2020; Mehta et al. 2020; Agrawal and Regulwar 2016).

The study area is located in Gujarat; the state average precipitation is 150 cm. The most of the rivers are floods occur during rainy seasons. Purna River is an important west-flowing seasonal river; this river watershed is located at south east part of Gujarat and north west part of Maharashtra (Mehta et al. 2021; Kumara and Mehtab 2020). This river originates from Western Ghats; this river is frequently flooded and carries huge flood water due to heavy rains in monsoon, but this river has inadequate capacity to carry the peak flood discharge (Kumara and Mehtab 2020). For flood forecasting is used 1D steady flow model in HEC-RAS. This model gives approximate flood water depth at short time by computation. The 1D steady flow model is used for hydraulic structure, inundation canal and flood wall design (Ahmad et al. 2016; Demir and Kisi 2016). This paper presents the use of a HEC-RAS approach for river modelling facilities.

### *11.1.1 Objectives of Study*

Objectives of this study is to examine and assess the adequacy of a part of Purna River reach of Navsari region which is nearly 13.5 km and to carry out 1D steady flow analysis using Hydrologic Engineering Center River Analysis System software.

## 11.2 Study Area

Purna River is the non-perennial river in Gujarat state. This river originated from Saputara hill ranges in near the Chinchai village in Maharashtra. Zankhri River is the main tributary of this river. Navsari town is situated at the left bank of the Purna River, this river delta ends into the Arabian Sea (Mehta et al. 2021).

The Purna River is westside flowing river and has its rises in the Sahyadri ranges of the Western Ghats in Dang district and its total length 180 km before joining with Gulf of Khambhat (Arabian Sea). This river watershed lies between 20° 43′–21° 05′ N latitudes and 72° 43′–73° 57′ E longitudes. This river watershed area is 2435 km$^2$, out of which is nearly 2377 km$^2$ lies in Gujarat and 58 km$^2$ lies in Maharashtra. The river catchment is bounded by Western Ghats separating Purna and Tapi rivers in east, the ridge separating Purna and Ambica rivers in south, Arabian Sea in west and the ridge separating Purna and Mind Hola on north. Important tributaries are Girra River, Zankhari River and Damas khadi. The maximum, minimum temperatures observed at CWC site in Mahuva it varies from 34 to 44 °C and 26 to 10 °C, respectively. The average precipitation 122 cm in this region (Mehta et al. 2021; Kumara and Mehtab 2020).

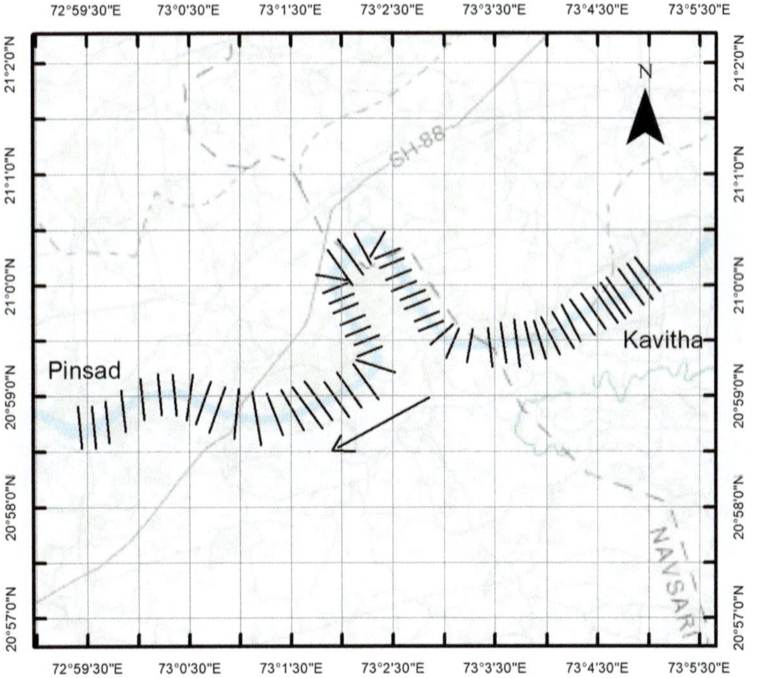


Fig. 11.1 Study area

In this study area, Purna River near Navsari is taken to carry out one-dimension steady flow analysis of Purna river basin of the Navsari region using HEC-RAS. This project work represents a process of one-dimensional steady flow analysis of the river by using computation. This study reach consists of 55 cross-sections, distance between two cross section 250 mitres at stream line. Approximately, 13.5 km length of Purna River is covered which is shown in Fig. 11.1. The upstream village is Kavitha and downstream village is Pinsad (Mehta and Yadav 2020).

## 11.3 Data Collection

The Hydrological data of Purna River at site Mahuva of is collected from Irrigation Department (Navsari) which include.

**Table 11.1** Past flood data

| Sr. No. | Year | Discharge (cumecs) |
|---|---|---|
| 1 | 2004 | 8836 |
| 2 | 2005 | 5437 |
| 3 | 2006 | 3273 |
| 4 | 2007 | 3058 |
| 5 | 2008 | 1853 |
| 6 | 2014 | 1543 |

### *11.3.1 Cross Sectional Data*

This spacing between each cross section is 250 m. It consists of 55 cross-sections. Approximately, 13.5 km length of Purna River is covered. The cross-section data consist of station-elevation. In addition to this, it also consists of data of left bank and right bank R.L.

### *11.3.2 Past Flood Event*

As discussed in previous section, peak discharge was used for analysis of flood. The data were collected from Irrigation Department, Navsari. Details of six flood events is shown in Table 11.1.

## 11.4 Methodology

The methodology taken in this study has been shown through the flowchart which is shown in Fig. 11.2.

### *11.4.1 HEC-RAS Analysis Procedure*

The Geometric data are added to cross section details which contain station, elevation, right over bank, left over bank and channel length, Manning's constant, left bank and right bank (Ingale and Shetkar 2017; Parhi et al. 2012).

Then from steady flow analysis menu, apply boundary condition and peak discharge parameters. After that save steady flow analysis file. Then, from hydraulic design function apply peak discharge and slope then after applying geometry it gives directly the water surface elevation height (Timbadiya et al. 2011, 2014).

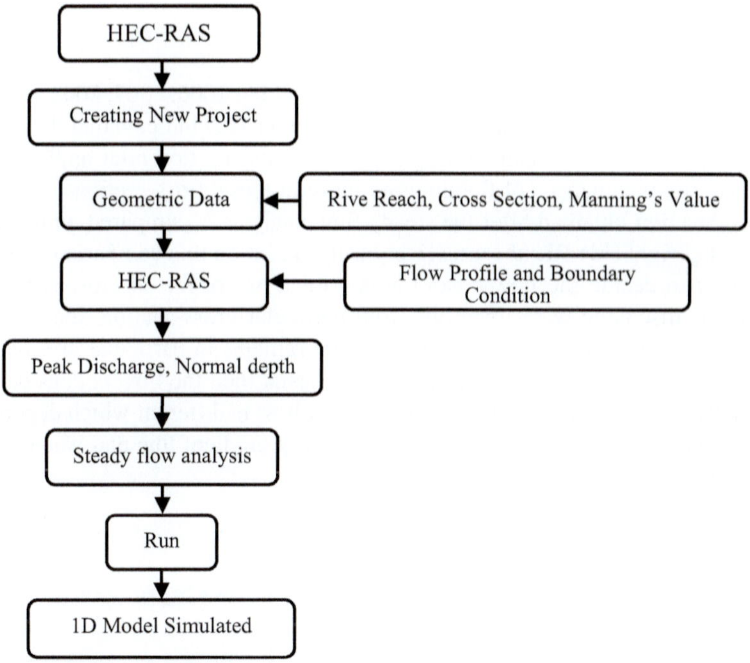


**Fig. 11.2** Flowchart of methodology

- Create a new HEC-RAS project. Create a new river geometry editor window. Give name of the Project as shown in Fig. 11.3.
- Create a new cross section. Paste the surveyed station/elevation points into the new cross section and then add the location of the left and right bank station.
- Enter the Manning's value for upstream reach. In this paper, value of '*n*'. The value of '*n*' can be taken according to bed material of the river reach as 0.03 or 0.04.

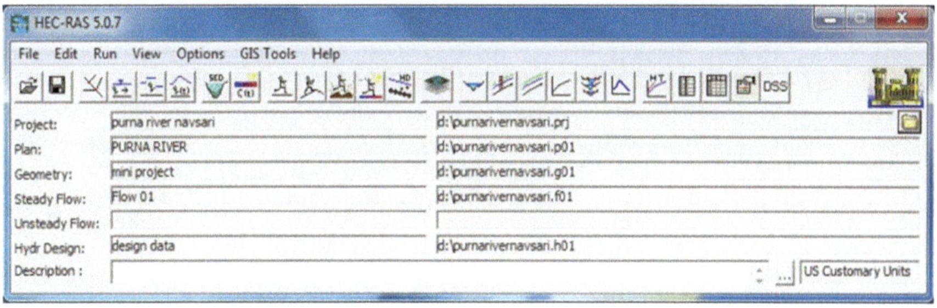


**Fig. 11.3** Main window of HEC-RAS

## 11.5 Result and Discussion

After giving imputes discharge for the years 2004, 2005, 2006, 2007, 2008 and 2014, the computed sections are obtained. Figures 11.4 and 11.5 indicate that the water is overtopped from the left bank and right bank of cross section after applying peak discharge. So, the chances of flooding are more in these cross sections. The water surface elevation obtained after the steady flow analysis is compared with existing left side and right side RL of Purna river basin. Based on this comparison analysis, water level to decide the flow condition is critical or not for all cross sections of river. If the discharge water overtops from particular cross-section, then the flow condition is high critical. If the discharge water remains in particular cross-section with sufficient remaining carrying capacity of basin, then the flow is considered as moderately critical. For all year, the condition of flow is different which depends on water discharge of that particular year. The energy gradient line and water surface line is obtained after computing the analysis.

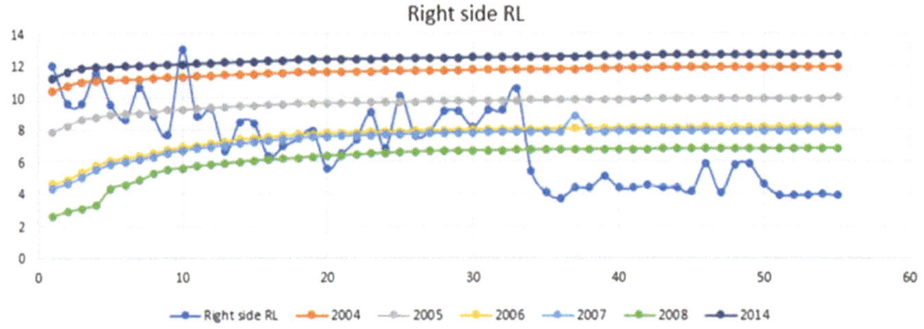


**Fig. 11.4** Graphical representation between existing and computed levels for flood events (right bank) of Purna River

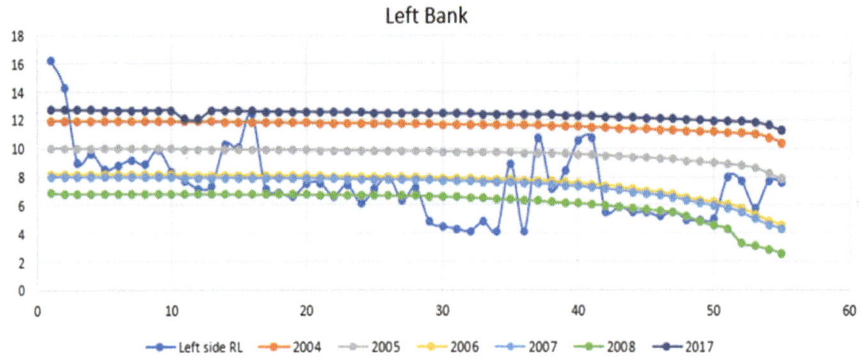


**Fig. 11.5** Graphical representation between existing and computed levels for flood events (left bank) of Purna River

By the analysis, it is found that cross-section no. 34–55 have less carrying capacity as compared to other cross-sections.

After giving imputes discharge for the years 2004, 2005, 2006, 2007, 2008 and 2014, the computed sections are obtained.

## 11.6 Conclusion

From flood analysis, the sections at which water overtop the existing level are found critical so it require restoration work and needs to be raised. With increased stream flows at different cross-sections of Purna river basin in the future, the chances of high magnitude flooding events are likely to increase under future climatic change in the river basin system. So, it is recommended that construction of levees, embankment and necessary flood gates in floodplain area. It is strongly recommended that no new construction is allowed in floodplain area. In our analysis, we found that in the year 2004, 2005 and 2014 the highest flood level has been reached. It is recommended that the storm drain outlets should be provided with flood gates to prevent entry of flood water in the study area. It is also recommended that there should be restriction in encroachment of width of the river because river sections are sensitive to high floods, encroachment will result in flooding of study region. It is strongly recommended that the sections, which water overtop over the existing embankment or retaining wall need to be raised. This study also recommends improving carrying capacity of Purna River so that it will minimize flood effects.

**Acknowledgements** Authors are thankful to the Navsari irrigation department for providing data of Purna River and Principal, Dr. S. & S. S. Ghandhy Government Engineering College, Surat for granting permission to carry out research work.

## References

Agrawal R, Regulwar DG (2016) Flood analysis of Dhudhana river in upper Godavari basin using HEC-RAS. Int J Eng Res 1:188–191

Ahmad HF, Alam A, Bhat MS, Ahmad S (2016) One dimensional steady flow analysis using HECRAS–a case of River Jhelum, Jammu and Kashmir. Eur Sci J 12:340–350

Demir V, Kisi O (2016) Flood hazard mapping by using geographic information system and hydraulic model: Mert River, Samsun, Turkey. Adv Meteorol

Ingale H, Shetkar RV (2017) Flood analysis of Wainganga River by using HEC-RAS model. Int J Sci Eng Technol 6(7):211–215

Khattak MS, Anwar F, Saeed TU, Sharif M, Sheraz K, Ahmed A (2016) Floodplain mapping using HEC-RAS and ArcGIS: a case study of Kabul River. Arab J Sci Eng 41(4):1375–1390

Kumara YV, Mehtab DJ (2020) Water productivity enhancement through controlling the flood inundation of the surrounding region of Navsari Purna river, India. Water Prod J. https://doi.org/10.2203/WPJ.2021.264752.1024

Mehta DJ, Yadav SM (2020) Hydrodynamic simulation of River Ambica for riverbed assessment: a case study of Navsari Region. In: Advances in water resources engineering and management. Springer, Singapore, pp 127–140

Mehta DJ, Ramani MM, Joshi MM (2013a) Application of 1-D HEC-RAS model in design of channels. Methodology 1(7):4–62

Mehta D, Yadav SM, Waikhom S (2013b) Geomorphic channel design and analysis using HEC-RAS hydraulic design functions. Paripex Int J Glob Res Anal 2(4):90–93

Mehta DJ, Yadav SM, Waikhom SI (2017) HEC-RAS flow analysis in the River Tapi. In: Proceedings of the 37th IAHR world congress, Kuala Lumpur, Malaysia, vol 16

Mehta D, Yadav SM, Waikhom S, Prajapati K (2020) Stable channel design of Tapi River using HEC-RAS for Surat region, vol 91. Springer, Cham, pp 25–36

Mehta DJ, Eslamian S, Prajapati K (2021) Flood modelling for a data-scare semi-arid region using 1-D hydrodynamic model: a case study of Navsari Region. Model Earth Syst Environ 1–11

Parhi PK, Sankhua RN, Roy GP (2012) Calibration of channel roughness for Mahanadi River, (India) using HEC-RAS model. J Water Resour Prot 4(10):847–850

Patel KB, Sanjay Y (2019) One dimensional unsteady flow analysis using HEC-RAS modelling approach for flood in Navsari City. In: Proceedings of recent advances in interdisciplinary trends in engineering & applications (RAITEA)

Timbadiya PV, Patel PL, Porey PD (2011) Calibration of HEC-RAS model on prediction of flood for lower Tapi River, India. J Water Resour Prot 3(11):805

Timbadiya PV, Patel PL, Porey PD (2014) One-dimensional hydrodynamic modelling of flooding and stage hydrographs in the lower Tapi River in India. Curr Sci 708–716

# Chapter 12
# Hydrological Parameter Estimation for Water Balance Study Using SWAT Model

**Sushmita Gouraha and Ishtiyaq Ahmad**

**Abstract** In recent years water demad has been increased due to a large population base, continued population growth and climate change uncertainties. To deal with issues of water management, quantification and estimation of different hydrological components are important. This study investigates hydrological parameter estimations using a semi-distributed physical-based model, the Soil and Water Assessment Tool (SWAT). The model was tested on a monthly basis for simulating the rate of streamflow, using rainfall and other climatic parameters in the basin. The watershed boundaries, sub-basins, slope, soil and land use maps, and streams were generated using Geographic Information System (GIS). Sensitivity of parameters was checked by *p*-value and *t*-stat. To check uncertainty in hydrological parameter, model was calibrated and validated for streamflow using the SWAT-Calibration and Uncertainty Program (SWAT-CUP). Model efficiency for calibration and validation was checked by different statistical parameters. Model uncertainty was analyzed by *P* factor and *R* factor. Results indicated $R^2$, NSE, and PBIAS were 0.62, 0.62, and 3.92, respectively, for the calibration period and also it was satisfactory with performance indicators $R^2$, NSE, and PBIAS as 0.64, 0.64, and 11.4 for the validation period. The study would be very useful for water resources community to take managerial actions in the watershed area.

**Keywords** Water balance · SWAT · SWAT-CUP · Sensitivity analysis · Calibration · Validation

---

S. Gouraha (✉)
Department of Hydrology, Indian Institute of Technology Roorkee, Roorkee 247667, India
e-mail: s_gouraha@hy.iitr.ac.in

I. Ahmad
Department of Civil Engineering, National Institute of Technology Raipur, Raipur 492001, India



P. Kumar et al. (eds.), *Water Resources Management and Sustainability*,
Advances in Geographical and Environmental Sciences,
https://doi.org/10.1007/978-981-16-6573-8_12

## 12.1 Introduction

Water is an important natural resource, which needs proper management for the survival of life on the earth, the improper and unmanaged use of water in India has led to decreasing groundwater resources, different land use/land cover planning and management provides the basis for better understanding of ecosystem to take a required decision and to make the decision for policymakers (Gupta et al. 1999). Water balance study is the application of the principle of mass conservation (Ghandhari and Alavi Moghaddam 2011), as there are different hydrological component involved in water budget analysis it is more practical to estimate hydrological component by hydrological modeling (Arnold and Allen 1996) through which water for the different hydrological component is quantified. There are various computer-based hydrological model available which are useful for watershed management (Strauch et al. 2012). To access the influence of climate change, topography and land use hydrologic models are very effective (Patel and Srivastava 2013). There are numerous physical-based watershed models available, among them the SWAT model developed by USDA has been used in this study. The Soil and Water Assessment Tool (SWAT) was developed to predict the effect of managerial practices on water quality, sediment yield, and pollution loading in watershed (Arnold et al. 1998), it has been applied many times in the study of water budget caused by changing climate (Leta et al. 2017; Cuceloglu et al. 2017; Zhou et al. 2011). SWAT model has also been used to study hydrological element in the agricultural area (Cao et al. 2018). Different input parameters were generated in Geographical Information System (GIS), which is a computer-based program which is used to map, analyze, integrate, transform, and manage the spatial or geographic data to solve complex planning and management operations. SWAT-Calibration and Uncertainty Program (SWAT-CUP) is a semi-automated method that has been used for sensitivity analysis, calibration, and validation of the model (Arnold et al. 2012). This model was used for calibration validation and uncertainty analysis of the model.

A good management practice with efficient planning may help for the survival of our water resources, for that computation of different hydrological components is important, hence water balance study has been performed, Andhiyarkore watershed of Chhattisgarh state was selected for the study. About 80% of the watershed area belongs to Kawardha district, which is facing problem of the water scarcity and was declared as a drought-prone area in recent years.

## 12.2 Materials and Method

### *12.2.1 Description of the Study Area*

The study area is a part of the Sheonath river basin, which tributary of the Mahanadi river basin in Chhattisgarh state, it covers an area of 2268.42 km$^2$, it extends from

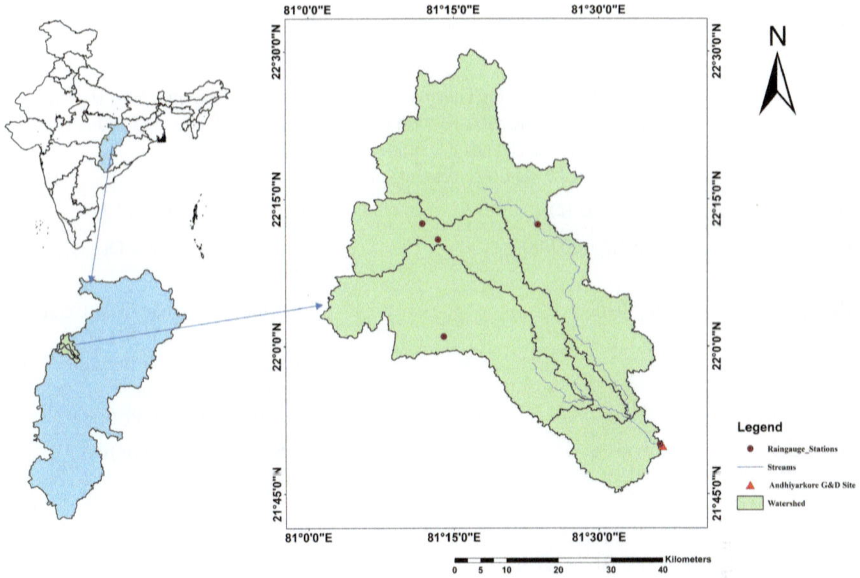


**Fig. 12.1** Location of Andhiyarkore watershed in India, Chhattisgarh

21° 45′ 5.83″ N to 22° 30′ 14.73″ N latitude and 81° 1′ 14.12″ E to 81° 36′ 38.87″ E longitude as shown in Fig. 12.1, Andhiyarkore is a Gauge and discharge measuring station of Central Water Commission (CWC), situated in the Bemetara district of Chhattisgarh state, which was used as an outlet for the study. The watershed covers parts of Kawardha, Bemetara, and Mungeli districts of Chhattisgarh state. The area receives an average annual rainfall of 895.46 mm. The mean annual temperature in the area is 33 °C, the climate in the area is dry sub-humid, the altitude varies from 224 to 979 m above MSL. There are mostly agricultural and forest area. In recent years, majority of the districts in Chhattisgarh state has been declared as drought-affected areas. There are 21 districts among 27 districts of Chhattisgarh facing the drought problem. Figure 12.1 is representing location map for the study area.

### *12.2.2 Data Used*

Digital elevation model, Land use—land cover, soil, and meteorological data such as rainfall, temperature, wind speed, relative humidity, and observed discharge datasets were collected from different sources and used for hydrological modeling in SWAT. Details of data sets are listed in Table 12.1.

**Table 12.1** Description of spatial datasets used for Andhiyarkore basin

| Data | Description | Source |
|---|---|---|
| DEM | Advanced Spaceborne Thermal Emission and Reflection Radiometer (ASTER) Global Digital Elevation Model (GDEM) of 30 m resolution | NASA and Japan's Ministry of Economy, Trade, and Industry (METI) |
| LULC | Scale 1:50,000 | Bhuvan, ISRO (2011–12) |
| Soil | Scale 1:5,000,000 | Food and Agriculture Organization (FAO), U.N. |
| Metrologic data | Daily data | Precipitation—State Data Center, Raipur, C.G. (1985–2013)<br>Wind speed, relative humidity, temperature and solar radiation—CFSR (1985–2013) |
| Discharge | Daily data | Central Water Commission, Bhubaneswar, Orissa (1985–2013) |

#### 12.2.2.1 Digital Elevation Model

DEM is a three-dimensional representation of the terrain, which represents elevations of the surface. The Digital Elevation Model (DEM) helps to understand catchment response, it helps to understand flow behavior and flow pattern in the watershed. In this study, Advanced Spaceborne Thermal Emission and Reflection Radiometer (ASTER) Global Digital Elevation Model (GDEM) of 30 m resolution, released by NASA and Japan's Ministry of Economy, Trade, and Industry (METI) has been used to analyze topography and to delineate watershed (Fig. 12.2). Watershed was divided into 5 number of sub-basins, and 40 number of Hydrological Response Units (HRUs). Percentage Slope was calculated using DEM, it was classified into four classes as shown in Fig. 12.3.

#### 12.2.2.2 Land Cover/Land Use

For hydrological study land use play an important role, land use data shows that how much region is covered by agricultural practice, forest, urbanization, etc., which is important to find runoff from that area. LULC data is downloaded from BHUVAN (LISS-III), National Remote Sensing Centre (NRSC). Most of the watershed is an agricultural and forest area (Fig. 12.4; Table 12.2).

Soil data are collected from the Food and Agriculture Organization of United Nations. Data is provided at 10 km spatial resolution soil database is a 30 arc-second raster database contained within 1:5,000,000 scale. Attribute data included with raster maps are different soil parameters like organic carbon, PH, water storage, soil,

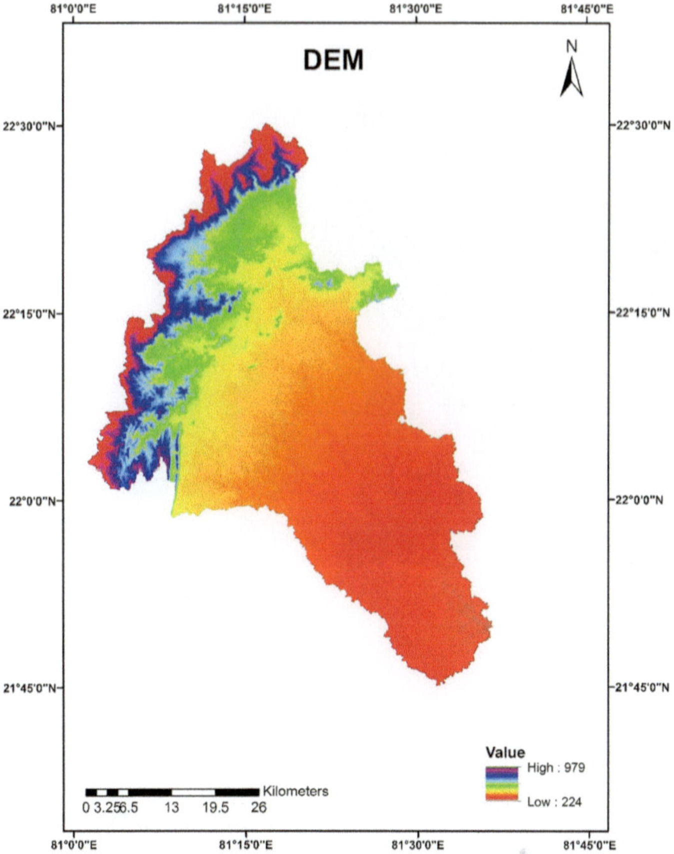


**Fig. 12.2** Digital Elevation Model of study area (ASTER Global Digital Elevation Model)

and clay fraction, etc. 3 groups of soil fall in this area, namely I-Bc-Lc-3714, I-bc-3735, and Lf92-1a-3791, in which basic variant is chromic vertisols and Lithosols (Fig. 12.5).

#### 12.2.2.3 Hydrometeorological Data

Daily rainfall data of six stations are used from 1985 to 2013 for hydrological modeling in SWAT many hydrological parameters as rainfall, temperature, wind speed, relative humidity, and solar radiation, are used, using these data weather generator database is prepared and used. Daily rainfall data of 6 stations are used from 1985 to 2013 which is collected from the state data center, Department of Water Resources, Raipur (CG) and other hydrological data are downloaded from the climate forecasting system reanalysis (CFSR) database which is simulated by National Centers for Environmental Prediction (NCEP) and Texas A & M University, United States.

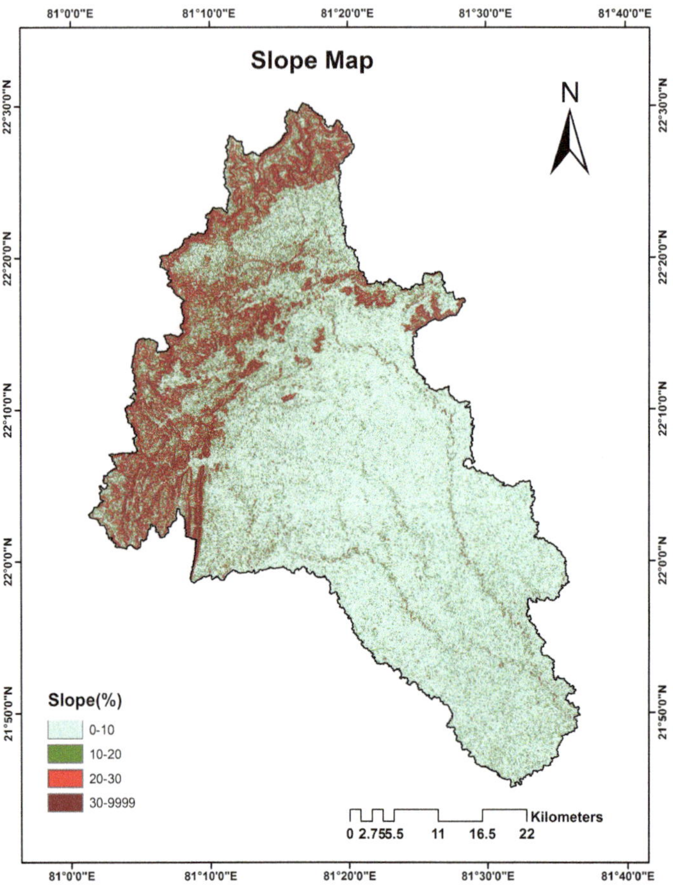


**Fig. 12.3** Percentage slope map of study area (ArcSWAT)

No of CFSR stations present in the Andhiyarkore watershed is 4. Daily discharge data (1985–2013) were collected from the Central Water Commission, Bhubaneswar, Orissa for Andhiyarkore gauging station.

## *12.2.3 SWAT Model Description*

Soil and water assessment tool (SWAT) is a physical-based, basin-scale continuous model and can be used to predict agricultural land management impacts on the hydrological regime of a watershed through simulation of variable soil, land use, and management conditions over long periods (Rahman et al. 2013). SWAT is a useful Geographic Information System (GIS) based decision support tool which has been used successfully for many watersheds around the world (Addis et al. 2016).

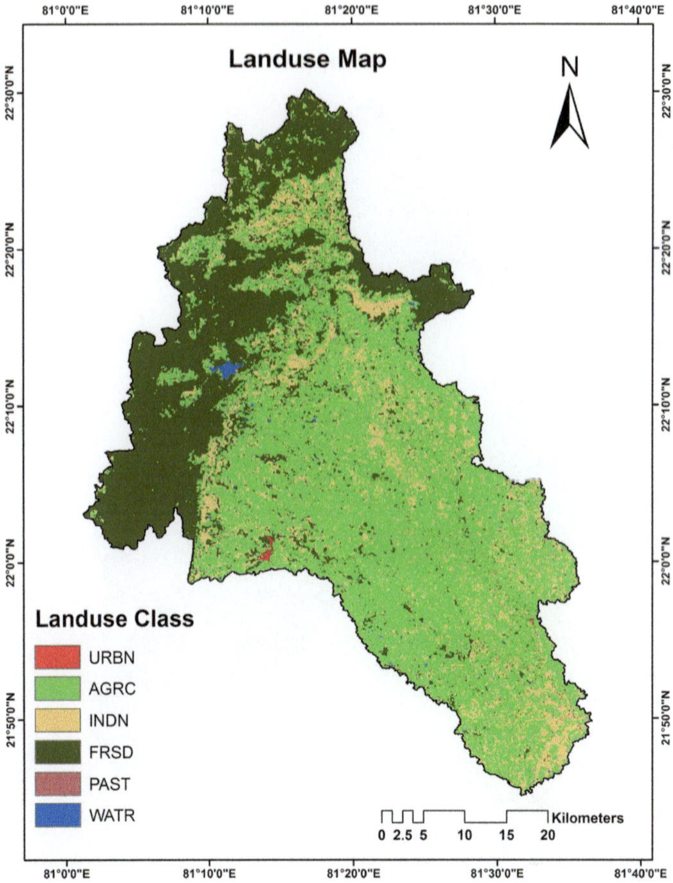


**Fig. 12.4** Land use map of study area (Bhuvan, ISRO (2011–12))

**Table 12.2** Land use/land cover (LULC) % area distribution (Bhuvan, ISRO (2011–12))

| S. No. | LULC type | Percentage of total area |
|---|---|---|
| 1 | Urban area | 0.13 |
| 2 | Agricultural area | 51.89 |
| 3 | Grass | 16.47 |
| 4 | Water | 0.33 |
| 5 | Pasture | 0.12 |
| 6 | Deciduous forest | 31.06 |

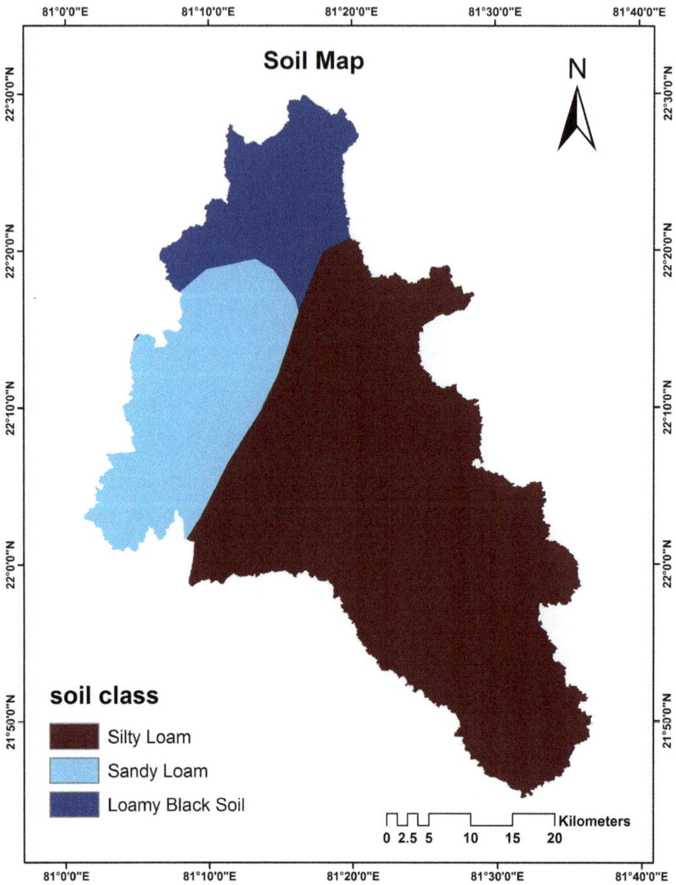


**Fig. 12.5** Soil map of study area (Food and Agriculture Organization (FAO), U.N.)

In the current study DEM was used for watershed delineation and also to draw drainage patterns of the land. Watershed was divided into five sub-basins. SWAT divides hydrology into two phases, the first phase land phase includes all vertical exchanges such as evaporation, infiltration, transpiration, percolation, and horizontal exchange as horizontal hypodermic flows. The watershed was divided into sub-basins which were further divided into Hydrologic Response Units (HRUs) according to similar soil, land use, and slope (Arnold et al. 1998). HRU is the smallest unit in the modeling. SWAT model simulates hydrological parameters for each HRU using the water balance equation. The water balance equation is as follows (Kundu et al. 2017)

$$\mathrm{SW}_t = \mathrm{SW}_o + \sum_{t=1}^{t} (R_t - Q_t - \mathrm{ET}_t - P_t - \mathrm{QR}_t) \tag{12.1}$$

To calculate $SW_t$ and $SW_o$ = soil water content at the beginning and end of the time period for which water balance equation is used (mm), $R_t$ = Rainfall for that day (mm), $Q_t$ = Surface Runoff for the day (mm), $ET_t$ = Evapotranspiration on day (mm), $P_t$ = Percolation on the day (mm), $QR_t$ = Lateral flow on the day (mm).

SWAT default method to calculate surface runoff is Soil Conservation Service Curve Number (SCS-CN) method has been used for this study. Potential evapotranspiration has been calculated by Hargreaves method. SWAT provides two methods for estimating runoff into the river system: the variable storage method developed by Williams (1969) and the method developed by Muskingum method by McCarthy (1938). In this study, the Muskingum method was chosen, which is the most commonly used routing mechanism in the literature.

### *12.2.4 Calibration and Uncertainty Analysis*

SUFI-2 Algorithm in SWAT-CUP has been used for model calibration and uncertainty analysis SWAT-CUP is specially developed by Abbaspour et al. (2007) to interface with the SWAT model (Abbaspour et al. 2007), it is capable of analyzing a large number of parameters. *t*-test and *p*-value are used to check the sensitivity of parameters. The higher the absolute value of *t*-stat and the smaller the value of *p*-value, the more sensitive is the parameter (Abbaspour et al. 2007). For calibration and validation semi-automated method SUFI-2 algorithm in SWAT-CUP was used (Abbaspour 2013). The model was calibrated for streamflow at one discharge gauging station Andhiyarkore, of Central Water Commission. Model calibration was performed for the year 1988–2008 (21 years) and validation for the year 2009–2013 (5 years).

### *12.2.5 Performance Indices*

The model performance for calibration and validation is checked by various statistical parameters *p* factor, *r* factor, Standard deviation, NSE, $R^2$, and PBIAS. Formulations to find these indices are described below

- Bias determination

$$\text{PBIAS} = 100 \times \frac{\sum_{i=1}^{n}(X_m - X_s)_i}{\sum_{i=1}^{n} X_{m,i}} \tag{12.2}$$

- Coefficient of determination

$$R^2 = \frac{\left[\sum_i \left(X_{m,i} - \overline{X}_m\right)\left(X_{s,i} - \overline{X}_s\right)\right]^2}{\sum_i \left(X_{m,i} - \overline{X}_m\right)^2 \sum_i \left(X_{s,i} - \overline{X}_s\right)^2} \tag{12.3}$$

- Nash Sutcliff efficiency

$$\mathrm{NSE} = 1 - \frac{\sum_i (X_m - X_s)_i^2}{\sum_i \left(X_{m,i} - \overline{X}_m\right)^2} \tag{12.4}$$

where,
$X$ = Variable value,
$m$ and $s$ = Measured and Simulated values,
$i$ = $i$th number of measured value and bar shows the average value.

## 12.3 Results and Discussions

### *12.3.1 SWAT Model Calibration*

SWAT is a physically based semi-distributed model, calibration of SWAT can be performed for gauged watersheds (Arnold and Allen 1996). The monthly flow rate at Andhiyarkore gauge and discharge station of CWC Bhubaneshwar was collected for the period 1988–2009 were used for model calibration. To achieve a certain level of model performance, calibration is important. Andhiyarkore was selected as an outlet and was taken for consideration to assign parameters of sub-basins. Rainfall data for six rain gauge stations were used and a rainfall data file was prepared. The weather generator database was generated using all meteorological data as was used as an input for SWAT model simulation. Priestley–Tayler's method was used to compute Evapotranspiration. SWAT default SCS-CN method was used for the estimation of Runoff.

#### 12.3.1.1 Simulation of Streamflow Rate Using Pre-calibrated ArcSWAT Model

Initially, pre-calibrated ArcSWAT outputs were compared with observed discharge at Andhiyarkore station. The hydrograph shows that model was overpredicting values of streamflow. Results indicate that model performance is not satisfactory for pre-calibration run and adequate calibration is required (Fig. 12.6).

#### 12.3.1.2 Parameters Used for Model Calibration

Sensitivity analysis was performed for calibration and validation of the model on monthly time steps. SWAT-CUP model was used for sensitivity analysis. Many parameters affect the performance of watershed sensitivity analysis is performed for 13 parameters, which are selected according to the objective function of streamflow

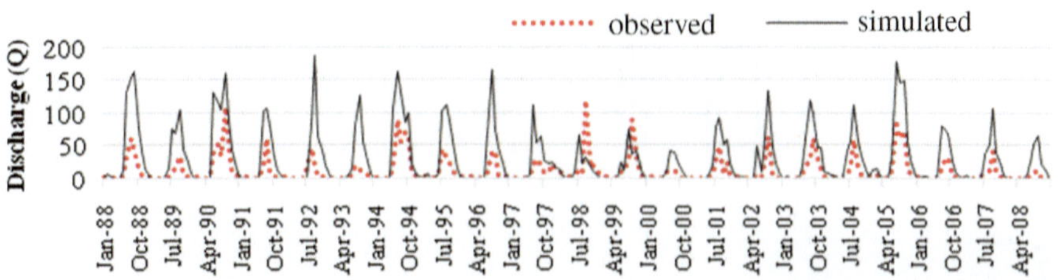


**Fig. 12.6** Graphical comparison of observed and simulated monthly flow before calibration of model for calibration period (1988–2008)

**Table 12.3** Results of sensitivity analysis (SWAT-CUP)

| Parameters | Description | *P*-value | *t*-stat | Rank |
|---|---|---|---|---|
| CN2 | Curve number for moisture condition 2 | 0 | 32.69 | 1 |
| GWQMN | Threshold depth of water in the shallow aquifer required for return flow to occur | 0 | 11.07 | 2 |
| GW_DELAY | Groundwater delay | 0.01 | 2.42 | 3 |
| SURLAG | Surface runoff lag time | 0.16 | 1.40 | 4 |
| SLSUBBSN | Average slope length | 0.24 | 1.18 | 5 |
| ALPHA_BF | Baseflow alpha factor | 0.43 | −0.78 | 6 |
| SOL_AWC | Available water capacity of the soil layer | 0.44 | 0.77 | 7 |
| SOL_K | Saturated hydraulic conductivity | 0.46 | −0.76 | 8 |
| ESCO | Soil evaporation compensation factor | 0.56 | 0.58 | 9 |
| SOL_BD | Moist bulk density | 0.61 | 0.51 | 10 |
| CH_K2 | Effective hydraulic conductivity in main channel alluvium | 0.66 | 0.44 | 11 |
| GW_REVAP | Groundwater "revap" coefficient | 0.80 | −0.26 | 12 |
| HRU_SLP | Average slope steepness | 0.98 | −0.02 | 13 |

measurement sensitivity of the model is checked by *p*-value and *t*-stat, the parameters are more sensitive for larger *t*-stat values. *P*-values are used to determine the significance of the sensitivity where the parameter becomes significant if the *p*-values is close to zero (Khalid et al. 2016). CN2, GWQMN, and GW_DELAY were found the most sensitive parameters for the watershed (Table 12.3).

#### 12.3.1.3 Simulation of Monthly Streamflow Rate

ArcSWAT model was calibrated against monthly streamflow at Andhiyarkore gauging station for years 1988–2013 with consideration of 3 years (1985–87) as a warm up period. *P*-factor and *R* factor were found 0.65 and 0.43, respectively, for the calibrated model. It can be observed in Fig. 12.7. The time to peak for observed

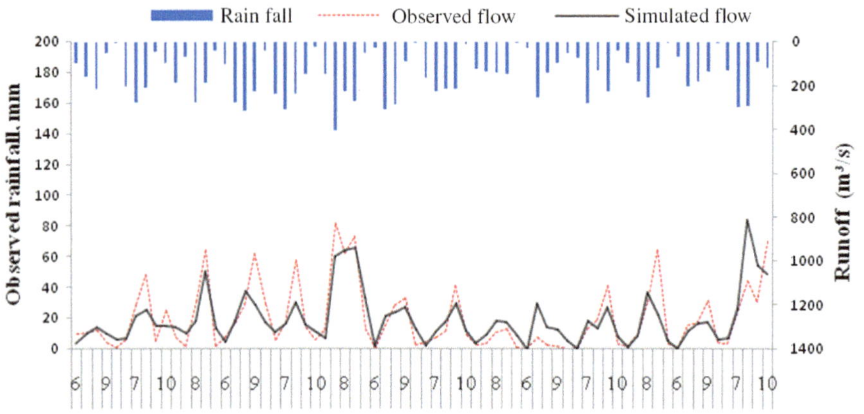


**Fig. 12.7** Graphical comparison of observed and simulated discharge with rainfall

and simulated flow matches well with peaks of rainfall in the watershed. Figure 12.8 is representing a scattered plot of observed flow and simulated flow for the watershed, with a coefficient of correlation as 0.62.

Model performance for calibration was checked by other statistical parameters, and it was found as standard deviation, coefficient of correlation ($R^2$), Nash–Sutcliffe efficiency (NSE), index of agreement, and ($d$) and percentage of bias (PBIAS) summary of statistics are tabulated in Table 12.4, model performance was found good with $R^2$ 0.62, NSE 0.62, PBIAS 3.92, and index of agreement as 0.991.

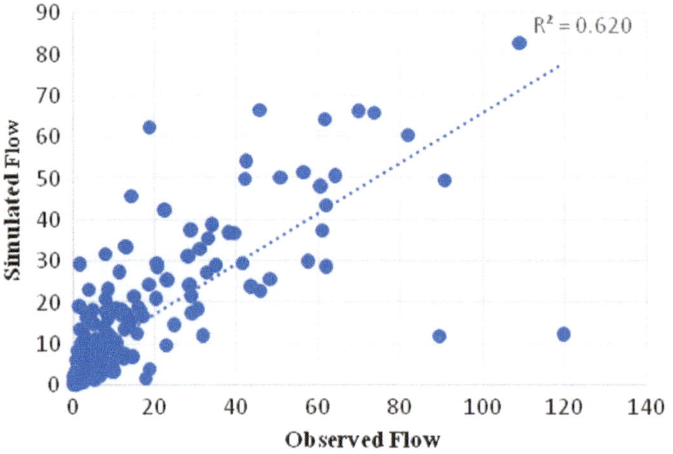


**Fig. 12.8** Scattered plot of observed and simulated monthly streamflow rate for model calibration period (1988–2008)

**Table 12.4** Summary of statistics for observed and simulated streamflow for calibration and validation period

| | Pre-calibrated | | Calibrated | | Validated | |
|---|---|---|---|---|---|---|
| | Observed | Simulated | Observed | Simulated | Observed | Simulated |
| Count | 252 | 252 | 252 | 252 | 60 | 60 |
| Mean ($m^3$/s) | 11.49 | 30.27 | 11.49 | 11.94 | 8.29 | 9.04 |
| Maximum ($m^3$/s) | 119.53 | 188 | 119.53 | 82.68 | 69.73 | 83.89 |
| SD | 19.74 | 43.23 | 19.74 | 15.32 | 15.34 | 15.42 |
| $R^2$ | 0.60 | | 0.62 | | 0.64 | |
| NSE | −2.3 | | 0.62 | | 0.65 | |
| *d* | 0.927 | | 0.991 | | 0.90 | |
| PBIAS | 163.31 | | 3.92 | | 11.4 | |
| RSR | 1.56 | | 0.45 | | 0.54 | |

## *12.3.2 Model Validation*

After calibrating the model, validation of the model has been performed for five years (2009–2013). Validation results show that the modeled monthly flow of stream matches well with observed stream flows. Figure 12.9 is representing the relation between the peak event of modeled and observed streamflow with rainfall, it matches well. Magnitude of the simulated monthly streamflow rate was lower for the years 2010, 2011, and 2012 and higher for the years 2009 and 2013.

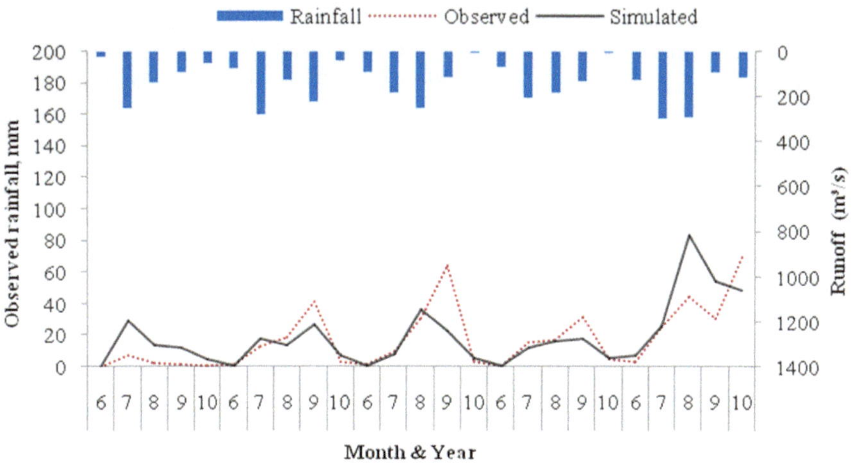


**Fig. 12.9** Graphical comparison of observed and simulated monthly stream flow rate for model validation (2009–2013)

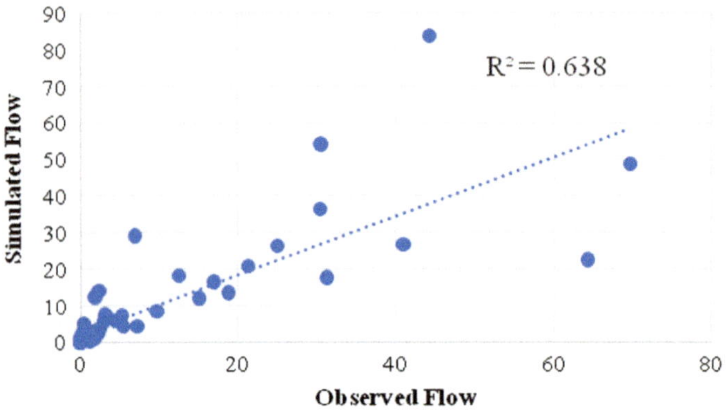


**Fig. 12.10** Scattered plot of Simulated and observed monthly flow during validation period (2009–2013)

Figure 12.10 is representing a scattered plot between observed and simulated streamflow for the validation period, they show good co-relation with the coefficient of variation as 0.64, NSE 0.65, and PBIAS 11.4 with the index of agreement as 0.9. Other statistics for the validation period are listed in Table 12.4.

### *12.3.3 Evaluation of Model Performance*

Model performance in calibration and validation has been tabulated in Table 12.4 which is representing a summary of all statistical parameters obtained. Moriasi et al. (2007) have given criteria for model evaluation for streamflow on monthly time steps, which is summarized in Table 12.5.

Model performance was found very good while analyzing PBIAS and RSR for model calibration, it was found good for PBIAS and RSR for model validation. While analyzing NSE model behavior was found satisfactory for calibration and good for validation.

**Table 12.5** General performance rating for monthly time step (Moriasi et al. 2007)

| Model performance | RSR | NSE | PBIAS |
|---|---|---|---|
| Very good | $0 < RSR < 0.5$ | $NSE > 0.75$ | $PBIAS < \pm 10$ |
| Good | $0.5 < RSR < 0.6$ | $0.65 < NSE < 0.75$ | $\pm 10 < PBIAS < \pm 15$ |
| Satisfactory | $0.6 < RSR < 0.7$ | $0.50 < NSE < 0.65$ | $\pm 15 < PBIAS < \pm 25$ |
| Unsatisfactory | $RSR > 0.7$ | $NSE < 0.50$ | $PBIAS > \pm 25$ |

**Table 12.6** Parameters representing uncertainty associated with the model prediction

| Parameters | After calibration | Validation |
|---|---|---|
| *P* factor | 0.65 | 0.62 |
| *R* factor | 0.43 | 0.44 |

## *12.3.4 Uncertainty Analysis*

There were total 6 iterations of SUFI-2 algorithm with 500 number of simulations for each iteration. *P* factor and *R* factor are important to analyze uncertainty for the model prediction, Table 12.5 represent parameters which represent uncertainty associated with model prediction using *P* factor and *R* factor. *P* factor was found 0.65 and *R* factor was found 0.43 for the calibrated model. During validation, *P* factor was 0.62 and *R* factor was 0.44. It is observed that *p*-value closer to 1 and *R* factor closer to 0 gives good relation between observed and simulated model result (Bekele and Nicklow 2007). It is observed that uncertainty is not very high in model prediction (Table 12.6).

## *12.3.5 Hydrological Parameter Estimation*

Calibration ensures that good correlation exists between observed and model-predicted flow, it also ensures that hydrological parameters associated with water budget are also in a reasonable range. Figure 12.11 shows the average annual value of all hydrological elements obtained from the calibrated model. It is observed that

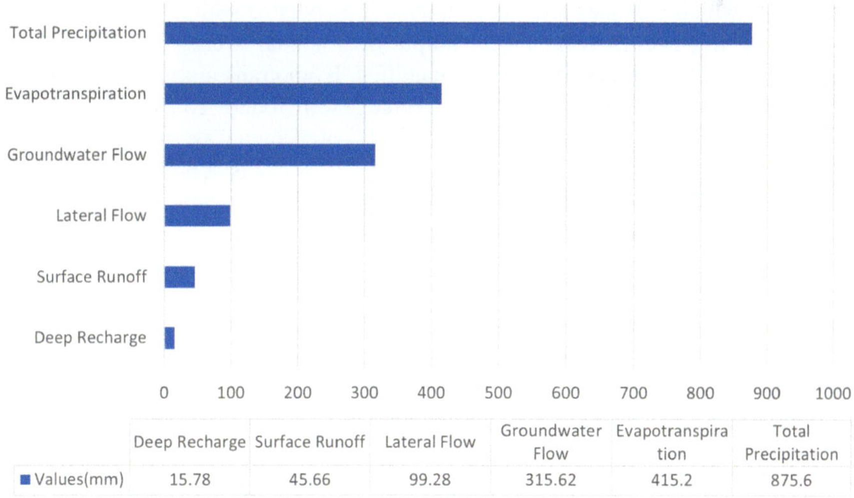


**Fig. 12.11** Average annual values of hydrological elements in mm for period 1988–2013

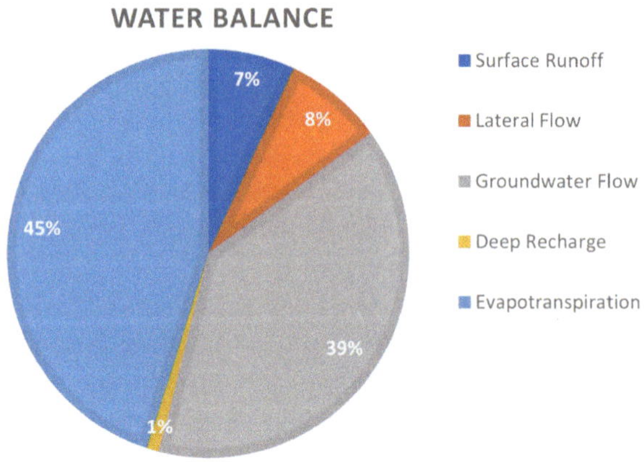


**Fig. 12.12** Percentage of hydrological component relative to precipitation

the average annual rainfall in the area was 875.6 mm. Figure 12.12 is representing the percentage distribution of all hydrological components. It is can be observed that a major portion of precipitation is distributed to evapotranspiration (45%) and Groundwater Flow (39%), whereas there is very least amount of water contributing to deep recharge and surface runoff.

## 12.4 Conclusion

The present study used the SWAT model to successfully simulate streamflow on a monthly scale for the Andhiyarkore watershed. All statistical parameters indicate the model was satisfactory calibrated and validated. *R* factor and *P* factor were satisfactory which indicates less uncertainty was associated with model simulation.

By analyzing twenty years model output data annual average contribution to different hydrological parameters was quantified. Evapotranspiration, groundwater flow, Lateral flow, surface runoff, and deep recharge were found 415.2 mm, 315.2 mm, 99.28 mm, 45.66 mm, and 18.78 mm, respectively out of annual average precipitation of 875.6 mm. It is observed that a major amount of water is contributed to evapotranspiration and groundwater flow as most of the part of catchment is agricultural land and forest area. These results for hydrological parameter estimation could be valuable for water resources management of the Andhiyarkore watershed.

**Acknowledgements** The authors would like to express deep thanks and acknowledge the support provided by various sources from where the data have been collected, State data Centre (WRD), Raipur, hydrometeorology department, Raipur, and National Remote Sensing Centre (NRSC) Hyderabad.

## References

Abbaspour KC (2013) Swat-cup 2012. In: SWAT calibration and uncertainty program—a user manual

Abbaspour KC, Vejdani M, Haghighat S, Yang J (2007) SWAT-CUP calibration and uncertainty programs for SWAT. In: MODSIM 2007 international congress on modelling and simulation, modelling and simulation society of Australia and New Zealand, pp 1596–1602

Addis HK, Strohmeier S, Ziadat F, Melaku ND, Klik A (2016) Modeling streamflow and sediment using SWAT in Ethiopian Highlands. Int J Agric Biol Eng 9(5):51–66

Arnold JG, Allen PM (1996) Estimating hydrologic budgets for three Illinois watersheds. J Hydrol 176(1–4):57–77

Arnold JG, Srinivasan R, Muttiah RS, Williams JR (1998) Large area hydrologic modeling and assessment part I: model development 1. JAWRA J Am Water Resour Assoc 34(1):73–89

Arnold JG, Moriasi DN, Gassman PW, Abbaspour KC, White MJ, Srinivasan R, Santhi C, Harmel RD, Van Griensven A, Van Liew MW, Kannan N (2012) SWAT: model use, calibration, and validation. Trans ASABE 55(4):491–1508

Bekele EG, Nicklow JW (2007) Multi-objective automatic calibration of SWAT using NSGA-II. J Hydrol 341(3–4):165–176

Cao Y, Zhang J, Yang M, Lei X, Guo B, Yang L, Zeng Z, Qu J (2018) Application of SWAT model with CMADS data to estimate hydrological elements and parameter uncertainty based on SUFI-2 Algorithm in the Lijiang river basin, China. Water 10(6):742

Cuceloglu G, Abbaspour KC, Ozturk I (2017) Assessing the water-resources potential of Istanbul by using a soil and water assessment tool (SWAT) hydrological model. Water 9(10):814

Ghandhari A, Alavi Moghaddam SMR (2011) Water balance principles: a review of studies on five watersheds in Iran. J Environ Sci Technol 4:465–479

Gupta HV, Sorooshian S, Yapo PO (1999) Status of automatic calibration for hydrologic models: comparison with multilevel expert calibration. J Hydrol Eng 4(2):135–143

Khalid K, Ali MF, Rahman NFA, Mispan MR, Haron SH, Othman Z, Bachok MF (2016) Sensitivity analysis in watershed model using SUFI-2 algorithm. Procedia Eng 162:441–447

Kundu S, Khare D, Mondal A (2017) Individual and combined impacts of future climate and land use changes on the water balance. Ecol Eng 105:42–57

Leta OT, El-Kadi AI, Dulai H (2017) Implications of climate change on water budgets and reservoir water harvesting of Nuuanu area watersheds, Oahu, Hawaii. J Water Resour Plann Manag 143(11):05017013

McCarthy GT (1938) The unit hydrograph and flood routing, Conference of North Atlantic Division. US Army Corps of Engineers, New London, CT. US Engineering

Moriasi DN, Arnold JG, Van Liew MW, Bingner RL, Harmel RD, Veith TL (2007) Model evaluation guidelines for systematic quantification of accuracy in watershed simulations. Trans ASABE 50(3):885–900

Patel DP, Srivastava PK (2013) Flood hazards mitigation analysis using remote sensing and GIS: correspondence with town planning scheme. Water Resour Manage 27(7):2353–2368

Rahman K, Maringanti C, Beniston M, Widmer F, Abbaspour K, Lehmann A (2013) Streamflow modeling in a highly managed mountainous glacier watershed using SWAT: the Upper Rhone River watershed case in Switzerland. Water Resour Manage 27(2):323–339

Strauch M, Bernhofer C, Koide S, Volk M, Lorz C, Makeschin F (2012) Using precipitation data ensemble for uncertainty analysis in SWAT streamflow simulation. J Hydrol 414–415:413–424

Williams JR (1969) Flood routing with variable travel time or variable storage coefficients. Trans ASAE 12(1):100–103

Zhou G, Wei X, Wu Y, Liu S, Huang Y, Yan J, Zhang D, Zhang Q, Liu J, Meng Z, Wang C, Chu G, Liu S, Tang X, Liu X (2011) Quantifying the hydrological responses to climate change in an intact forested small watershed in Southern China. Glob Change Biol 17(12):3736–3746

# Chapter 13
# Groundwater Potential Mapping Using Maximum Entropy

**Arnold R. Salvacion**

**Abstract** Maximum entropy (MaxEnt) is a widely used general-purpose machine learning approach for species distribution modeling. However, in recent years, other researchers have used MaxEnt in other areas such as disease risk mapping, flooding risk assessment, and fire hazard analysis among others. This study demonstrates the use of MaxEnt to map groundwater potential in the province of Marinduque, the Philippines using groundwater wells location and different environmental variables. These environmental variables include elevation, slope, topographic wetness index, drainage density, distance from faults, distance from rivers, rainfall during the wettest month, and annual rainfall. Based on the results, elevation and annual rainfall were the variables with the highest contribution in predicting the groundwater potential in the province.

**Keywords** GIS · Groundwater · MaxEnt · Marinduque · Philippines

## 13.1 Introduction

Groundwater is the subsurface water in the fractures of rock formations, soil pore spaces, natural voids, or aquifers (Tang et al. 2017; Beckie 2013). It is a critical part of the water systems that constitute 30% of the world's water resource providing for small and large communities across the globe (Flores 2014; Alley 2009; Ajami 2020). Groundwater is a valuable natural resource for human life and several infrastructures for human survival and development (Chen et al. 2018; Li et al. 2013). Around 50% of drinking water and 40% of irrigation came from groundwater, supporting approximately 20 billion people worldwide (Beckie 2013; Ajami 2020; Waller and Yitayew 2016). In addition, discharge from groundwater supplies springs, lakes, rivers, wetlands, and the ocean (Beckie 2013; Shrestha and Pandey 2016).

---

A. R. Salvacion (✉)
Department of Community and Environmental Resource Planning, College of Human Ecology, University of the Philippines Los Baños College, 4031 Laguna, Philippines
e-mail: arsalvacion@up.edu.ph

 
P. Kumar et al. (eds.), *Water Resources Management and Sustainability*,
Advances in Geographical and Environmental Sciences,
https://doi.org/10.1007/978-981-16-6573-8_13

However, despite its importance, it has not been appropriately managed, and limited information about groundwater exist (Ajami 2020; Shrestha and Pandey 2016).

Mapping spatial patterns of groundwater occurrence are crucial for water resource management (Chen et al. 2019; Miraki et al. 2019). Delineation of potential groundwater zones is a critical input for successful groundwater identification, management, and protection programs (Chen et al. 2018; Naghibi and Pourghasemi 2015; Jha et al. 2010). Traditional approaches of groundwater exploration such as test drilling, stratigraphy analysis, and geophysical techniques are costly, uneconomical, and time-consuming (Jha et al. 2010; Rahmati et al. 2016; Oh et al. 2011). In addition, some of these methods traditional methods do not account for the diverse factors affecting groundwater occurrence (Oh et al. 2011). Recently, several researchers have applied different statistical; and data mining techniques in mapping groundwater potential based on the locations of well, spring, and qanat (Chen et al. 2018; Naghibi and Pourghasemi 2015; Kordestani et al. 2019; Guru et al. 2017). These methods include frequency ratio (Guru et al. 2017; Naghibi et al. 2015b; Lee et al. 2019), weights-of-evidence (Chen et al. 2018; Khoshtinat et al. 2019), index entropy (Naghibi et al. 2015b; Al-Abadi et al. 2016), classification regression tree (Naghibi et al. 2015a; Zhao et al. 2016), random forest (Rahmati et al. 2016; Naghibi et al. 2015a; Kim et al. 2019), and boosted regression tree (Naghibi et al. 2015a; Kim et al. 2019) were among the widely used method for potential groundwater assessment (Naghibi and Pourghasemi 2015; Kordestani et al. 2019; Guru et al. 2017).

### *13.1.1 MaxEnt*

Maximum entropy (MaxEnt) is a general-purpose machine learning method with a simple and precise mathematical design, which is well suited for modeling species geographic distribution based on presence-only data (Phillips et al. 2006). According to Phillips et al. (2006), MaxEnt has many advantages over other modeling methods. Presence-only and environmental data for the entire study region were the only requirements for running MaxEnt. MaxEnt can handle both categorical and continuous input data. Also, MaxEnt can make a spatial prediction from incomplete data (Rahmati et al. 2016; Medley 2010; Moreno et al. 2011). In addition, MaxEnt has an efficient deterministic algorithm and performs better over other methods even with a small sample size of presence data (Wisz et al. 2008). MaxEnt estimates a target probability distribution subject to a set of constraints with respect to incomplete information regarding the target distribution based on maximum entropy (i.e., closest to uniform distribution) (Phillips et al. 2006). In the case of mapping groundwater potential, the model starts with a uniform distribution and performs several iterations (until no further improvement in spatial prediction) using different geo-environmental factors to estimate the probability of occurrence of groundwater source (i.e., wells, springs, or qanat) (Rahmati et al. 2016; Phillips et al. 2006; Merow et al. 2013). A detailed description of MaxEnt can be found elsewhere (Rahmati et al. 2016; Phillips et al. 2006; Merow et al. 2013; Elith et al. 2011).

In the case of groundwater potential mapping, only a few researchers have applied MaxEnt to their study. Rahmati et al. (2016) used MaxEnt to predict the groundwater potential of Mehran Region, Iran. Golkarian and Rahmati (2018) used MaxEnt to determine factors affecting groundwater availability in Gonabad Plain, also in Iran. However, results from Rahmati et al. (2016) showed MaxEnt outperformed random forest predicting groundwater potential.

### 13.1.2 Geo-Environmental Factors of Groundwater Potential

Several geo-environmental factors influence the occurrence, movement, and productivity of groundwater (Rahmati et al. 2016; Oh et al. 2011; Mukherjee 1996). These factors include topography, groundwater table distribution, lithology, geological structure, fracture density, aperture and connectivity of fractures, secondary porosity, drainage pattern, slope, landform, and land use, climatic conditions, and the interrelationship among these factors (Rahmati et al. 2016; Oh et al. 2011; Mukherjee 1996). However, over the past years, others have explored other geo-environmental data set to map groundwater potential. These factors include soil texture (Chen et al. 2018, 2019; Naghibi and Pourghasemi 2015; Oh et al. 2011; Lee et al. 2019; Khoshtinat et al. 2019; Al-Abadi et al. 2016; Kim et al. 2019; Al-Fugara et al. 2020; de Souza et al. 2019; Al-Abadi and Shahid 2015; Ibrahim-Bathis and Ahmed 2016; Thapa et al. 2017), topographic wetness index (TWI) (Chen et al. 2018, 2019; Naghibi and Pourghasemi 2015; Rahmati et al. 2016; Oh et al. 2011; Naghibi et al. 2015b; Al-Abadi et al. 2016; Kim et al. 2019; Golkarian and Rahmati 2018; Nampak et al. 2014; Razandi et al. 2015), distance to river/stream (Chen et al. 2018, 2019; Naghibi and Pourghasemi 2015; Rahmati et al. 2016; Oh et al. 2011; Naghibi et al. 2015b; Khoshtinat et al. 2019; Kim et al. 2019; Golkarian and Rahmati 2018; Razandi et al. 2015), drainage density (Rahmati et al. 2016; Guru et al. 2017; Naghibi et al. (2015a, b); Golkarian and Rahmati 2018; Ibrahim-Bathis and Ahmed 2016; Thapa et al. 2017; Nampak et al. 2014; Razandi et al. 2015), and distance to fault (Naghibi and Pourghasemi 2015; Khoshtinat et al. 2019; Al-Abadi et al. 2016; Naghibi et al. 2015a; Golkarian and Rahmati 2018; Al-Fugara et al. 2020). Soil texture controls the amount of surface water that can infiltrate and recharge groundwater affecting the storage and yield of the aquifer system (Chen et al. 2018; Al-Abadi and Shahid 2015). As infiltration increases, groundwater storage increases (Ibrahim-Bathis and Ahmed 2016). Grain shape, grain size, void ratio, adsorbed water, impurities, and the degree of saturation affect soil permeability and effective porosity (Thapa et al. 2017). Clay and silt soils exhibit low permeability compared with fine sand and loamy soils (Thapa et al. 2017). Meanwhile, TWI connects the upslope area as a measure of water flowing toward a certain point (local slope) or a measure of subsurface lateral transmissivity (Oh et al. 2011; Nampak et al. 2014; Beven and Kirkby 1979). TWI influence the size and zoning of saturated areas, thereby affecting the occurrence of springs (Chen et al. 2019; Nampak et al. 2014; Pourtaghi and Pourghasemi 2014). The higher TWI indicates higher groundwater potential (Chen et al. 2019; Nampak et al.

2014; Pourtaghi and Pourghasemi 2014). Areas closer to rivers, especially those with permanent or prolonged flows, have higher groundwater potential (Oh et al. 2011; Golkarian and Rahmati 2018). Drainage density is the ratio of the total length of stream segment of all orders per unit area and indicates the closeness of spacing of stream channels (Ibrahim-Bathis and Ahmed 2016; Razandi et al. 2015; Singh et al. 2014). High drainage density causes lower infiltration and increased surface runoff resulting in lower groundwater potential (Razandi et al. 2015; Kumar et al. 2007). Faults are fractures and cracks on the soil and rocks where water can easily infiltrate; thus, close distance to fault can result in higher groundwater potential (Khoshtinat et al. 2019).

### *13.1.3 Objective of the Chapter*

This chapter aims to demonstrate the use of MaxEnt for delineating groundwater potential in the province of Marinduque, Philippines. The study used locations of wells and different publicly available environmental data.

## 13.2 Methodology

### *13.2.1 Study Site*

Marinduque is an island province approximately 188 km south of the country's capital (Manila) is considered the geographic center and heart of the Philippines (Salvacion 2017, 2019, 2020; Salvacion and Magcale-Macandog 2015). The province is around 96 thousand hectares and consist of 6 municipalities (Salvacion and Magcale-Macandog 2015; Salvacion 2016). The terrain of Marinduque is predominantly rolling to steep slope mountainous regions in the inner part and low-lying coastal areas in the outer portion (Salvacion 2016).

### *13.2.2 Data*

Locations of wells were extracted from the website (http://www.nwrb.gov.ph/index.php/products-and-services/water-permittees) of the National Water Resource Board of the Philippines. In the case of Marinduque, there are around 27 registered groundwater wells distributed around the province (Fig. 13.1). Meanwhile, environmental variables (Table 13.1) were downloaded or digitized from existing publicly available data sets. The elevation data of the province was extracted from the advanced spaceborne thermal emission and reflection radiometer (ASTER) global digital elevation

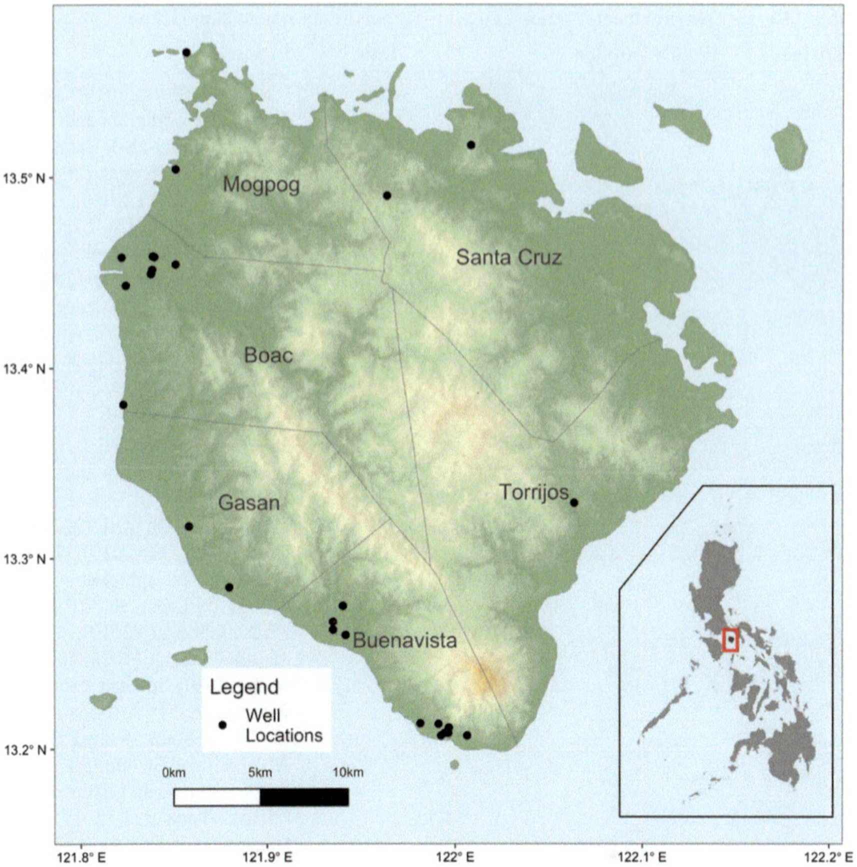


**Fig. 13.1** Location map of Marinduque, Philippines showing well locations

model (GDEM) (https://asterweb.jpl.nasa.gov/gdem.asp) (Abrams 2000; Abrams et al. 2015). This elevation data was the input to generate other topographic features such as slope, topographic wetness index (TWI), and drainage density (Salvacion 2016). The river network map of Marinduque was from the Philippine GIS data clearing house (http://philgis.org/), while the existing fault line map of the province was from the Philippine Institute of Volcanology and Seismology hazard maps (http://www.phivolcs.dost.gov.ph/index.php/gisweb-hazard-maps). Calculations of the distances to the river networks and fault lines were from the digitized maps of rivers and fault lines of the province. The landcover data was extracted from Philippine Geoportal (http://www.geoportal.gov.ph/). Rainfall data for Marinduque was from the latest WorldClim 2.0 database (Fick and Hijmans 2017). Soil texture data (% sand, % clay, and % silt) for Marinduque were from SoilGrid (https://soilgrids.org/) database (Hengl et al. 2017). All the environmental data was set or resampled to 30 m resolution following the spatial resolution of digital elevation data.

**Table 13.1** Geo-environmental variables used for groundwater potential modeling

| Variables | Description | Unit | References |
|---|---|---|---|
| Elevation | Elevation | masl | Chen et al. (2018, 2019), Miraki et al. (2019), Naghibi and Pourghasemi (2015), Rahmati et al. (2016), Oh et al. (2011), Kordestani et al. (2019), Naghibi et al. (2015a, b), Lee et al. (2019), Khoshtinat et al. (2019), Al-Abadi et al. (2016), Al-Fugara et al. (2020), de Souza et al. (2019), Al-Abadi and Shahid (2015), Thapa et al. (2017), Nampak et al. (2014), Razandi et al. (2015) |
| Slope | Slope | ° | Chen et al. (2018, 2019), Miraki et al. (2019), Naghibi and Pourghasemi (2015), Rahmati et al. (2016), Oh et al. (2011), Kordestani et al. (2019), Guru et al. (2017), Naghibi et al. (2015a, b), Lee et al. (2019), Khoshtinat et al. (2019), Al-Abadi et al. (2016), Kim et al. (2019), Golkarian and Rahmati (2018), Al-Fugara et al. (2020), de Souza et al. (2019), Al-Abadi and Shahid (2015), Ibrahim-Bathis and Ahmed (2016), Thapa et al. (2017), Nampak et al. (2014), Razandi et al. (2015) |
| TWI | Topographic wetness index | – | Chen et al. (2018, 2019), Naghibi and Pourghasemi (2015), Rahmati et al. (2016), Oh et al. (2011), Naghibi et al. (2015b), Al-Abadi et al. (2016), Kim et al. (2019), Golkarian and Rahmati (2018), Nampak et al. (2014), Razandi et al. (2015) |
| Drainage density | Drainage density | $km/km^2$ | Rahmati et al. (2016), Guru et al. (2017), Naghibi et al. (2015a, b), Golkarian and Rahmati (2018), Ibrahim-Bathis and Ahmed (2016), Thapa et al. (2017), Nampak et al. (2014), Razandi et al. (2015) |

(continued)

**Table 13.1** (continued)

| Variables | Description | Unit | References |
|---|---|---|---|
| Landcover | Landcover | – | Chen et al. (2018, 2019), Miraki et al. (2019), Naghibi and Pourghasemi (2015), Rahmati et al. (2016), Kordestani et al. (2019), Guru et al. (2017), Naghibi et al. (2015a, b), Lee et al. (2019), Khoshtinat et al. (2019), Al-Abadi et al. (2016), Kim et al. (2019), Golkarian and Rahmati (2018), Ibrahim-Bathis and Ahmed (2016), Thapa et al. (2017), Nampak et al. (2014) |
| % Sand | Sand content (50–2000 μm) mass fraction | % | Chen et al. (2018, 2019), Naghibi and Pourghasemi (2015), Oh et al. (2011), Lee et al. (2019), Khoshtinat et al. (2019), Al-Abadi et al. (2016), Kim et al. (2019), Al-Fugara et al. (2020), de Souza et al. (2019), Al-Abadi and Shahid (2015), Ibrahim-Bathis and Ahmed (2016), Thapa et al. (2017), Nampak et al. (2014) |
| % Silt | Silt content (2–50 μm) mass fraction | % | Chen et al. (2018, 2019), Naghibi and Pourghasemi (2015), Oh et al. (2011), Lee et al. (2019), Khoshtinat et al. (2019), Al-Abadi et al. (2016), Kim et al. (2019), Al-Fugara et al. (2020), de Souza et al. (2019), Al-Abadi and Shahid (2015), Ibrahim-Bathis and Ahmed (2016), Thapa et al. (2017), Nampak et al. (2014) |
| % Clay | Clay content (0–2 μm) mass fraction | % | Chen et al. (2018, 2019), Naghibi and Pourghasemi (2015), Oh et al. (2011), Lee et al. (2019), Khoshtinat et al. (2019), Al-Abadi et al. (2016), Kim et al. (2019), Al-Fugara et al. (2020), de Souza et al. (2019), Al-Abadi and Shahid (2015), Ibrahim-Bathis and Ahmed (2016), Thapa et al. (2017), Nampak et al. (2014) |

(continued)

**Table 13.1** (continued)

| Variables | Description | Unit | References |
|---|---|---|---|
| Rainfall | Total annual rainfall | mm/year | Chen et al. (2019), Miraki et al. (2019), Lee et al. (2019), Khoshtinat et al. (2019), Zhao et al. (2016), Al-Fugara et al. (2020), de Souza et al. (2019), Ibrahim-Bathis and Ahmed (2016), Thapa et al. (2017), Nampak et al. (2014), Razandi et al. (2015) |
| Distance to fault | Euclidean distance to fault | m | Naghibi and Pourghasemi (2015), Khoshtinat et al. (2019), Al-Abadi et al. (2016), Naghibi et al. (2015a), Golkarian and Rahmati (2018), Al-Fugara et al. (2020), Al-Abadi and Shahid (2015) |
| Distance to river | Euclidean distance to river | m | Chen et al. (2018, 2019), Naghibi and Pourghasemi (2015), Rahmati et al. (2016), Oh et al. (2011), Naghibi et al. (2015b), Khoshtinat et al. (2019), Kim et al. (2019), Golkarian and Rahmati (2018), Razandi et al. (2015) |

### *13.2.3 Groundwater Potential Mapping*

MaxEnt modeling was done via the R-software (Ihaka and Gentleman 1996) using the *ENMeval* package for model parameter optimization (Muscarella et al. 2014). In addition, the random k-fold method was used to partition the training and test data set for cross-validation. On the other hand, geo-environmental variables in the predictor with permutation importance equal to zero per model run were removed (Zeng et al. 2016). This procedure was repeated until no more environmental predictor has permutation importance equal to zero. Groundwater potential map was generated using the remaining geo-environmental predictors. The predicted groundwater potential map was classified into quantile following the methodology of Rahmati et al. (2016).

## 13.3 Results and Discussion

### *13.3.1 Variable Selection and Model Optimization*

Out of the 11 geo-environmental variables, five (5) remained in the final MaxEnt model (Table 13.2). These variables include elevation (Fig. 13.2), annual rainfall (Fig. 13.3), landcover (Fig. 13.4), % clay (Fig. 13.5), and % silt (Fig. 13.6). On average, the final model AUCs were 0.89 and 0.85 for the training and test data,

**Table 13.2** Permutation importance of geo-environmental variables used in the final MaxEnt model

| Variables | Permutation importance |
|---|---|
| Elevation | 69.1 |
| Rainfall | 21.8 |
| Landcover | 4.8 |
| % Clay | 2.9 |
| % Silt | 1.3 |

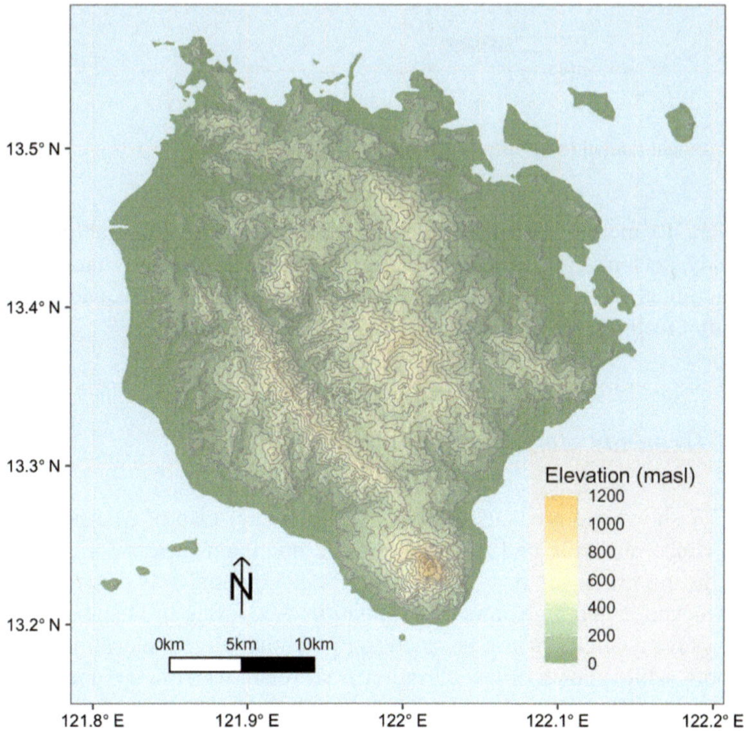


**Fig. 13.2** Elevation map of Marinduque, Philippines

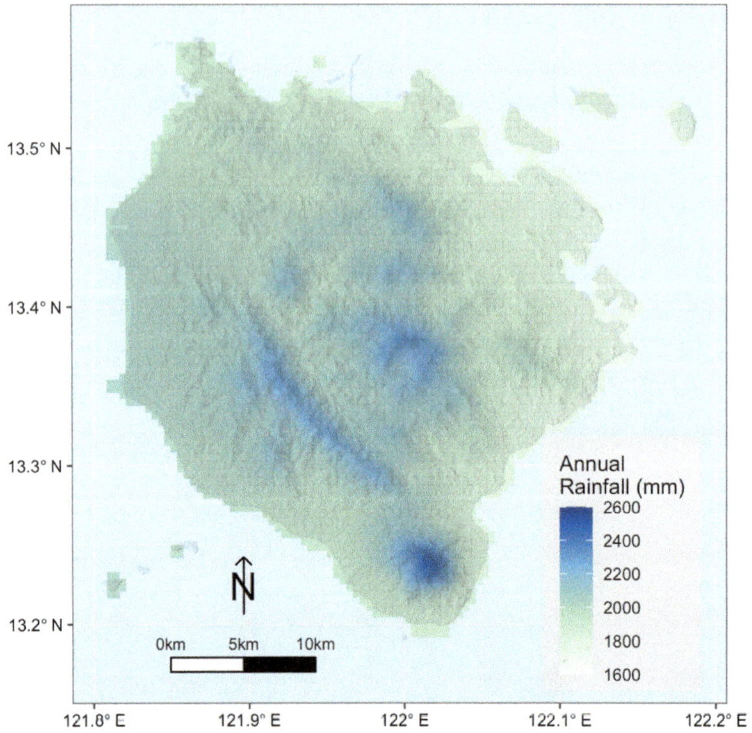


**Fig. 13.3** Annual rainfall (mm) map of Marinduque, Philippines

respectively. Its means that the MaxEnt model for groundwater potential mapping, in this study, performs very well with respect to the training and test data (Guru et al. 2017; Naghibi et al. 2015b). Lastly, the final MaxEnt model used a quadratic feature combination technique with a regularization multiplier of 1.5.

## *13.3.2 Groundwater Potential Map*

Figure 13.7 shows the predicted groundwater potential map of Marinduque using MaxEnt. Since elevation and rainfall were the dominant factors for groundwater potential in the province, very high groundwater potential was observed in areas with low elevation and high annual precipitation. According to Thapa et al. (2017), elevation plays a crucial role in groundwater potential. Higher elevation areas have more runoff, while plainer or low elevation areas retain water longer and have more recharge and higher infiltration rate (Miraki et al. 2019; Thapa et al. 2017; Manap

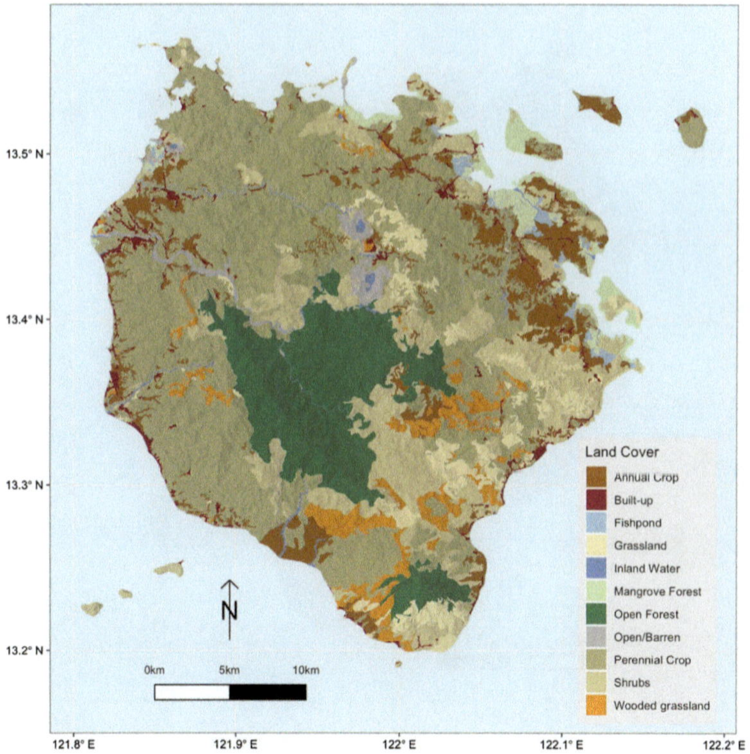


**Fig. 13.4** Landcover map of Marinduque, Philippines

et al. 2014). In addition, elevation also influences other factors (i.e., rainfall, soil, and vegetation) that affect groundwater potential (Al-Abadi et al. 2016; Al-Abadi and Shahid 2015; Aniya 1985). Likewise, rainfall is one of the most important sources and determinants of the amount of water that will recharge the groundwater system via percolation (Thapa et al. 2017; Nampak et al. 2014; Razandi et al. 2015). Most of the low elevation and high rainfall areas in Marinduque were in the western portion of the province. The highest proportion of very high groundwater potential was located in Boac (33%), while the highest proportion of areas with low groundwater potential was in the municipality of Torrijos (35%). Table 13.3 summarizes the distribution and percentage of groundwater potential for each town in Marinduque.

## 13.4 Conclusion

Mapping potential zones is a crucial step in the development of the groundwater resource management program. Groundwater potential zone map can provide important information necessary to formulate programs for monitoring, protection, and

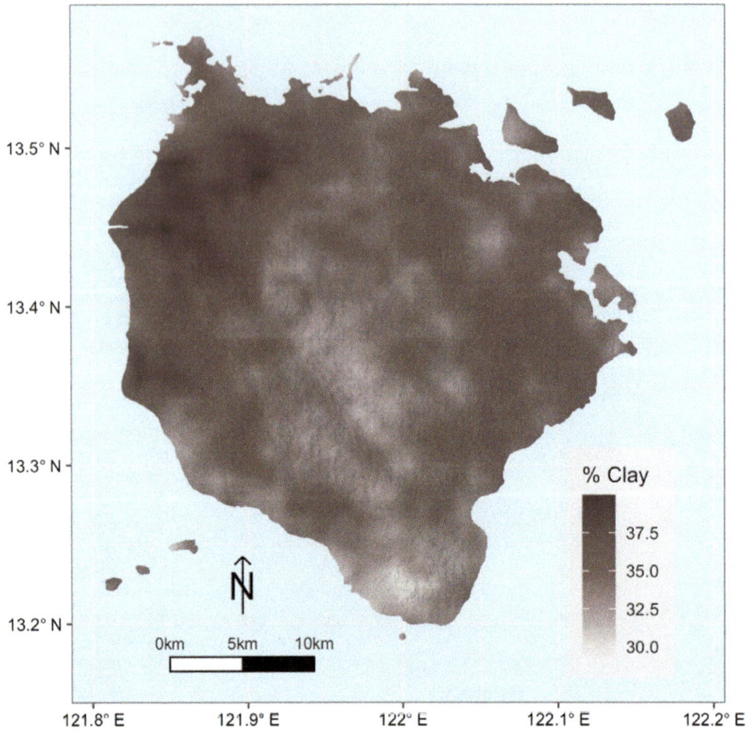


**Fig. 13.5** Percent (%) clay map of Marinduque, Philippines

management of groundwater resources of an area. Using a machine learning approach such as MaxEnt with publicly available geospatial data of groundwater determinants (e.g., elevation, rainfall, soil, and landcover) can provide a fast and cheap alternative to traditional approaches for groundwater assessment. This paper demonstrated the use of the above-mentioned techniques to delineate groundwater potential in Marinduque, Philippines. The result of this paper can guide the local government and concerned institutions in the province to start or develop a comprehensive groundwater resource management plan. Meanwhile, the approach used in this study can also be applied elsewhere since most of the data used in this study were available online, and there is a rich resource on the internet on implementing the MaxEnt model.

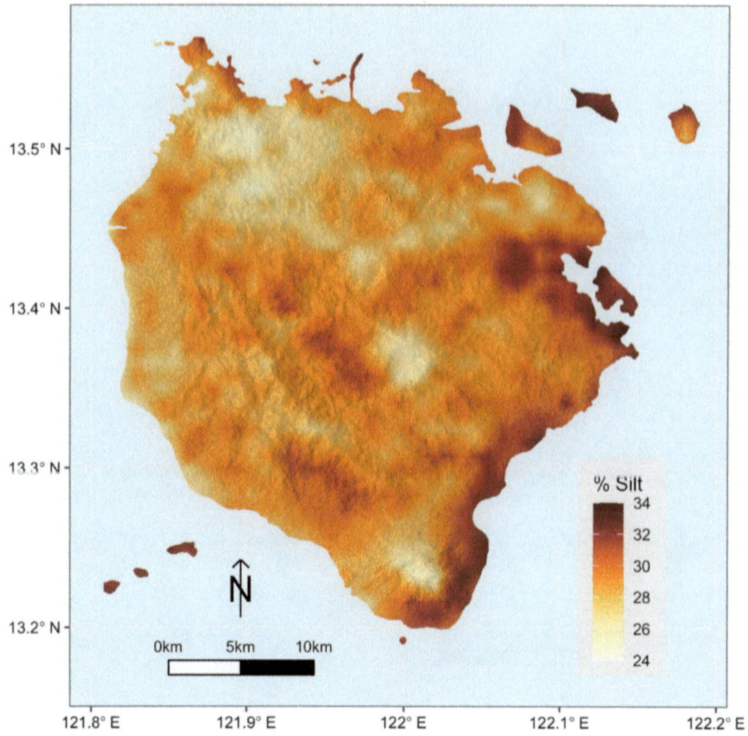


**Fig. 13.6** Percent (%) silt map of Marinduque, Philippines

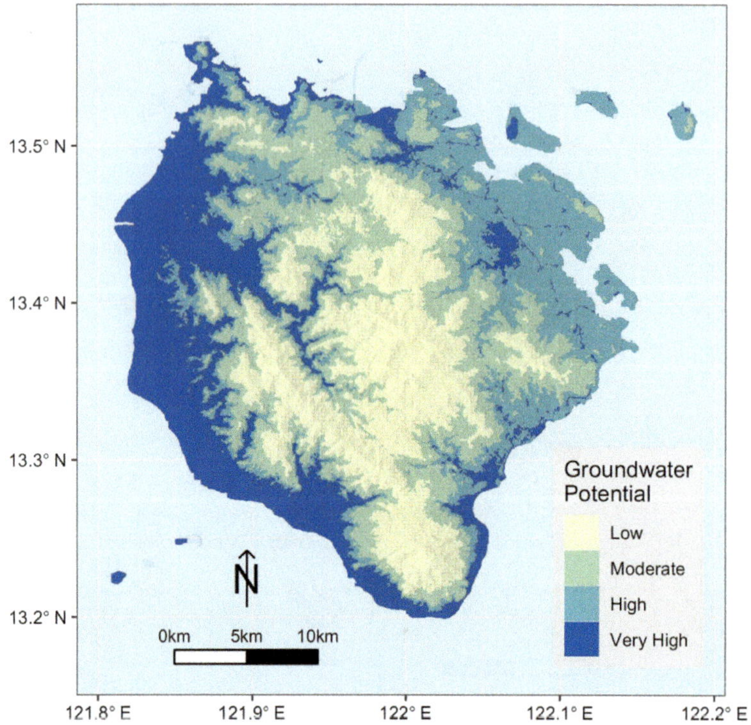


**Fig. 13.7** Groundwater potential map of Marinduque, Philippines

**Table 13.3** Area (ha) and proportion (%) of different groundwater potential class per municipality of Marinduque, Philippines

| Municipality | Groundwater potential | | | |
|---|---|---|---|---|
| | Low | Moderate | High | Very high |
| Boac | 3953 (22) | 4287 (24) | 2078 (12) | 7410 (42) |
| Buenavista | 2836 (35) | 1466 (18) | 688 (9) | 3075 (38) |
| Gasan | 2985 (27) | 1981 (18) | 930 (8) | 5219 (47) |
| Mogpog | 524 (5) | 3482 (36) | 2371 (25) | 3216 (34) |
| Santa Cruz | 3477 (15) | 6061 (27) | 10,977 (49) | 1957 (9) |
| Torrijos | 8520 (42) | 5017 (25) | 5251 (26) | 1418 (7) |

## References


Abrams M (2000) The advanced spaceborne thermal emission and reflection radiometer (ASTER): data products for the high spatial resolution imager on NASA's Terra platform. Int J Remote Sens 21:847–859. https://doi.org/10.1080/014311600210326

Abrams M, Tsu H, Hulley G, Iwao K, Pieri D, Cudahy T, Kargel J (2015) The advanced spaceborne thermal emission and reflection radiometer (ASTER) after fifteen years: review of global products. Int J Appl Earth Obs Geoinf 38:292–301. https://doi.org/10.1016/j.jag.2015.01.013

Ajami H (2020) Encyclopedia of geology. In: Reference module in earth systems and environmental sciences. Elsevier. https://doi.org/10.1016/B978-0-12-409548-9.12388-7

Al-Abadi AM, Shahid S (2015) A comparison between index of entropy and catastrophe theory methods for mapping groundwater potential in an arid region. Environ Monit Assess 187:576. https://doi.org/10.1007/s10661-015-4801-2

Al-Abadi AM, Al-Temmeme AA, Al-Ghanimy MA (2016) A GIS-based combining of frequency ratio and index of entropy approaches for mapping groundwater availability zones at Badra–Al Al-Gharbi–Teeb areas, Iraq. Sustain Water Resour Manag 2:265–283. https://doi.org/10.1007/s40899-016-0056-5

Al-Fugara A, Pourghasemi HR, Al-Shabeeb AR, Habib M, Al-Adamat R, Al-Amoush H, Collins AL (2020) A comparison of machine learning models for the mapping of groundwater spring potential. Environ Earth Sci 79:206. https://doi.org/10.1007/s12665-020-08944-1

Alley WM (2009) Ground water. In: Likens GE (ed) Encyclopedia of inland waters. Academic Press, Oxford, pp 684–690. https://doi.org/10.1016/B978-012370626-3.00015-6

Aniya M (1985) Landslide-susceptibility mapping in the Amahata river basin, Japan. Ann Assoc Am Geogr 75:102–114. https://doi.org/10.1111/j.1467-8306.1985.tb00061.x

Beckie RD (2013) Groundwater. In: Reference module in earth systems and environmental sciences. Elsevier. https://doi.org/10.1016/B978-0-12-409548-9.05923-6

Beven KJ, Kirkby MJ (1979) A physically based, variable contributing area model of basin hydrology. Hydrol Sci Bull 24:43–69

Chen W, Li H, Hou E, Wang S, Wang G, Panahi M, Li T, Peng T, Guo C, Niu C, Xiao L, Wang J, Xie X, Ahmad BB (2018) GIS-based groundwater potential analysis using novel ensemble weights-of-evidence with logistic regression and functional tree models. Sci Total Environ 634:853–867. https://doi.org/10.1016/j.scitotenv.2018.04.055

Chen W, Tsangaratos P, Ilia I, Duan Z, Chen X (2019) Groundwater spring potential mapping using population-based evolutionary algorithms and data mining methods. Sci Total Environ 684:31–49. https://doi.org/10.1016/j.scitotenv.2019.05.312

de Souza E, Pontes LM, Fernandes Filho EI, Schaefer CEGR, dos Santos EE (2019) Spatial and temporal potential groundwater recharge: the case of the Doce River Basin, Brazil. Rev Bras Ciênc Solo 43:e0180010. https://doi.org/10.1590/18069657rbcs20180010

Elith J, Phillips SJ, Hastie T, Dudík M, Chee YE, Yates CJ (2011) A statistical explanation of MaxEnt for ecologists. Divers Distrib 17:43–57. https://doi.org/10.1111/j.1472-4642.2010.00725.x

Fick SE, Hijmans RJ (2017) WorldClim 2: new 1-km spatial resolution climate surfaces for global land areas. Int J Climatol 37:4302–4315. https://doi.org/10.1002/joc.5086

Flores RM (2014) Co-produced water management and environmental impacts, Chap 8. In: Flores RM (ed) Coal and coalbed gas. Elsevier, Boston, pp 437–508. https://doi.org/10.1016/B978-0-12-396972-9.00008-2

Golkarian A, Rahmati O (2018) Use of a maximum entropy model to identify the key factors that influence groundwater availability on the Gonabad Plain, Iran. Environ Earth Sci 77:369. https://doi.org/10.1007/s12665-018-7551-y

Guru B, Seshan K, Bera S (2017) Frequency ratio model for groundwater potential mapping and its sustainable management in cold desert, India. J King Saud Univ Sci 29:333–347. https://doi.org/10.1016/j.jksus.2016.08.003

Hengl T, de Jesus JM, Heuvelink GBM, Gonzalez MR, Kilibarda M, Blagotić A, Shangguan W, Wright MN, Geng X, Bauer-Marschallinger B, Guevara MA, Vargas R, MacMillan RA, Batjes NH, Leenaars JGB, Ribeiro E, Wheeler I, Mantel S, Kempen B (2017) SoilGrids250m: global gridded soil information based on machine learning. PLoS ONE 12:e0169748. https://doi.org/10.1371/journal.pone.0169748

Ibrahim-Bathis K, Ahmed SA (2016) Geospatial technology for delineating groundwater potential zones in Doddahalla watershed of Chitradurga district, India. Egypt J Remote Sens Space Sci 19:223–234. https://doi.org/10.1016/j.ejrs.2016.06.002

Ihaka R, Gentleman R (1996) R: a language for data analysis and graphics. J Comput Graph Stat 5(3):299–314. https://doi.org/10.1080/10618600.1996.10474713

Jha MK, Chowdary VM, Chowdhury A (2010) Groundwater assessment in Salboni Block, West Bengal (India) using remote sensing, geographical information system and multi-criteria decision analysis techniques. Hydrogeol J 18:1713–1728. https://doi.org/10.1007/s10040-010-0631-z

Khoshtinat S, Aminnejad B, Hassanzadeh Y, Ahmadi H (2019) Application of GIS-based models of weights of evidence, weighting factor, and statistical index in spatial modeling of groundwater. J Hydroinform 21:745–760. https://doi.org/10.2166/hydro.2019.127

Kim J-C, Jung H-S, Lee S (2019) Spatial mapping of the groundwater potential of the geum river basin using ensemble models based on remote sensing images. Remote Sens 11:2285. https://doi.org/10.3390/rs11192285

Kordestani MD, Naghibi SA, Hashemi H, Ahmadi K, Kalantar B, Pradhan B (2019) Groundwater potential mapping using a novel data-mining ensemble model. Hydrogeol J 27:211–224. https://doi.org/10.1007/s10040-018-1848-5

Kumar PKD, Gopinath G, Seralathan P (2007) Application of remote sensing and GIS for the demarcation of groundwater potential zones of a river basin in Kerala, southwest coast of India. Int J Remote Sens 28:5583–5601. https://doi.org/10.1080/01431160601086050

Lee S, Hyun Y, Lee M-J (2019) Groundwater potential mapping using data mining models of big data analysis in Goyang-si, South Korea. Sustainability 11:1678. https://doi.org/10.3390/su11061678

Li S-Q, Li Y-X, Liu ZH (2013) Hebei rural groundwater contamination and integrated control. J Cangzhou Norm Univ 29:8–10

Manap MA, Nampak H, Pradhan B, Lee S, Sulaiman WNA, Ramli MF (2014) Application of probabilistic-based frequency ratio model in groundwater potential mapping using remote sensing data and GIS. Arab J Geosci 7:711–724. https://doi.org/10.1007/s12517-012-0795-z

Medley KA (2010) Niche shifts during the global invasion of the Asian tiger mosquito, Aedes albopictus Skuse (Culicidae), revealed by reciprocal distribution models. Glob Ecol Biogeogr 19:122–133. https://doi.org/10.1111/j.1466-8238.2009.00497.x

Merow C, Smith MJ, Silander JA (2013) A practical guide to MaxEnt for modeling species' distributions: what it does, and why inputs and settings matter. Ecography 36:1058–1069. https://doi.org/10.1111/j.1600-0587.2013.07872.x

Miraki S, Zanganeh SH, Chapi K, Singh VP, Shirzadi A, Shahabi H, Pham BT (2019) Mapping groundwater potential using a novel hybrid intelligence approach. Water Resour Manag 33:281–302. https://doi.org/10.1007/s11269-018-2102-6

Moreno R, Zamora R, Molina JR, Vasquez A, Herrera MT (2011) Predictive modeling of microhabitats for endemic birds in South Chilean temperate forests using Maximum entropy (Maxent). Ecol Inform 6:364–370. https://doi.org/10.1016/j.ecoinf.2011.07.003

Mukherjee S (1996) Targetting saline aquifer by remote sensing and geophysical methods in a part of Hamirpur-Kanpur, India. Hydrol J 19:1867–1884

Muscarella R, Galante PJ, Soley-Guardia M, Boria RA, Kass JM, Uriarte M, Anderson RP (2014) ENMeval: an R package for conducting spatially independent evaluations and estimating optimal model complexity for Maxent ecological niche models. Methods Ecol Evol 5:1198–1205. https://doi.org/10.1111/2041-210X.12261

Naghibi SA, Pourghasemi HR (2015) A comparative assessment between three machine learning models and their performance comparison by bivariate and multivariate statistical methods in

groundwater potential mapping. Water Resour Manag 29:5217–5236. https://doi.org/10.1007/s11269-015-1114-8
Naghibi SA, Pourghasemi HR, Dixon B (2015a) GIS-based groundwater potential mapping using boosted regression tree, classification and regression tree, and random forest machine learning models in Iran. Environ Monit Assess 188:44. https://doi.org/10.1007/s10661-015-5049-6
Naghibi SA, Pourghasemi HR, Pourtaghi ZS, Rezaei A (2015b) Groundwater qanat potential mapping using frequency ratio and Shannon's entropy models in the Moghan watershed, Iran. Earth Sci Inform 8:171–186. https://doi.org/10.1007/s12145-014-0145-7
Nampak H, Pradhan B, Manap MA (2014) Application of GIS based data driven evidential belief function model to predict groundwater potential zonation. J Hydrol 513:283–300. https://doi.org/10.1016/j.jhydrol.2014.02.053
Oh H-J, Kim Y-S, Choi J-K, Park E, Lee S (2011) GIS mapping of regional probabilistic groundwater potential in the area of Pohang City, Korea. J Hydrol 399:158–172. https://doi.org/10.1016/j.jhydrol.2010.12.027
Phillips SJ, Anderson RP, Schapire RE (2006) Maximum entropy modeling of species geographic distributions. Ecol Model 190:231–259. https://doi.org/10.1016/j.ecolmodel.2005.03.026
Pourtaghi ZS, Pourghasemi HR (2014) GIS-based groundwater spring potential assessment and mapping in the Birjand Township, southern Khorasan Province, Iran. Hydrogeol J 22:643–662. https://doi.org/10.1007/s10040-013-1089-6
Rahmati O, Pourghasemi HR, Melesse AM (2016) Application of GIS-based data driven random forest and maximum entropy models for groundwater potential mapping: a case study at Mehran Region, Iran. CATENA 137:360–372. https://doi.org/10.1016/j.catena.2015.10.010
Razandi Y, Pourghasemi HR, Neisani NS, Rahmati O (2015) Application of analytical hierarchy process, frequency ratio, and certainty factor models for groundwater potential mapping using GIS. Earth Sci Inform 8:867–883. https://doi.org/10.1007/s12145-015-0220-8
Salvacion AR (2016) Terrain characterization of small island using publicly available data and open-source software: a case study of Marinduque, Philippines. Model Earth Syst Environ 2:1–9. https://doi.org/10.1007/s40808-016-0085-y
Salvacion AR (2017) Exploring determinants of child malnutrition in Marinduque Island, Philippines. Hum Ecol 45:853–863. https://doi.org/10.1007/s10745-017-9951-0
Salvacion AR (2019) Mapping land limitations for agricultural land use planning using fuzzy logic approach: a case study for Marinduque Island, Philippines. GeoJournal. https://doi.org/10.1007/s10708-019-10103-4
Salvacion AR (2020) Spatial pattern and determinants of village level poverty in Marinduque Island, Philippines. GeoJournal 85:257–267. https://doi.org/10.1007/s10708-018-9944-6
Salvacion AR, Magcale-Macandog DB (2015) Spatial analysis of human population distribution and growth in Marinduque Island, Philippines. J Mar Isl Cult 4:27–33. https://doi.org/10.1016/j.imic.2015.06.003
Shrestha S, Pandey VP (2016) Groundwater as an environmental issue in Asian cities, Chap 1. In: Shrestha S, Pandey VP, Shivakoti BR, Thatikonda S (eds) Groundwater environment in Asian cities. Butterworth-Heinemann, pp 1–13. https://doi.org/10.1016/B978-0-12-803166-7.00001-5
Singh P, Gupta A, Singh M (2014) Hydrological inferences from watershed analysis for water resource management using remote sensing and GIS techniques. Egypt J Remote Sens Space Sci 17:111–121. https://doi.org/10.1016/j.ejrs.2014.09.003
Tang Y, Zhou J, Yang P, Yan J, Zhou N (2017) Groundwater. In: Tang Y, Zhou J, Yang P, Yan J, Zhou N (eds) Groundwater engineering. Springer, Singapore, pp 1–34. https://doi.org/10.1007/978-981-10-0669-2_1
Thapa R, Gupta S, Guin S, Kaur H (2017) Assessment of groundwater potential zones using multi-influencing factor (MIF) and GIS: a case study from Birbhum district, West Bengal. Appl Water Sci 7:4117–4131. https://doi.org/10.1007/s13201-017-0571-z
Waller P, Yitayew M (2016) Groundwater. In: Waller P, Yitayew M (eds) Irrigation and drainage engineering. Springer International Publishing, Cham, pp 169–183. https://doi.org/10.1007/978-3-319-05699-9_10

Wisz MS, Hijmans RJ, Li J, Peterson AT, Graham CH, Guisan A (2008) NCEAS predicting species distributions working group: effects of sample size on the performance of species distribution models. Divers Distrib 14:763–773. https://doi.org/10.1111/j.1472-4642.2008.00482.x

Zeng Y, Low BW, Yeo DCJ (2016) Novel methods to select environmental variables in MaxEnt: a case study using invasive crayfish. Ecol Model 341:5–13. https://doi.org/10.1016/j.ecolmodel.2016.09.019

Zhao Y, Li Y, Zhang L, Wang Q (2016) Groundwater level prediction of landslide based on classification and regression tree. Geod Geodyn 7:348–355. https://doi.org/10.1016/j.geog.2016.07.005

# Chapter 14
# Application of Remote Sensing and GIS in Floodwater Harvesting for Groundwater Development in the Upper Delta of Cauvery River Basin, Southern India

**Kongeswaran Thangaraj and Sivakumar Karthikeyan**

**Abstract** People living in semiarid areas with inadequate rainfall are frequently affected by water scarcity. Upper delta region of Cauvery River Basin (CRB) in southern India was selected to search suitable areas for floodwater harvesting to induce artificial recharge that improves the groundwater level. The aim of this study is floodwater harvesting based on the technical design and identification of the appropriate locations for artificial recharge structures. Remote sensing and Geographic Information System (GIS) were used to produce the flood hazard map and recommend suitable areas for floodwater harvesting. Thematic layers were prepared and overlaid to determine the flood vulnerable zones and suitable recharge structures were identified based on the hazard map. Burrowing and flooding are the most favorable artificial recharge structures should be implemented in all parts of CRB, whereas battery wells near to the river banks should be built to improve the groundwater level. Hydrologists, decision-makers, and planners will use this appropriate map to quickly identify the locations with the greatest potential for flood water collection. This study concludes that geospatial technology becomes very effective for flood vulnerable zone mapping, floodwater harvesting, and suggesting management plans to improve groundwater level for sustainable development.

**Keywords** Floodwater · Harvesting · GIS · Semi-arid · Sustainable and development

## 14.1 Introduction

Drought is one of the periodically happened phenomena of the Indian agriculture system because water requirement has amplified manifold because of the increasing population, agricultural and commercial expansions (Belal et al. 2014). It affects not best the national food security however also the reasons for discomfort to human

K. Thangaraj · S. Karthikeyan (✉)
Department of Geology, Alagappa University, Karaikudi, Tamil Nadu 630003, India



P. Kumar et al. (eds.), *Water Resources Management and Sustainability*,
Advances in Geographical and Environmental Sciences,
https://doi.org/10.1007/978-981-16-6573-8_14

existence and livestock. Drought is an insidious natural phenomenon that results from an anticipated (or) ordinary deficiency of precipitation so that when it is miles spread over a season or longer period, the amount of precipitation is insufficient to meet the needs of human activities and the environment (Wilhite and Buchanan Smith 2005; Hussain and Javadi 2016).

A qualitative and quantitative appraisal about droughts in the Indian state of Tamil Nadu were analyzed and identified in nine drought-prone districts such as Coimbatore, Pudukottai, Thirunelveli, Ramanathapuram, Madras, Madurai, Salem, Dharmapuri, North Arcot, and Thiruchirapalli (Nathan 1998). It has become the well-understood fact that the monsoon brings enough rainfall in catchment regions which leads to either flooding along river banks or simply let the turbulent water gush into the sea rapidly to avoid flooding and related problems. On the contrary, during the summer season, we suffer from a drought situation due to chronic water deficit because of the ever-increasing water demand for several purposes (Bariweni et al. 2012). While moving aloof from river flood to different water bodies by constructing small channel networks with artificial recharge structures that help for floodwater harvesting. Yet, another fact is also not a new one for this state that to divert major part of monsoon water to suitable areas where groundwater recharge can happen naturally or by artificial means or to store in periodically de-silted lakes, reservoirs, and tanks and strengthening of the river, lake, and canal bunds which was done promptly in the olden days by several kings. Nowadays, floodwater harvesting has been dealt with seriously as an essential practice to cope with today's water needs as well as to minimize groundwater abstraction along river bank sides and to improve the groundwater quality of the region by maintaining minimal flow in the river canals (Ramachandran et al. 2020; Sivakumar et al. 2016).

The use of Geospatial technology is rapidly developing among several international researchers in groundwater potential zones identification, managing the surface water resources, appropriate sites selection for artificial recharge systems and in groundwater recharge estimations (Chandra et al. 2019; Cabrera and Lee 2019; Bagyaraj et al. 2013; Scanlon et al. 2002). Geospatial technology is comprised of remotely sensed data with Geographic Information System (GIS) that is cost-effective and reliable (Selvam et al. 2015; Srivastava et al. 2012; Machiwal et al. 2011). Thus, a geospatial model study has been conducted in parts of upper Cauvery delta of Tiruchirappalli district, located in central Tamil Nadu to demonstrate how to harvest flood water to cater to the water requirements during summer and also to groundwater level depletion which leads to sustainable agriculture development.

### 14.1.1 Study Area

The upper delta of the Cauvery River Basin (CRB) is an active agriculture zone throughout the year. It has highly fertile soil and enough groundwater resources for its productivity. However, in the last two decades, gradual decrease of rainfall and groundwater levels were recorded in this region which will adversely affect

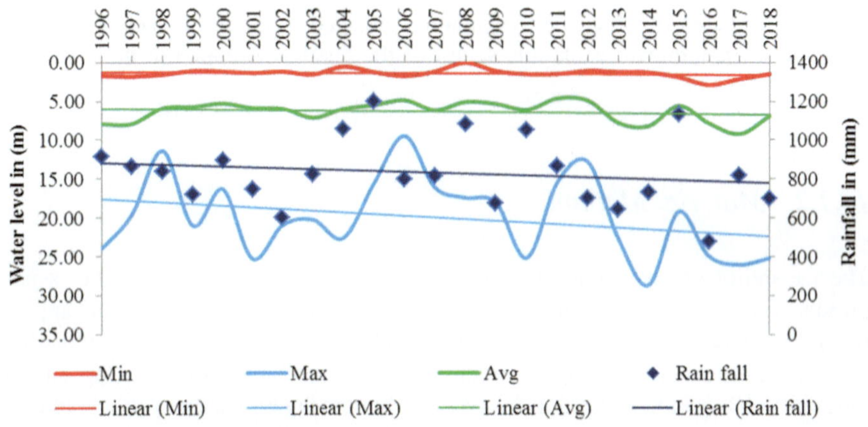


**Fig. 14.1** Shown gradual decrease of rainfall and groundwater level. *Source* CGWD and IMD

agriculture productivity (Fig. 14.1). The study area has been buffered for 10 km on either side of the Cauvery River which is considered as an active area during the monsoon and is a linear strip along the river Cauvery covers approximately 1865 km$^2$ (Fig. 14.2). The area covers in 6 sheets of 1:50,000 scale Survey of India Toposheet (58 J/1, 5, 9, 13, 58 I/4, and 8), contained within the following Latitude N 10° 45′–11° 05′ and Longitude E 78° 10′–78° 55′. The study area climate is tropical with the lowest temperature of 20.6 °C and the highest temperature of noted as 37.2 °C. The total area of Tiruchirappalli District is about 5114 km$^2$ and its population as per Census

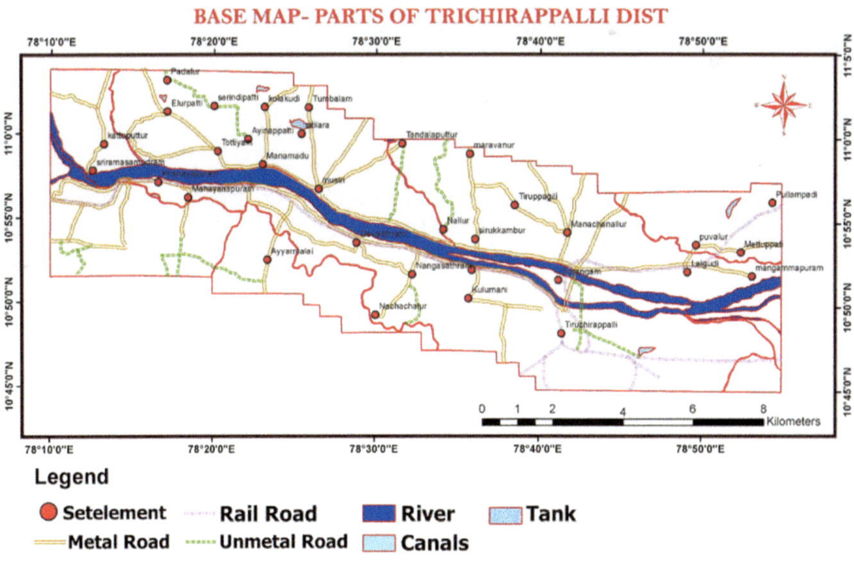


**Fig. 14.2** Study area map

2011 is 2,722,290. In which 70% population engaged in agriculture actvities and it is the most predominant economic sector of the district. The area of net cropping land is 1412.82 $km^2$.

### 14.1.2 *Material Method*

The capability of geospatial technologies was applied to understand the issues present within the study area for mapping flood hazard zones, recommending appropriate techniques for floodwater harvesting, and to recommend acceptable remedial measures to attenuate flooding in Tiruchirapalli district. Survey of India Toposheets indexed 58 J/1, 5, 9, 13 and 58 I/4, 8 used as the bottom layers to prepare the avoidance map and relief map. The thematic layers of lithology, lineament, geomorphology, drainage, and land use/land cover were prepared from the satellite imagery IRS (Indian Remote Sensing) 1C LISS (Linear Imaging Self-Scanning Sensor)–III, Path–101, Row–65 (Sivakumar et al. 2017). The ASTER (Advanced Spaceborne Thermal Emission and Reflection Radiometer) data of DEM (Digital Elevation Model) information was set as base to prepare the relief map to spot the low-lying surfaces within the study area (Jayaprakash et al. 2016). To generate the flood plain map, MODIS (Moderate Resolution Imaging Spectroradiometer) information was used. In order to generate the final flood prone zone map in the GIS, both the flood plain map and the low-lying spot map are combined after the realm is categorized as extreme, moderate, and low flood zone (Fig. 14.3).

## 14.2 Flood Vulnerable Zone Mapping

### 14.2.1 *One Meter Contour Map*

The one-meter contour map was produced with the aid of and geographic information system spatial analyst tools by using ASTER data and interpolated the contours for spot heights. All the 20 m interval contour's height details in the form of triangulation points, benchmarks, and spot heights are available in 1:50,000 scales Toposheets that were digitized separately (Sharma et al. 2009). Then the contour height was converted into point data and integrated with spot heights derived from the ASTER data. The Earth Explorer website (United States Geological Survey—USGS) from downloaded Advanced ASTER. The study area has been vectorized and integrated with the above point data of spot heights. From this final integrated point data, a one-meter contour map was generated to find out the low-lying area to delineate the possible flood vulnerable zones (Fig. 14.4a).

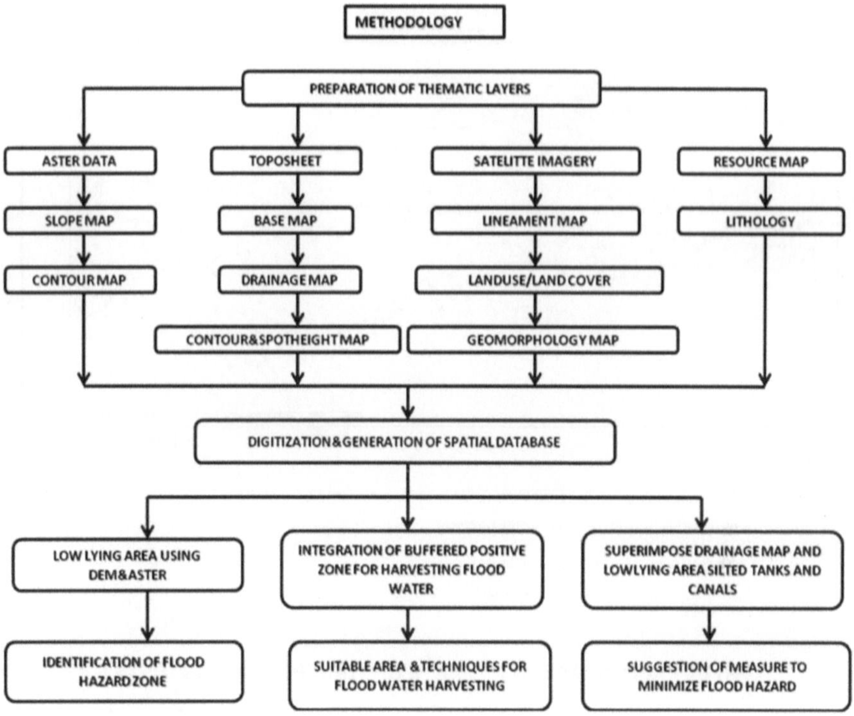


**Fig. 14.3** Flow chart

### *14.2.2 Slope Map*

The slope is a very vital aspect for flood water harvesting that if the slope is higher degree there is a risk for greater runoff and less infiltration and therefore automatically high erosion in the study region (Maina and Raude 2016). The slope map has been produced the usage of ASTER data and classified into 3 classes that are 0–3, 4–5, and 5–50°. The maximum part of the study region is located to be under the Very Shallow Sloping category. The slope map shows the following 3 different slope categories obtained (Fig. 14.4b).

### *14.2.3 Flood Vulnerable Zones*

Using the one-meter contour map, all the low-lying areas concerning the local height of the river bed on either bank and every adjacent place have been digitized as flood vulnerable polygons. Thus, the GIS-based layout has been generated showing the flood vulnerable zones.

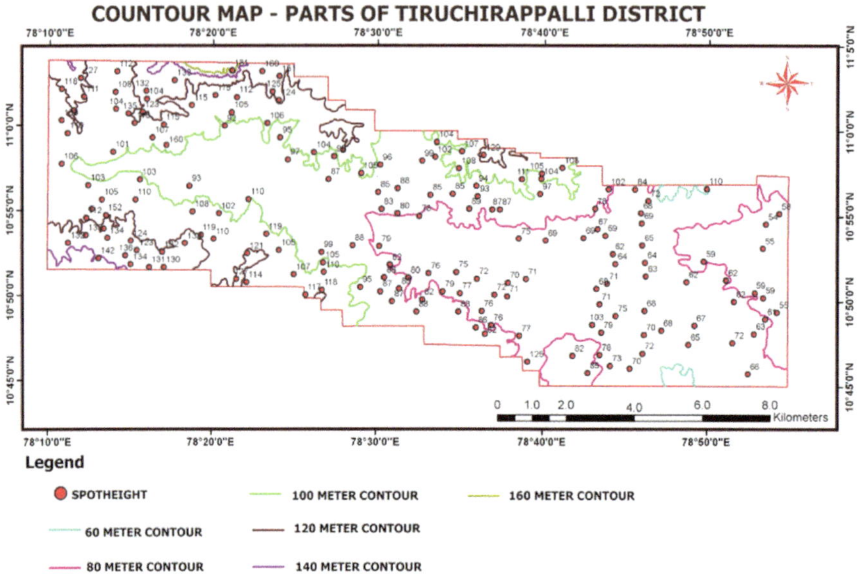


(a)

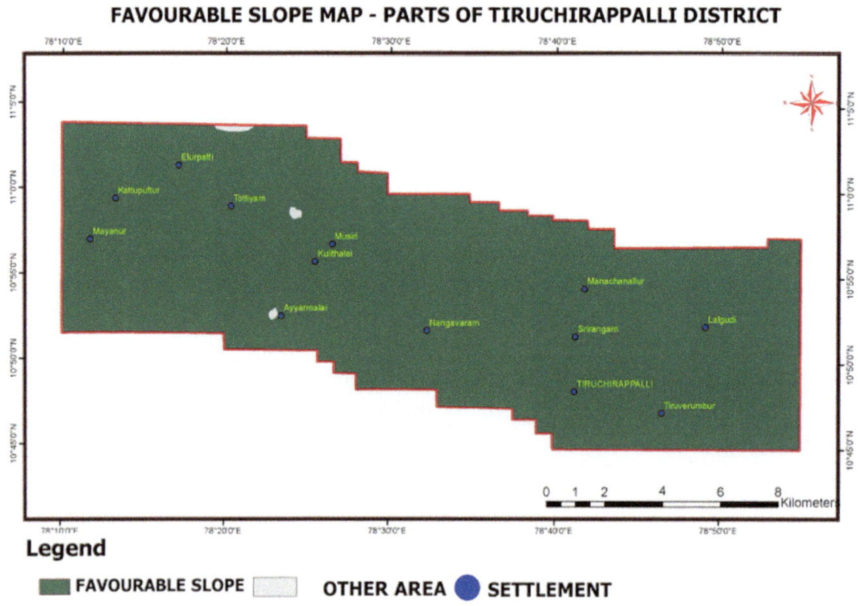


(b)

**Fig. 14.4** **a** Contour map. **b** Favorable slope map of the study area

### 14.2.4 Validation of Flood Vulnerable Zone

The flood zones have been mapped for the past 9 years with 36 spectral bands with MODIS (Moderate Resolution Imaging Spectroradiometer) data ranging in wavelength from 0.4 to 14.4 μm and at varying spatial resolutions (2 bands at 250 m, 5 bands at 500 m, and 29 bands at 1 km). This flood map has been used to verify the commonality of flood polygons prepared in this present study. It is identified that though the web-published flood map is on a very small scale, there is a very good matching of polygons between them (Fig. 14.5a, b).

## 14.3 Floodwater Harvesting and Management—Data Base Generation

### 14.3.1 Lithology

The Lithology map was developed from the Tiruchirappalli district resource map produced by Geological Survey of India (GSI). The following are the geological age-wise different lithology classes that available in the present study area; Recent Alluvium, Pink Migmatites, Fissile Hornblende Gneiss, Granite, Garnet Granulite, Hornblende Biotite Gneiss, and Charnockites (Fig. 14.6a).

### 14.3.2 Lineament

Linear ground features such as fractures, joints, faults, litho contacts, and bedding are expressed together as lineamants. These linear structures influences on infiltraion, conductive and hydrostatic strain development on sloping ground surface (Nagarajan et al. 2010). 88% of the detected landslides have associated with proximity of major faults zones of about 250 m (Gökceoglu and Aksoy 1996). In order to create the lineamant map for the study area, IRS 1C, 1D LISS III satellite imagery was visually interpreted. Linear, rectilinear, and curvilinear characteristics originating in tectonics are derived from satellite data, and these lineaments usually display variations in satellite data in textural, soil tonal, relief, drainage, flower linearity, and curvilinear links. There are faults, joints, and limits between stratigraphic formations in geology. On satellite data, the lineaments are observed by deliberating various terrain characteristics such as drainage association, directly in-stream lines, valley, vegetation-tone variations are ordinary lineaments' geomorphic expressions. N–S, NE–SW, NW–SE, and E–W lineaments are defined in this analysis area (Fig. 14.6b).

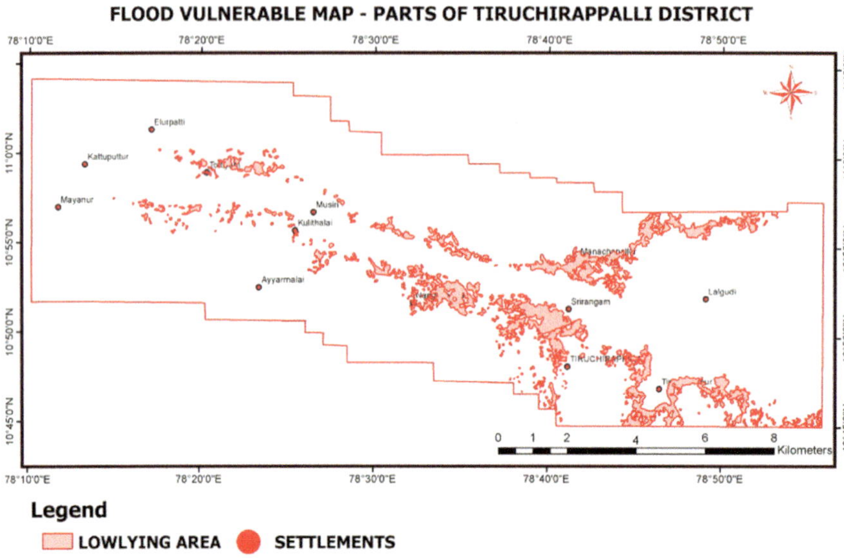


(a)

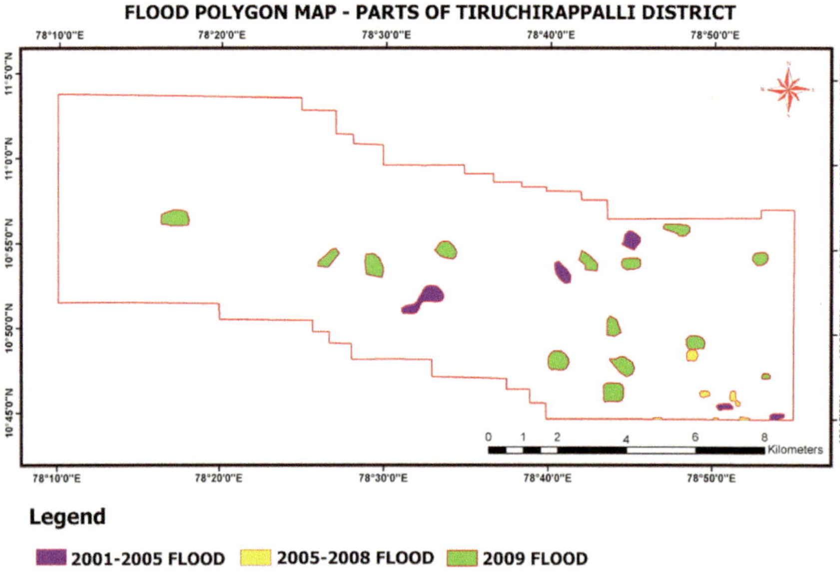


(b)

**Fig. 14.5** **a** Flood vulnerable map. **b** Flood polygon map of the study area

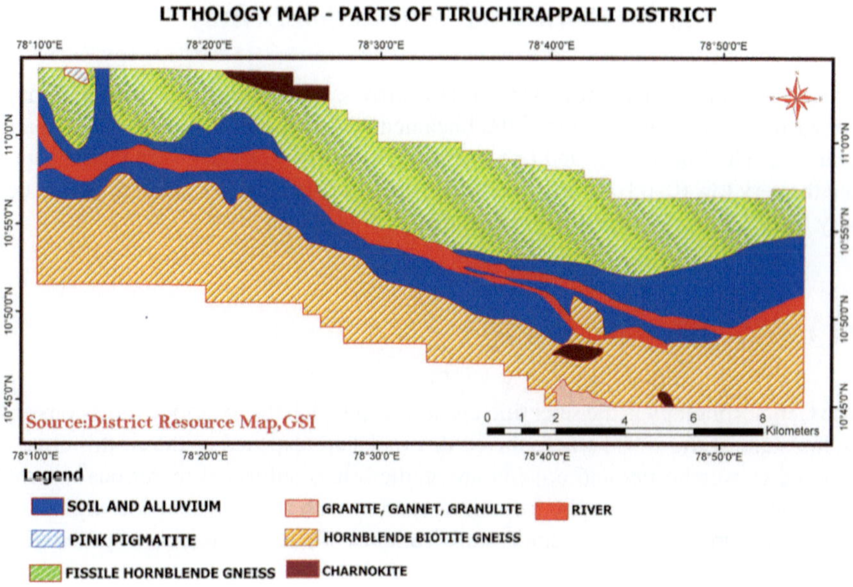


(a)

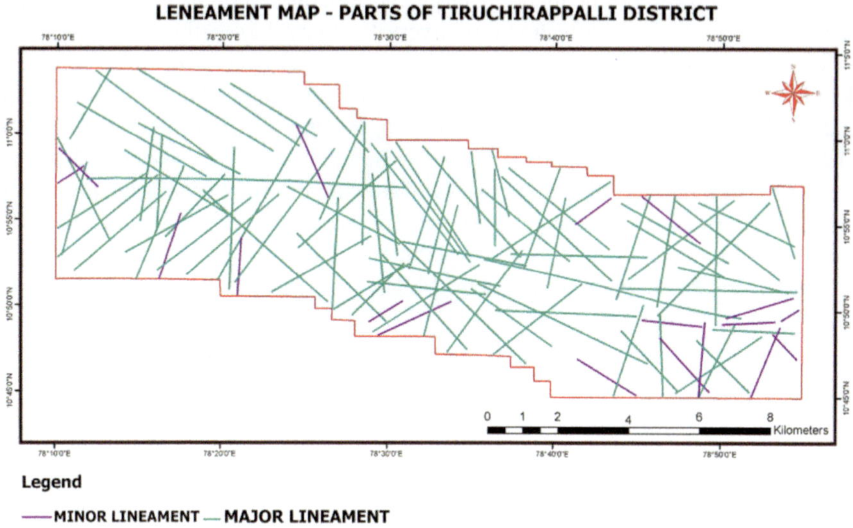


(b)

**Fig. 14.6** **a** Lithology map. **b** Lineament map of the study area

### 14.3.3 Lineament Density and Intersection Density

The line map had a length of 140 m × 140 m and was superimposed over the grid map. For each grid, the total length of the lineaments and intersections was counted and tabulated. The line density and line intersection density contours were then mapped and the very low (0–0.67), low (0.68–1.32) medium (1.33–1.99) and high (2.0–2.66) very high (2.67–3.32) zones were demarcated (Fig. 14.7a, b).

### 14.3.4 Geomorphology Map

The Geomorphology is the scientific discipline in which the exterior expressions and architecture of the world where the earth's rocks are exposed or the earth's crust is exposed to weathering and erosion are studied. It is subjected to various external morphodynamics and morpho tectonic processes such as horizontal convergent and weathering process due to temperature variants, biotic interference, hydrological interference, the destructive and constructive processes, and the related interactive processes, the wind erosion, volcanic eruptions, geothermal processes, etc. Hence, the artwork of geomorphic mapping has simplest expanded for the closing six decades particularly the geomorphic mapping has gained terrific importance after the advent of modern remote sensing technology. Every landform has distinct physical properties and subsurface characteristics. In the existing study, a try made to prepare a detailed geomorphologic map on a 1:50,000 scale using IRS-1D data (Kongeswaran and Karikalan 2015, 2017; Karikalan and Kongeswaran 2015; Kongeswaran 2019). The photo recognition elements like tone, texture, shape, size, associated features, etc., have been utilized in delineating the following different landforms present in the study area such as Inselbergs, Residual Hill, Bajada, Rocky pediments, Shallow pediments, Moderate pediments, Deep pediments, Colluvial fill Gullies, Flood plain (younger), and Flood plain (older) (Fig. 14.8a).

### 14.3.5 Drainage Map

The drainage map was developed from the Survey of India (SOI) toposheet (58 J/1, 5, 9, 13, 58 I/4, 5) by digitizing possible rivers, streams, and tanks. Cauvery River, Kolidam River (Colleroon), Kudamurutti River, Vennar River were the major rivers flowing in the area. Cauvery river flows W–E direction then divided as Cauvery River and Kolidam river. Vennar river flows NW–SE direction and joins with Cauvery river. Most of the tanks are existing in the upper region of the study region and lower parts of the NW–SE area (Thirumalai et al. 2015) (Fig. 14.8b).

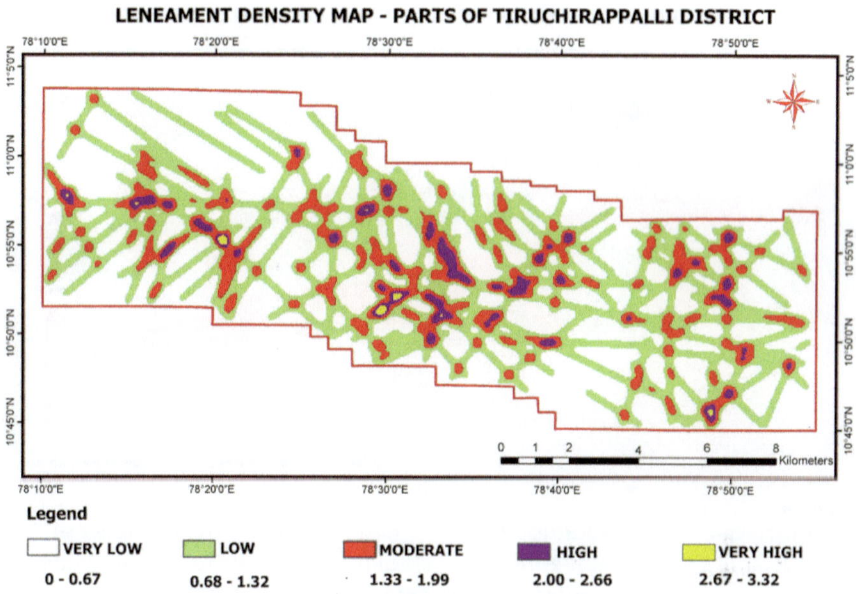


(a)

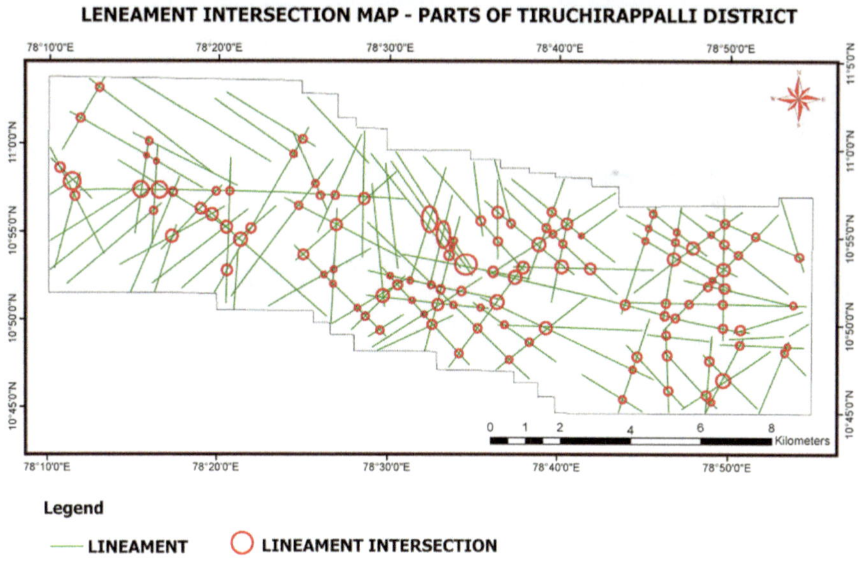


(b)

**Fig. 14.7** **a** Lineament density map. **b** Lineament intersection map of the study area

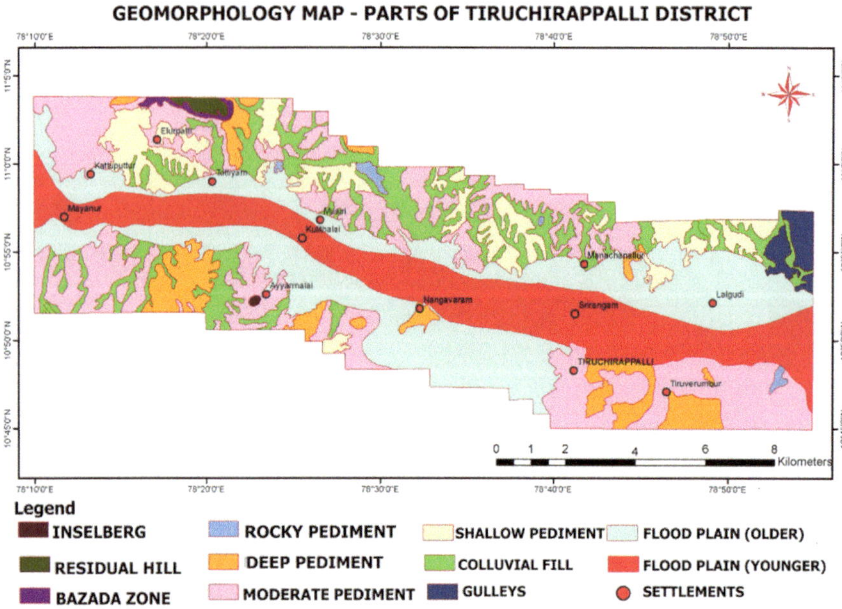


(a)

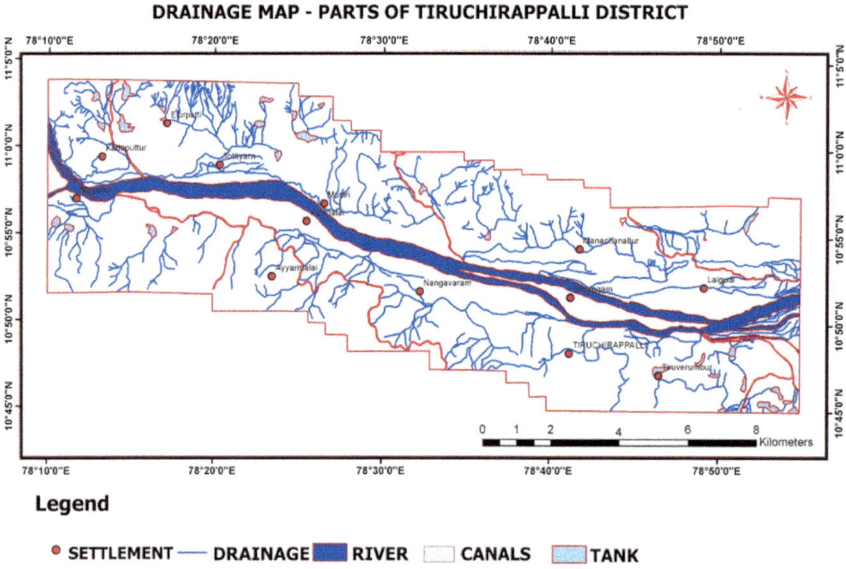


(b)

**Fig. 14.8** **a** Geomorphology map. **b** Drainage map of the study area

### 14.3.6 Drainage Density

The entire study area was girded with the 1 m value of each. The drainage map was superimposed over the grid plotted in the corresponding grid center and contoured using Surfer software. As a drainage density diagram, these contours have been designated. The low, medium–high, and very high zones were demarcated after the anomalous values were removed and the image was generated in the ArcGIS environment (Fig. 14.9a).

### 14.3.7 Land Use/Land Cover Map

The Land use refers to the activities of man and different uses carried on land, land cover refers to natural vegetation, water sources, rock/soil, artificial cover, and other results attributable to land transformation. The classification of land use of the defined area using remotely sensed data may provide useful information on the interrelationship between land use and land cover. For the planners of various development activities, a systematic and detailed collection of land use/land cover map data is required. Since land use/land cover has a direct or indirect effect on soil erosion, the land use/land cover of the investigating area must be identified. Due to its synaptic view and temporal data capability, the role of remote sensing data in providing such information has been well defined (Kongeswaran and Karikalan 2016a, b, 2019; Venkata Ramireddy et al. 2015). The standard classification of land use/land cover defined by National Remote Sensing Agency (NRSA) is followed in the present research. The land use/land cover map was prepared using IRS 1-D imagery on a 1:50,000 scale. Settlements, Wet Crop, Dry Crop, Fallow Land and Plantations, Evergreen Green Thick, Land with Scrub, and Land without Scrub are the various land use and land cover groups of the study area (Fig. 14.9b).

### 14.3.8 Flood Water Harvesting and Management—Identification of Suitable Area to Divert Floodwater

The various thematic maps are prepared using Arc GIS 10.2 as a lithology map, geomorphology map, land use land cover map, slope map, and these separate thematic maps have been incorporated. From the integrated map it is dissolved into two categories that are 0 and 1, 0 is defined as an unfavorable area to divert the floodwater and 1 is defined as a favorable area to divert the floodwater. Likewise, the different thematic maps are integrated into different levels as mentioned below (Gnanachandrasamy et al. 2018).

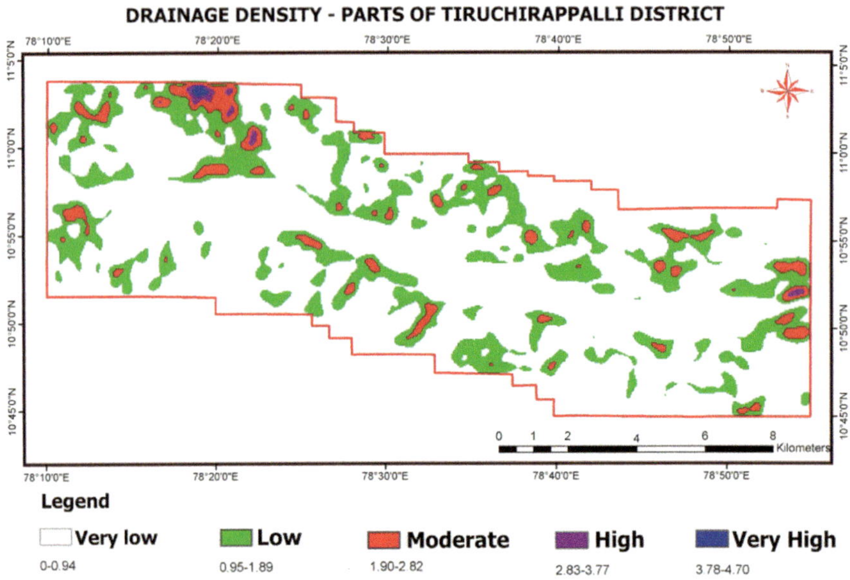


(a)

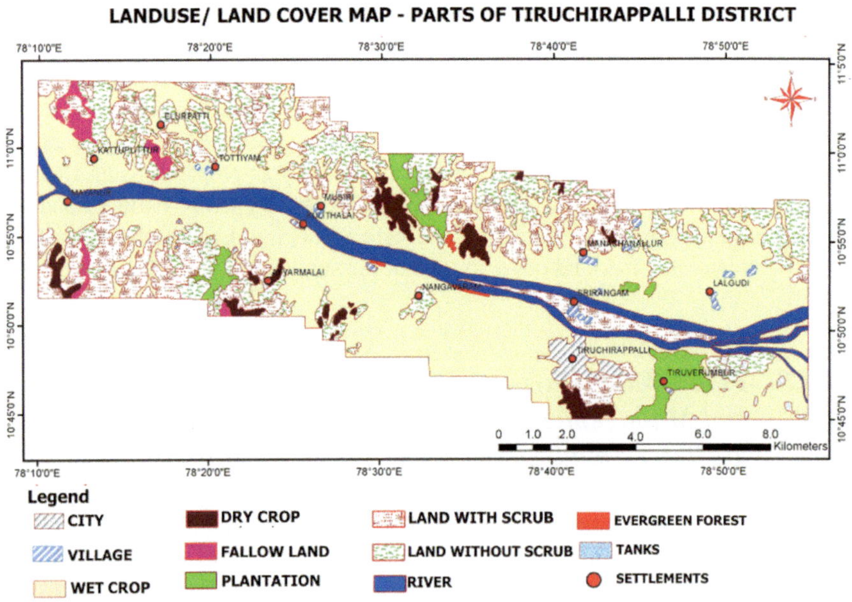


(b)

**Fig. 14.9** **a** Drainage density map. **b** Land use/land cover map of the study area

### 14.3.9 Porous and Pervious Lithology

In this level I, the lithology map was dissolved into two categories by taking the features like weathered granitic rocks, soil, and alluvium, etc. are taken as favorable zones for diverting floodwater and remaining areas such as compact hard unaltered exposed rock as unfavorable zones. The dissolved output map is kept separately for further analysis (Musthafa et al. 2017).

### 14.3.10 Favorable Geomorphology and Level I Integration

Geomorphology map was dissolved into two categories by taking the features like Deep pediment, moderate pediment, colluvial fill, flood plain (older, younger) are taken as favorable zones for diverting floodwater and remaining area such as shallow pediment, rocky pediment, gullies as unfavorable zones. The dissolved output map of geomorphology is combined with a Porous and pervious map (Fig. 14.10a). This map is also kept separately for further analysis.

### 14.3.11 Favorable Slope and Level II Integration

In this level III, the slope map was dissolved into two categories by taking features like 0–2 and 3–5° are taken as favorable zones for diverting floodwater and remaining areas such as 6–50° as unfavorable zones. With the dissolved output map of Slope, the level I output map is pooled. This map is also kept separately for further analysis (Fig. 14.10b).

### 14.3.12 Favorable Land Use Land Cover and Level III Integration

In this level IV, the Land use/Land cover map was dissolved into two categories by taking features like fallow land, scrubland, are taken as favorable zones for diverting floodwater and remaining cropland settlement area are as unfavorable zones. The dissolved output map of Landuse/Landcover is combined with a level II output map. This map is also kept separately for further analysis (Fig. 14.11a).

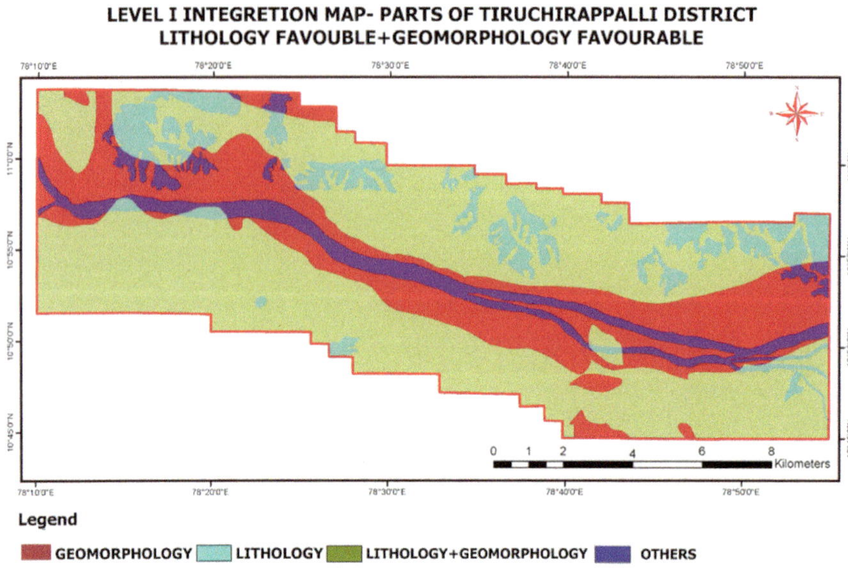


(a)

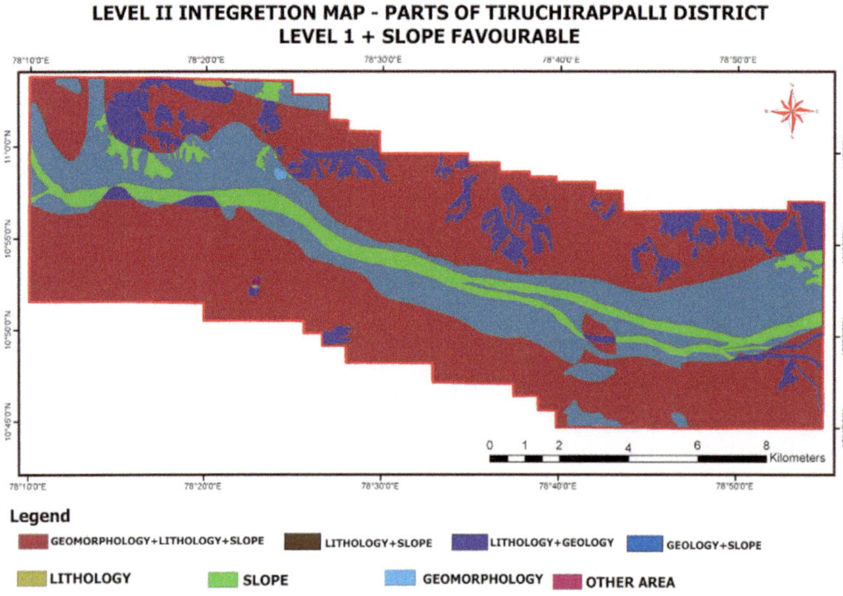


(b)

**Fig. 14.10** **a** Level I integration map. **b** Level II integration map of the study area

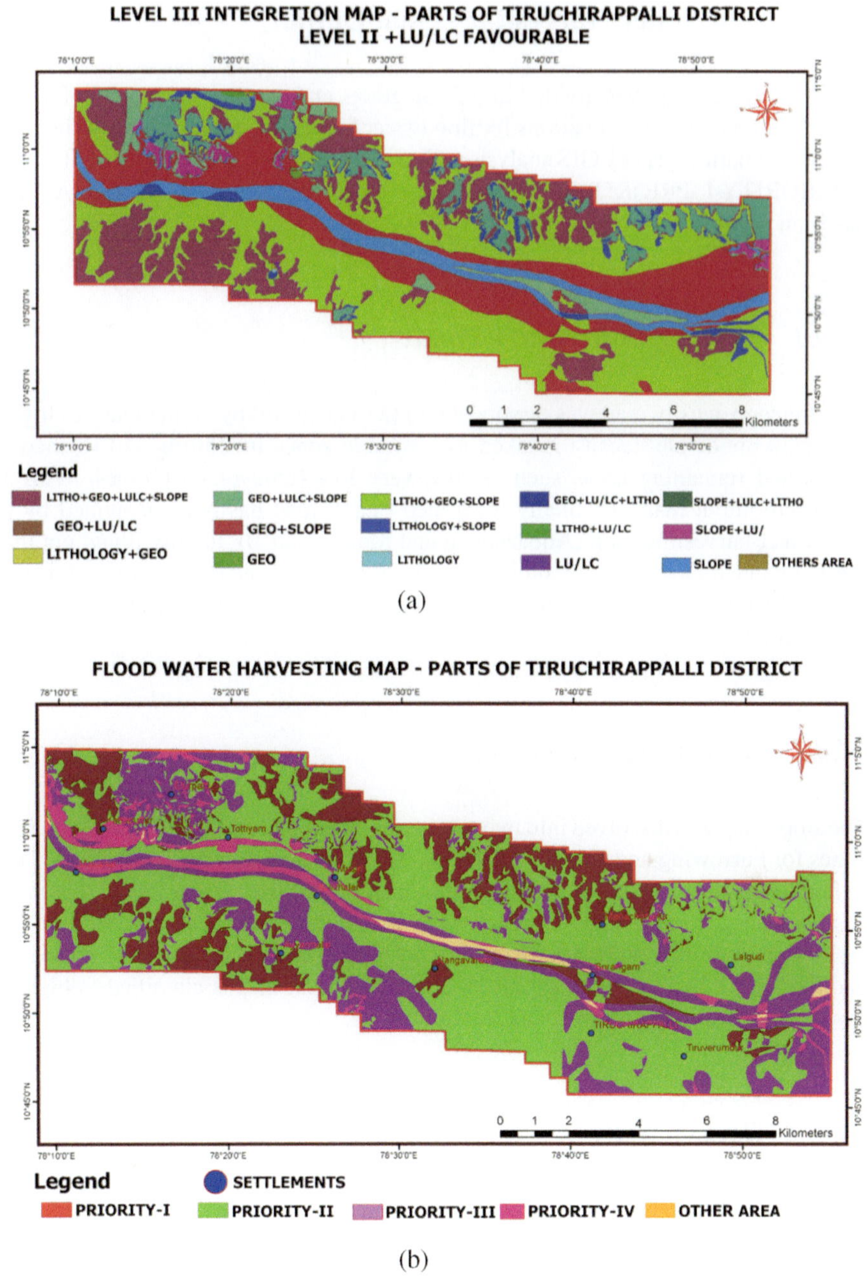


**Fig. 14.11** **a** Level III integration map. **b** Flood water harvesting map of the study area

## 14.4 Flood Water Harvesting (Final Output)

Level III output map was divided into 5 categories and the final output map for the identification of suitable locations for floodwater harvesting was produced using the various thematic layers by GIS analysis with the results which are given in Fig. 14.11b as PRIORITY-I, PRIORITY-II, PRIORITY-III, and PRIORITY-IV, another AREA has been derived.

### *14.4.1 Pitting and Recharge Ponds*

The drainage density map was dissolved into two categories by taking features high, very high, moderate density is taken as favorable zones for Pitting and Recharge Ponds and remaining areas such as low, very low (unfavorable) zones. With a dissolved output map, the finally integrated output map has been generated by a floodwater harvesting map (Anbazhagan and Manivel 2010). Finally found out the areas suitable for harvesting floodwater through the Pitting and Recharge Ponds technique. The final output map is classified as highly favorable, moderate favorable, and least favorable.

### *14.4.2 Burrowing and Flooding*

The slope map was dissolved into two categories by taking features 0–2° as favorable zones for Furrowing and Flooding and the remaining area such as 3–5 and 6–50° as unfavorable zones and dissolved the output map accordingly. Then finally this map has been integrated with the floodwater harvesting map and found out the suitable areas for harvesting floodwater to improve the groundwater level through Furrowing and Flooding (Asare-Kyei et al. 2015). The final output map is classified as highly favorable, moderate favorable, and least favorable.

### *14.4.3 Battery Wells*

The maps of lineament density and lineament intersection have been dissolved into two categories by taking features high, very high, moderate density as favorable zones because they are highly porous and previous zones and can receive and infiltrate maximum flood water very quickly. Hence, in these areas, several boreholes can be drilled. The remaining areas such as low, very low are considered as unfavorable zones and a dissolved output map is generated. Then finally the output map and floodwater harvesting map are integrated to find the favorable areas to harvest the flood water through the Battery Wells technique (Chowdhury and Nag 2017; Kumari

2018) (Fig. 14.12a). The final output map is classified and presented as highly favorable, moderate favorable, and least favorable zones for harvesting floodwater during monsoon using a Battery Wells technique.

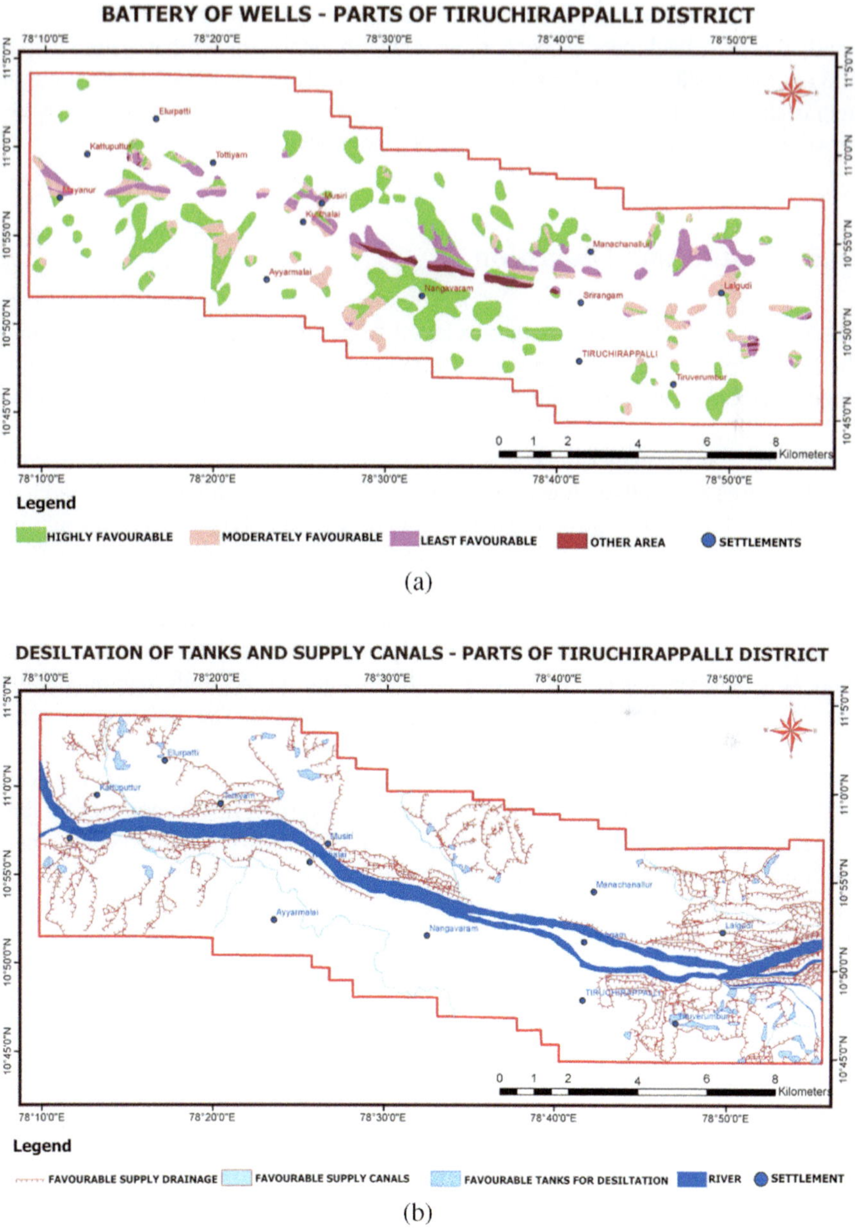


**Fig. 14.12** **a** Battery wells. **b** Desiltation of tanks and supply canals maps for the study area

### 14.4.4 Desiltation of Tanks and Supply Canals

The drainage map was dissolved into two categories by taking the tanks and canals as favorable zones for desiltation of tanks and supply canals and remaining areas such as drainage are kept as unfavorable zones and a dissolved output map produced. Then finally this dissolved map has been integrated with the floodwater harvesting map to find out the Desiltation of Tanks and Supply Canals (Fig. 14.12b). The final output map is classified as highly favorable, moderate favorable, and least favorable zones for adopting the desiltation technique before the monsoon.

## 14.5 Results and Discussion

The various thematic maps were developed from GIS techniques. Among which the Geomorphology is a very important parameter for deciding zones for harvesting floodwater. Geomorphologically, the area has covered with Inselbergs, Residual hills, Colluvial fill, rocky sediment, shallow pediment, moderate pediment, deep pediment, older flood plain, younger flood plain, etc. Flood vulnerable zones are identified in the study area shows there are several such areas seen throughout the study area on either side of the river Cauvery. There are a lot of such polygons seen surrounding the Tiruchirappalli city and Tiruverumbur area and also on the north-east as an E–W strip north of Lalgudi area.

To preserve these areas from flooding, several canals have to be laid starting from the upstream side, aligned out of these low-lying areas toward the outreaches where there is actual water need. To harvest the flood water in the study region, suitable sites were identified by integrating various geo-parameters. The high priority zones are seen in little elevated and on either border sides of the study area for harvesting floodwater. These areas have to be connected by laying canals from the upstream areas in the western side of the river Cauvery for diverting floodwater, storing, and recharging the ground. A separate study can lead to identifying suitable alignment for constructing such canals. Then the suitable areas have been identified for implementing different floodwater harvesting techniques again using precise conditions existing in the study area. There are so many areas identified as highly suitable for harvesting floodwater through the Flooding and Furrowing technique in the study region. Similarly, several regions were found favorable for adopting the Battery Wells technique. Hence, the supply canals and tanks available in the study area have to be de-silted periodically before monsoon so that the floodwater can be easily diverted without any obstruction in the canals.

The map prepared was by using flood vulnerable maps and drainage maps. Overlay the flood vulnerable map and drainage map and finally find out to minimize the flood area. There is one big canal to drain the floodwater in the southern part of the study area, so these forms of canal construction help to mitigate the flood automatically in the future. Maps showing different techniques for recharging the groundwater in the

**Table 14.1** Identification of suitable area to divert floodwater

| Layers | Favorable zone | Unfavorable zone |
|---|---|---|
| Lithology | Soil, alluvium, and weathered gneiss | Compact hard unaltered exposed Charnockite rocks |
| Geomorphology | Deep pediment, moderate pediment, colluvial fill, flood plain (older, younger) | Shallow pediment, rocky pediment, gulley's |
| Slope | 0–2 and 3–5° | 6–50° |
| Land use/land cover | Fallow land, scrub land | Crop land settlement |

**Table 14.2** Flood water harvesting in favorable zone and unfavorable zone

| Flood water harvesting technique | Thematic layer | Favorable zone | Unfavorable zone |
|---|---|---|---|
| Pitting and recharge ponds | Drainage density | High, very high, moderate density | Low, very low density |
| Burrowing and flooding | Slope map | 0–2° | 3–5 and 6–50° |
| Desiltation of tanks and supply canals | Drainage map | Tanks and canals | Drainage |
| Battery of wells | Lineament density | High, very high, moderate density | Low, very low density |

priority areas lead to an idea that more canals need to be laid to bring the monsoonal flood water to these areas. Further studies are warranted to lay new canals that have to be constructed to divert the floodwater during monsoon to link the tanks, lakes, and reservoirs in nearby villages and to floodwater harvesting areas (Tables 14.1 and 14.2).

## 14.6 Conclusion

The current study has shown that satellite data is very useful in the preparation of various thematic maps such as Lineament, Geomorphology, Slope, Drainage, and Land use and Land Cover. Even with minor info, it provides fast and precise data for easy mapping on a GIS platform. All of the above different thematic maps were digitized, GIS data-based using ArcGIS 10.2, and analyzed using ENVI 4.5 digital image processing software. The GIS tool provides a variety of scenarios that are generated and studied before the finalizing of plans for implementation. From the analysis of thematic layers, the suitable intersection areas for flood water harvesting have been delineated and mapped. Further, suitable areas for adopting different floodwater harvesting cum recharge techniques are also be identified. High favorable zone for pitting and recharge ponds was found in the upstream direction of the river banks. All the parts of CRB are mostly favorable for burrowing and flooding whereas battery wells can be constructed very nearer to the river banks where the lineament density is more. Distillation of tanks with feeder canals is favorable in some parts upstream and downstream directions. This study concludes that remote sensing, digital image processing, and GIS becomes very effective technology for flood vulnerable zone mapping, floodwater harvesting, and suggesting management plans to improve the quantity as well as the quality of the groundwater. Further studies are warranted to delineate several suitable canal alignments for diverting flood water to water starving outer areas.

**Acknowledgements** This work was supported by Department of Geology, Faculty of Science, Alagappa University, Karaikudi, Tamil Nadu, India. This article has been written with the financial support of RUSA—Phase 2.0 grant sanctioned vide letter No. 24-51/2014-U, Policy (TNMulti-Gen), Department of Education, Government of India, Dated. 09.10.2018.

**Conflicts of Interest** No conflict of interest was declared by the authors.

## References

Anbazhagan G, Manivel M (2010) Targeting groundwater potential zones in hard rock terrain using remote sensing and GIS. Recent Trends Water Res Remote Sens Gen Perspect 11(2):8–24

Asare-Kyei D, Forkuor G, Venus V (2015) Modeling flood hazard zones at the sub-district level with the rational model integrated with GIS and remote sensing approaches. Water 7(7):3531–3564

Bagyaraj M, Ramkumar T, Venkatramanan S, Gurugnanam B (2013) Application of remote sensing and GIS analysis for identifying groundwater potential zone in parts of Kodaikanal Taluk, South India. Front Earth Sci 7(1):65–75

Bariweni PA, Tawari CC, Abowei JFN (2012) Some environmental effects of flooding in the Niger Delta Region of Nigeria. Int J Fish Aquat Sci 1(1):35–46

Belal AA, El-Ramady HR, Mohamed ES, Saleh AM (2014) Drought risk assessment using remote sensing and GIS techniques. Arab J Geosci 7(1):35–53

Cabrera JS, Lee HS (2019) Flood-prone area assessment using GIS-based multi-criteria analysis: a case study in Davao Oriental, Philippines. Water 11(11):2203

Chandra S, Auken E, Maurya PK, Ahmed S, Verma SK (2019) Large scale mapping of fractures and groundwater pathways in crystalline hardrock by AEM. Sci Rep 9(1):1–11

Chowdhury P, Nag SK (2017) Delineating ground water potential zones in Khatra block, Bankura district, West Bengal using MCE, RS and GIS techniques. J Appl Geol Geophys 5(4):45–60

Gnanachandrasamy G, Zhou Y, Bagyaraj M, Venkatramanan S, Ramkumar T, Wang S (2018) Remote sensing and GIS based groundwater potential zone mapping in Ariyalur district, Tamil Nadu. J Geol Soc India 92:484–490

Gökceoglu C, Aksoy H (1996) Landslide susceptibility mapping of the slopes in the residual soils of the Mengen region (Turkey) by deterministic stability analyses and image processing techniques. Eng Geol 44(1–4):147–161

Hussain MS, Javadi AA (2016) Assessing impacts of sea level rise on seawater intrusion in coastal aquifer with sloped shoreline boundary. J Hydro-Environ Res 11:29–41

Jayaprakash M, Sivakumar K, Muthusamy S, Krishnamurthy RR, Edward Patterson JK (2016) Shrinking of Vann Island, Gulf of Mannar, SE coast of India: assessing the impacts. Nat Hazards 84(3):1529–1538

Karikalan R, Kongeswaran T (2015) Morphological changes caused by post-tsunami in the Region of Nagapattinam Coast, Tamil Nadu, east coast of India. J Adv Res Geo Sci Remote Sens 2(3–4):85–93

Kongeswaran T (2019) Geoinformatics study on coastal geomorphological landforms and shoreline changes between Cuddalore and Nagapattinam, south east coast of Tamil Nadu, India. Ph.D. thesis, Alagappa University, Karaikudi, Tamil Nadu, India, pp 76–125

Kongeswaran T, Karikalan R (2015) Studies on remote sensing and geographical information system applications on coastal geomorphological landforms from Portonova to Coleroon river mouth, South Arcot, Tamil Nadu, east coast of India. Int J Geomat Geosci 5(4):544–554

Kongeswaran T, Karikalan R (2016a) Land use and land cover changes in the Gulf of Mannar using GIS techniques. J Ocean Sci 1:8–15

Kongeswaran T, Karikalan R (2016b) Mapping of shoreline changes in between Devipattinam and Kilakkarai, Tamil Nadu, southeast coast of India. J Ocean Sci 2:12–19

Kongeswaran T, Karikalan R (2017) A study of coastal geomorphological features changes in part of east coast from Cuddalore to Nagapattinam, Tamil Nadu using remote sensing & GIS techniques. In: Geospatial technologies for rural development. The Gandhigram Rural Institute—Deemed University, Dindigul, pp 139–143

Kongeswaran T, Karikalan R (2019) Assessment of shoreline changes between Cuddalore and Nagapattinam coast, east coast of Tamil Nadu, India using geospatial techniques. Disaster Adv 12(2):28–36

Kumari L (2018) Delineation of potential groundwater zone using RS and GIS: a review. Int J Curr Microbiol Appl Sci 7(12):196–203

Machiwal D, Jha MK, Mal BC (2011) Assessment of groundwater potential in a semi-arid region of India using remote sensing, GIS and MCDM techniques. Water Resour Manage 25(5):1359–1386

Maina CW, Raude JM (2016) Assessing land suitability for rainwater harvesting using geospatial techniques: a case study of Njoro Catchment, Kenya. Appl Environ Soil Sci 1–9

Musthafa M, Thirukumaran V, Kalaivanan K, Suresh M (2017) Delineation of groundwater potential zones studies using geospatial technology in lower Tamirabharani river basin, southern India. Int J Recent Sci Res 8(12):22361–22369

Nagarajan R, Rajmohan N, Mahendran U, Senthamilkumar S (2010) Evaluation of groundwater quality and its suitability for drinking and agricultural use in Thanjavur city, Tamil Nadu, India. Environ Monit Assess 171(1–4):289–308

Nathan KK (1998) Droughts in Tamil Nadu: a qualitative and quantitative appraisal, vol 10(3). Drought Network News, Drought-National Drought Mitigation Center at Digital Commons@University of Nebraska, Lincoln

Ramachandran A, Sivakumar K, Shanmugasundharam A, Sangunathan U, Krishnamurthy RR (2020) Evaluation of potable groundwater zones identification based on WQI and GIS techniques in Adyar River basin, Chennai, Tamil Nadu, India. Acta Ecol Sin 1–11. https://doi.org/10.1016/j.chnaes.2020.02.006

Scanlon BR, Healy RW, Cook PG (2002) Choosing appropriate techniques for quantifying groundwater recharge. Hydrogeol J 10:18–39

Selvam S, Dar FA, Magesh NS, Singaraja C, Venkatramanan S, Chung SY (2015) Application of remote sensing and GIS for delineating groundwater recharge potential zones of Kovilpatti Municipality, Tamil Nadu using IF technique. Earth Sci Inform 9(2):137–150. https://doi.org/10.1007/s12145-015-0242-2

Sharma A, Tiwari KN, Bhadoria PBS (2009) Measuring the accuracy of contour interpolated digital elevation models. J Indian Soc Remote Sens 37:139–146

Sivakumar K, Priya J, Muthusamy S, Saravanan P, Jayaprakash M (2016) Spatial diversity of major ionic absorptions in groundwater: recent study from the industrial region of Tuticorin, Tamil Nadu, India. EnviroGeoChem Acta 3(1):138–147

Sivakumar K, Muthusamy S, Jayaprakash M, Mohana P, Sudharson ER (2017) Application of post classification in landuse & landcover strategies at north Chennai industrial area. J Adv Res Geo Sci Remote Sens 4(3 & 4):1–13

Srivastava VK, Giri DN, Bharadwaj P (2012) Study and mapping of ground water prospect using remote sensing, GIS and geoelectrical resistivity techniques—a case study of Dhanbad district, Jharkhand, India. J Indian Geophys Union 16(2):55–63. Retrieved from http://www.j-igu.in/Archives/16-2/paper3_srivastava.pdf

Thirumalai P, Anand PH, Sudha D, Manivel P, Senthilkumar R (2015) Sand quarry locations on the segment of Kollidam river and its impact on rural environment using spatial information technology. Int J Recent Sci Res 6(3):3239–3244

Venkata Ramireddy P, Padma GV, Balayerikala Reddy N (2015) Identification of groundwater recharge zones and artificial recharge structures for part of Tamil Nadu, India—a geospatial approach. Int J Eng Sci Res Technol 4(7):999–1009

Wilhite DA, Buchanan Smith M (2005) Drought as hazard: understanding the natural and social context. In: Wilhite DA (ed) Drought and water crises: science, technology, and management issues. CRC Press, Taylor & Francis Group, Florida, pp 3–29

# Chapter 15
# Strategic Evaluation of Water Quality Monitoring Network Using GIS-AHP Model in a Large River System

**Preeti Rajput, Manish Kumar Sinha, and Gaurav Kant Nigam**

**Abstract** Mahanadi River and its tributaries are passes through the various important cities of Chhattisgarh and Odisha state in India. It serves as the valuable source of water for its domestic, agricultural and industrial need. To monitor river water quality, analytical hierarchy process-based water quality index (WQI) has been adopted to reveal the general status of water quality spatio-temporally. Along the length of the river variation of quality is modelled by interpolation in GIS platform. It is found that the 13.11% of observed stations located nearby major cities of Chhattisgarh and Odisha are extremely bad in general status. The water is of these locations cannot be used as drinking source even after the conventional treatment and disinfection. 4.91% of location lie under very bad status with controlled waste disposal. 22.95% of location identify as bad status of its quality. 54.09% falls for the medium status and the remaining 4.91% stations are having good class of general status that can be used as drinking source after conventional treatment. This study results the overall WQI in Mahanadi River basin and also gives a strategic evaluation of various water quality monitoring network in a large river system.

**Keywords** Analytical hierarchy process · Cluster analysis · Mahanadi River system · Sustainable use · Water quality index · Water quality parameters

P. Rajput
Department of Civil Engineering, Government Engineering College, Raipur, Chhattisgarh 492001, India

M. K. Sinha (✉)
Department of Environmental and Water Resources Engineering, UTD, Chhattisgarh Swami Vivekanand Technical University, Bhilai, Chhattisgarh 491107, India

G. K. Nigam
Krishi Vigyan Kendra—Korba, Indira Gandhi Krishi Vishwavidyalaya, Raipur 492012, India

 
P. Kumar et al. (eds.), *Water Resources Management and Sustainability*, Advances in Geographical and Environmental Sciences,
https://doi.org/10.1007/978-981-16-6573-8_15

## 15.1 Introduction

The pollution of water depends on its sources as it board water into the water body. Mukherjee (2017) says that pollution creates due to unmanaged activity in system. Urban evolution changes regional conditions of environment and effectually alters local settings and in specifically the relative flow of toxic waste into streams, thereby resentfully changing the status of water body. Duan (2016) highlights the issues of surface water contamination due to wastewater disposal is an intense problem in developing countries, where unplanned small and medium industry and densely populated localities with unsewered areas exist. This is created by high rates of migration into cities due to urbanization. Mattos (2012) does the study of rainfall and the resulting surface runoff transport. There are different types of materials of chemical and biological root to the adjacent collecting water body. This results in that the water body is essentially altered from its innate state. These operations do not permit the system itself to be modelled by simple mathematical modelling approach. As the effect of which spread over the other sectors, consequently makes system unsustainable by increase in vulnerability. Apart from its impact on environment, the footprint goes to the economy also. According to Karamouz (2008), the most ruinous economic aftermath from water pollution is the expense of water treatment for human use for different purpose, loss of commercial fish husbandry and lessening the value of real estate properties. Hou (2014) focusses on the effect of urban runoff, which will outturn both short-term and long-term transformation in collecting waters noted to habitat vulnerability and chemical toxicity. This in trend will consequently bring transformation to aquatic communities such as increased fatality of biota and destructive revision to breeds assortment and profusion. Therefore, though stormwater runoff events are episodic, what is of serious concern is the shock pollutant load on collecting waters resulting from an urban stormwater runoff event. The study of Mirrasooli (2017) reveals that the good quality of river water indicates the sustainable use of water resources. It is highly important to maintain the good quality of river water because it affects the ecosystem further. The river stretches nearby urban area is vulnerable to pollution due to untreated effluents being dumped in the river, and consequently, quality of river water has deteriorated significantly. According to Sutadian (2015), the quality of water judged by the large number of variables belongs from its chemical, physical and biological property which represent the status of water body. These large number of parameters with their different scale of measurement makes it tough to understand the status report by the public. To overcome this problem, Klimas (1996) suggests water quality indexing approach which has been used worldwide for evaluating river water quality reasonably. WQI is a single number which is the cumulative effects of properties, extracted from the parameters with the different scale of measurements by bring them into a single platform. The first WQI proposed in 1965 and since then it has been identified as one of the most efficient ways to know the water quality status. The study of Wu (2015) gives the idea about various indices that have been developed in the past by many agencies and researcher with the objective to refine the existing indices. However, there is no WQI till date accepted globally for defining the water quality status. And also, WQI is not capable to report the water quality for all uses nor capable to provide the complete information, it only express general status of water body. Therefore, it takes the continuous attention to develop WQI that is congenial to local and regional area. In Reda et al. (2015) and Ning and Chang (2002), they have done their work for different river basin using analytical hierarchy process (AHP)-based modelling for evaluation of water quality monitoring system. Based on the literature review, this study is an attempt to develop WQI for Mahanadi River basin with the application of AHP.

According to Sutadian (2017), quality of water reflects the health of river. Water quality degradation is a big challenge for decision makers. Because it is difficult to quantify the overall quality due to large number of parameters. This concern has been solved by WQI approach. WQI consolidates the complexity of factors affecting water quality in a single number.

Shoba and Shobha (2014) prove the fact that due to industrial revolution and urbanization vulnerability of rivers has increased. The waste generated from the industry and urban area dumped into the river either directly or indirectly. The study of Zi (2010) reveals that the design of WQI depends on the intended use of water body and the data availability. Apart from that the redundancy of parameters, the selected set of parameters involves the regress subjective judgement. To conclude the status, the set of selected parameters has to standardize either by linear interpolation rescaling, categorical scaling or comparison with the permissible limits. And, aggregated according to the relative importance. There are many WQI developed with time and requirements as CCMEWQI (Khan et al. 2004) developed by the Canadian Council of Ministers of the Environment as a tool to assess and report water quality information to both management institutions and the public. The CCMEWQI gives the liberty to choose the water quality parameters, which helps the index users to modify and adopt the index according to local water quality issues. It provides a simple mathematical concept for the computation of final index value. The NSFWQI (Brown 1977) is one of the earliest WQIs, which was developed during the early 1970s. The Bascarón index (House and Newsome 1989) was developed specifically for Spain.

The fundamental concept for the assessment of river water quality is the compliance of water quality objectives (Hunsaker and Levine 1995). This concept is applied for all the considered water quality parameter. But, this type of analysis fails to give sufficient information on the general status of water quality. In Sutadian et al. (2015), it shows the statistical analysis between considered parameters. This type of assessment provides the information about the water quality even though there is a smaller number of parameters to be monitored. The management of river water quality is one of the important environmental challenges in front of the users. The objective of this study is to develop a concept based on multi-criteria analysis (MCA).

## 15.2 Study Area

This study has been done on Mahanadi River system in India. The area of interest for this study circumscribed within 80° 30′ E–86° 50′ E longitudes and 19° 20′ N–23° 35′ N latitudes. The total catchment area of the basin is 141,600 km$^2$ with maximum elevation of 877 m and minimum of 193 m (Sinha and Rajput 2020). The tributaries of Mahanadi roam around Chhattisgarh and Orissa and in some parts of neighbouring state. Dadhwal et al. (2010). The total length of the river is about 851 km. In this study, 66 measuring station locations (observed by Central Pollution Control Board, India) were selected shown in Fig. 15.1 for water quality sampling. Selected locations are surrounded by the urban area influenced by industrial activities which makes these locations vulnerable to the river water pollution. According to Panda (2006), Mahanadi provides housing for various large industries along with aluminium and thermal plants. All the industrial waste and sewage dump to Mahanadi daily. Also, there are practice that the entire town's sewage is discharged directly into the river which consists of faecal sludge, grey water from toilets, kitchen waste and several other contaminated forms of liquid waste.

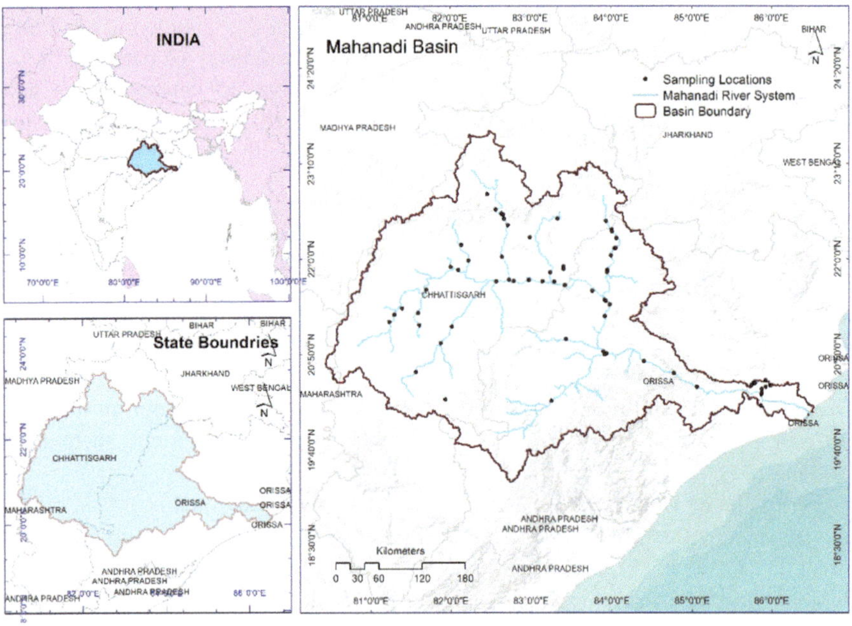


**Fig. 15.1** Location map of the Mahanadi River system indicating sampling points in the basin

## 15.3 Material and Methods

The computation of WQI and its representation is a tedious work as it involves the users need and the extent of use. The methodology used in this work is depicted in Fig. 15.2. WQI converts the specifically shortlist parameters of water property into a unique index value that represent the quality of river water at any location. To develop the WQI as per Sutadian (2017), four steps can be followed, which are the selection of variables for representing the water property, converting them to a common single scale and obtain sub-index value, assigning weight to each parameter according to the relative importance and addition of sub-indices to generate the unique index value.

### *15.3.1 Water Quality Parameter Characterization*

The selection of variables is crucial step in the computation of WQI. Stehling (1988) gives the suggestion that the water quality parameters were selected to identify the

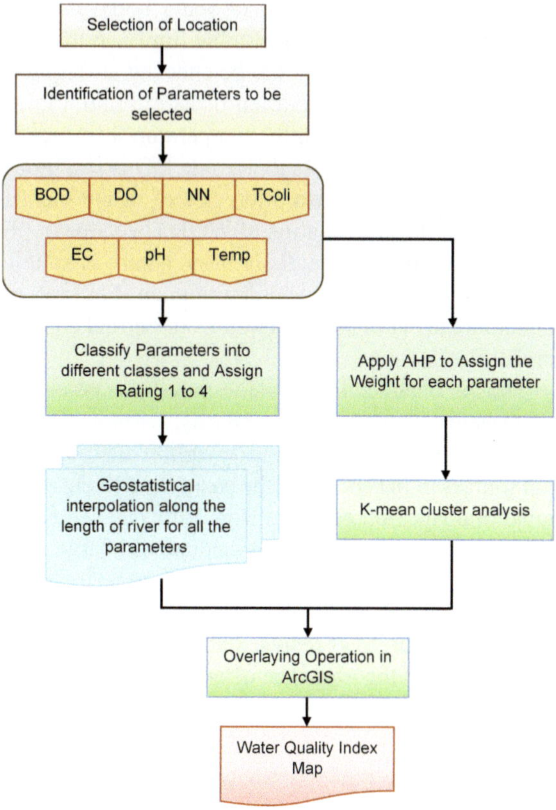


**Fig. 15.2** Flow chart of the methodology adopted to compute WQI

most affected parameter due to pollution load on the water body. These parameters were selected on the basis of their pollution intimation property for the different types of uses of water body. And on the basis of the data values, which show anomalies as compared to expected values. The general status of water quality in river has been identified by considering the complexity of different factors affecting the water quality. To take into account, the following variables of water properties, namely temperature (temp), dissolve oxygen (DO), biological oxygen demand (BOD), pH, nitrite + nitrate (NN), conductivity (EC) and total coliform (Tcoli), have been selected to consider all the three physical, chemical and biological properties in this study to describe the status of the River of Mahanadi basin.

The temperature is the parameters which affect the solubility of oxygen in water and control the decay and growth rates. The rate of reactions in the water body between the chemicals and biological activities is affected by temperature in a considerable scale (Miles and Water 2012).

Every process is set to make at a certain temperature level in nature. Any biological and chemical change have some process that has positive or negative impact on the environment. The deviation of temperature from normal scale is triggered by anthropogenic activities. Effluents that are released from industry and urban land feed and the water body spreads thermal pollution (Weber et al. 2006). Dissolved oxygen represents the status of water body. Its concentration represents pollution load on the system. Its permissible limit is 6 mg/l or more. DO changes drastically over a day therefore it should be checked at frequent interval of day and night time with the specific consideration of its minimum level. Also, the dilution of oxygen in water will depends on the temperature of water body, as higher the temperature lowers the DO level, at high altitude DO is low. For the estimation of chemical and biological action as adsorption rate of heavy metals in the river, pH is taken into account for the computation of WQI. The value of pH should lie in between 6.5 and 8.5. pH measures the acidic or basic nature of solution. For any biological function in a system, pH is a critical determinant. For biological integrity, it is necessary to maintain the pH as the deviation from normal condition may interfere the metabolic process.

Drastic changes in pH value may have potential calamitous influence on river ecosystems. During photosynthesis, carbon dioxide consumes and consequently increases pH. During respiration carbon dioxide release, thus decreasing pH. Potential sources of pH change at the basin are effluents from wastewater plants, industries, mine drainage and effluents from informal settlements. Nitrate + nitrite is used as WQI indicator to check the immediate discharge of organic waste to the river. If NN goes beyond 1.2 mg/l, it will be objectionable to use that water. The biological oxygen demand concentration effectively affects the dissolve oxygen concentration, therefore it is considered as representative of river status according to aquatic life. BOD is an indicator of the existence of organics that are susceptible to biooxidation. When surplus biodegradable organic substance comes into aquatic ecosystems and encounters bacterial putrefaction, huge amounts of DO absorb creating an imbalance which leads to caustic deterioration to the ecosystem. Biological oxygen demand should be 2 mg/l or less in five days at 20 degree centigrade. Conductivity is the capacity of water to regulate electric current. It depends on the salt concentration of water. Salts in the water body come from the industrial discharge or release of illegal effluent into the body. The sudden change in conductivity provides fair indication of the change in water quality. Fluctuations in conductivity values in a catchment may be experienced seasonally due to rainfall influences also. To know the domestic point load and agricultural non-point load, total coliform has been considered in this study. Total coliform has the permissible limit of 50 or less MPN/100 ml (Junior et al. 2014).

### *15.3.2 WQI Computation*

To convert the parameters into a representative unit which reveals the comprehensive quality of water for different uses, it needs to bring the parameters of different scale into a common scale. To standardize the parameters of different scale, rescaling is done by assigning the rate value to the parameters according to the permissible limits for the general water quality standard provided by the Central Pollution Control Board (CPCB), GOI. Parameters are divided into various classes and rating of one to four is provided as listed in Table 15.1. Rating of one is used to represent most departed range of water quality parameter value and four is used for the most ideal range. On this reasoning, actual parameter values are converted into a value having common scale for all parameters. Then, weighted overlay method applied. In this study, weights are allotted to the parameters by employing the AHP. The weights are computed through pair-wise comparison matrices of water quality parameters by comparing several choices.

### *15.3.3 Application of AHP Method*

The paired comparison of parameters with their relative importance is depicted in Table 15.2. The intensity of importance is assigned from 1 to 9. Where 1 is assigned for representing the equally important, 3 moderately important, 5 strongly important, 7 very strongly important and 9 extremely important. The values 2, 4, 6, 8 are the intermediate scale between two judgement. The judgement is the then checked by

**Table 15.1** Rating of individual parameters according to class

| S. No. | Parameters | Weight (%) | Range | Rating |
|---|---|---|---|---|
| 1 | Biological oxygen demand mg/l | 38.18 | < 2.0 | 4 |
| | | | 2.0–2.5 | 3 |
| | | | 2.5–3.0 | 2 |
| | | | > 3.0 | 1 |
| 2 | Dissolve oxygen mg/l | 22.35 | < 6.0 | 1 |
| | | | 6.0–7.0 | 2 |
| | | | 7.0–8.0 | 3 |
| | | | > 8.0 | 4 |
| 3 | pH | 14.11 | 6.5–7.5 | 4 |
| | | | 7.5–8.5 | 3 |
| | | | 8.5–9.5 | 2 |
| | | | 9.5–10 | 1 |
| 4 | Nitrate + nitrite mg/l | 10.11 | < 1.2 | 4 |
| | | | 1.2–2.1 | 3 |
| | | | 2.1–3.0 | 2 |
| | | | > 3.0 | 1 |
| 5 | Temperature | 6.97 | 24.5–26.5 | 4 |
| | | | 26.5–28.5 | 3 |
| | | | 28.5–30.5 | 2 |
| | | | 30.5–32.5 | 1 |
| 6 | Total coliform MPN/100 ml | 4.94 | < 50 | 4 |
| | | | 50–500 | 3 |
| | | | 500–5000 | 2 |
| | | | > 5000 | 1 |
| 7 | Conductivity micromhos/cm | 3.34 | < 2000 | 4 |
| | | | 2000–2250 | 3 |
| | | | > 2250 | 1 |

computing the consistency of paired matrix. The comparison matrix is then normalized, and the priority vectors have been computed. These priority vectors represent the weight of parameters respective to their relative importance. To check the consistency of paired matrix, principal eigenvector is computed (Harker and Vargas 1987). Saaty (1980) gives the equation to measure the consistency, called degree of consistency. The ratio of degree of consistency and random consistency index is called consistency ratio (CR). The random consistency index is based on the size of the matrix formed by the selected parameters. The maximum size of the matrix could be ten. Its values vary according to the size and it may lie between 0 and 1.49. If CR is less than or equal to 10%, inconsistency is tolerable. If CR is greater than 10%, it needs

Table 15.2 Comparison matrix and relative weight of WQI parameters

| Para. | BOD | DO | pH | NN | Temp | Tcoli | EC | Weight(%) |
|---|---|---|---|---|---|---|---|---|
| BOD | 1 | 2 | 3 | 4 | 6 | 7 | 9 | 38.18 |
| DO | 1/2 | 1 | 2 | 2 | 3 | 5 | 7 | 22.35 |
| pH | 1/3 | 1/2 | 1 | 2 | 2 | 3 | 4 | 14.11 |
| NN | 1/4 | 1/2 | 1/2 | 1 | 2 | 2 | 3 | 10.11 |
| Temp | 1/6 | 1/3 | 1/2 | 1/2 | 1 | 2 | 2 | 6.97 |
| Tcoli | 1/7 | 1/5 | 1/3 | 1/2 | 1/2 | 1 | 2 | 4.94 |
| EC | 1/9 | 1/7 | 1/4 | 1/3 | 1/2 | 1/2 | 1 | 3.34 |

to amend the tendentious judgement. The WQI is then computed by aggregating the sub-index value. The maximum value of WQI could be 4.0 and the minimum value 1.0. Therefore, to classify the status of water body, WQI values are divided into four equal parts by normal distribution as 1.0–1.75 very bad (water source conditions are departed natural or desirable level) status, 1.75–2.50 bad (water source can be used for industrial cooling and controlled waste disposal) status, 2.50–3.25 medium (water source can be used for promulgation of aquatic life) status, 3.25–4.00 good (water source can be used for drinking water after traditional method of purification) status.

### 15.3.4 Water Quality Mapping Using GIS Application

To represent the river water quality situation and its variation along the length of the river, these computations are performed in ArcGIS platform. The individual parameter maps are prepared by converting point valued quality data into spatially varying water quality with the help of interpolation. Interpolation is the technique of estimation of pixel values of unsampled point based on known pixel value of surrounding sampled point. In GIS, interpolation is done in two ways as the geostatistical and the deterministic way. Geostatistical interpolation is based on the statistical behaviour of known sampled data series. Mathematically, sound technique to interpolate parameters changing under environmental process. The deterministic interpolation works on the concept of degree of similarity. In this type of interpolation, a smooth surface is created by using surrounding points and mathematical function to assign values to the unknown point. In present work, geostatistical interpolation technique has been used to identify the spatial variation of water quality parameters along the river stretch. This technique helps to understand the pluming behaviour of pollutants, which suggest the overall water quality of river.

After getting the interpolated maps of all the seven parameters, reclassification is done according to the range of parameters into four different classes and assigns new values from 1 to 4. Then, weighted overlaying is performed by stacking the maps with its associated weight determined by AHP technique. The resultant map is representing the WQI map of Mahanadi River.

### *15.3.5 Application of K-Mean Clustering*

To support the AHP analysis, K-mean clustering technique is used in this study. K-mean clustering technique is the algorithm of unsupervised learning. With this algorithm, the monitoring sites are grouped in four clusters. All the four clusters represent four classes of very bad, bad, medium and good status of WQI as identified in AHP analysis. These clusters are formed on the basis of water quality parameters observed in different locations. Each cluster having the locations of water quality monitoring station of similar water quality characteristics.

## 15.4 Results and Discussion

The sustainablc dcvclopment is the need of this hour. Management of good water quality status is one of the crucial domains of development. The objective of which is to optimally utilize the available water resources and maintaining its quality with social and economic development. The objective was to establish the WQI for the determination of general water quality status of the Rivers of Mahanadi basin. This was accomplished by the application of AHP with GIS to the water quality parameters.

### *15.4.1 Data Analysis and Mapping of Water Quality Parameters*

The statistical analysis of water quality parameters and its variation along the length of the river gives the overview on the general character of the river. The individual parameters stats were listed in Table 15.3. The graphical variation of water quality parameters along the river in selected 66 locations is depicted in Fig. 15.3.

Biological Oxygen Demand (BOD): BOD is the parameter that gives the idea about the immediate discharge of organic waste to the river system. This parameter is departed from its normal standard range which affects the DO level and they are by aquatic life. Biological oxygen demand has its lowest value 1.2 mg/l and highest

Table 15.3 Statistical characterization parameters used in the computation of WQI

| Parameters | Min. | Max. | Mean | SD | CV |
|---|---|---|---|---|---|
| BOD | 1.2 | 8.0 | 2.4 | 1.4 | 57.1 |
| DO | 3.1 | 7.4 | 5.8 | 0.9 | 15.4 |
| Temp. | 27.0 | 44.0 | 37.6 | 4.8 | 12.9 |
| EC | 163.0 | 50,910.0 | 2808.0 | 9804.6 | 349.16 |
| pH | 7.5 | 10.0 | 8.3 | 0.3 | 4.2 |
| Tcoli | 300.0 | 160,000.0 | 39,846.9 | 55,125.9 | 138.3 |
| NN | 0.8 | 29.0 | 3.8 | 4.2 | 107.2 |

value 8.0 mg/l which are beyond the permissible range when tested for five days at 20 degree centigrade.

Dissolve Oxygen (DO): To estimate the pollutant load on the river system, DO is considered as it affects the aquatic life. Dissolved oxygen stabilizes the biodegradable substances with the help of living organisms. In this study, DO varies with minimum value of 3.1 mg/l to maximum value of 7.4 mg/l in selected locations along the length of Mahanadi River but ideally DO should be more than 6 mg/l.

Temperature (Temp.): It is considered for the knowledge of decay and growth rates in river system. Therefore, it is one of the important parameters for river water quality simulation. Temperature is the parameter which affects the acceptability of inorganic and chemical contaminants in water body. In this study, its value varies from 27 to 44°. Temperature will alter the odour, colour and taste of water.

pH: The strength of acidity or alkalinity of any solution sample is represented by the pH. There is no health-based guideline for the pH value. It is required for evaluation of chemical and biological processes such as adsorption rate of heavy metals in river. It has range of 7.5 to 10 along the Mahanadi River stretch.

Electrical Conductivity (EC): It is directly related to total dissolved solids (TDS). Concentrations of TDS in water depend on different geological regions due to the differences in the solubility of minerals. For domestic use, it is recommended that water should have TDS less than 500 mg/l. If it founds more than 1000 mg/l, it imparts unpleasant taste and makes inadequate for other uses. In Mahanadi River, EC value varies with minimum of 163 mhos/cm to maximum of 50,910 mhos/cm along the length. For the Mahanadi River system, this parameter is well within the permissible limit.

Nitrate Nitrogen (NN): Naturally, the nitrate concentration is low in water bodies but due to the humans, it appears in the water body in considerable amount. Nitrate reaches the water body by leaching or runoff from agricultural land or contamination from human or animal wastes as a consequence of the oxidation of ammonia. As per WHO standard for drinking water, it should not more than 10 mg/l. In this study, nitrate varies from 0.8 to 29 mg/l.

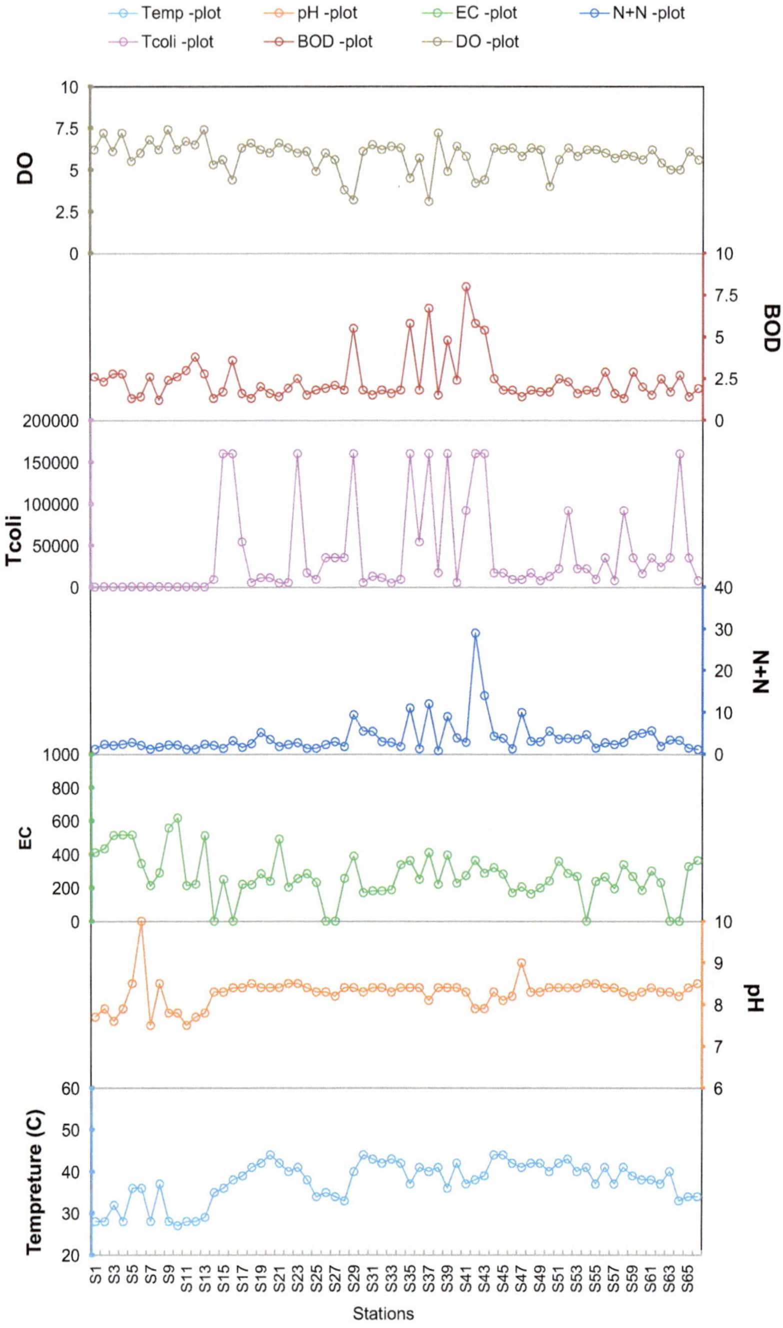


**Fig. 15.3** Station-wise water quality parameters distribution in 66 sites

Total Coliform (Tcoli): The values of this variable depart from river water quality standards in the study area. For the determination of agriculture pollution load and domestic point and non-point loads, total and faecal coliform used to identify location. As per the standards of CPCB, the permissible value of total coliform is 50 or less MPN/100 ml but the resultant map has the minimum value of 300 MPN/100 ml and maximum value of 160,000 MPN/100 ml. This map indicates the situation that would not be acceptable to have for sustainable use of river system.

### 15.4.2 Result of AHP Application on Spatial Distribution of River Water Quality

The percentage influence amongst the parameters is computed and it was found according to the relative significance of the parameters. The computation suggests that BOD has the maximum weight of 38.18%. The DO, pH, NN, Temp, Tcoli and EC have weights of 22.35%, 14.11%, 10.11%, 6.97%, 4.94% and 3.34%, respectively. The resultant maps of selected water qualities are depicting the variation along the length of the river in Figs. 15.4, 15.5, 15.6, 15.7, 15.8, 15.9 and 15.10 obtained from spatial interpolation. These maps are overplayed with its weight obtained from AHP

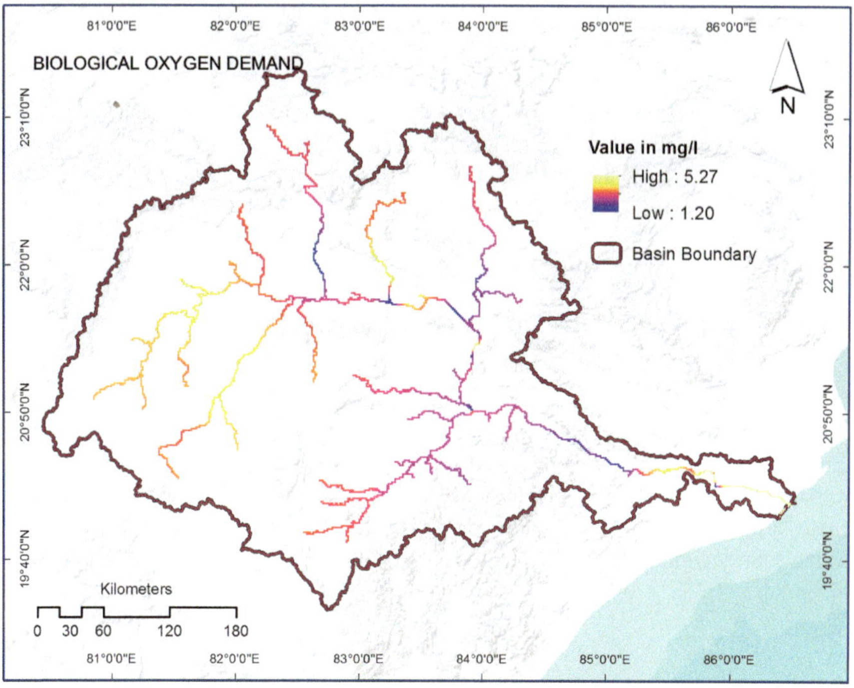


**Fig. 15.4** Spatial variation of biological oxygen demand in Mahanadi River

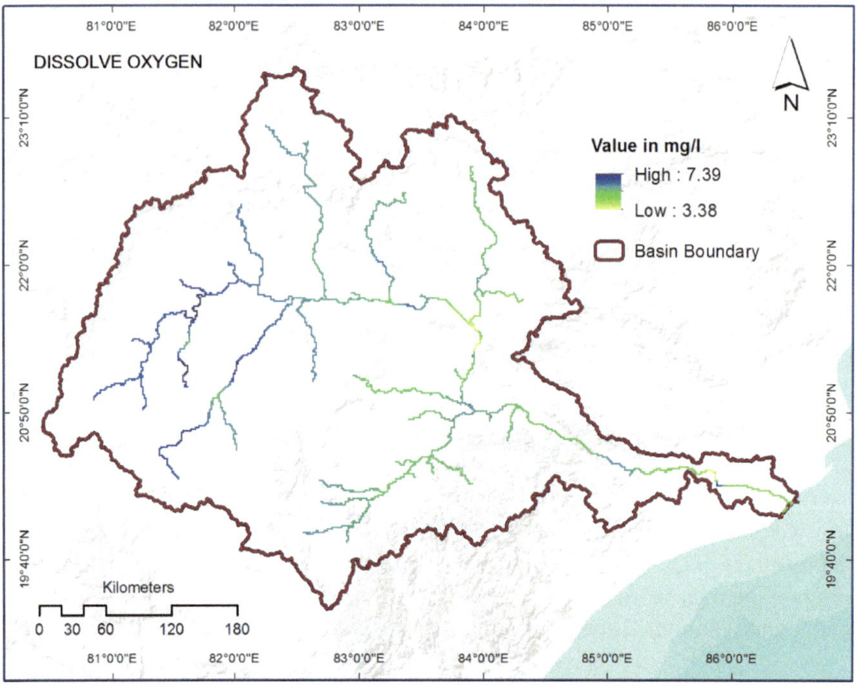


**Fig. 15.5** Spatial variation of dissolved oxygen in Mahanadi River

analysis. The resultant map is showing the variation of WQI along the river length in Fig. 15.11. On the detail observation of WQI Map, the total length of river is classified according to the good, medium, bad and very bad status with respect to WQI value as listed in Table 15.4.

### *15.4.3 Results of K-Mean Cluster Analysis*

To check the consistency of methodology adopted in this study, K-mean cluster analysis was done. The result of cluster analysis shows that all the four clusters have same members as the members in WQI category with the error of 18.16%, obtained from the AHP analysis. The statistical properties of water quality parameters for each cluster are listed in Table 15.5. These stats reveal the limiting values of parameters for the four designated water quality standards as per obtained by AHP-WQI.

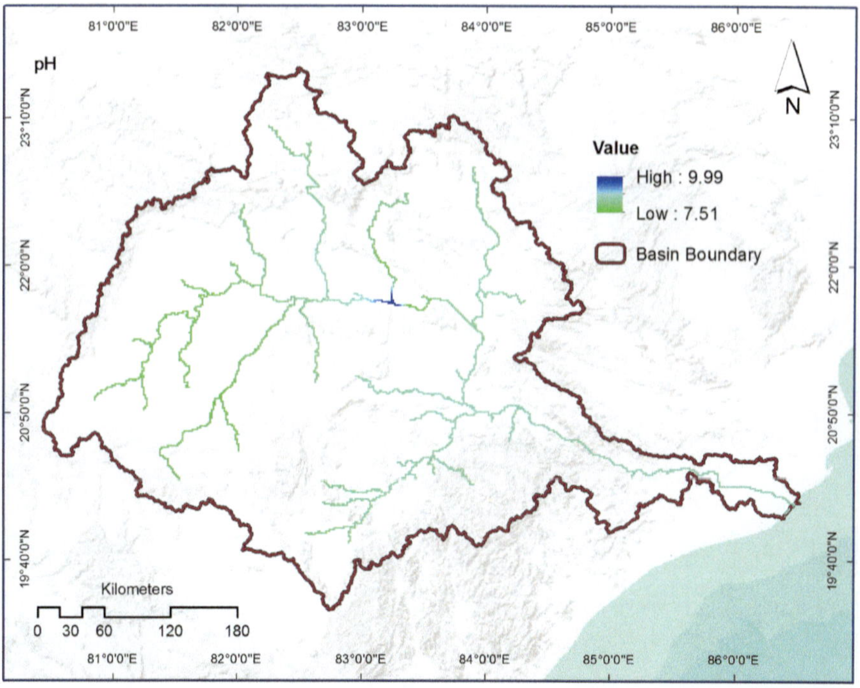


**Fig. 15.6** Spatial variation of pH value in Mahanadi River

These clusters reveal the location of water quality measuring site with respective water quality status. The resultant map is shown in Fig. 15.12. Figure shows the clusters of measuring sites along the length of river. The lowest value calculated is 1.38, which belongs to very bad situation and the highest value of 3.27 which suggests the good condition. On the analysis of result, it was found that 21.21% of the location is in very bad situation, 27.27% bad, 48.51% medium and 3.03% is in good situation.

On the investigation, it was found that the greatest threat to the Mahanadi comes from coal mines, industry, thermal power station and urban local bodies. The chemical compounds in industrial waste and sewage are poisonous and toxic, and it is having long-term effect on rivers ecological balance. And the problem intensifies when people continue to use this water for daily purpose which contaminates land as well. The measuring stations located at Kharun River upstream side of the Raipur city is of medium status whereas downstream side of it has bad status. From this observation, it can be conclude that the Kharun River water gets deteriorated after passing the Raipur city. This result represents that the water of Kharun River is not in good practice of sustainable use. Also, it can be seen that the Kelo River passes nearby

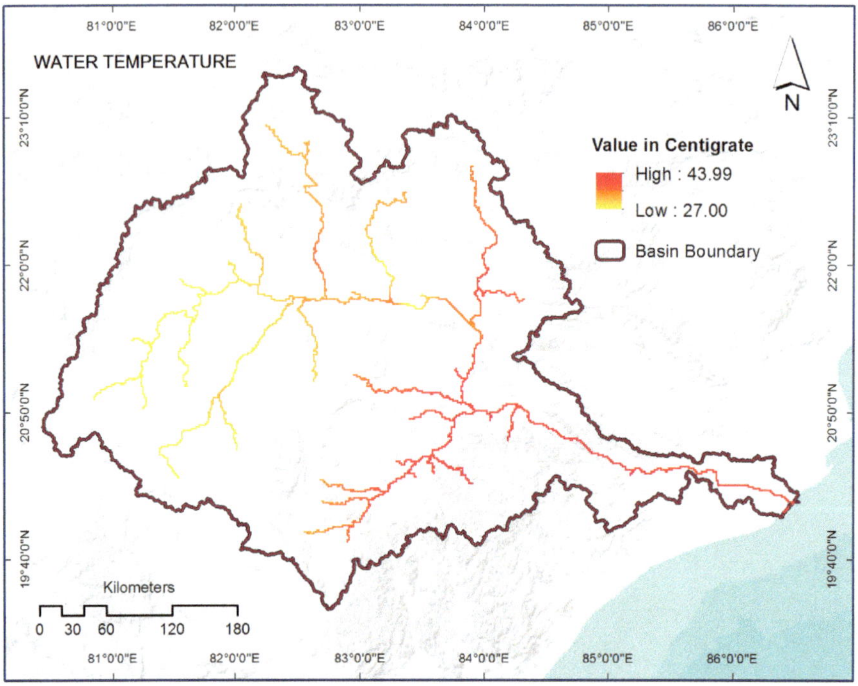


**Fig. 15.7** Spatial variation of maximum temperature in Mahanadi River

Raigarh city has its general water quality in bad status. This may be due to the dense industrial Sproule in this area. Similarly, the Seonath River at Sigma, Mahanadi at Rajim and at interstate boundary have its general quality in bad status. Hence, the resultant map can be used as the base information map for the implementation of management measuring strategy.

## 15.5 Conclusion

Water is one of the fundamental lives supporting element for all living organisms on the earth and for them the quality of water is a consistent matter of concern for the perpetual transformation. Conservation of quantity, quality and authenticity of water resources is the primary responsibility of humans for sustainable development. It is

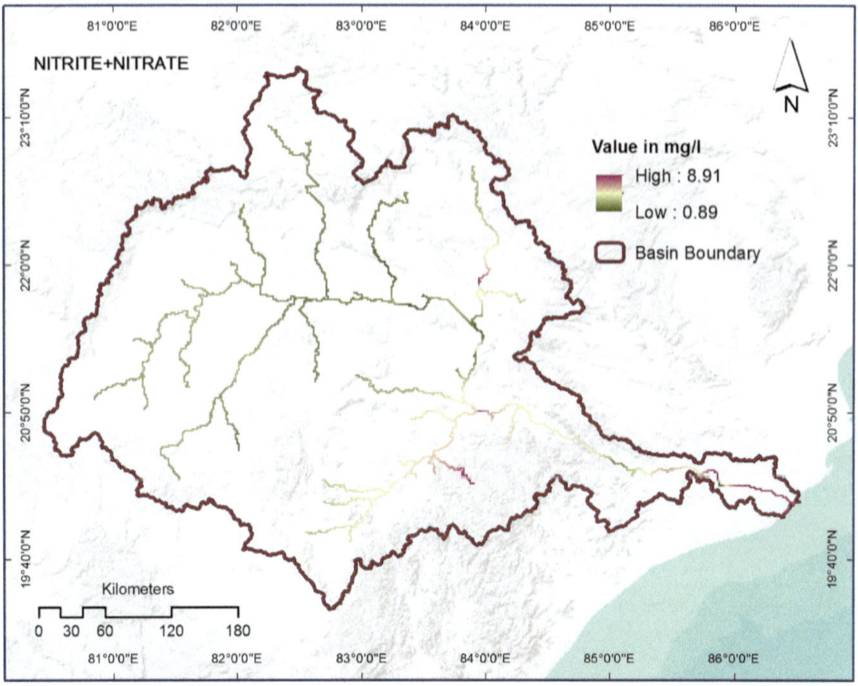


**Fig. 15.8** Spatial variation of nitrate + nitrite in Mahanadi River

suggested from this study that the maintenance of water quality should be the responsibility of users. And the management of water resources should be the responsibility of government authorities. Availability of fresh water for humans, animals and plants should not be compromised at any cost as it is a basic need. Although water quality problems have been addressed and action plans proposed in the national legislature, most of these plans have not been fully implemented for various reasons. Illegal dumping of waste, effluent discharge into rivers, direct discharge of storm water and releases of polluted mine drainage are common in the Mahanadi River basin. Most of industries were built without comprehensive analyzes of long-term impact on the catchments and without consultation with the affected surrounding communities. At some location, it is concluded that there are point sources of contamination which claim their standard of effluent discharge is as per norms. Therefore, it is suggested bring a body who legally monitor the water quality at regular interval. Making policy for setting the standard to regulate usage could be the possible step towards control-

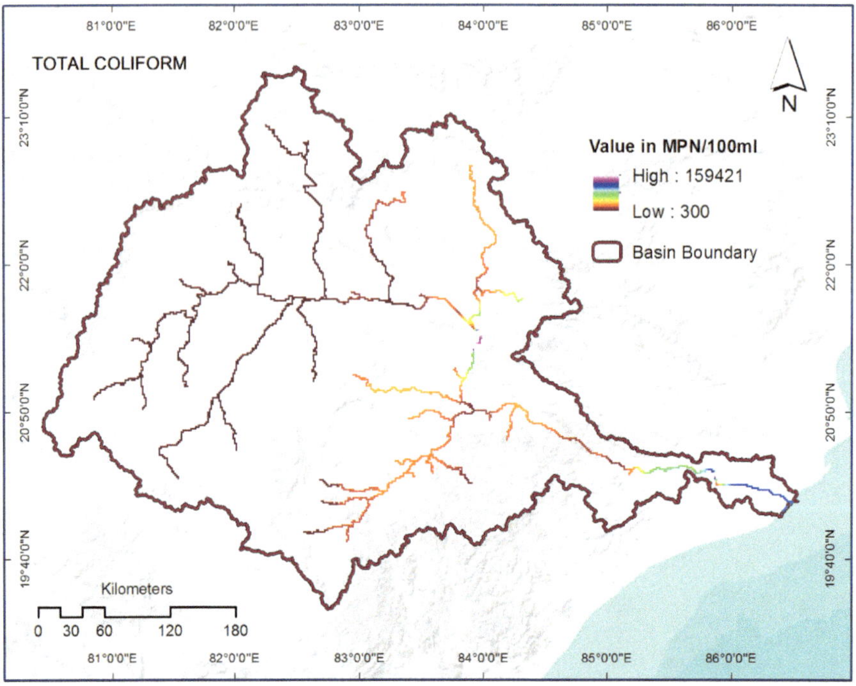


**Fig. 15.9** Spatial variation of total coliform in Mahanadi River

ling damage by pollution. The rollback centres for recycling of resultant waste water from small industry and urban activity should be activated under the pollution prevention policy. Formulating strategy for the provision of good sanitation in the area of informal settlements. Applying the use of modern forming practices in which reprocess the manure and produce biofuels can cut short some amount of pollution load. It is concluded that because the location specific factors are influencing water quality, its quantification is extremely difficult. Chemical processes exert a strong influence on water quality characteristics. This study was an attempt to assess the sustainability of the Rivers of Mahanadi basin. To monitor the quality index-based approach, it has been applied so that the different locations can be compared in same scale of general status. The AHP provides a system in this study to integrate the individual parameters defining different properties of water resulted from environmental pollution .

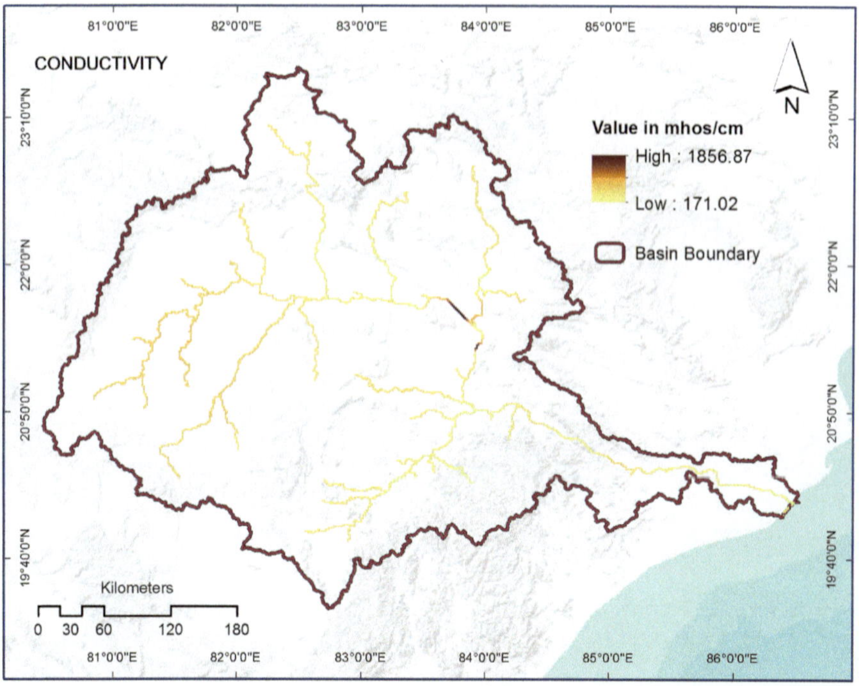


**Fig. 15.10** Spatial variation of conductivity in Mahanadi River

AHP helps in representing and quantifying problem parameter by giving percentage influence. For the Mahanadi basin, it was the biological oxygen demand which was departed from its desirable limits in most of the station followed by dissolved oxygen and nitrite + nitrate. These three parameters are representing heavy organic pollution load. So, it is necessary to monitor the disposal of waste water at river site. The outcome of this study can be useful for interpretation of footprint of regional developmental activity due to pro industry policy. And the spatial variation of water quality index value along the river suggests the location of possible sites of water treatment plant.

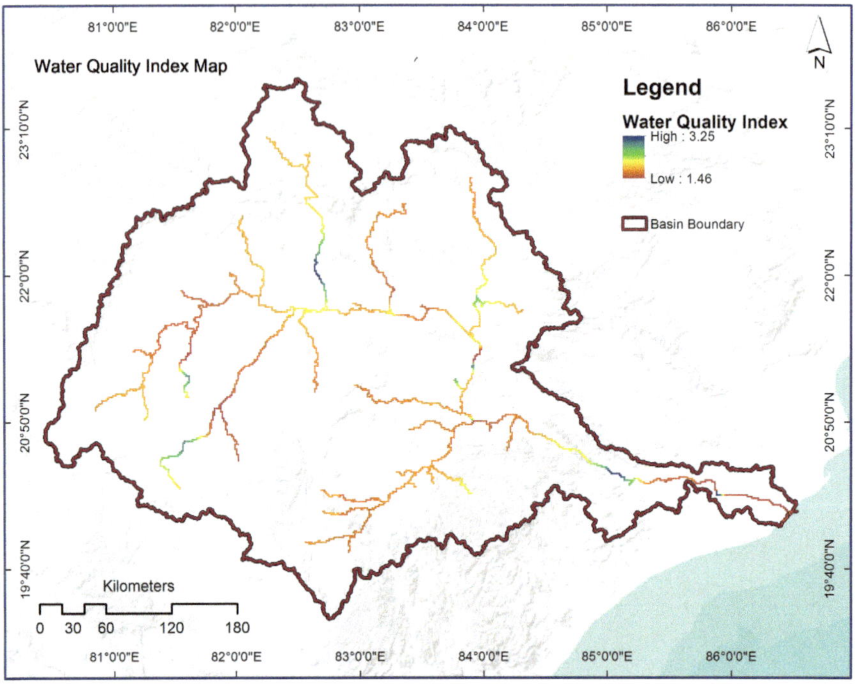


**Fig. 15.11** Spatial variation of resulting WQI map for Mahanadi River

**Table 15.4** Statistical characterization of Mahanadi River length based on WQI

| S. No. | Class name | Range | Length of river (in km) | Length of river (in %) (%) |
|---|---|---|---|---|
| 1 | Very bad | 1.00–1.75 | 230 | 7 |
| 2 | Bad | 1.75–2.50 | 523 | 17 |
| 3 | Medium | 2.50–3.25 | 1713 | 55 |
| 4 | Good | 3.25–4.00 | 647 | 21 |

**Table 15.5** Statistical result of K-Mean cluster analysis

| Parameters | Cluster 1 | Cluster 2 | Cluster 3 | Cluster 4 | Over all |
|---|---|---|---|---|---|
| BOD | 2.6–3.8<br>(2.9 ± 0.5) | 2.3–2.4<br>(2.4 ± 0.1) | 1.2–8<br>(2.4 ± 1.5) | 2.8–2.8<br>(2.8 ± 0) | 1.2–8<br>(2.4 ± 1.4) |
| DO | 6.2–6.8<br>(6.48 ± 0.3) | 7.2–7.4<br>(7.3 ± 0.1) | 3.1–7.2<br>(5.68 ± 0.8) | 7.2–7.4<br>(7.3 ± 0.1) | 3.1–7.4<br>(5.84 ± 0.9) |
| Temp. | 27–28<br>(27.8 ± 0.45) | 28–28<br>(28 ± 0.01) | 32–44<br>(39.1 ± 3.23) | 28–29<br>(28.5 ± 0.7) | 27–44<br>(37.6 ± 4.87) |
| Cond | 212–616<br>(334.6 ± 178.7) | 434–556<br>(495 ± 86.3) | 163–50,910<br>(3186 ± 10,512) | 511–516<br>(513.5 ± 3.53) | 163–50,910<br>(2808 ± 9804) |
| pH | 7.5–7.8<br>(7.64 ± 0.13) | 7.8–7.9<br>(7.85 ± 0.07) | 7.6–10<br>(8.37 ± 0.28) | 7.8–7.9<br>(7.85 ± 0.07) | 7.5–10<br>(8.28 ± 0.35) |
| Tcoli | 300–460<br>(406 ± 76.02) | 300–500<br>(400 ± 141.4) | 350–160,000<br>(46,078 ± 56,898) | 300–300<br>(300 ± 0) | 300–160,000<br>(39,847 ± 55,125) |
| N + N | 1.2–2.2<br>(1.4 ± 0.45) | 2.2–2.4<br>(2.3 ± 0.14) | 0.8–29<br>(4.19 ± 4.2) | 2.4–2.4<br>(2.4 ± 0) | 0.8–29<br>(3.87 ± 4.14) |

*Note* Minimum–Maximum
(Mean ± Standard Deviation)

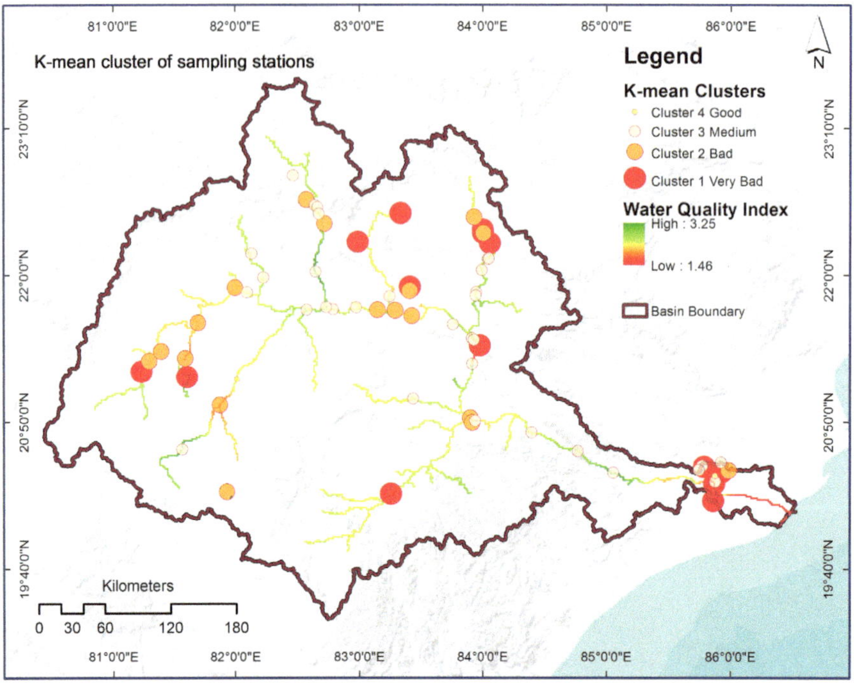


**Fig. 15.12** Location of K-mean cluster members for Mahanadi River system

**Acknowledgements** Authors would like to acknowledge Central Pollution Control Board (CPCB), ENVIS Centre and National Water Quality Monitoring Programme (NWMP) data portal for making water quality data freely available which was essential for this research.

## References

Brown RM (1977) National sanitation foundation keeps services, standards current. J Environ Health 39(5):382–384

Dadhwal V, Aggarwal S, Mishra N (2010) Hydrological simulation of Mahanadi river basin and impact of land use/land cover change on surface runoff using a macro scale hydrological model

Duan W et al (2016) Water quality assessment and pollution source identification of the Eastern Poyang Lake basin using multivariate statistical methods. Sustainability 8(2):133

Harker PT, Vargas LG (1987) The theory of ratio scale estimation: Saaty's analytic hierarchy process. Manage Sci 33(11):1383–1403

Hou D et al (2014) A real-time, dynamic early-warning model based on uncertainty analysis and risk assessment for sudden water pollution accidents. Environ Sci Pollut Res 21(14):8878–8892

House MA. Newsome DH (1989) Water quality indices for the management of surface water quality. In: Ellis JB (ed) Urban discharges and receiving water quality impacts. Pergamon, p 159–173

Hunsaker CT, Levine DA (1995) Hierarchical approaches to the study of water quality in rivers. Bioscience 45(3):193–203

Junior HLDS, Silva GLD, Silva VLD (2014) Qualitative analysis of the presence of emerging contaminants in water supplies for human use: a case study of the Guilherme de Azevedo reservoir in Caruaru (PE, Brazil). Int J Adv Oper Manage 6(2):101–109

Karamouz M et al (2008) Design of river water quality monitoring networks: a case study. Environ Model Assess 14(6):705

Khan AA, Paterson R, Khan H (2004) Modification and application of the Canadian council of ministers of the environment water quality index (CCME WQI) for the communication of drinking water quality data in newfoundland and labrador. Water Quality Res J 39(3):285–293

Klimas AA (1996) Methodology for mapping shallow groundwater quality in urbanized areas: a case study from Lithuania. Environ Geol 27(4):320–328

Mattos RD et al. Evaluation of an indicator for water yield in a watershed of Alto Rio Grande Region, State of Minas Gerais, Brazil. Engenharia Agrícola 32(4):698–707

Miles EJ (2012) Water quality monitoring: guidelines on the application of environmental impact assessment methodology as a decision making tool to improve data quality

Mirrasooli E, Ghorbani R, Molaei M (2017) Ecological health evaluation of Ziarat River using water quality index, Golestan Province, Iran. Open J Ecol 7(13):631

Mukherjee A, Sen S, Paul S (2017) A deviation from standard quality approach for characterisation of surface water quality. Int J Sustain Dev Plann 12(1):30–41

Ning S-K, Chang N-B (2002) Multi-objective, decision-based assessment of a water quality monitoring network in a river system. J Environ Monit 4(1):121–126

Panda UC et al (2006) Application of factor and cluster analysis for characterization of river and estuarine water systems–a case study: Mahanadi River (India). J Hydrol 331(3–4):434–445

Reda M et al (2015) Application of multi criteria analysis technique in surface water quality management

Saaty TL (1980) The analytical hierarchy process, planning Priority. Resource Allocation. RWS Publications, USA

Shoba G, Shobha G (2014) Water quality prediction using data mining techniques: a survey. Int J Eng Comput Sci 3(06)

Sinha MK, Rajput P (2020) Geospatial evaluation of drought resilience in sub-basins of Mahanadi river in India. Water Supply

Stehling F (1988) Environmental quality indices: problems, concepts, examples. Physica-Verlag HD, Heidelberg

Sutadian AD et al (2015) Development of river water quality indices—a review. Environ Monit Assess 188(1):58

Sutadian AD et al (2017) Using the analytic hierarchy process to identify parameter weights for developing a water quality index. Ecol Ind 75:220–233

Sutadian AD et al (2015) Use of exploratory data analysis for cost effective monitoring of water quality data. In: 36th hydrology and water resources symposium: the art and science of water. Engineers Australia

Weber S, Khan S, Hollender J (2006) Human risk assessment of organic contaminants in reclaimed wastewater used for irrigation. Desalination 187(1):53–64

Wu J et al (2015) On the sensitivity of entropy weight to sample statistics in assessing water quality: statistical analysis based on large stochastic samples. Environ Earth Sci 74(3):2185–2195

Zi T (2010) Assessment on water ecosystem services in the Songhua River Basin by AHP Method. In: 2010 4th international conference on bioinformatics and biomedical engineering

# Chapter 16
# Application of RS and GIS for Determination of Various Criteria Causing Drying of Kanari River System

**Ayushi Trivedi, Manoj Kumar Awasthi, and Malay Singh**

**Abstract** Uncontrolled ground water extraction is the basic cause and foremost problem of ceasing effluent discharge into the water stream. To revive Kanari River in Jabalpur district using various technical approaches and determining the root causes of drying of the same is the main objective of this study. The technical advance toward the study was integration of LULC map by ERDAS imagine and ArcGIS which subsequently aimed of assessing the spatial and quantum recharge requirement for revival of river. The LULC classification was done on the grounds of satellite imagery of Landsat 5 thematic mapper (TM) from 1990, Landsat 8 enhanced thematic mapper (ETM) data from 2004, 2009, and Sentinel 2B data from 2019 that were used for visual assessment of development and land use trends within the area. The runoff was also calculated by the integration of ArcGIS and SCS-CN method which finally depicted the runoff in the area kept on increasing as rainfall kept on increasing following a linear trend which predicted that the infiltration opportunity decrease and ultimately resulting in decreased infiltration rate causing decreased baseflow resulted in drying of river.

**Keywords** Thematic map · LULC · SCS-CN method · Satellite imagery

## 16.1 Introduction

Despite of certainty ample amount of money is spent on rehabilitating our rivers across the globe, the science of restoration is still inconclusive. It is known that this is an integrated disciplinary perspective system requiring consolidated space of knowledge into a single empirical and conceptual framework (Pickett et al. 1994; Sinha et al. 2012). On that account, a systematic scientific approach is needed to resolve the upcoming challenges in river restoration which is multidisciplinary that is process based and prophetic. On a particular region, a river system may vary from upstream to downstream or from catchment to catchment. However, a better

A. Trivedi (✉) · M. K. Awasthi · M. Singh
Department of Soil and Water Engineering, Jawaharlal Nehru Krishi Vishwa Vidyalaya, Jabalpur, India

 
P. Kumar et al. (eds.), *Water Resources Management and Sustainability*, Advances in Geographical and Environmental Sciences,
https://doi.org/10.1007/978-981-16-6573-8_16

knowledge in morphology and process of river is accessible at numerous scale (Sinha et al. 2013). The first and fundamental order data set of river management programs are obtained by the delineation of geomorphic conditions in a river system. Rivers go through everlasting adaptation for internal alterations, thus a river system never remains in a fixed state morphologically, for instance change in one channel due to alterations in the confluent channel, or due to external movement or changes in the basin's land cover (Brierley et al. 2006; Brierley and Fryirs 2005). Rivers go through everlasting adaptation for internal alterations, thus a river system never remains in a fixed state morphologically, for instance change in one channel due to amendments in the joining channel, or due to extraneous movement or change in the basin's land cover (Hohensinner et al. 2018). Analyzing the temporal perspective, it is observed that river adjustments reflect deferred counter to the previous advents and cumulative counter to recent advents (Brierley and Fryirs 2005). Boosting or curbing riverine activities or dynamics is somewhat related to universal channel adjustments overlapped by man-made activities. Modifications in the geomorphological characteristics or composition of river reach can undoubtedly influence or alter its capability to assist the habitat availability and ecological functions of a fluvial system (Hohensinner et al. 2018). Furthermore, riverine ecosystems, exceptionally, hinge on interruptions that create single portion of the system on typical basis. Forecasts utilizing mathematical models or tools (or any other scale models in laboratories) are generally used, yet there is still possibility to master about their efficiency or accuracy and a lot to upgrade on the model concepts. Lange et al. (2015) adopted a design which are model-based for reviving an urban river Panke situated in Berlin, Germany (Lange et al. 2015). This unfamiliar technique linked high resolution 2D hydraulic modeling including habitat modeling and using river-ecological expert ability in a continual way. Numerous approaches were proposed for habitat modeling then the habitat suitability maps were refined for the habitat and fish appropriateness for benthos was determined by involving clusters having distinct hydraulic prepossession. The advance basically model-based for intensification provided numerous antique advices on recent trends of shortcomings in the morphology of river, preferences for new form habitats creation, and quantitative learning on increment of applicable sites to be expected (Lange et al. 2015).

LULC classification performs an evident role in advancement or improvement of particular region. Precise and relevant knowledge of LULC are remarkably necessary for investigating diversified socio-ecological interests. This knowledge is critical in various operations including sustainable development, rural management, urban land planning, and agriculture. Remote sensing technology is broadly utilized for estimating LULC changes. Remote sensing data gathered using satellite images was extensively utilized for fetching LULC-related knowledge (Babykalpana et al. 2010; Saadat et al. 2011). There are numerous methods introduced basically for LULC classification with the advances and development of satellite handling and remote sensing technology (Sekertekin et al. 2017; Mishra et al. 2014).

Various observations and stream gaging data in various parts of the world and especially in India show that there is significant decrease in the non-monsoon river water flow notably in all river systems. This unusual recession of flow and drying of rivers in post-monsoon period is recognized by small hydrologic units (watersheds and sub-catchments). The influenced areas particularly are watersheds and sub-catchments of rainfed areas, the major irrigation projects, downhill areas of degraded forest with diminishing soil cover, ground water overdraft areas, river stretches in the downstream which accommodate higher water uplift, underground mines, or deep open cast. The drying of rivers or flow depletion is broadly recognized firstly near the origin and then subsequently in the large hydrologic units. Rivers are suffering from flow degradation because of numerous possible reasons, including the water usage for agriculture and dam installation. But in most of the cases, the depleted flow is mainly due to climate change including increasing evaporation due to high temperature range and altering rainfall patterns. Diminishing runoff is elevating burden on freshwater resources all over the world and in India, because of high demand rate for water due to increasing population. Freshwater being a fundamental and core resource, the downward trends of the same are of great interest. Depleted river flow conclusively influences the world's climate. If less amount of freshwater is allowed in the oceans, they become more salty, which simultaneously effect temperature-driven ocean circulation patterns and salinity consequently playing integral role in climate regulation. The Food and Agriculture Organization (FAO) predicted that international agricultural production will elevate by 60% by the year 2050 for compensating growing population demand and dynamic consumption patterns. This signifies accelerated agriculture, which recommends that more water will be required for irrigation and other practices, more amount of fertilizer will be enforced to fields (resulting in more nutrient runoff), and burden will grow to focus on undeveloped parts of the remaining floodplains or catchment. The emerging challenges in recent era are reviving rivers to their natural state which is not attainable in most of the situations, creating balance between multiple functions of the river, scale and complexity, increasing unpredictability for future scenario, assuring the sustainability of river restoration, scientific approaches for river revival which draw the attention to the respective study.

## 16.2 Study Area

### *16.2.1 Kanari River Watershed*

The present study is based on reviving of Kanari River originating from Ghutehi hill region located at Ghutehi village (Sihora Tehsil) in Jabalpur district of Madhya Pradesh (23° 33′ 36.73″ N and 80° 06′ 57.49″ E) at an elevation of about 490 m. Geographically, the origin of the river is positioned about 62 km to the northeast of Jabalpur. The location was tracked using Survey of India Toposheet no. 64A/2 (OSM No. F44C2), 64A/3 (OSM No. F44C3) and 55M/15 (OSM No. F44B15) (Fig. 16.1). The Kanari River lies under the subtropical region having an average mean sea level (MSL) of 399–401 m. Further, this river finally conjoined Suhar River (23°

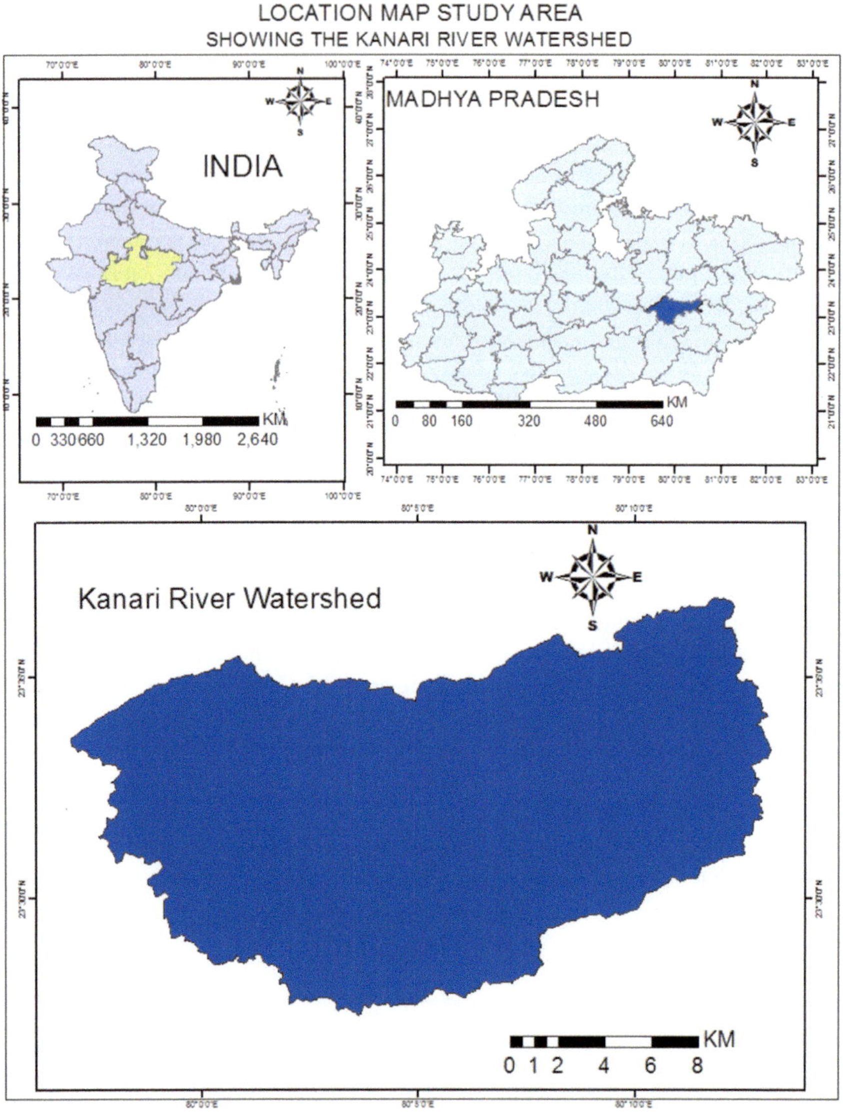


**Fig. 16.1** Location map of the study area

29′ 18.68″ N and 79° 59′ 09.20″ E) in Budhi village (Majholi Tehsil) of Jabalpur (Madhya Pradesh).

The Kanari River flowing through the Sihora and Majholi Tehsil of Jabalpur is connecting about 38 g Panchayats of the area. Previously, the originating site was the hills of Gurji, Silodi, Darshani, Dinai, Khamaria, and Khudawal but gradually due to various human interference activities and climate change scenarios, the origin of the river shifted by approximately 1.07 km.

Some of the salient features associated with the Kanari River watershed are that the river flows through the Sihora and Majhauli Tehsil of Jabalpur district. The water course of the Kanari River is 63 km long including Bah Nala which progressively conjoins the river at 23° 29′ 07.65″ N and 80° 04′ 13.45″ E. The catchment area of the river is 298 $km^2$. The study area involves 38 g Panchayats and 29 micro watersheds.

### 16.2.2 Climate

The climate of the region is mild, and generally can be classified as warm and temperate. The area receives more precipitation during the winters than in summer. The average annual temperature and precipitation of the area are about 24.6 °C and 1277 mm, respectively. The precipitation is lowest in the month of April, with an average of about 6 mm. The maximum amount of precipitation is received in the month of July and August, with an average of about 419 mm. The hottest month of the year of the area is May having an average temperature of 33.4 °C. While in the month of December and January, the average nominal temperature falls to 16.5 °C. The data collected during the period of 1982–2012 depicted that the difference in precipitation between the driest and wettest months is 413 mm/in. The maximum amount of precipitation is received in the month of July having an average of about 244.9 mm. The wind velocity is maximum in the month of June with an average of about 5 km/h.

### 16.2.3 Topography

The study area is surrounded by highlands having long narrow plains extending northeast and southwest. The study area is located in Jabalpur district, falling under the Mahakoshal region, situated between the watershed divide of Son and Narmada, but most of the area lies within the valley of the Narmada falling 30 ft. over a rocky ledge (the Dhuandhar, or misty shoot). The area has gentle rolling topography with an elevation difference of 1–2 m.

### 16.2.4 Soil

The plain of the study area which creates an offshoot from the valley of Narmada covers its southern and western portions by rich alluvial deposit including black cotton soil. The soil of the study area is black cotton soil, and water table depth is adequate near the surface.

## 16.3 Methodology

### *16.3.1 Data Products*

**The types of data products used in the process are as follows**

For this study, satellite image of Sihora Tehsil, Jabalpur District, covering Kanari River watershed was acquired for four epochs: 1990, 2004, 2009 and 2019 from global land cover facility (GLCF) an earth science data interface (Table 16.1).

### *16.3.2 Software and Tools Used*

Software and tools such as (a) ERDAS IMAGINE, (b) ArcGIS 10.3.1, (c) Google Earth Pro, (d) ArcInfo, (e) Microsoft Excel, and (f) Microsoft Word were used for the study. ERDAS IMAGINE was used for the image mosaic, unsupervised classification, layer stacking, AOI preparation, and accuracy assessment. ArcGIS 10.3.1 was used for watershed and subwatershed delineation, complementing the image display, various clipping features, and processing of the data. Google Earth Pro was used for validating the various features present in the study area by using sync function. ArcInfo was used for geographic input, processing, and output production. Microsoft Excel and Microsoft Word were used for graph and chart making and manuscript preparation.

### *16.3.3 Watershed and Subwatershed Delineation*

#### 16.3.3.1 Importing of DEM Data to ArcMap

The preliminary or foremost step was delineation of watershed and subwatershed. Previously, the delineation was carried out physically using the technique of area upstream from a certain outlet point or manually using paper maps. Currently, the same is borned out digitally in a GIS environment using ArcGIS (ESRI) desktop and Goggle Earth Pro software. For delineating Kanari River watershed, DEM data were used for determining area of interest. This DEM data were downloaded using USGS Earth Explorer.

**Table 16.1** Description of various satellite images used

| Image ID | Status | WRS:P/R | [UTM_ZONE]/[LATITUDE_BAND]/[GRID_SQUARE] | Acquisition date | Data set | Producer | Spatial resolution | Type |
|---|---|---|---|---|---|---|---|---|
| LT514404 41990008I SP00 | Online | 2:144/044 | – | 31st Nov 1990 | Thematic mapper (TM) | U.S. Geological Survey (USGS) | 30 * 30 m | GeoTIFF |
| LT514404 42004335B KT00 | Online | 2:144/044 | – | 31st Nov 2004 | Thematic mapper (TM) | U.S. Geological Survey (USGS) | 30 * 30 m | GeoTIFF |
| LT514404 42009332 KHC00 | Online | 2:144/044 | – | 31st Nov 2009 | Thematic mapper (TM) | U.S. Geological Survey (USGS) | 30 * 30 m | GeoTIFF |
| S2B_MSIL 1C_201903 20T050649 _N0207_R 019_T44Q MM_2019 0320T1026 53 | Online | – | Military grid reference system/52.3140232/2B | 31st Nov 2019 | Multispectral sensor (MSS) | U.S. Geological Survey (USGS) | 10 * 10 m | GeoTIFF |

#### 16.3.3.2 Delineation of Kanari River Watershed

The DEM grid was set as input surface raster for processing. Fill tool from the Hydrology toolbox (ArcGIS 10.3.1) was utilized for removing any sort of imperfections or (sinks) in the digital elevation model (DEM). The output surface raster was also added to the working directory. The flow direction tool was used for creating flow direction grids. The grid processor was used for finding the lowest neighboring cell from the center (for every 3 * 3 cell neighborhood). The working directory was also updated with output surface raster. Each of the specified number in the listed matrix below depicts the flow direction—that is, on the assumption that the center cell flows northeast, its value will be assumed 128; if they flows due north, its value will be 64, etc. For calculating the flow in each call, the flow accumulation tool was used and simultaneously the upstream cells was identified that flow in each downslope cell (Fig. 16.2). Nevertheless, each cell's flow accumulation value was estimated by the number of cells flowing upstream based on topography and landscape. The flow accumulation layer was now symbolized for enhancing the visual quality of pinpoint areas having high flow accumulation. Outlet (pour) points were created by providing the latitude and longitude value in ArcCatalog window by using start and stop editing dialog box (viz. 23° 29′ 18.68″ N and 79° 59′ 09.20″ E). The tool, namely snap pour point, was applied for accomplishment of two things, that is, snaping of the pour point created to the nearest cell of high flow accumulation for accounting any error at the time of placement and to convert the pour points into the raster format for watershed tool input. Eventually, the watershed was delineated using Hydrology tool in ArcCatalog window.

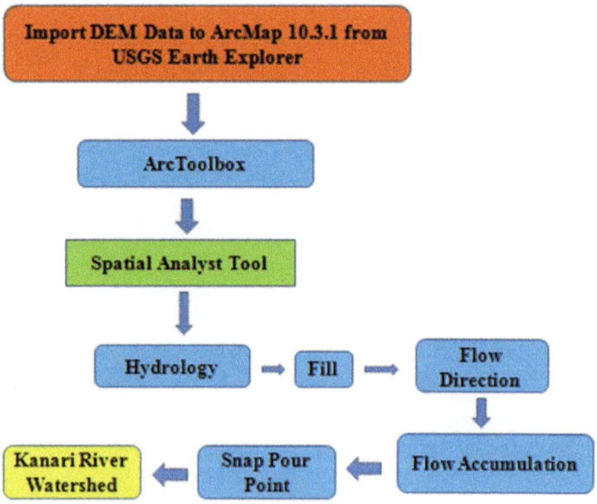


**Fig. 16.2** Delineation of Kanari River watershed

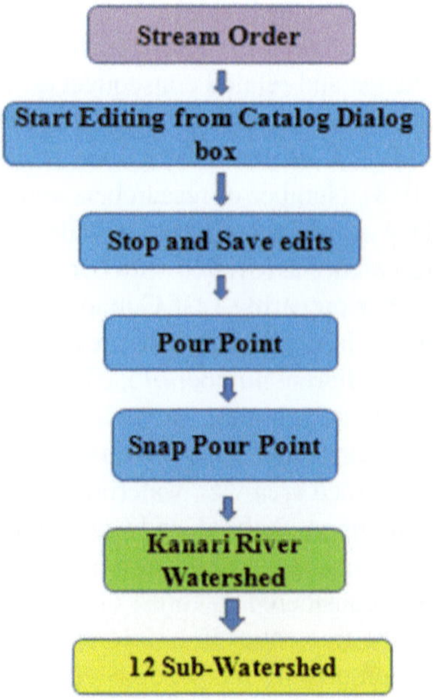


**Fig. 16.3** Delineation of Kanari River subwatershed

#### 16.3.3.3 Delineation of Kanari River Subwatershed

Stream link function was used for creating a grid of stream segments (or links) having unique identification (Fig. 16.3). This provide the clarification that whether a link is a head link, or it is a conjunction between two links. In a particular link, all the cell must have the similar grid code that is specified to that link. The DEM used for deriving raster representation was simultaneously converted in vector format representation. Strahler stream order function from hydrology tool was eventually used to calculate the number of stream order present in the catchment area. It is of great importance to delineate or determine watersheds draining to a specific points on the stream network, for example, stream gages. Using spatial analyst tools of ArcGIS 10.3.1 followed by Hydrology and snap pour point, the subwatersheds were delineated.

#### 16.3.3.4 LULC Preparation

Identification, determination, and detection of digital change are the activity that assists in discovering the changes linked with the land use land cover properties with instance to multi-temporal geo-registered RS data. Identification of temporal changes (normal variation) that is generally not possible with other technologies are quite feasible. LULC change detection has many utilizations such as land use changes, rate of coastal change, afforestation/deforestation, urban sprawl, and other progressive changes through temporal and spatial analysis tools and techniques such as RS and geographic information system (GIS) in conjunction with digital image processing techniques. For supporting LULC, land use change models (LUCM) mostly treated as an efficient tool for analyzing the root causes of land use land cover change and related consequences. In addition, the LULC model is also effective in anticipating the post or future scenario and analyzing spatial, temporal distribution of LULC utilizing the gained observation from last or previous years (Islam et al. 2018). Number of researchers and scientists, such as Singh et al. (2001a, b, c, d, e), Dey and Singh (2003), Okada et al. (2001), and Dey et al. (2004), conducted researches associated with ocean, surface, and atmospheric variables or parameters.

For preparing LULC map using satellite images, types of classification scheme which basically define or prescribe the LULC classes were taken into consideration. The different number of LULC classes were decided on the basis of requirement of a definite program for peculiar implication or operation (Arora and Mathur 2001; Saha et al. 2005). Eight primary LULC classes were selected for mapping the unified watershed area viz.; waterbody, agricultural land, forest/mountain, wasteland, open land, mines, habitat, and vegetation. Area under dry, irrigated or rainfed agricultural land was enclosed as agricultural land mean while space occupied by agroforestry was considered as forest class because of analogous spectral and spatial response of the tree cover in agroforestry systems to open forests. After preparing classification system, most extensively used image classification system or approach, i.e., maximum likelihood classification was used for mapping all types of land use/cover classes (Fig. 16.4). Before selecting the training samples, satellite imagery was analyzed empirically; Toposheet of the watershed and Google Earth images were also investigated carefully and simultaneously unsupervised classification then into 50 classes. There were many places where land use/land cover was different from actual land use/land cover. So, in order to correct it and for improving the accuracy assessment, AOI was drawn for each and every wrong land use/land cover.

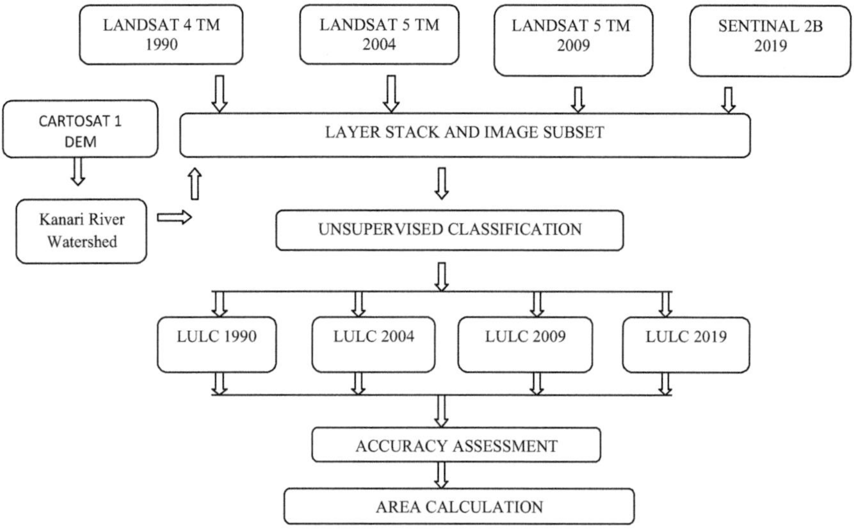


**Fig. 16.4** Flowchart for LULC preparation

## 16.4 Results and Discussion

### *16.4.1 Delineation of Kanari River Watershed and Subwatershed*

Firstly, the SRTM DEM was processed using ArcGIS. The basin of Kanari River was delineated utilizing the Hydrology tool of ArcGIS (ESRI), ArcHydro tools. Numerous automatic watershed delineation techniques and tools generated different areal extents of the basin and stream networks using DEMs. These steps viz., fill, flow direction, flow accumulation, watershed outlet (pour point), stream order, watershed, pour point, and subwatershed are depicted in Fig. 16.5.

The drainage pattern observed in the Kanari basin is homogeneous in nature and having the drainage density of 0.93 $km^{-1}$. Stream ordering was accomplished using drainage network and DEM in ArcGIS using Hydrology tool as depicted in Fig. 16.6. Drainage network is utilized for watershed delineation and for proposing numerous soil conservation measures and water harvesting structures.

### *16.4.2 LULC Preparation*

Multi-temporal and multi-spectral LULC map consisting of eight major (main) classes: waterbody, agricultural land, forest/mountain, wasteland, open land, mines, habitat, and vegetation of 1990, 2004, 2009 and 2019 are depicted in Figs. 16.7 and 16.8. Obtained spatial and spectral distribution pattern of LULC using unsupervised classification are depicted in Table 16.2.

Area wise comparison of different years on the basis of LULC map prepared is illustrated in Fig. 16.9 showing devastated decrease in water bodies and forest land.

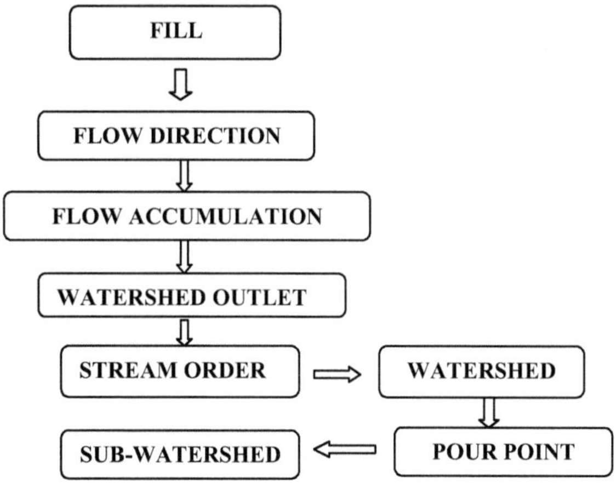


**Fig. 16.5** Delineation of Kanari River watershed

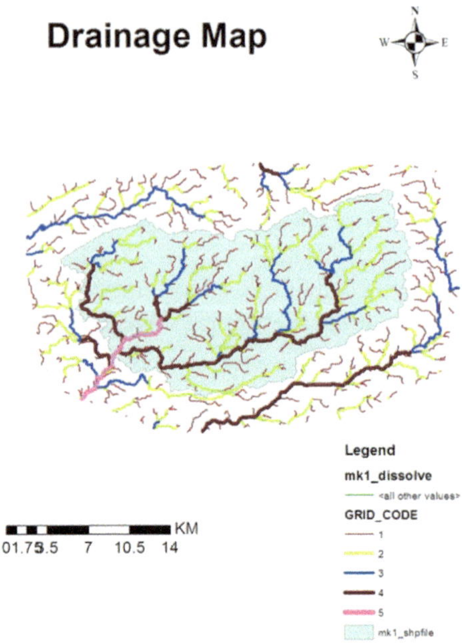


Fig. 16.6 Drainage map of Kanari River watershed

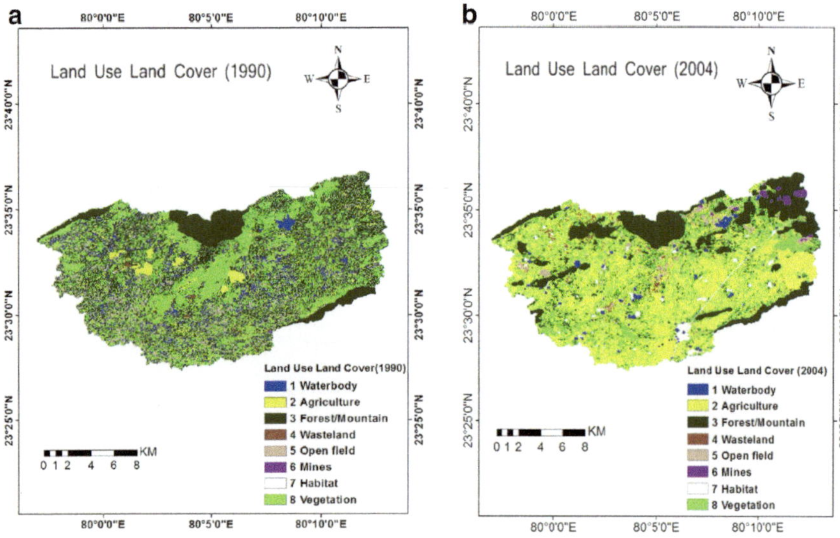


Fig. 16.7 LULC map for the year 1990 and 2004

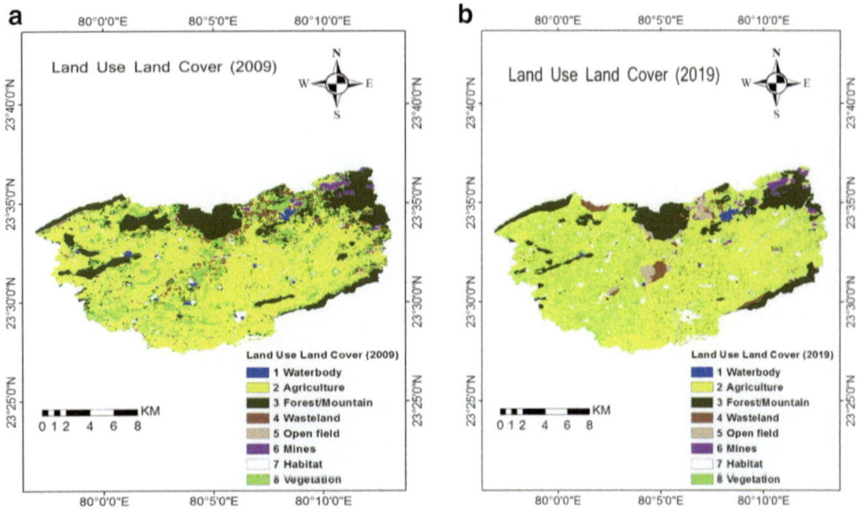


**Fig. 16.8** LULC map for the year 2009 and 2019

The graphical comparison of all the eight classes on the basis of areal distribution obtained from LULC map was also analyzed and is shown in Fig. 16.10 illustrating rising and falling linear trend of different parameters. The classified maps depicted that in the year 1990 area under different classes viz.; agricultural land was 9.80%, wasteland was 0.40%, mining activity was 0.154%, habitat area covered 0.493%, dense forest and vegetation covering major portion of the watershed, and covered about 40.06% and 28.52% space, respectively. While in the year 2004, about 43.09% of the space was covered by agricultural land against 9.80% space in 1990 clearly noticed an increment in cultivated land area. Waste land, mines, and habitat area covered 1.16%, 0.879%, and 2.57%, respectively while agriculture had greatest share of 43.09% over open forest 25.65%. The water body share was about 1.339% of the total geographical area for the year 2004 (Table 16.2; Fig. 16.11).

The classified image for the year 2009 depicted that above 50% of the space was occupied by agricultural land, whereas open forest share was 21.84%. Vegetation occupied 10.34%; habitat area covered 2.68% while waste land and water bodies occupied 2.88% and 0.89% (less than 1%) area, respectively. In 2019, the area under agricultural field has fairly grown and covered 61.86% collectively whereas forest lands depleted to 14.76%. Mines and habitat covered 1.205% and 5.40% area, respectively, resulting in depletion of waterbody to minimal amount that is only 0.625% (< 1%) (Table 16.2).

**Table 16.2** LULC distribution in Kanari river watershed of Jabalpur

| LULC classes | 1990 | | 2004 | | 2009 | | 2019 | |
|---|---|---|---|---|---|---|---|---|
| | Area (ha) | Area (%) | Area (ha) | Area (%) | Area (ha) | Area (%) | Area (ha) | **Area (%)** |
| Waterbody | 2366.45 | 7.910807 | 400.68 | 1.339433 | 268.38 | 0.897168 | 187.11 | 0.62549 |
| Agriculture | 2933.7 | 9.807068 | 12,891.8 | 43.09601 | 16,981.5 | 56.76747 | 18,507.1 | 61.8674 |
| Forest/mountain | 11,984.7 | 40.06366 | 7675.83 | 25.65954 | 6535.08 | 21.84612 | 4417.92 | 14.76867 |
| Wasteland | 119.827 | 0.40057 | 347.58 | 1.161925 | 862.11 | 2.881948 | 1205.64 | 4.030335 |
| Openfield | 3771.41 | 12.60745 | 1212.84 | 4.054404 | 1052.01 | 3.516765 | 591.75 | 1.978161 |
| Mines | 46.1703 | 0.154343 | 263.16 | 0.879718 | 316.44 | 1.057828 | 360.63 | 1.20555 |
| Habitat | 147.763 | 0.493957 | 770.67 | 2.576273 | 803.79 | 2.68699 | 1617.8 | 5.408145 |
| Vegetation | 8533.41 | 28.52634 | 6377.58 | 21.31962 | 3094.83 | 10.34571 | 3030.39 | 10.13029 |
| Total | 29,914.14 | | 29,914.14 | | 29,914.14 | | 29,914.14 | |

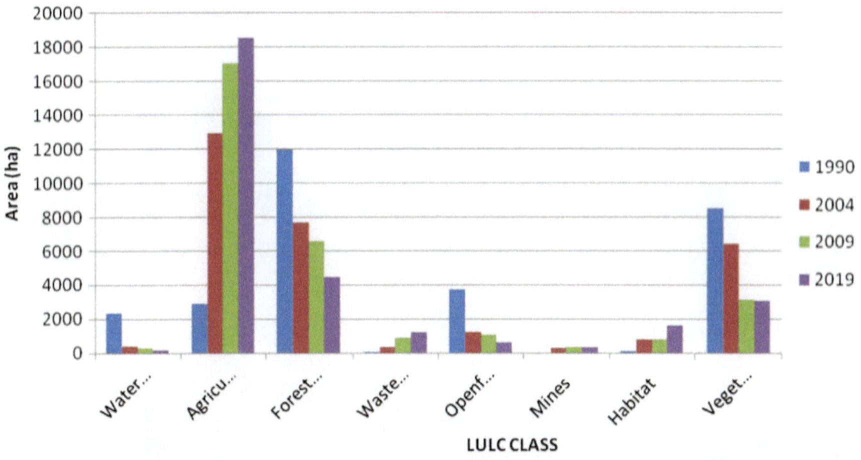


**Fig. 16.9** LULC area wise comparison of different years

## *16.4.3 Field Survey and Accuracy Assessment*

The ground truthing was performed at different levels involving personal interview with the resident of the area for validating the path of the river and determination of causes of flow cessation. Some of the photograph of the field survey and ground truthing are illustrated in Fig. 16.12.

Kappa accuracy assessment which is one of the most generously used approaches was selected for assessing the accuracy of classified map of the year 1990, 2004, 2009, and 2019. Accuracy assessment was also performed using satellite imagery having high resolution of Planet scope 100 stratified random points with minimum 12 points in each class were generated. User's accuracy and individual producer's accuracy of all the classes are also presented.

**Accuracy assessment for the year 1990**

The overall classification accuracy for the obtained classified image of 1990 was determined and was 63% having overall Kappa coefficient of 0.5734 (Tables 16.3 and 16.4).

**Accuracy assessment for the year 2004**

The overall classification accuracy for the obtained classified image of 2004 was determined and was 83% having overall Kappa coefficient of 0.8016 (Tables 16.5 and 16.6).

**Accuracy assessment for the year 2009**

The overall classification accuracy for the obtained classified image of 2009 was determined and was 90% having overall Kappa coefficient of 0.8690 (Tables 16.7 and 16.8).

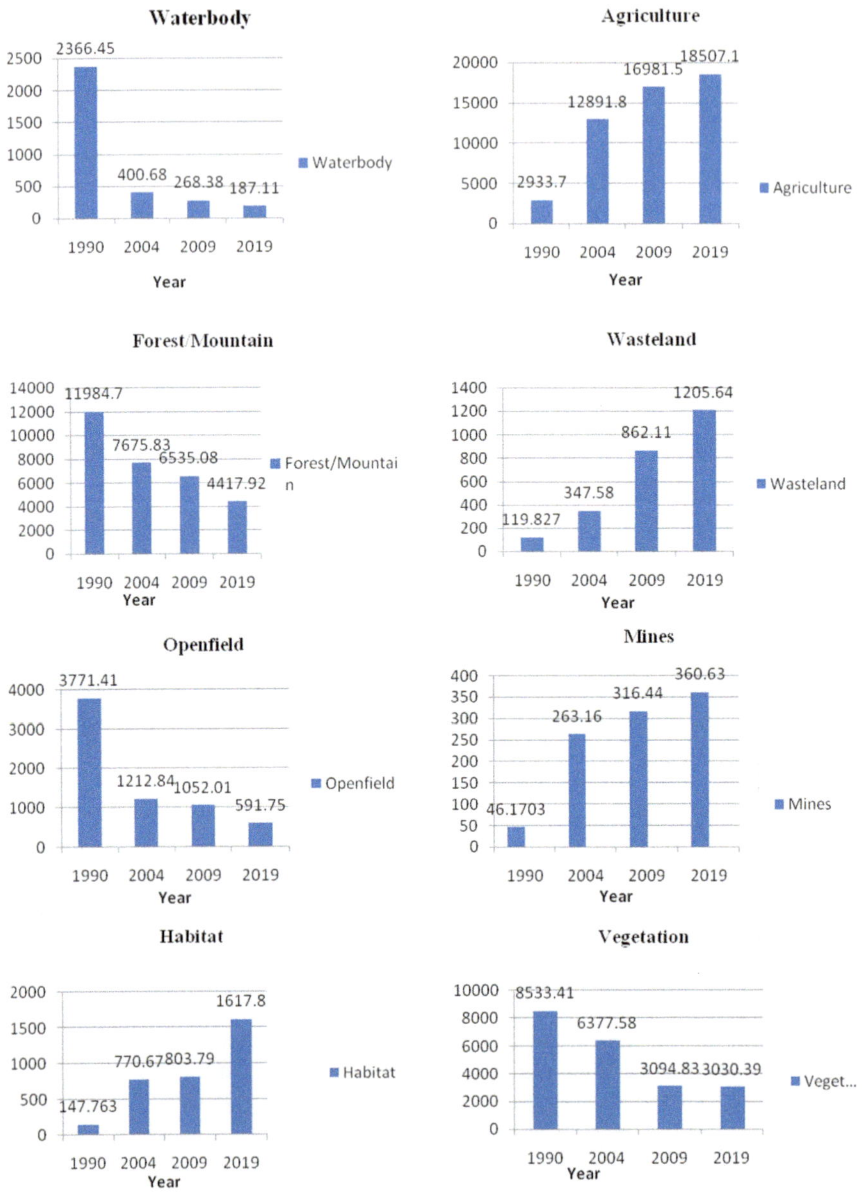


**Fig. 16.10** LULC area wise comparison of different years of eight classes

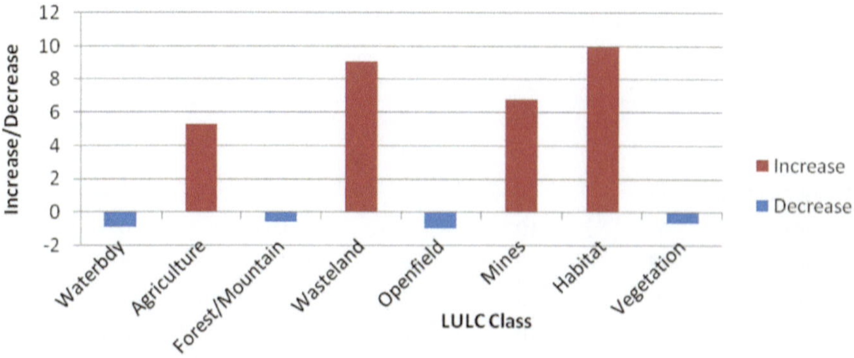


**Fig. 16.11** Increase/decrease in area (ha) of eight classes

**Fig. 16.12** Photograph captured during ground truthing and accuracy assessment

**Table 16.3** Classification accuracy

| LULC class | Reference total | Classified total | Number correct | Producers accuracy (%) | Users accuracy (%) |
|---|---|---|---|---|---|
| Waterbody | 9 | 12 | 7 | 77.78 | 58.33 |
| Agriculture | 16 | 12 | 7 | 43.75 | 58.33 |
| Forest/mountain | 17 | 18 | 8 | 47.06 | 44.44 |
| Wasteland | 6 | 10 | 6 | 100.00 | 60.00 |
| Openfield | 13 | 12 | 7 | 53.85 | 58.33 |
| Mines | 9 | 10 | 8 | 88.89 | 80.00 |
| Habitat | 11 | 10 | 8 | 72.73 | 80.00 |
| Vegetation | 19 | 16 | 12 | 63.16 | 75.00 |

**Table 16.4** Overall Kappa statistics

| LULC class | Kappa |
|---|---|
| Waterbody | 0.5421 |
| Agriculture | 0.504 |
| Forest/mountain | 0.3307 |
| Wasteland | 0.5745 |
| Openfield | 0.5211 |
| Mincs | 0.7802 |
| Habitat | 0.7753 |
| Vegetation | 0.6914 |

**Table 16.5** Classification accuracy

| LULC class | Reference total | Classified total | Number correct | Producers accuracy (%) | Users accuracy (%) |
|---|---|---|---|---|---|
| Waterbody | 9 | 10 | 9 | 100.00 | 90.00 |
| Agriculture | 26 | 23 | 18 | 69.23 | 78.26 |
| Forest/mountain | 13 | 13 | 13 | 100.00 | 100.00 |
| Wasteland | 9 | 11 | 8 | 88.89 | 72.73 |
| Openfield | 7 | 10 | 6 | 85.71 | 60.00 |
| Mines | 8 | 10 | 8 | 100.00 | 80.00 |
| Habitat | 12 | 11 | 11 | 91.67 | 100.00 |
| Vegetation | 16 | 12 | 10 | 62.50 | 83.33 |

**Accuracy assessment for the year 2019**

The overall classification accuracy for the obtained classified image of 2019 was determined and was 87.34% having overall Kappa coefficient of 0.8390 (Tables 16.9 and 16.10).

**Table 16.6** Overall Kappa statistics

| LULC class | Kappa |
|---|---|
| Waterbody | 0.8901 |
| Agriculture | 0.7062 |
| Forest/mountain | 1 |
| Wasteland | 0.7003 |
| Openfield | 0.5699 |
| Mines | 0.7826 |
| Habitat | 1 |
| Vegetation | 0.8016 |

**Table 16.7** Classification accuracy

| LULC class | Reference total | Classified total | Number correct | Producers accuracy (%) | Users accuracy (%) |
|---|---|---|---|---|---|
| Waterbody | 3 | 5 | 3 | 100.00 | 60.00 |
| Agriculture | 42 | 42 | 38 | 90.48 | 90.48 |
| Forest/mountain | 17 | 18 | 17 | 100.00 | 94.44 |
| Wasteland | 7 | 6 | 6 | 85.71 | 100.00 |
| Openfield | 8 | 9 | 7 | 87.50 | 77.78 |
| Mines | 5 | 5 | 5 | 100.00 | 100.00 |
| Habitat | 7 | 6 | 5 | 71.43 | 83.33 |
| Vegetation | 11 | 9 | 9 | 81.82 | 100.00 |

**Table 16.8** Overall Kappa statistics

| LULC class | Kappa |
|---|---|
| Waterbody | 0.5876 |
| Agriculture | 0.8358 |
| Forest/mountain | 0.9331 |
| Wasteland | 1 |
| Openfield | 0.7585 |
| Mines | 1 |
| Habitat | 0.8208 |
| Vegetation | 1 |

**Runoff estimation using integration of SCS-CN method and ArcGIS**

The methodology applied for runoff calculation is depicted in Fig. 16.13. The soil conservation service curve number approach is usually utilized empirical methods to determine the direct runoff from a drainage basin (USDA 1972) in the Kanari basin. The surface storage is combined with infiltration losses by the relation of

**Table 16.9** Classification accuracy

| LULC class | Reference total | Classified total | Number correct | Producers accuracy (%) | Users accuracy (%) |
|---|---|---|---|---|---|
| Waterbody | 6 | 10 | 4 | 79.67 | 79.00 |
| Agriculture | 59 | 23 | 19 | 82.20 | 82.61 |
| Forest/mountain | 4 | 13 | 2 | 87.98 | 96.87 |
| Wasteland | 3 | 11 | 2 | 86.67 | 18.18 |
| Openfield | 4 | 10 | 1 | 87.00 | 69.00 |
| Mines | 3 | 10 | 3 | 100.00 | 87.00 |
| Habitat | 6 | 11 | 2 | 79.87 | 91.87 |
| Vegetation | 19 | 12 | 7 | 83.84 | 58.33 |

**Table 16.10** Overall Kappa statistics

| LULC class | Kappa |
|---|---|
| Waterbody | 0.7617 |
| Agriculture | 0.7758 |
| Forest/mountain | 1 |
| Wasteland | 0.8754 |
| Openfield | 1 |
| Mines | 0.7976 |
| Habitat | 0.8976 |
| Vegetation | 0.9321 |

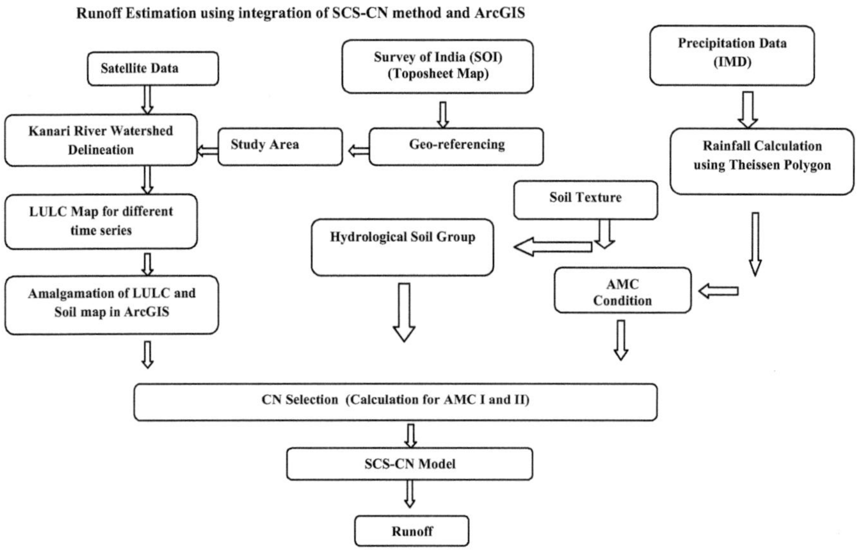


**Fig. 16.13** Methodology for runoff calculation

$$Q = (P - \mathrm{Ia})^2/P - \mathrm{Ia} + S \tag{16.1}$$

Here,

$Q$ gathered/collected runoff (mm),
$P$ effective rainfall depth (mm),
Ia initial abstraction (mm) (surface storage, interception, and infiltration preceding to runoff in the entire watershed). The empirical relation was obtained in the terms of Ia and is depicted below.

The empirical relationship is

$$\mathrm{Ia} = 0.3S \tag{16.2}$$

The potential maximum retention ($S$) for Indian conditions is given by

$$S = \left(\frac{25{,}400}{\mathrm{CN}}\right) - 254 \tag{16.3}$$

where

CN curve number (source: USDA 1972).

Rewriting the equation as

$$Q = (P - 0.3S)^2/(P + 0.7S) \tag{16.4}$$

The runoff from the entire area was calculated in significance of value of CN using Eqs. 16.3 and 16.4.

The SCS curve number serves the determination of capability of soils to accommodate infiltrated water in regards of land use/ land cover (LULC) and antecedent soil moisture condition (AMC) (Amutha and Porchelvan 2009) (Table 16.11). U.S soil conservation service (SCS) soils distributed the soil into four hydrologic soil groups, namely group $A$, $B$, $C$, and $D$, in regards of probable rate of runoff and final infiltration. CN, $S$, and $P$ calculated for various years are shown in Table 16.12.

**Table 16.11** Antecedent soil moisture classes (AMC) group

| AMC group | Soil characteristics | Five-day antecedent rainfall in mm | |
|---|---|---|---|
| | | Dormant season | Growing season |
| I | Wet condition | < 13 | < 36 |
| II | Average condition | 13–28 | 36–53 |
| III | Heavy condition | > 28 | > 53 |

**Table 16.12** Hydrological and quantitative calculations in the watershed

| AMC | CN | | | | *S* | | | | *P* > 0.3S | | | |
|---|---|---|---|---|---|---|---|---|---|---|---|---|
| | 1990 | 2004 | 2009 | 2019 | 1990 | 2004 | 2009 | 2019 | 1990 | 2004 | 2009 | 2019 |
| I | 67.75 | 71.33 | 73.48 | 75.06 | 120.88 | 102.06 | 91.64 | 84.38 | 36.26 | 30.61 | 27.49 | 25.3 |
| II | 82.76 | 85.05 | 86.37 | 87.32 | 52.91 | 44.63 | 40.06 | 36.88 | 15.86 | 13.38 | 12.01 | 11.0 |
| III | 92.00 | 93.02 | 93.68 | 94.16 | 22.07 | 19.05 | 17.10 | 15.74 | 6.62 | 5.71 | 5.13 | 4.24 |

**Table 16.13** Aimed weighted curve number (WCN) at Kanari River watershed for the year 1990

| Land use cover | Soil type (HSG) | Area in ha | Area in $m^2$ | CN | Area | Area * CN | Weighted curve number (WCN) |
|---|---|---|---|---|---|---|---|
| Waterbody | D | 2265.4 | 22,654,000 | 100 | 0.075757 | 7.575728 | AMC I—67.75 AMC II—82.76 AMC III—92.00 |
| Agriculture | D | 2941.15 | 29,411,500 | 90 | 0.098355 | 8.851954 | |
| Forest/mountain | D | 12,028.5 | 120,285,000 | 77 | 0.402245 | 30.97289 | |
| Wasteland | D | 119.827 | 1,198,270 | 94 | 0.004007 | 0.376671 | |
| Openfield | D | 3780.12 | 37,801,200 | 84 | 0.126411 | 10.61853 | |
| Mines | D | 46.7092 | 467,092 | 95 | 0.001562 | 0.14839 | |
| Habitat | D | 149.38 | 1,493,800 | 86 | 0.004995 | 0.429606 | |
| Vegetation | D | 8572.31 | 85,723,100 | 83 | 0.286667 | 23.79334 | |
| Total | | | 299,033,962 | | | 82.76711 | |

It was found from the calculations (SCS method), annual surface runoff depth (465.59 mm) for the year 1990 are multiplied by watershed area ($A$ = 299,033,962 $m^2$) provides the total average runoff volume as (139,228,136.1 $m^3$). The result for the year 1990 is depicted in Table 16.13.

It was found from result of the calculations (SCS method), annual surface runoff depth (597.21 mm) for the year 2004 is multiplied by watershed area ($A$ = 299,033,962 $m^2$) provides the total average runoff volume as (178,807,886.9 $m^3$). The result for the year 2004 is depicted in Table 16.14.

It was found from the calculations (SCS method), annual surface runoff depth (737.44 mm) for the year 2009 are multiplied by watershed area ($A$ = 299,033,962 $m^2$) provides the total average runoff volume as (220,599,996.4 $m^3$). The result for the year 2009 is depicted in Table 16.15.

It was found from the calculations (SCS method), annual surface runoff depth (1319.24 mm) for the year 2019 are multiplied by watershed area ($A$ = 299,033,962 $m^2$) provides the total average runoff volume as (361,883,866 $m^3$). The result for the year 2019 is depicted in Table 16.16.

The runoff varies 465.59–1319.24 mm (1990–2019) as shown in Table 16.17. The amount of rainfall varies between 1136.7 and 2466.9 mm in the Kanari River watershed as shown in Table 16.17. The calculated average annual runoff is found to

**Table 16.14** Aimed weighted curve number (WCN) at Kanari River watershed for the year 2004

| Land use cover | Soil type (HSG) | Area in ha | Area in $m^2$ | CN | Area | Area * CN | Weighted curve number (WCN) |
|---|---|---|---|---|---|---|---|
| Waterbody | D | 400.68 | 4,006,800 | 100 | 0.013383 | 1.33827 | AMC I—71.33 AMC II—85.05 AMC III—93.02 |
| Agriculture | D | 12,891.8 | 128,918,000 | 90 | 0.430586 | 38.75272 | |
| Forest/mountain | D | 7675.83 | 76,758,300 | 77 | 0.256373 | 19.74069 | |
| Wasteland | D | 347.58 | 3,475,800 | 94 | 0.011609 | 1.091261 | |
| Openfield | D | 1212.84 | 12,128,400 | 84 | 0.040509 | 3.402742 | |
| Mines | D | 263.16 | 2,631,600 | 95 | 0.00879 | 0.835006 | |
| Habitat | D | 770.67 | 7,706,700 | 86 | 0.02574 | 2.213671 | |
| Vegetation | D | 6377.58 | 63,775,800 | 83 | 0.213011 | 17.67992 | |
| Total | | | 299,033,962 | | | 85.05428 | |

**Table 16.15** Aimed weighted curve number (WCN) at Kanari River watershed for the year 2009

| Land use cover | Soil type (HSG) | Area in ha | Area in $m^2$ | CN | Area | Area * CN | Weighted curve number (WCN) |
|---|---|---|---|---|---|---|---|
| Waterbody | D | 268.38 | 2,683,800 | 100 | 0.008972 | 0.897168 | AMC I—73.48 AMC II—86.37 AMC III—93.68 |
| Agriculture | D | 16,981.5 | 169,815,000 | 90 | 0.567675 | 51.09072 | |
| Forest/mountain | D | 6535.08 | 65,350,800 | 77 | 0.218461 | 16.82152 | |
| Wasteland | D | 862.11 | 8,621,100 | 94 | 0.028819 | 2.709031 | |
| Openfield | D | 1052.01 | 10,520,100 | 84 | 0.035168 | 2.954083 | |
| Mines | D | 316.44 | 3,164,400 | 95 | 0.010578 | 1.004936 | |
| Habitat | D | 803.79 | 8,037,900 | 86 | 0.02687 | 2.310812 | |
| Vegetation | D | 3094.83 | 30,948,300 | 83 | 0.103457 | 8.586939 | |
| Total | | | 299,033,962 | | | 86.3752 | |

be 779.87 mm and average runoff volume for the period of 29 years is 225,129,971.3 $m^3$. The rainfall runoff relationship showed in Fig. 16.14 for Kanari watershed. Rainfall-runoff variation in Kanari watershed is showed in Figs. 16.15 and 16.16, respectively depicting rising linear trend in aspect of both the parameters simultaneously decreasing opportunity time and hence infiltration decreasing baseflow of the area.

**Table 16.16** Aimed weighted curve number (WCN) at Kanari River watershed for the year 2019

| Land use cover | Soil type (HSG) | Area in ha | Area in $m^2$ | CN | Area | Area * CN | Weighted curve number (WCN) |
|---|---|---|---|---|---|---|---|
| Waterbody | D | 187.11 | 1,871,100 | 100 | 0.006254 | 0.625402 | AMC I—75.06 AMC II—87.32 AMC III—94.16 |
| Agriculture | D | 18,507.1 | 185,071,000 | 90 | 0.618587 | 55.67284 | |
| Forest/mountain | D | 4417.92 | 44,179,200 | 77 | 0.147666 | 11.37028 | |
| Wasteland | D | 1205.64 | 12,056,400 | 94 | 0.040298 | 3.787983 | |
| Openfield | D | 591.75 | 5,917,500 | 84 | 0.019779 | 1.661422 | |
| Mines | D | 360.63 | 3,606,300 | 95 | 0.012054 | 1.145112 | |
| Habitat | D | 1617.8 | 16,178,000 | 86 | 0.054074 | 4.650352 | |
| Vegetation | D | 3030.39 | 30,303,900 | 83 | 0.101289 | 8.406963 | |
| Total | | | 299,033,962 | | | 87.32035 | |

**Table 16.17** Annual average runoff depth and volume

| Year | Rainfall in mm | Runoff in mm | Volume ($m^3$) = (Runoff * area) | Runoff Co. = (RO/RF) |
|---|---|---|---|---|
| 1990 | 1136.7 | 465.59 | 139,228,136.1 | 0.4 |
| 2004 | 1220.38 | 597.22 | 178,807,886.9 | 0.48 |
| 2009 | 1225.1 | 737.44 | 220,599,996.4 | 0.6 |
| 2019 | 2466.9 | 1319.25 | 361,883,866 | 0.49 |
| Average | 1512.27 | 779.8750954 | 225,129,971.3 | 0.49 |

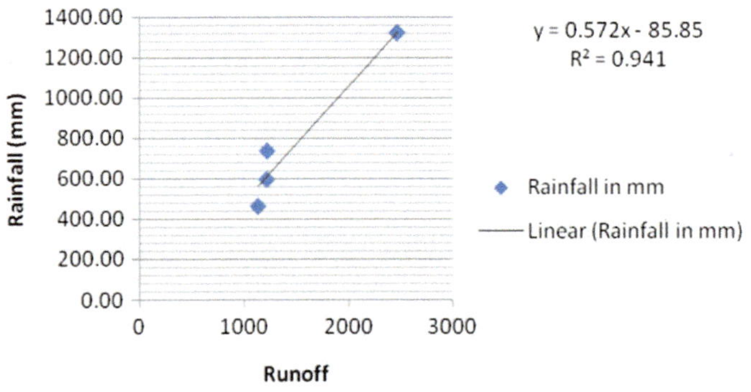


**Fig. 16.14** Scatter plot between the rainfall and calculated runoff

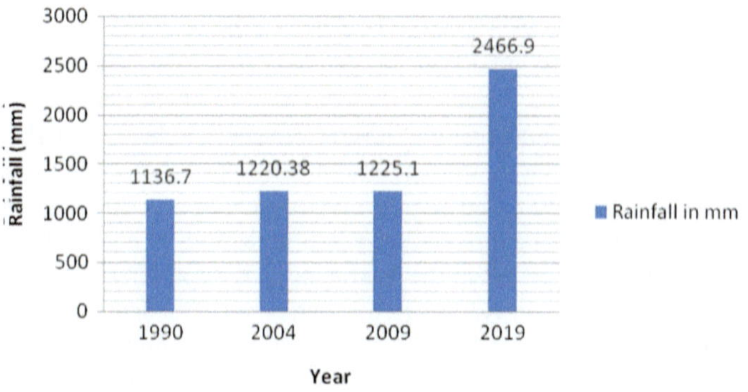


**Fig. 16.15** Rainfall variation in Kanari watershed

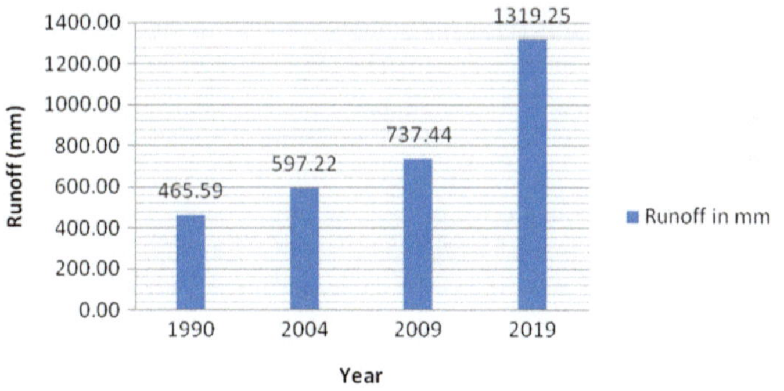


**Fig. 16.16** Runoff variation in Kanari watershed

## 16.5 Conclusion

Remote sensing (RS) and geographic information system (GIS) technology is one of the broadly or extensively applied science used for detection analysis of land use and land cover change. The detection of land use change criterion for Kanari River watershed was performed using downloaded satellite images and was further classified utilizing ERDAS imagine 2014 software. The observations of the study finally concluded that there were significant change in land use land cover over the period of 29 years in Kanari River watershed. There was significant rise in agriculture land by about 53.08% and a downfall in water bodies by about 92.07% from the year 1990–2019. The classification scheme utilized eight LULC classes representing urban/built-up areas (habitation), vegetation, forest, wasteland, open field, water features, agriculture field, and mined area for the year 1990, 2004, 2009 and 2019. The classification method used for this project produced an overall accuracy of 93.4% which finally interpreted contrasting percentage of various classes that resulted in drying of the river. The result indicated that the habitation and mining activities was increased by 98.3% and 67.2% (form the year 1990 to 2019), respectively, which consequently enhanced the groundwater overdraft. Further analysis showed that the forest (63.2% decrease), vegetation (64.64% decrease), and water body (91.7% decrease) acreage were converted into waste land (90.61% increase), agriculture field (52.92% increase) resulted in shifting of the origin of the river by 1.07 km obtained during ground truthing. The forest area was chiefly converted into habitat land as a result of incremented population leading to high requirement rate in the area. This was associated to the continued extension of settlement and cultivated over period of years in Kanari River watershed. The science about the land use change is pre-requisite for land use planning and administration activities in Kanari River watershed. Therefore, this study revealed that there was an increase of mined area and habitat which was the root cause of drying of the river. Curve number and soil conservation service model were utilized in the study by analyzing soil map and land use map described in ArcGIS, as an input data. In the current study, the methodology applied for the tenacity or determination of runoff utilizing combined SCS and GIS approach can also be applied in any other watershed for the conservation measures. The fruitful soil and water conservation measures need be designed and enforced in the watersheds for grasping or gathering runoff volume and soil loss. Antecedent moisture condition of the soil imitates a significant role in SCN curve number method accounted for estimating runoff depth.The runoff in the study area was also calculated by the integration of ArcGIS and SCS-CN method which finally depicted the runoff in the area kept on increasing as rainfall kept on increasing following a linear trend which predicted that the infiltration opportunity decrease and ultimately resulting in decreased infiltration rate causing decreased baseflow resulted in drying of river. Concluding, soil conversations service –curve number technique is handily proven as a superior method, consuming little time and facilitating handling of big data set as well as identification of suitable site for artificial recharge structures. Various questioner surveys were also conducted which indicated that direct uplifting of water from the stream for irrigation purpose is root cause of the problem.

## References

Amutha R, Porchelvan P (2009) Estimation of surface runoff in Malattar sub-watershed using SCS-CN method. J Soc Remote Sens 37(2):291–304

Arora KM, Mathur S (2001) Multi-source classification using artificial neural network in a rugged terrain. Geocarto Int 16(3):37–44

Babykalpana Y, ThanushKodi K (2010) Supervised/unsupervised classification of LULC using remotely sensed data for Coimbatore city, India. Int J Comput Appl 2(7):26–30

Brierley GJ, Fryirs KA (2005) Geomorphology and river management: applications of the river styles framework. Blackwell, Oxford, pp 398

Dey S, Singh RP (2003) Surface latent heat flux as an earthquake precursor. Nat Hazards Earth Syst Sci 3:749–755

Dey S, Sarkar S, Singh RP (2004) Anomalous changes in column water vapor after Gujarat earthquake. Adv Space Res 33:274–278

Dimyati MUH, Mizuno K, Kobayashi S, Kitamura T (1996) An analysis of landuse/cover change in Indonesia. Int J Remote Sens 17(5):931–944

USDA Soil Conservation Service (1972) National engineering handbook. Hydrology section 4. Chapters 4–10. USDA, Washington, D.C.

Herold M, Gardner ME, Robert DA (2003) Spectral resolution requirements for mapping urban areas. IEEE Trans Geosci Remote Sens 41:1907–1919

Hohensinner S, Hauer C, Muhar S (2018) River morphology, channelization, and habitat restoration. In: Riverine ecosystem management. Springer, pp 41–65

Islam K, Rahman MF, Jashimuddin M (2018) Modeling land use change using cellular automata and artificial neural network: the case of Chunati wildlife sanctuary, Bangladesh. Ecol Indic 88:439–453

Kiruki HM, van der Zanden EH, Malek Z, Verburg PH (2017) Land cover change and woodland degradation in a charcoal producing semi-arid area in Kenya. Land Degr Dev 28:472–481

Lange C, Schneider M, Mutz M, Haustein M, Halle M, Seidel M, Sieker H, Wolter C, Hinkelmann R (2015) Model-based design for restoration of a small urban river. J Hydro-Environ Res 9:226–236

Mishra VN, Kumar P, Gupta DK, Prasad R (2014) Classification of various land features using risat-1 dual polarimetric data. Int Arch Photogramm Remote Sensing Spatial Inf Sci:833–837

Muhar S, Januschke K, Kail J (2016) Evaluating good practice cases for river restoration across Europe: context, methodological framework, selected results and recommendations. Hydrobiologia 769:3–19

Niquisse S, Cabral P, Rodrigues Â, Augusto G (2017) Ecosystem services and biodiversity trends in Mozambique as a consequence of land cover change. Int J Biodivers Sci Ecosyst Serv Manage 13:297–311

Okada Y, Mukai S, Singh RP (2004) Changes in atmospheric aerosol parameters after Gujarat earthquake of January 26 2001. Adv Space Res 33:254–258

Pauleit S, Ennos R, Golding Y (2005) Modeling the environmental impacts of urban land use and land cover change—a study in Merseyside, UK. Landsc Urban Plan 71:295–310

Petit C, Scudder T, Lambin E (2001) Quantifying processes of land-cover change by remote sensing: resettlement and rapid land-cover changes in south-eastern Zambia. Int J Remote Sens 22(17):3435–3456

Read JM, Lam NSN (2002) Spatial methods for characterising land cover and detecting land-cover changes for the tropics. Int J Remote Sens 23(12):2457–2474

Saadat H, Adamowski J, Bonnell R, Sharifi F, Namdar M, Ale-Ebrahim S (2011) Land use and land cover classification over a large area in Iran based on single date analysis of satellite imagery. ISPRS J Photogramm Remote Sens 66:608–619

Saha AK, Arora MK, Csaplovics E, Gupta RP (2005) Land cover classification using IRS LISS III image and DEM in a rugged terrain: a case study in Himalayas. Geocarto Int 20(2):33–40

Sekertekin A, Marangoz AM, Akcin H (2017) Pixel based classification analysis of land use land cover using sentinel-2 and landsat-8 data. Int Arch Photogramm Remote Sensing Spatial Inf Sci:91–93

Shalaby A, Tateishi R (2007) Remote sensing and GIS for mapping and monitoring land cover and landuse changes in the Northwestern coastal zone of Egypt. Appl Geogr 27(1):28–41

Singh RP, Sahoo AK, Bhoi S, Kumar MG, Bhuiyan CS (2001) Ground deformation of Gujarat earthquake of 26 January. J Geol Soc India 58:209–214

Singh RP, Bhoi S, Sahoo AK, Raj U, Ravindran S (2001) Surface manifestations after the Gujarat earthquake. Curr Sci 81:164–166

Singh RP, Bhoi S, Sahoo AK (2001) Significant changes in the ocean parameters after the Gujarat earthquake. Curr Sci 80:1376–1377

Singh R, Simon B, Joshi PC (2001) Estimation of surface latent heat fluxes from IRSP4/MSMR satellite data. Proc Indian Acad Sci 110:231–238

Singh RP, Bhoi S, Sahoo AK (2002) Changes observed on land and ocean after Gujarat earthquake 26 January 2001 using IRS data. Int J Remote Sens 23:3123–3128

Sinha R, Jain V, Tandon SK (2012) River systems and river science in India: major drivers and challenges. Earth system processes and disaster management. Springer, pp 67–90
Sinha R, Jain V, Tandon S (2013) River systems and river science in India: major drivers and challenges. Earth system processes and disaster management, pp 67–90
USDA Soil Conservation Service (1974) USDA-SCS Soil Survey of Travis County, Texas. College Station, Texas. Texas Agricultural Experiment Station, and Washington, D.C.
Tolessa T, Senbeta F, Kidane M (2017) The impact of land use/land cover change on ecosystem services in the central highlands of Ethiopia. Ecosyst Serv 23:47–54
Trivedi A, Pyasi SK, Galkate RV (2019) Impact of climate change using trend analysis of rainfall, RRL AWBM toolkit, synthetic and arbitrary scenarios. Curr J Appl Sci Technol:1–18
Verburg PH, Veldkamp A, de Koning GHJ, Kok K, Bouma J (1999) A spatial explicit allocation procedure for modelling the pattern of land use change based upon actual land use. Ecol Model 116:45–61
Vitousek PM, Mooney HA, Lubchenco J, Melillo JM (1997) Human domination of Earth's ecosystems. Science 277(5325):494–499
Youssef M, Ali BA (2017) Revival of forgotten rivers through recreating the cultural promenade: a case study of the revival of beirut river, lebanon. Sustain Dev Plann 226:725–737
Zende AM, Devkate S, Atal KR (2015) Selection of suitable sites for water harvesting structures in chand watershed using geospatial techniques. In: 20th international conference on hydraulics. IIT Roorkee, pp 1–8

# Chapter 17
# Impacts of Climate Variability on Urban Rainfall Extremes Using Statistical Analysis of Climatic Variables for Change Detection and Trend Analysis

**Manish Kumar Sinha, Klaus Baier, Rafig Azzam, Mukesh Kumar Verma, and Sunil Kumar**

**Abstract** Importance of extreme weather events under climate variability on design rainfall to design a hydraulic structure is very important to mitigate the risk of disasters. This study yields an approach to identify the changes in Intensity–Duration–Frequency (IDF) using statistical trend analysis and change detection test on climatic parameters of Raipur City, a capital of Chhattisgarh state in India. Extreme events are often described by their expected frequency of recurrence; hence frequency analysis of shorter duration (15 min) rainfall has been conceded using Gumbel's Extreme Value-I distribution. The results of change-point analysis indicated a climate shift nearly in year 1993, that has been taken as standard change-point which has split the data series and three different periods. Results concluded the temperature was the main factor which was causing the effect of changing rainfall pattern in Raipur city. Frequency analysis results Intensity–Duration–Frequency (IDF) curve for Raipur city which was influenced by changing climate pattern in Raipur urban area. Design intensities of rainfall were corrected using results of trend analysis and change-point test. The result attributes the effect of urbanization, rapid development of industrial and commercial activity which altered natural ecology and environment in Raipur city.

**Keywords** Change-point detection · Climate shift · Climate variability · Intensity–duration–frequency curve · Trend analysis · Raipur city

---

M. K. Sinha (✉) · K. Baier · R. Azzam
Department of Engineering Geology and Hydrogeology, RWTH Aachen University, 52064 Aachen, Germany

M. K. Verma
Civil Engineering, Chhattisgarh Swami Vivekanand Technical University, Bhilai, Newai, Chhattisgarh 491107, India

S. Kumar
Council of Scientific & Industrial Research, National Environmental Engineering Research Institute, Nagpur 440020, India



P. Kumar et al. (eds.), *Water Resources Management and Sustainability*, Advances in Geographical and Environmental Sciences,
https://doi.org/10.1007/978-981-16-6573-8_17

## 17.1 Introduction

Changes in social and biogeophysical environment and is modulated by the inherent dynamics of natural systems is a key driver of change and variability in the climate state, or climate dynamics (IPCC 2007). The impacts of urbanization on climate change are converging in unsafe manner. Urban cities are major contributors to climate change, although they occupied less than 2% of the earth's surface, cities consume 78% of the world's energy and yield more than 60% of all $CO_2$ and other greenhouse gases, mainly via generation of energy, automobile and industries use (UN-Habitat 2011). Simultaneously, towns and cities are vulnerable to climate change. Urban population in developing and emerging countries grows at higher rates compared to countryside. In the last fifty years, South Asia's total population doubled while its urban population increased fivefold. The twentieth century has witnessed the phenomenon of rapid urbanization (Baier et al. 2014). By year 1900, 13% of the world's population was urban, and according to UNEP (2011), the urban population increased from 220 million in 1900 to 732 million in 1950 (29% of the world's population). By 2007, 50% of the world populations were living in cities. According to latest predictions, 4.9 billion people, or 60% of the world's population, are expected to be urban dwellers by 2030 (Strohschön et al. 2009; UNEP 2011). 1/3rd of India's population is already been urban, it is necessary to acquire information on growth patterns of cities and how they impact the living environment (Farooq and Ahmad 2008). And now India has decided to develop 98 smart cities by investing 1.9 billion USD (Praharaj et al. 2017). Raipur city is one of them and still developing since 2001 after became capital of Chhattisgarh in India. A new city has been proposed as Naya Raipur besides the old Raipur City. This results expansion of the city in all directions as the large-scale urban sprawl and urban land use change. Today, Raipur city is an important regional, commercial and industrial hub for steel, power, coal and aluminium industries. Raipur city has the largest market of steel and iron in India and among the richest cities of India. The city witnessed a high growth rate in population which may be a cause of extreme weather events in and around the city (Narain 2017).

In recent practice, non-stationarity in rainfall trend causing uncertainty in design rainfall events. Hence, the present study has been conceded at Raipur city to show how the climate variability due to human intervention has become more common in India and how to cope with this using a methodology adopted in this study. The changes in climatic parameter trend may be determined by Global Circulation Models (GCMs) or Downscaled Regional Circulation Models (RCMs) (Ghosh and Mujumdar 2006) or statistical analysis of historical time series (Kioutsioukis et al. 2010). In India, Australia and countries of United States, where usually wet climate has been seen are used to follow urban storm water management practices to make their city resilient to flood (Smith 1999). To mitigate and manage these floods, Intensity–Duration–Frequency (IDF) curve is being used since 30s (Kimball 1938). Climate change impact studies related to urban areas have focused generally on flood risk or storm water management. IDF curves are frequency analysis of shorter duration rainfall events used by scientists and engineers to create flood frequency estimates (Kimball 1938). According to Chandra et al. (2015), the optimum design of urban infrastructure is important to have a reliable estimate of extreme rainfall intensity. Chandra et al. (2015) have used Bayesian analysis to deal with application of Global Circulation Model (GCM). Application of statistical tools for extremes event forecast, not only in meteorology but also in several hydrology disciplines, has led to dissatisfaction (De Paola et al. 2014). Several examples abound in gross underestimation/overestimation of designs based on engineered standard procedure that eventually cause catastrophic damage. Cheng and AghaKouchak (2014) used Gumle's extreme value type I distribution to predict non-stationarity in ground-based observations of extremes rainfall events (annual maximum series) from the US- NOAA Atlas 14.

Bai (1997), Bates et al. (2012), Bisai et al. (2014), Khan et al. (2014) performed change-point tests, namely, Pettitt's test (PeT), Buishand's range test (BRT), von Neumann ratio test (vNRT) and standard normal homogeneity test (SNHT) to identify single change-point in the data series. Multiple change-point methods are sometimes more realist but in case of climatic scenario in Raipur city, it has already been proposed by Jaiswal et al. (2015) for application of single change-point in this study area that has significant results. Moreover, these methods were also been used by Fealy and Sweeney (2005), Gallagher et al. (2013), Gao et al. (2011), Jandhyala et al. (2010), Khan et al. (2014), Lee and Kim (2014), Lund and Reeves (2002), Menne and Williams (2005); Pohlert (2017), Pranuthi et al. (2014) predict single change-point in climatological variables. Bisai et al. (2014), Dhorde and Zarenistanak (2013), Gadgil and Dhorde (2005), Jain and Kumar (2012), Karmeshu (2012) and Mavromatis and Stathis (2010) analysed mean daily/monthly/annual temperature of one or more station using non-parametric linear regression (LRT), Mann Kendall (MKT) and Spermann rho test (SRT) to identify change in data series. These tests result positive as well negative trends at 95% confidence interval for an urban area which was well correlated with recorded historical time series. Many researchers have identified historical time series of evaporation, temperature, and rainfall for change-point detection and statistical trend investigation to study possible impact of climate variability and impact of industrialization and anthropogenic activities (Barrow et al. 2004; Boroujerdy 2008; Buhairi 2010; Croitoru et al. 2011; Karpouzos et al. 2010; Serra et al. 2001; Tabari and Marofi 2010; Zarenistanak et al. 2014).

Present paper is a part of research work undergoing at RWTH Aachen University, Department of Engineering Geology and Hydrogeology (LIH). In this project, 'Interaction between urbanization and water management' at Indian, Indonesian and Chinese urban city has focused (Azzam et al. 2014; Christ et al. 2012; Putra and Baier 2008; Sinha et al. 2016; Strohschön et al. 2009; Wakode et al. 2014). An Indian City Raipur is one of them, which is the capital city of Chhattisgarh state in Indian (Agrawal 2013; CRISIL 2014). This paper explains an approach to detect the change in IDF relations under climate variability. Also, the statistical methods of trend detection and change-point analysis have been used for climatic variables including evaporation (EP), minimum temperature (MinT), maximum temperature (MaxT), relative humidity (RH), wind speed (WS), sunshine hour (SS) and rainfall (RF) time series of Raipur city.

## 17.2 Study Area and Data Used

### 17.2.1 Description of Study Area

Raipur city is the capital and largest city of the Indian state Chhattisgarh. It was part of the state Madhya Pradesh, until the 1 November 2000, when 16 districts of Madhya Pradesh were formed into the new state Chhattisgarh (Agrawal 2013). Raipur Municipal area extends between 21° 12' N–21° 18' N and 81° 33′–81° 41' E (Fig. 17.1). Raipur Municipal Corp. (RMC) is the core town forming Raipur Standard Urban Area (SUA) situated in Raipur district. The Mahanadi River flows to the east of the Raipur city and one of its tributary Kharun River is on west. The city situated at the bank of Kharun River with its land merges at north up to Chota Nagpur Plateau and south up to Deccan Plateau. The city has tropical wet and dry climate where temperature remains moderate throughout the year. March and June are extremely hot, whereas the temperature in April–May sometimes rises above 48 °C. The city experiences the average annual rainfall about 1200 mm of rainfall, 90% contributed in late June to early October by the monsoon season. Winters in Raipur, temperature can fall to 8 °C (mostly at night), although it is mild in November to January.

The whole of Raipur city is part of the open Chhattisgarh plain, which is thickly populated and closely cultivated. The climate of the study area is sub-tropical with three distinct seasons where temperatures remain moderate for most of the year.

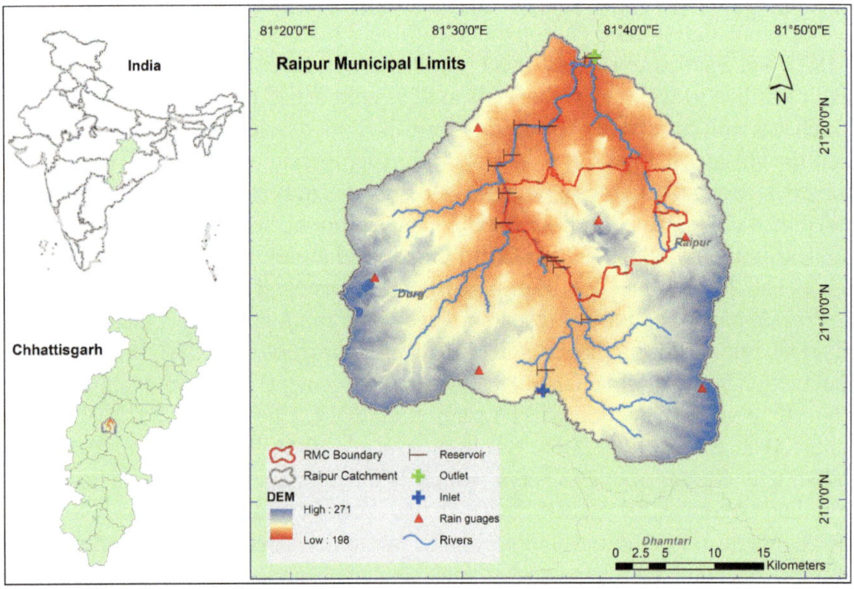


**Fig. 17.1** Location of study area indicating rain gauge location (designated delta shape is Raipur catchment and red boundary is Raipur city municipal boundary besides bank of Kharun River)

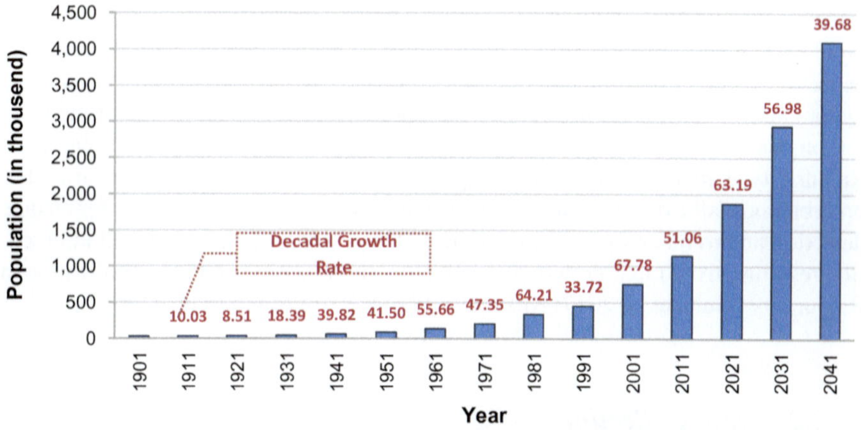


**Fig. 17.2** Population growth of Raipur City in thousands, modified after Census-of-India (2011b) and GIZ et al. (2011)

May is found hottest month and December is the coldest. The average elevation is about 295 m above mean sea level. The city occupies the North-western part of Raipur district. Raipur city is situated in the fertile plains of Chhattisgarh state (Sinha et al. 2016). The city has one of the richest bio-diversity habitats and above all, over 200 non-timber forest produces, with tremendous potential for value addition. There are mega industries in steel, aluminium and cement already functioning in the near vicinity of Raipur municipal corporation area even before Chhattisgarh was declared a separate state from Madhya Pradesh. After the new state was formed, the city is also gradually emerging as an educational hub (Bodhankar 2002; Sinha et al. 2002).

Population statics from Census of India (Census-of-India 2011b) and City Sanitation planning survey (GIZ et al. 2010) has been collected and field survey has been done during this study. Figure 17.2 shows population of RMC area up to census 2011 and forecasted population by RMC (2010). The city urban area has extended from 25.17 km$^2$ in 1971 to 64.29 km$^2$ in 2001. The change in sprawl of the municipal corporation of Raipur city has come due to the inclusion of the part of the eight villages, namely Telibandha (2.00 km$^2$), Mowa (2.00 km$^2$), Labandih (2.00 km$^2$), Amlidih (1.50 km$^2$), Bhanpur (1.20 km$^2$), Khamtarai (0.30 km$^2$), Tatibandh (0.25 km$^2$) and Mathpurena (0.01 km$^2$), vide Government of Madhya Pradesh, notification dated 21st Feb 1977 (Madhuri et al. 2012). It clearly indicated that a portion of eight villages had, in fact, been merged with the city and had become a part of the expanding urban sprawl. Number of town's has also increased from 3 in 1971 to 20 in 1991 (Census-of-India 2011a). Total population of Raipur city, according to 2001 census, is 758,891 persons but the decadal population growth marked quiet uneven pattern varying from 10.03% during 1901–1911 to 47.35% in 1961–71 and 51.13% in 1991–2001. The population projection indicates that the RMC would accommodate a population of 1.8 million in 2021 and 4.1 million in 2041(Fig. 17.2). A recent study has been carried out by Guha et al. (2017) explained how the city has grown and sprawl in all directions due to industrial and economic rationality. The rapid growth and industrialization have increased Raipur city land use demand and hence the needs of housing and infrastructure land use have been rise. This was the major cause for increased urban heat in Raipur city centre which is one of the main reason for change in climatic pattern and impact for extreme weather scenario in Raipur city. This is the motivation why this study of impacts of climate variability on urban rainfall extremes using statistical analysis of climatic variables for change detection and trend analysis due to climate change and/or human-induced intervention resulting from rapid growth of population, urbanization and industrialization in Raipur city (Guha et al. 2017; Khan and Jhariya 2016; Madhuri et al. 2012).

### *17.2.2 Data Collection*

Climatic parameters, including daily and monthly rainfall (RF), minimum temperature (MinT), maximum temperature (MaxT), relative humidity (RH), sunshine hour (SS), wind speed (WS) and evapotranspiration (EP) from year 1971 to 2016 were collected from observatory of Climatic Department, Indira Gandhi Krishi Vishwavidyalaya Raipur (Chhattisgarh), India, used in this analysis (Table 17.1).

Shorter duration rainfall has been digitized from a histogram of daily recorded rain gauge station which is situated at Indian Climatic Department at Lalpur. 15 min shorter duration incremental rainfall has been derived from daily cumulative chart. Lalpur rain gauge station and station at IGKV have aerial distance of 6 kms. Hence a similar areal distribution has been assumed throughout the study area. To check the correctness and consistency of the data as the time series for rainfall data is 46 years i.e. (1971–2016) 5 more rain gauge stations data has been taken from state data centre, water resources department Chhattisgarh, Raipur which is shown in Table 17.2.

If the condition relevant to the recording of a rain gauge station has undergone a significant change during the period of record, inconsistency would arise in the rainfall data of that station. This inconsistency would be felt from the time the significant change took place. The check for inconsistency of record is done by the double-mass curve technique. A group of 7 stations are averaged and plotted with accumulated annual rainfall of Raipur_IGKV station and Raipur at Lalpur station. The graphs reveal no such changes are found within the records and the data is free from error (Fig. 17.3).

## 17.3 Methodology Applied

The adopted methodology for this study covers analysis of climatic variables on a monthly basis, statistical analysis of these variables, change-point test, trend detection and climate uncertainty on Intensity–Duration–Frequency relation for Raipur city.

**Table 17.1** Climatic parameter statistic of EP, MaxT, MinT, RH, SS and WS as mean monthly, annual and seasonal series while the RF is cumulative monthly, annual and seasonal

| Period | EP (mm) | | | |
|---|---|---|---|---|
| | Min | Max | Mean | SD |
| Jan | 2.39 | 3.71 | 3.05 | 0.36 |
| Feb | 3.59 | 6.18 | 4.34 | 0.51 |
| Mar | 4.89 | 8.65 | 6.62 | 0.86 |
| Apr | 7.89 | 12.10 | 9.95 | 1.07 |
| May | 7.83 | 17.78 | 12.17 | 1.87 |
| Jun | 4.85 | 11.69 | 8.77 | 1.85 |
| Jul | 2.79 | 6.95 | 4.35 | 0.79 |
| Aug | 2.66 | 5.11 | 3.55 | 0.44 |
| Sep | 2.85 | 5.10 | 3.74 | 0.40 |
| Oct | 2.97 | 5.37 | 3.72 | 0.43 |
| Nov | 2.66 | 5.12 | 3.26 | 0.42 |
| Dec | 2.30 | 5.16 | 2.92 | 0.55 |
| Annual | 4.75 | 6.22 | 5.54 | 0.35 |
| Summer | 27.21 | 44.86 | 37.51 | 4.14 |
| Rainy | 12.91 | 20.69 | 15.35 | 1.39 |
| Winter | 11.72 | 17.81 | 13.56 | 1.13 |
| Period | MaxT (°C) | | | |
| | Min | Max | Mean | SD |
| Jan | 24.11 | 30.25 | 27.50 | 1.24 |
| Feb | 27.60 | 33.52 | 30.39 | 1.38 |
| Mar | 32.47 | 37.70 | 35.18 | 1.38 |
| Apr | 36.63 | 42.63 | 39.74 | 1.33 |
| May | 37.05 | 44.23 | 41.90 | 1.61 |
| Jun | 31.31 | 41.33 | 37.24 | 2.23 |
| Jul | 29.05 | 34.29 | 31.15 | 1.12 |
| Aug | 28.71 | 31.31 | 30.04 | 0.74 |
| Sep | 29.21 | 32.81 | 31.07 | 0.78 |
| Oct | 29.17 | 33.65 | 31.14 | 0.94 |
| Nov | 27.39 | 31.85 | 29.46 | 1.11 |
| Dec | 25.31 | 29.75 | 27.53 | 1.10 |
| Annual | 31.11 | 33.94 | 32.69 | 0.49 |
| Summer | 142.73 | 161.57 | 154.06 | 3.96 |
| Rainy | 118.21 | 128.69 | 123.39 | 2.42 |

(continued)

**Table 17.1** (continued)

| Period | MaxT (°C) | | | |
|---|---|---|---|---|
| | Min | Max | Mean | SD |
| Winter | 109.76 | 120.67 | 114.88 | 2.90 |
| Period | MinT (°C) | | | |
| | Min | Max | Mean | SD |
| Jan | 8.36 | 14.29 | 11.39 | 1.58 |
| Feb | 10.45 | 17.73 | 14.01 | 1.43 |
| Mar | 15.53 | 20.67 | 17.92 | 1.22 |
| Apr | 20.27 | 25.88 | 22.74 | 1.11 |
| May | 23.57 | 30.33 | 26.51 | 1.33 |
| Jun | 23.74 | 28.55 | 26.15 | 1.22 |
| Jul | 22.33 | 25.79 | 24.24 | 0.76 |
| Aug | 21.89 | 25.56 | 24.15 | 0.76 |
| Sep | 22.42 | 25.84 | 24.03 | 0.73 |
| Oct | 18.25 | 24.75 | 20.91 | 1.17 |
| Nov | 11.77 | 20.49 | 15.13 | 2.02 |
| Dec | 8.17 | 15.73 | 11.25 | 1.70 |
| Annual | 18.31 | 21.46 | 19.87 | 0.75 |
| Summer | 85.86 | 104.60 | 93.31 | 3.78 |
| Rainy | 87.60 | 101.42 | 93.34 | 2.86 |
| Winter | 42.83 | 58.94 | 51.79 | 4.06 |
| Period | RH (%) | | | |
| | Min | Max | Mean | SD |
| Jan | 18.68 | 57.35 | 35.88 | 8.57 |
| Feb | 14.54 | 57.32 | 32.38 | 8.19 |
| Mar | 13.39 | 38.32 | 23.20 | 6.17 |
| Apr | 9.70 | 35.37 | 18.89 | 4.98 |
| May | 10.84 | 40.87 | 20.17 | 5.88 |
| Jun | 27.13 | 65.77 | 45.51 | 9.50 |
| Jul | 56.84 | 80.84 | 71.84 | 5.12 |
| Aug | 66.23 | 82.94 | 76.25 | 3.50 |
| Sep | 58.03 | 81.23 | 70.40 | 4.80 |
| Oct | 41.32 | 68.87 | 54.75 | 7.04 |
| Nov | 27.40 | 58.57 | 39.72 | 7.39 |

(continued)

**Table 17.1** (continued)

| Period | RH (%) | | | |
|---|---|---|---|---|
| | Min | Max | Mean | SD |
| Dec | 15.97 | 67.23 | 34.40 | 7.49 |
| Annual | 36.87 | 51.48 | 43.62 | 2.74 |
| Summer | 75.47 | 151.75 | 107.76 | 17.49 |
| Rainy | 249.40 | 301.61 | 273.23 | 13.21 |
| Winter | 96.13 | 190.53 | 142.38 | 19.01 |
| Period | SS (h) | | | |
| | Min | Max | Mean | SD |
| Jan | 2.93 | 9.77 | 7.82 | 1.55 |
| Feb | 2.90 | 10.35 | 8.50 | 1.38 |
| Mar | 2.38 | 10.03 | 8.63 | 1.41 |
| Apr | 2.64 | 10.40 | 8.85 | 1.42 |
| May | 3.10 | 9.92 | 8.43 | 1.24 |
| June | 2.65 | 10.06 | 5.45 | 1.94 |
| Jul | 1.22 | 9.83 | 3.82 | 2.35 |
| Aug | 0.68 | 9.49 | 3.93 | 2.29 |
| Sep | 2.78 | 9.92 | 5.91 | 1.81 |
| Oct | 3.21 | 9.71 | 7.82 | 1.34 |
| Nov | 2.57 | 10.04 | 7.96 | 1.57 |
| Dec | 2.76 | 9.73 | 7.71 | 1.62 |
| Annual | 6.15 | 8.18 | 7.07 | 0.47 |
| Summer | 17.88 | 37.93 | 31.36 | 3.69 |
| Rainy | 14.18 | 37.72 | 21.48 | 5.96 |
| Winter | 23.39 | 38.05 | 32.00 | 3.48 |
| Period | WS (kmph) | | | |
| | Min | Max | Mean | SD |
| Jan | 0.85 | 4.27 | 2.88 | 0.88 |
| Feb | 1.81 | 5.84 | 3.72 | 1.03 |
| Mar | 2.26 | 6.94 | 4.63 | 1.13 |
| Apr | 3.88 | 9.77 | 6.60 | 1.40 |
| May | 4.50 | 11.18 | 8.23 | 1.45 |

(continued)

**Table 17.1** (continued)

| Period | WS (kmph) | | | |
|---|---|---|---|---|
| | Min | Max | Mean | SD |
| Jun | 6.71 | 14.74 | 10.92 | 1.79 |
| Jul | 6.32 | 14.36 | 10.30 | 2.06 |
| Aug | 3.55 | 13.62 | 8.91 | 2.37 |
| Sep | 1.43 | 9.36 | 5.62 | 1.70 |
| Oct | 1.46 | 5.19 | 3.26 | 0.85 |
| Nov | 0.88 | 4.38 | 2.78 | 0.88 |
| Dec | 0.74 | 7.99 | 2.55 | 1.10 |
| Annual | 3.66 | 7.80 | 5.87 | 1.13 |
| Summer | 21.08 | 39.17 | 30.38 | 4.86 |
| Rainy | 16.29 | 38.47 | 28.09 | 6.01 |
| Winter | 5.46 | 19.94 | 11.94 | 3.41 |
| Period | RF (mm) | | | |
| | Min | Max | Mean | SD |
| Jan | 0.00 | 104.40 | 13.57 | 21.48 |
| Feb | 0.00 | 78.20 | 13.88 | 14.94 |
| Mar | 0.00 | 51.20 | 12.75 | 14.70 |
| Apr | 0.00 | 83.40 | 16.99 | 18.24 |
| May | 0.00 | 87.40 | 24.20 | 22.21 |
| Jun | 41.30 | 713.40 | 186.48 | 133.75 |
| Jul | 71.80 | 715.50 | 339.50 | 147.81 |
| Aug | 141.50 | 553.40 | 329.98 | 117.86 |
| Sep | 25.60 | 605.50 | 197.90 | 118.52 |
| Oct | 0.00 | 154.20 | 47.12 | 43.06 |
| Nov | 0.00 | 112.20 | 12.57 | 22.99 |
| Dec | 0.00 | 72.60 | 6.00 | 14.19 |
| Annual | 709.10 | 1716.70 | 1200.94 | 269.12 |
| Summer | 77.50 | 731.30 | 240.41 | 137.35 |
| Rainy | 473.10 | 1389.60 | 914.50 | 254.81 |
| Winter | 0.90 | 192.20 | 46.02 | 40.10 |

Following statistical tests were applied on RF, MaxT, MinT, WS, RH, SS and EP on monthly, annual and seasonal (rainy, winter and summer) basis for determination of change-point and trend detection for identification of temporal climate change impact on IDF relation. The methods for different statistical tests for climatic parameters are explained in next section.

**Table 17.2** Gauging station location, gauge type and the climatic parameter with data scale

| S. No. | Station | Lat. | Long. | Operated by | Gauging type | NRRG | RRG | Parameters | Data Scale |
|---|---|---|---|---|---|---|---|---|---|
| 1 | Raipur | 21° 12′ 00″ | 81° 39′ 00″ | IMD | Tipping bucket rain gauge | | X | RF | 15 min |
| 2 | Raipur_IGKV | 21° 14′ 00″ | 81° 42′ 00″ | IMD | Full Climatic Station | | X | RF, MaxT, MinT, RH, WS, SS, EP | Daily |
| 3 | Kharun_Amdi | 21° 05′ 56″ | 81° 34′ 37″ | SWRD | – | X | | RF | Daily |
| 4 | Patharidih | 21° 20′ 28″ | 81° 35′ 45″ | SWRD | – | X | | RF | Daily |
| 5 | Otebandh | 21° 20′ 00″ | 81° 31′ 00″ | SWRD | – | X | | RF | Daily |
| 6 | Kendri | 21° 06′ 00″ | 81° 44′ 00″ | SWRD | – | X | | RF | Daily |
| 7 | Bhilai | 21° 12′ 00″ | 81° 25′ 00″ | SWRD | – | X | | RF | Daily |

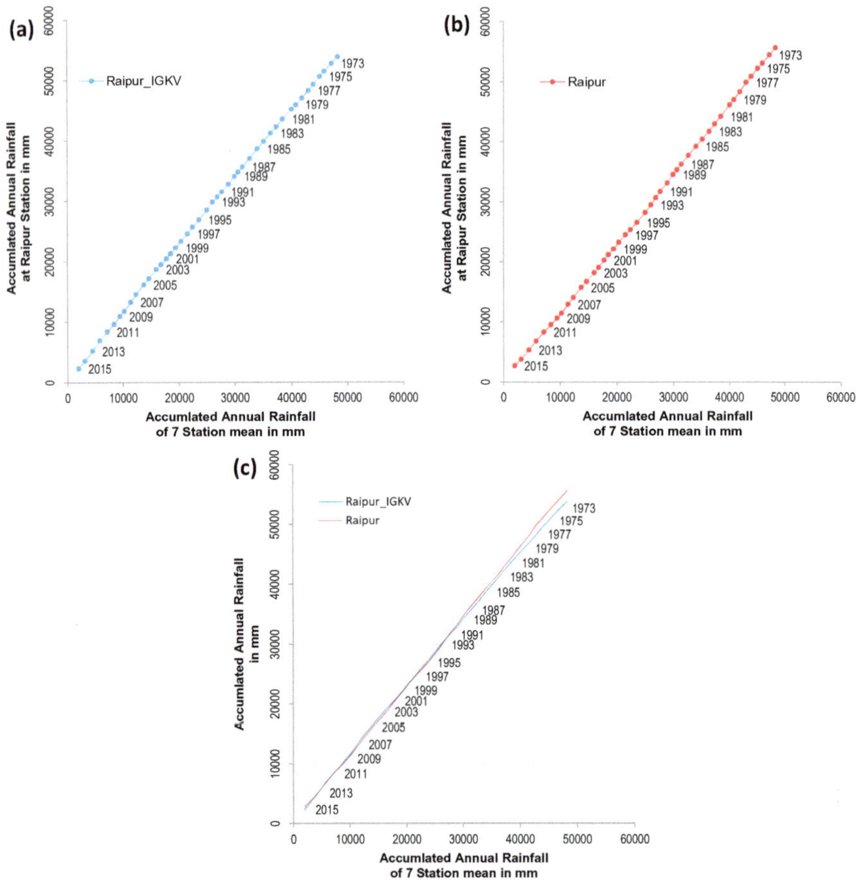


**Fig. 17.3** Double-Mass curve of annual rainfall **a** Raipur_IGKV staion, **b** Raipur at Lalpur station and **c** graph is overlapping curves

## *17.3.1 Methods for Change-Point Test and Trend Analysis*

The change-point test is used to determine/estimate the point of change from the time scale data series from where significant change has occurred. PeT, SNHT, BRT and vNRT have been applied in the present data for identification of change-point in climatic variables (Appendix 1). The time step values Xi of the testing variables X are independent and identically distributed under null hypothesis and the series are assumed as homogeneous. Meanwhile under alternative hypothesis, PeT, SNHT and BRT consider the series composed of break in the mean and assumed as non-uniform under critical values of change-point tests. Similarly, in the present study, the trend analysis for climatic parameters is examined using LRT, MKT and SRT, as described in Appendix 2.

### 17.3.2 Frequency Analysis in Rainfall Data

The characteristics of shorter durations point rainfall data can be examined through frequency analysis. In present study, we have used ground-based point data rainfall to perform frequency analysis. The IDF is a tool for statistically summarizing regional rainfall records and is often used in municipal storm water management and other engineering design applications. The IDF analysis starts by collecting time series records of different durations (here 15–30–45–60 min, 1.5–2–3–4–5–6–7–8–10–11–12–16–24 h used to smoothen the curve). Then time serried data has cumulated to extract annual maximum of the data series for each duration. The annual max. Data is then fit to a probability plot, in order to estimate rainfall counts. Most widely accepted probability distribution used is the Gumbel Extreme Value type I distribution [also used by Prodanovic and Simonovic (2008)] and is therefore adopted in this study.

Continuous rainfall data is used for designing storm intensities. 15 min duration rainfall intensities were analysed and probability plot has been plotted to compute maximum intensities for a particular duration like (15, 30, 45 min, … 0.24 h). To draw a smooth curve, 17 different durations have been selected. For the rainfall computed above, IDF curves were generated to determine the intensity of the rainfall event at various time durations for different return periods. These curves were generated for return periods of 2, 5, 10, 25, 50, 100 years. Before preparing the curve, goodness of fit test was applied for the available rainfall data. This test was carried out for Gumbel's Distribution. It was found out that the rainfall data follows Gumbel's extreme Value Type-1 Distribution. The values obtained from the goodness of fit test were compared with chai-squad and Kolmogorov–Smirnov table to check the significance level of the test performed. The rainfall ($X_T$) corresponding to a specific return period ($T$) using the Gumbel's extreme value distribution is given by (Ward, 1989):

$$X_t = \mu + \sigma K_T \tag{17.1}$$

where $X_t$ is the rainfall intensity of the $T$th year storm event, and frequency factor ($KT$) that is a function of return period $T$ and can be computed using Eq. 17.2. $\mu$ is mean and $\sigma$ is the standard deviation of the annual maximum series. The frequency factor $KT$ is calculated using the equation:

$$K_T = -\frac{\sqrt{6}}{\pi}\left[0.5772 + \ln\left(\ln\left(\frac{T}{T-1}\right)\right)\right] \tag{17.2}$$

## 17.4 Results and Discussion

Present study constitutes in three parts, first various change-point test PeT, BRT, SNHT and vNRT have been performed to identify a change-point in monthly, annual and seasonal historical data series of seven climatic parameters. Second, after change-point test three different climatic scenario (time series) (P1 whole series of a variables 1971–2016, P2 covers time series up to 1993 and P3 covers time series from 1994 to 2016) have been proposed to apply trend analysis test namely, LRT, MKT and SRT. After these two steps with climatic parameters, the third step was to find the impact of this climate shift on IDF relations of Raipur city has been performed. Various methods have been used to derive IDF curve for Raipur city during this study namely Gumbel's Extreme Value distribution, Normal distribution, Log Pearson Type III distribution, Log-normal distribution (two parameter), Gamma distribution (two parameter) which has been explained in another report of this project, which is beyond the scope of this paper. In this paper, only a well-established Gumbles's Extreme Value distribution has been presented which shows a significant change in IDF relation under climatic parameter uncertainty to analyse the impact of climate shift in Raipur city.

### *17.4.1 Discussion on Change-Point Analysis Results*

To perform change-point test on average monthly, annual and seasonal (summer, rainy and winter) series of MaxT, MinT, WS, RH, SS, and EP have been analysed while the rainfall is analysed as cumulative and average series with the use of PeT, BRT, SNHT and vNRT. The result of various tests and acceptance and rejection of null hypothesis for all climatic parameters with their static has been tabulated in Table 17.3. To identify a significant change-point for a particular parameter from the whole series, a well-known approach has been used. According to (Wijngaard et al. 2003), a series may be considered as change-point/inhomogeneous (CP); if more than two tests out of all tests rejects the null hypothesis. Similarly, no change-point/homogeneous (HG); if none or one test out of all tests rejects the null hypothesis and doubtful series (DF); if two tests out of all tests rejects the null hypothesis at 5% significant level.

The result indicating the change-point test results of various climatic parameters have been tabulated in Table 17.4, and final result of homogeneity test is tabulated in Table 17.5. The graphical representations of historical time series of various climatic variables indicating change-point have been represented in Fig. 17.4.

The change-point analysis on EP confirms that April and Annual series have a change-point while Rainy series should be examined using another test while all other series are homogeneous or doubtful in nature at 95% confidence level. The behaviour of MaxT has indicated a notable change-point in the series of November and December at 99% confidence level exists, which yields a change-point in winter series also. MinT of February–March–April–May–June–July–August–September–annual–summer–rainy and winter season marked a notable change-point, while other series can be taken as homogeneous in nature at 95% and 99% confidence limit. RH confirms the change-point may occur in 1993 in April series, while rest of the series are either homogeneous or doubtful. WS presents notable change-points in 1990–1994 at all series except April and Annual series at 95% confidence level. Wind speed

**Table 17.3** Change-point test results of various climatic parameters

| EP (mm) | | | |
|---|---|---|---|
| Period | PeT | | |
| | Statistic | Break | Year of shift |
| Jan | 141 | No | – |
| Feb | 159 | No | – |
| Mar | 212 | No | – |
| Apr | 317 | Yes | 1993 |
| May | 193 | No | – |
| Jun | 138 | No | – |
| Jul | 244 | No | – |
| Aug | 191 | No | – |
| Sep | 117 | No | – |
| Oct | 89 | No | – |
| Nov | 215 | No | – |
| Dec | 195 | No | – |
| Annual | 287 | Yes | 1993 |
| Summer | 248 | No | – |
| Rainy | 227 | No | – |
| Winter | 201 | No | – |
| EP (mm) | | | |
| Period | SNHT | | |
| | Statistic | Break | Year of shift |
| Jan | 3.87 | No | – |
| Feb | 2.99 | No | – |
| Mar | 7.67 | No | – |
| Apr | 10.97 | Yes | 1993 |
| May | 7.79 | No | – |
| Jun | 9.31 | Yes | 1972 |
| Jul | 6.53 | No | – |
| Aug | 12.90 | Yes | 1971 |
| Sep | 12.05 | Yes | 1971 |
| Oct | 14.80 | Yes | 1971 |
| Nov | 20.55 | Yes | 1971 |
| Dec | 16.95 | Yes | 1993 |
| Annual | 7.08 | No | – |
| Summer | 12.87 | Yes | 1972 |
| Rainy | 15.11 | Yes | 1971 |
| Winter | 14.54 | Yes | 1971 |

(continued)

**Table 17.3** (continued)

| EP (mm) | | | |
|---|---|---|---|
| Period | BRT | | |
| | Statistic | Break | Year of shift |
| Jan | 5.46 | No | – |
| Feb | 4.94 | No | – |
| Mar | 7.81 | No | – |
| Apr | 11.36 | Yes | 1993 |
| May | 5.53 | No | – |
| Jun | 4.27 | No | – |
| Jul | 7.89 | No | – |
| Aug | 7.52 | No | – |
| Sep | 6.07 | No | – |
| Oct | 4.05 | No | – |
| Nov | 4.93 | No | – |
| Dec | 4.92 | No | – |
| Annual | 9.04 | Yes | 1993 |
| Summer | 8.07 | No | – |
| Rainy | 8.56 | Yes | 1983 |
| Winter | 4.73 | No | – |

| EP (mm) | | |
|---|---|---|
| Period | vNRT | |
| | Statistic | Break |
| Jan | 1.69 | No |
| Feb | 2.02 | No |
| Mar | 1.79 | No |
| Apr | 1.41 | Yes |
| May | 1.63 | No |
| Jun | 1.56 | No |
| Jul | 2.15 | No |
| Aug | 1.67 | No |
| Sep | 1.40 | Yes |
| Oct | 1.61 | No |
| Nov | 1.51 | Yes |
| Dec | 1.30 | Yes |
| Annual | 1.59 | Yes |
| Summer | 1.24 | Yes |
| Rainy | 1.47 | Yes |
| Winter | 1.22 | Yes |

(continued)

**Table 17.3** (continued)

| EP (mm) | | |
|---|---|---|
| Period | Result | |
| | Nature | Year of shift |
| Jan | HG | – |
| Feb | HG | – |
| Mar | HG | – |
| Apr | CP | 1993 |
| May | HG | – |
| Jun | HG | – |
| Jul | HG | – |
| Aug | HG | – |
| Sep | DF | – |
| Oct | HG | – |
| Nov | DF | – |
| Dec | DF | – |
| Annual | CP | 1993 |
| Summer | DF | – |
| Rainy | CP | 1993 |
| Winter | DF | – |

| MaxT (°C) | | | |
|---|---|---|---|
| Period | PeT | | |
| | Statistic | Break | Year of shift |
| Jan | 117 | No | – |
| Feb | 141 | No | – |
| Mar | 91 | No | – |
| Apr | 166 | No | – |
| May | 93 | No | – |
| Jun | 104 | No | – |
| Jul | 112 | No | – |
| Aug | 148 | No | – |
| Sep | 119 | No | – |
| Oct | 163 | No | – |
| Nov | 285 | Yes | 1995 |
| Dec | 284 | Yes | 1993 |
| Annual | 153 | No | – |
| Summer | 190 | No | – |
| Rainy | 130 | No | – |
| Winter | 310 | Yes | 199 |

(continued)

**Table 17.3** (continued)

| MaxT (°C) | | | |
|---|---|---|---|
| Period | SNHT | | |
| | Statistic | Break | Year of shift |
| Jan | 2.31 | No | – |
| Feb | 3.04 | No | – |
| Mar | 1.95 | No | – |
| Apr | 2.91 | No | – |
| May | 4.15 | No | – |
| Jun | 7.21 | No | – |
| Jul | 4.18 | No | – |
| Aug | 4.64 | No | – |
| Sep | 1.89 | No | – |
| Oct | 4.46 | No | – |
| Nov | 9.87 | Yes | 1995 |
| Dec | 10.75 | Yes | 1993 |
| Annual | 10.49 | Yes | 1971 |
| Summer | 8.35 | Yes | 1971 |
| Rainy | 6.12 | No | – |
| Winter | 12.59 | Yes | 1993 |
| MaxT (°C) | | | |
| Period | BRT | | |
| | Statistic | Break | Year of shift |
| Jan | 5.03 | No | – |
| Feb | 5.30 | No | – |
| Mar | 3.44 | No | – |
| Apr | 5.79 | No | – |
| May | 3.50 | No | – |
| Jun | 3.59 | No | – |
| Jul | 3.80 | No | – |
| Aug | 5.54 | No | – |
| Sep | 3.82 | No | – |
| Oct | 6.49 | No | – |
| Nov | 10.27 | Yes | 1998 |
| Dec | 10.85 | Yes | 1995 |
| Annual | 6.84 | No | – |
| Summer | 5.37 | No | – |
| Rainy | 4.96 | No | – |
| Winter | 11.87 | Yes | 1993 |

(continued)

**Table 17.3** (continued)

| MaxT (°C) | | |
|---|---|---|
| Period | vNRT | |
| | Statistic | Break |
| Jan | 2.28 | No |
| Feb | 1.98 | No |
| Mar | 2.34 | No |
| Apr | 1.88 | No |
| May | 2.24 | No |
| Jun | 2.15 | No |
| Jul | 1.92 | No |
| Aug | 1.82 | No |
| Sep | 2.33 | No |
| Oct | 1.83 | No |
| Nov | 1.31 | Yes |
| Dec | 1.99 | No |
| Annual | 0.56 | No |
| Summer | 1.83 | No |
| Rainy | 0.99 | No |
| Winter | 1.37 | Yes |
| MaxT (°C) | | |
| Period | Result | |
| | Nature | Year of shift |
| Jan | HG | – |
| Feb | HG | – |
| Mar | HG | – |
| Apr | HG | – |
| May | HG | – |
| Jun | HG | – |
| Jul | HG | – |
| Aug | HG | – |
| Sep | HG | – |
| Oct | HG | – |
| Nov | CP | 1995 |
| Dec | CP | 1993 |
| Annual | HG | – |
| Summer | HG | – |
| Rainy | HG | – |
| Winter | CP | 1993 |

(continued)

**Table 17.3** (continued)

| MinT (°C) | | | |
|---|---|---|---|
| Period | BRT | | |
| | Statistic | Break | Year of shift |
| Jan | 125 | No | – |
| Feb | 275 | Yes | 1997 |
| Mar | 312 | Yes | 2000 |
| Apr | 285 | Yes | 2001 |
| May | 272 | Yes | 2000 |
| Jun | 261 | Yes | 2001 |
| Jul | 401 | Yes | 1995 |
| Aug | 395 | Yes | 1994 |
| Sep | 326 | Yes | 1994 |
| Oct | 204 | No | – |
| Nov | 189 | No | – |
| Dec | 251 | Yes | 2001 |
| Annual | 344 | Yes | 2000 |
| Summer | 354 | Yes | 2000 |
| Rainy | 371 | Yes | 1997 |
| Winter | 279 | Yes | 2001 |
| MinT (°C) | | | |
| Period | SNHT | | |
| | Statistic | Break | Year of shift |
| Jan | 3.22 | No | – |
| Feb | 8.65 | Yes | 1997 |
| Mar | 12.56 | Yes | 2000 |
| Apr | 10.10 | Yes | 2001 |
| May | 11.16 | Yes | 2000 |
| Jun | 10.71 | Yes | 2001 |
| Jul | 18.16 | Yes | 1995 |
| Aug | 16.25 | Yes | 1994 |
| Sep | 10.57 | Yes | 1994 |
| Oct | 6.89 | No | – |
| Nov | 4.41 | No | – |
| Dec | 7.28 | No | 2008 |
| Annual | 16.42 | Yes | 2000 |
| Summer | 16.83 | Yes | 2000 |
| Rainy | 11.97 | Yes | 1997 |
| Winter | 11.76 | Yes | 2001 |

(continued)

**Table 17.3** (continued)

| MinT (°C) | | | |
|---|---|---|---|
| Period | BRT | | |
| | Statistic | Break | Year of shift |
| Jan | 4.47 | No | – |
| Feb | 9.93 | Yes | 1997 |
| Mar | 11.51 | Yes | 2000 |
| Apr | 10.22 | Yes | 2001 |
| May | 10.91 | Yes | 2000 |
| Jun | 10.33 | Yes | 2001 |
| Jul | 14.56 | Yes | 1995 |
| Aug | 13.81 | Yes | 1994 |
| Sep | 10.81 | Yes | 2001 |
| Oct | 7.12 | No | – |
| Nov | 6.14 | No | – |
| Dec | 8.03 | No | 2008 |
| Annual | 13.12 | Yes | 2000 |
| Summer | 13.40 | Yes | 2000 |
| Rainy | 11.68 | Yes | 1997 |
| Winter | 10.03 | Yes | 2001 |

| MinT (°C) | | |
|---|---|---|
| Period | vNRT | |
| | Statistic | Break |
| Jan | 2.04 | No |
| Feb | 1.99 | No |
| Mar | 1.40 | Yes |
| Apr | 1.52 | Yes |
| May | 1.60 | No |
| Jun | 1.48 | Yes |
| Jul | 0.82 | Yes |
| Aug | 1.01 | Yes |
| Sep | 1.21 | Yes |
| Oct | 1.57 | No |
| Nov | 1.43 | Yes |
| Dec | 1.96 | No |
| Annual | 0.72 | Yes |
| Summer | 1.07 | Yes |
| Rainy | 0.98 | Yes |
| Winter | 1.13 | Yes |

(continued)

**Table 17.3** (continued)

| MinT (°C) | | |
|---|---|---|
| Period | Result | |
| | Nature | Year of shift |
| Jan | HG | – |
| Feb | CP | 1997 |
| Mar | CP | 2000 |
| Apr | CP | 2001 |
| May | CP | 2000 |
| Jun | CP | 2002 |
| Jul | CP | 1995 |
| Aug | CP | 1994 |
| Sep | CP | 1994 |
| Oct | HG | – |
| Nov | HG | – |
| Dec | HG | – |
| Annual | CP | 2000 |
| Summer | CP | 2000 |
| Rainy | CP | 1997 |
| Winter | CP | 2001 |

| RH (%) | | | |
|---|---|---|---|
| Period | PeT | | |
| | Statistic | Break | Year of shift |
| Jan | 198 | No | – |
| Feb | 153 | No | – |
| Mar | 190 | No | – |
| Apr | 265 | Yes | 1993 |
| May | 160 | No | – |
| Jun | 148 | No | – |
| Jul | 201 | No | – |
| Aug | 215 | No | – |
| Sep | 191 | No | – |
| Oct | 213 | No | – |
| Nov | 166 | No | – |
| Dec | 103 | No | – |
| Annual | 199 | No | – |
| Summer | 237 | No | – |
| Rainy | 158 | No | – |
| Winter | 173 | No | – |

(continued)

**Table 17.3** (continued)

| RH (%) | | | |
|---|---|---|---|
| Period | SNHT | | |
| | Statistic | Break | Year of shift |
| Jan | 8.48 | Yes | 1977 |
| Feb | 3.24 | No | – |
| Mar | 5.90 | No | – |
| Apr | 10.66 | No | – |
| May | 2.17 | No | – |
| Jun | 3.88 | No | – |
| Jul | 4.92 | No | – |
| Aug | 12.32 | Yes | 1976 |
| Sep | 7.29 | No | – |
| Oct | 6.12 | No | – |
| Nov | 3.85 | No | – |
| Dec | 3.84 | No | – |
| Annual | 8.09 | No | – |
| Summer | 6.47 | No | – |
| Rainy | 3.05 | No | – |
| Winter | 7.24 | No | – |
| RH (%) | | | |
| Period | BRT | | |
| | Statistic | Break | Year of shift |
| Jan | 7.34 | No | – |
| Feb | 4.43 | No | – |
| Mar | 7.58 | No | – |
| Apr | 9.41 | Yes | 1993 |
| May | 4.85 | No | – |
| Jun | 5.24 | No | – |
| Jul | 6.85 | No | – |
| Aug | 8.88 | Yes | 1979 |
| Sep | 7.35 | No | – |
| Oct | 7.51 | No | – |
| Nov | 6.50 | No | – |
| Dec | 4.82 | No | – |
| Annual | 7.99 | No | – |
| Summer | 8.59 | No | – |
| Rainy | 5.83 | No | – |
| Winter | 6.63 | No | – |

(continued)

**Table 17.3** (continued)

| RH (%) | | |
|---|---|---|
| Period | vNRT | |
| | Statistic | Break |
| Jan | 1.58 | No |
| Feb | 2.13 | No |
| Mar | 1.88 | No |
| Apr | 1.82 | Yes |
| May | 2.01 | No |
| Jun | 1.94 | No |
| Jul | 2.07 | No |
| Aug | 1.66 | No |
| Sep | 2.20 | No |
| Oct | 1.79 | No |
| Nov | 1.60 | No |
| Dec | 1.95 | No |
| Annual | 1.59 | No |
| Summer | 1.86 | No |
| Rainy | 2.36 | No |
| Winter | 1.29 | Yes |

| RH (%) | | |
|---|---|---|
| Period | Result | |
| | Nature | Year of shift |
| Jan | HG | – |
| Feb | HG | – |
| Mar | HG | – |
| Apr | CP | 1993 |
| May | HG | – |
| Jun | HG | – |
| Jul | HG | – |
| Aug | DF | – |
| Sep | HG | – |
| Oct | HG | – |
| Nov | HG | – |
| Dec | HG | – |
| Annual | HG | – |
| Summer | HG | – |
| Rainy | HG | – |
| Winter | HG | – |

(continued)

**Table 17.3** (continued)

| SS (h) | | | |
|---|---|---|---|
| Period | PeT | | |
| | Statistic | Break | Year of shift |
| Jan | 230 | No | – |
| Feb | 192 | No | – |
| Mar | 266 | Yes | 2000 |
| Apr | 203 | No | – |
| May | 151 | No | – |
| Jun | 238 | No | – |
| Jul | 316 | Yes | 1983 |
| Aug | 282 | Yes | 1990 |
| Sep | 250 | Yes | 1996 |
| Oct | 247 | No | – |
| Nov | 171 | No | – |
| Dec | 258 | Yes | 2000 |
| Annual | 439 | Yes | 1993 |
| Summer | 221 | No | – |
| Rainy | 346 | Yes | 1990 |
| Winter | 199 | No | – |
| SS (h) | | | |
| Period | SNHT | | |
| | Statistic | Break | Year of shift |
| Jan | 4.37 | No | – |
| Feb | 7.88 | No | – |
| Mar | 5.18 | No | – |
| Apr | 10.13 | No | – |
| May | 2.05 | No | – |
| Jun | 21.59 | Yes | 1978 |
| Jul | 33.09 | Yes | 1979 |
| Aug | 27.44 | Yes | 1979 |
| Sep | 11.67 | Yes | 1993 |
| Oct | 13.46 | Yes | 1974 |
| Nov | 6.93 | No | – |
| Dec | 2.05 | No | – |
| Annual | 22.71 | Yes | 1993 |
| Summer | 8.15 | No | – |
| Rainy | 26.33 | Yes | 1979 |
| Winter | 6.95 | No | – |

(continued)

**Table 17.3** (continued)

| SS (h) | | | |
|---|---|---|---|
| Period | BRT | | |
| | Statistic | Break | Year of shift |
| Jan | 4.84 | No | – |
| Feb | 7.30 | No | – |
| Mar | 6.19 | No | – |
| Apr | 8.66 | Yes | 1979 |
| May | 3.37 | No | – |
| Jun | 15.65 | Yes | 1978 |
| Jul | 14.25 | Yes | 1979 |
| Aug | 7.09 | Yes | 1979 |
| Sep | 5.62 | Yes | 1993 |
| Oct | 3.77 | No | – |
| Nov | 6.93 | No | – |
| Dec | 2.05 | No | – |
| Annual | 16.09 | Yes | 1995 |
| Summer | 6.59 | No | – |
| Rainy | 13.96 | Yes | 1979 |
| Winter | 6.65 | No | – |

| SS (h) | | |
|---|---|---|
| Period | vNRT | |
| | Statistic | Break |
| Jan | 1.24 | Yes |
| Feb | 0.93 | Yes |
| Mar | 0.92 | Yes |
| Apr | 1.01 | Yes |
| May | 1.57 | No |
| Jun | 1.00 | Yes |
| Jul | 0.42 | Yes |
| Aug | 0.68 | Yes |
| Sep | 1.30 | Yes |
| Oct | 1.17 | Yes |
| Nov | 1.23 | Yes |
| Dec | 1.61 | No |
| Annual | 0.73 | Yes |
| Summer | 1.01 | Yes |
| Rainy | 0.53 | Yes |
| Winter | 0.97 | Yes |

(continued)

**Table 17.3** (continued)

| SS (h) | | |
|---|---|---|
| Period | Nature | Year of shift |
| Jan | HG | – |
| Feb | HG | – |
| Mar | DF | – |
| Apr | DF | – |
| May | HG | – |
| Jun | CP | 1978 |
| Jul | CP | 1979 |
| Aug | CP | 1979 |
| Sep | CP | 1993 |
| Oct | DF | – |
| Nov | HG | – |
| Dec | HG | – |
| Annual | CP | 1993 |
| Summer | HG | – |
| Rainy | CP | 1979 |
| Winter | HG | – |

| WS (kmph) | | | |
|---|---|---|---|
| Period | PeT | | |
| | Statistic | Break | Year of shift |
| Jan | 491 | Yes | 1991 |
| Feb | 512 | Yes | 1990 |
| Mar | 521 | Yes | 1993 |
| Apr | 466 | Yes | 1992 |
| May | 367 | Yes | 1993 |
| Jun | 474 | Yes | 1994 |
| Jul | 428 | Yes | 1994 |
| Aug | 454 | Yes | 1994 |
| Sep | 414 | Yes | 1990 |
| Oct | 436 | Yes | 1990 |
| Nov | 492 | Yes | 1992 |
| Dec | 521 | Yes | 1991 |
| Annual | 526 | Yes | 1994 |
| Summer | 523 | Yes | 1993 |
| Rainy | 496 | Yes | 1994 |
| Winter | 528 | Yes | 1992 |

(continued)

**Table 17.3** (continued)

| WS (kmph) | | | |
|---|---|---|---|
| Period | SNHT | | |
| | Statistic | Break | Year of shift |
| Jan | 28.85 | Yes | 1987 |
| Feb | 30.97 | Yes | 1990 |
| Mar | 28.62 | Yes | 1993 |
| Apr | 24.82 | Yes | 1990 |
| May | 17.01 | Yes | 2002 |
| Jun | 25.57 | Yes | 1994 |
| Jul | 21.76 | Yes | 1994 |
| Aug | 23.20 | Yes | 1994 |
| Sep | 19.59 | Yes | 1990 |
| Oct | 21.31 | Yes | 1988 |
| Nov | 26.48 | Yes | 1992 |
| Dec | 20.50 | Yes | 1990 |
| Annual | 33.06 | Yes | 1991 |
| Summer | 32.09 | Yes | 1994 |
| Rainy | 28.47 | Yes | 1994 |
| Winter | 33.37 | Yes | 1991 |
| WS (kmph) | | | |
| Period | BRT | | |
| | Statistic | Break | Year of shift |
| Jan | 17.95 | Yes | 1991 |
| Feb | 18.96 | Yes | 1992 |
| Mar | 18.34 | Yes | 1993 |
| Apr | 17.03 | Yes | 1993 |
| May | 14.03 | Yes | 1993 |
| Jun | 17.32 | Yes | 1994 |
| Jul | 15.98 | Yes | 1994 |
| Aug | 16.50 | Yes | 1994 |
| Sep | 15.05 | Yes | 1990 |
| Oct | 15.46 | Yes | 1990 |
| Nov | 17.63 | Yes | 1992 |
| Dec | 15.39 | Yes | 1990 |
| Annual | 19.68 | Yes | 1992 |
| Summer | 19.41 | Yes | 1994 |
| Rainy | 18.28 | Yes | 1994 |
| Winter | 19.73 | Yes | 1991 |

(continued)

**Table 17.3** (continued)

| WS (kmph) | | |
|---|---|---|
| Period | vNRT | |
| | Statistic | Break |
| Jan | 0.49 | Yes |
| Feb | 0.49 | Yes |
| Mar | 0.44 | Yes |
| Apr | 0.85 | Yes |
| May | 1.04 | Yes |
| Jun | 0.76 | Yes |
| Jul | 0.85 | Yes |
| Aug | 0.67 | Yes |
| Sep | 1.07 | Yes |
| Oct | 0.93 | Yes |
| Nov | 0.60 | Yes |
| Dec | 1.03 | Yes |
| Annual | 0.19 | Yes |
| Summer | 0.32 | Yes |
| Rainy | 0.43 | Yes |
| Winter | 0.25 | Yes |

| WS (kmph) | | |
|---|---|---|
| Period | Nature | Year of shift |
| Jan | CP | 1991 |
| Feb | CP | 1990 |
| Mar | CP | 1993 |
| Apr | DF | – |
| May | CP | 1993 |
| Jun | CP | 1994 |
| Jul | CP | 1994 |
| Aug | CP | 1994 |
| Sep | CP | 1990 |
| Oct | CP | 1990 |
| Nov | CP | 1992 |
| Dec | CP | 1990 |
| Annual | DF | – |
| Summer | CP | 1994 |
| Rainy | CP | 1994 |
| Winter | CP | 1991 |

(continued)

**Table 17.3** (continued)

| RF (mm) | | | |
|---|---|---|---|
| Period | PeT | | |
| | Statistic | Break | Year of shift |
| Jan | 121 | No | – |
| Feb | 89 | No | – |
| Mar | 100 | No | – |
| Apr | 148 | No | – |
| May | 79 | No | – |
| Jun | 58 | Yes | 1972 |
| Jul | 196 | No | – |
| Aug | 177 | Yes | 1995 |
| Sep | 240 | No | – |
| Oct | 121 | No | – |
| Nov | 138 | No | – |
| Dec | 92 | No | – |
| Annual | 180 | Yes | 1993 |
| Summer | 94 | No | – |
| Rainy | 170 | No | – |
| Winter | 103 | No | – |
| RF (mm) | | | |
| Period | SNHT | | |
| | Statistic | Break | Year of shift |
| Jan | 2.49 | No | – |
| Feb | 2.44 | No | – |
| Mar | 2.94 | No | – |
| Apr | 7.87 | No | – |
| May | 1.65 | No | – |
| Jun | 3.64 | No | – |
| Jul | 9.22 | Yes | 2008 |
| Aug | 4.07 | Yes | 1995 |
| Sep | 4.48 | No | – |
| Oct | 7.80 | No | – |
| Nov | 2.22 | No | – |
| Dec | 1.59 | No | – |
| Annual | 4.76 | Yes | 1993 |
| Summer | 5.83 | No | – |
| Rainy | 5.31 | No | – |
| Winter | 2.25 | No | – |

(continued)

**Table 17.3** (continued)

| RF (mm) | | | |
|---|---|---|---|
| Period | BRT | | |
| | Statistic | Break | Year of shift |
| Jan | 4.10 | No | – |
| Feb | 4.04 | No | – |
| Mar | 4.46 | No | – |
| Apr | 6.48 | No | – |
| May | 3.46 | No | – |
| Jun | 2.91 | No | – |
| Jul | 8.06 | No | – |
| Aug | 6.48 | Yes | 1995 |
| Sep | 5.99 | No | – |
| Oct | 5.96 | No | – |
| Nov | 4.98 | No | – |
| Dec | 4.28 | No | – |
| Annual | 6.89 | No | – |
| Summer | 2.65 | No | – |
| Rainy | 2.70 | No | – |
| Winter | 3.84 | No | – |

| RF (mm) | | |
|---|---|---|
| Period | vNRT | |
| | Statistic | Break |
| Jan | 1.92 | No |
| Feb | 2.02 | No |
| Mar | 2.22 | No |
| Apr | 2.24 | No |
| May | 1.61 | No |
| Jun | 2.15 | No |
| Jul | 1.93 | No |
| Aug | 1.65 | No |
| Sep | 2.14 | No |
| Oct | 2.57 | No |
| Nov | 2.05 | No |
| Dec | 2.00 | No |
| Annual | 1.85 | Yes |
| Summer | 2.01 | No |
| Rainy | 1.97 | No |
| Winter | 2.01 | No |

(continued)

**Table 17.3** (continued)

| RF (mm) | | |
|---|---|---|
| Period | Nature | Year of shift |
| Jan | HG | – |
| Feb | HG | – |
| Mar | HG | – |
| Apr | HG | – |
| May | HG | – |
| Jun | HG | – |
| Jul | HG | – |
| Aug | CP | 1995 |
| Sep | HG | – |
| Oct | HG | – |
| Nov | HG | – |
| Dec | HG | – |
| Annual | CP | 1993 |
| Summer | HG | – |
| Rainy | HG | – |
| Winter | HG | – |

is the parameter that has been found to be very variable to analyse as its complex behaviour, which usually cause difficulty to find a change-point (Eslamian et al. 2009; Portnyagin et al. 2006). Change-point of SS identified in June-July–August and September at 99% confidence limit which produces change-point in Annual and Rainy series also, in most cases, the change-point for sunshine hour is varying from 1979 to 1993. The behaviour of historical time series of RF has evidence of existence of a significant change-point in the August and Annual series at 95% confidence level; the plot of Annual series of rainfall is shown in Fig. 17.4. From the result, it can be concluded that the mean rainfall has been increased after year 1990.

## *17.4.2 Result and Discussion on Trend Analysis*

Change-point test reveals that most of the series of climatic parameters are homogeneous and doubtful in nature; however, some of the parameters yield a significant change-point between 1990 and 2001 which may be due to human-induced factor as industrialization and increase in commercial activity as Raipur has been capitalized on year 2001 (Ambade 2015; Madhuri et al. 2012). After examining the change-point test results and dividing the data series into nearly equal length, the trend test of all climatic parameters have been carried out as P1 whole series of a variables

**Table 17.4** Corollary of change point test analysis of climatic parameters

| EP (mm) | | |
|---|---|---|
| Period | Nature | Year of shift |
| Jan | HG | – |
| Feb | HG | – |
| Mar | HG | – |
| Apr | CP | 1993 |
| May | HG | – |
| June | HG | – |
| Jul | HG | – |
| Aug | HG | – |
| Sep | DF | – |
| Oct | HG | – |
| Nov | DF | – |
| Dec | DF | – |
| Annual | CP | 1993 |
| Summer | DF | – |
| Rainy | CP | 1993 |
| Winter | DF | – |
| MaxT (°C) | | |
| Period | Nature | Year of shift |
| Jan | HG | – |
| Feb | HG | – |
| Mar | HG | – |
| Apr | HG | 1993 |
| May | HG | – |
| June | HG | – |
| Jul | HG | – |
| Aug | HG | – |
| Sep | HG | – |
| Oct | HG | – |
| Nov | CP | 1995 |
| Dec | CP | 1993 |
| Annual | HG | – |
| Summer | HG | – |
| Rainy | HG | – |

(continued)

**Table 17.4** (continued)

| MaxT (°C) | | |
|---|---|---|
| Period | Nature | Year of shift |
| Winter | CP | 1993 |
| MinT (°C) | | |
| Period | Nature | Year of shift |
| Jan | HG | – |
| Feb | CP | 1997 |
| Mar | CP | 2000 |
| Apr | CP | 2001 |
| May | CP | 2000 |
| June | CP | 2002 |
| Jul | CP | 1995 |
| Aug | CP | 1994 |
| Sep | CP | 1994 |
| Oct | HG | – |
| Nov | HG | – |
| Dec | HG | – |
| Annual | CP | 2000 |
| Summer | CP | 2000 |
| Rainy | CP | 1997 |
| Winter | CP | 2001 |
| RH (%) | | |
| Period | Nature | Year of shift |
| Jan | HG | – |
| Feb | HG | – |
| Mar | HG | – |
| Apr | CP | 1993 |
| May | HG | – |
| June | HG | – |
| Jul | HG | – |
| Aug | DF | – |
| Sep | CP | – |
| Oct | HG | – |
| Nov | HG | – |

(continued)

**Table 17.4** (continued)

| RH (%) | | |
|---|---|---|
| Period | Nature | Year of shift |
| Dec | HG | – |
| Annual | HG | – |
| Summer | HG | – |
| Rainy | HG | – |
| Winter | HG | – |
| SS (h) | | |
| Period | Nature | Year of shift |
| Jan | HG | – |
| Feb | HG | – |
| Mar | DF | – |
| Apr | DF | – |
| May | HG | – |
| June | CP | 1978 |
| Jul | CP | 1979 |
| Aug | CP | 1979 |
| Sep | CP | 1993 |
| Oct | DF | – |
| Nov | HG | – |
| Dec | HG | – |
| Annual | CP | 1993 |
| Summer | HG | – |
| Rainy | CP | 1979 |
| Winter | HG | – |
| WS (kmph) | | |
| Period | Nature | Year of shift |
| Jan | CP | 1990 |
| Feb | CP | 1991 |
| Mar | CP | 1993 |
| Apr | DF | – |
| May | CP | 1993 |
| June | CP | 1994 |
| Jul | CP | 1994 |
| Aug | CP | 1994 |
| Sep | CP | 1990 |
| Oct | CP | 1990 |
| Nov | CP | 1992 |

(continued)

**Table 17.4** (continued)

| WS (kmph) | | |
|---|---|---|
| Period | Nature | Year of shift |
| Dec | CP | 1990 |
| Annual | DF | – |
| Summer | CP | 1994 |
| Rainy | CP | 1994 |
| Winter | CP | 1991 |
| RF (mm) | | |
| Period | Nature | Year of shift |
| Jan | HG | – |
| Feb | HG | – |
| Mar | HG | – |
| Apr | HG | – |
| May | HG | – |
| June | HG | – |
| Jul | HG | – |
| Aug | HG | 1995 |
| Sep | HG | – |
| Oct | HG | – |
| Nov | HG | – |
| Dec | HG | – |
| Annual | HG | 1993 |
| Summer | HG | – |
| Rainy | HG | – |
| Winter | HG | – |

1971–2016, P2 covers time series up to 1993 and P3 covers time series from 1994 to 2016 series of a variable.

To identify trend in the data series, LRT, MKT and SRT have been applied in the climatic parameters time series. Tables 17.5 and 17.6 have been attributed with final results of all trend test parameters with their significance level.

Figure 17.5 shows annual trends in P1, P2 and P3 time periods of all climatic variables that show rising trend in annual MaxT, MinT, RH and RF while falling trend in annual WS, EP and SS. The MaxT of P1 period results a notable rise in trend of November–December–Rainy and winter series, P3 period also shows rising trend in October, Annual and winter series. This rise in MaxT may cause of urbanization, increasing industrialization and climate change in Raipur city after 1993, which is supported by the results of Guha et al. (2017). However, P2 period has no trend in any period. The plot of change in Annual series of MaxT has been shown in Fig. 17.5 and the result of Sen Slope of Mann–Kendall test has tabulated in Table

**Table 17.5** Corollary of trend analysis for various climatic parameters for design period P1, P2 and P3

| Period | P1 (whole series) | | | | | | |
|---|---|---|---|---|---|---|---|
| | MaxT | MinT | RH | WS | EP | SS | RF |
| Jan | 0.61 | 0.98 | 1.07 | – 5.59*** | 0.12 | – 2.37* | 0.34 |
| Feb | 1.00 | 2.05* | 1.14 | – 6.12*** | – 0.62 | – 1.78+ | 0.09 |
| Mar | – 0.34 | 2.53* | 2.36* | – 5.88*** | – 1.62 | – 2.73 | – 0.32 |
| Apr | – 1.47 | 1.63 | 2.25* | – 5.25*** | – 2.45* | – 1.53 | 1.08 |
| May | – 0.97 | 2.36* | 0.95 | – 4.37*** | – 1.96* | – 0.73 | – 0.14 |
| Jun | – 0.45 | 1.57 | 1.29 | – 5.62*** | – 1.01 | – 0.80 | 0.21 |
| Jul | 0.76 | 3.26** | 2.15* | – 5.06*** | – 2.36* | – 2.41* | 1.48 |
| Aug | 1.33 | 3.44*** | 1.69+ | – 5.25*** | – 1.45 | – 1.97* | – 1.23 |
| Sep | 0.76 | 2.42* | 1.71+ | – 4.68*** | – 0.47 | – 2.97** | 2.25* |
| Oct | 0.74 | – 0.56 | – 1.74+ | – 5.25*** | 0.17 | – 2.25* | – 0.26 |
| Nov | 2.95** | 1.25 | – 1.11 | – 5.77*** | 1.64 | – 1.34 | – 0.92 |
| Dec | 2.67** | 1.42 | – 0.33 | 0.19 | – 5.56*** | 1.21 | 2.56* |
| Annual | 0.47 | 2.61** | 1.74+ | – 7.42*** | – 2.75** | – 5.83*** | 0.62 |
| Summer | – 1.95+ | 2.75** | 2.35* | – 6.80*** | – 2.45* | – 2.46* | 0.80 |
| Rainy | 1.38 | 2.39* | 1.12 | – 6.34*** | – 1.62 | – 3.71*** | 0.62 |
| Winter | 2.67** | 2.23* | 0.74 | – 6.63*** | 0.98 | – 1.87+ | 0.17 |
| Period | P2 (up to 1993) | | | | | | |
| | MaxT | MinT | RH | WS | EP | SS | RF |
| Jan | 0.32 | – 0.11 | 0.69 | – 2.80** | 0.53 | 0.53 | 0.36 |
| Feb | 0.79 | – 0.53 | – 0.58 | – 3.54*** | 2.54* | 1.11 | 0.45 |
| Mar | – 0.69 | – 1.21 | 1.53 | – 1.90+ | 1.35 | – 0.11 | 0.45 |
| Apr | – 0.37 | – 2.32* | – 0.79 | – 1.74+ | 1.14 | – 1.74+ | 0.37 |
| May | – 0.32 | – 1.37 | – 0.90 | – 1.43 | 0.00 | – 0.21 | – 0.58 |
| Jun | – 0.74 | – 2.43* | 0.79 | – 1.37 | 0.00 | – 2.43* | – 0.05 |
| Jul | 1.32 | – 3.38*** | 0.85 | – 2.06* | – 1.51 | – 2.75** | – 0.32 |
| Aug | 0.37 | – 2.48* | 1.72+ | – 2.38* | – 1.75+ | – 2.11* | – 1.96 |
| Sep | – 0.16 | – 3.43*** | 1.53 | – 1.64 | – 0.53 | – 0.95 | – 0.174 |
| Oct | – 0.58 | – 3.54*** | – 1.06 | – 3.17** | – 0.32 | 0.95 | – 0.55 |
| Nov | 0.63 | – 0.85 | – 0.32 | – 3.51*** | – 0.48 | – 0.37 | 0.25 |
| Dec | – 0.16 | – 1.21 | 1.85+ | 1.06 | – 3.64*** | – 1.11 | – 0.69 |
| Annual | – 0.90 | – 3.38*** | 0.48 | – 4.54*** | 0.37 | – 2.96** | – 0.63 |
| Summer | – 1.53 | – 1.90+ | 0.21 | – 2.75** | 0.08 | – 1.43 | 0.00 |
| Rainy | 0.74 | – 4.44*** | 0.79 | – 3.86*** | – 1.93+ | – 1.64 | – 0.95 |
| Winter | 0.26 | – 1.64 | 0.53 | – 4.49*** | 1.16 | 0.74 | 0.45 |

(continued)

**Table 17.5** (continued)

| Period | P3 (1994–2016) | | | | | | |
|---|---|---|---|---|---|---|---|
| | MaxT | MinT | RH | WS | EP | SS | RF |
| Jan | 0.95 | 1.37 | − 0.48 | − 0.21 | − 0.69 | − 1.21 | − 0.65 |
| Feb | 1.43 | 1.64 | − 0.21 | − 0.11 | − 0.37 | **− 2.38*** | − 0.37 |
| Mar | − 0.37 | **2.96**** | 1.16 | − 0.71 | − 1.16 | **− 3.22**** | − 1.01 |
| Apr | 0.40 | **2.64**** | 0.11 | − 0.58 | 0.32 | − 0.69 | 1.37 |
| May | − 0.16 | **3.22**** | 0.05 | − 1.06 | − 1.06 | − 0.58 | 0.05 |
| Jun | 0.26 | 1.64 | 0.63 | **− 2.48*** | − 0.42 | 1.21 | 0.00 |
| Jul | 0.21 | **3.22**** | 0.50 | − 1.37 | − 0.16 | 1.16 | 1.00 |
| Aug | 0.95 | **2.80**** | − 1.43 | − 1.00 | 0.13 | **− 1.69+** | − 0.26 |
| Sep | 0.21 | **2.75**** | 0.00 | − 0.53 | − 0.11 | − 1.37 | **1.90+** |
| Oct | **1.74+** | 0.71 | **− 2.27*** | − 1.11 | 0.32 | − 0.74 | − 0.55 |
| Nov | 1.21 | 0.85 | − 1.00 | − 0.37 | 0.26 | − 0.63 | − 0.88 |
| Dec | 1.37 | **1.90+** | **− 2.27*** | 0.79 | 0.48 | 0.63 | **2.27*** |
| Annual | **1.74+** | **4.01***** | − 0.85 | **− 2.38*** | − 0.48 | **− 1.69+** | **2.25*** |
| Summer | 0.16 | **3.38***** | 0.26 | **− 2.38*** | − 0.69 | − 0.90 | 0.11 |
| Rainy | 0.85 | **2.80**** | − 1.21 | − 1.11 | 0.48 | − 0.05 | 1.21 |
| Winter | **2.22*** | **3.06**** | − 0.53 | − 0.48 | 0.05 | **− 2.22*** | − 0.21 |
| If a trend at | | | | | | | |
| Level of significance | 0.001 | 0.01 | 0.05 | 0.1 | | | |
| Notations | *** | ** | * | + | | | |

Bold shows level of significance at 0.001, 0.01, 0.05, 0.1 levels with respective notations ***, **, *, +

17.5. The analysis of MinT parameter at Raipur city shows a fall in trend of April–June–July–August–September–October–November–December–Annual–Rainy and summer in P2 period. While P3 period shows rise in trend of March–April–May–July–August–September–December–Annual–Rainy–summer and winter. The cause of this change may be similar to the impact seen in the analysis of MaxT. Although P1 period of MinT shows rising trend; which may be a cause of increase in urban heat. These two parameters were the most important factors to conclude the scenario of urban heat island at Raipur city, which causes significant climate change and changes the behaviour of rainfall pattern in Raipur city. Hence, overall rising trend may be concluded for minimum and maximum temperature from this analysis. RH at Raipur city indicates a rise in trend of August and December of period P2. P3 period shows falling trend in October and December while P1 period shows rise in trend of March–April–July–August–September–Annual and Rainy which may lead to conclusion of overall rising trend in RH at Raipur city. The result of WS parameter for all three periods P1, P2 and P3 reveals fall in trend of all monthly annual and seasonal series. EP parameter indicates randomly distributed during periods P1, P2

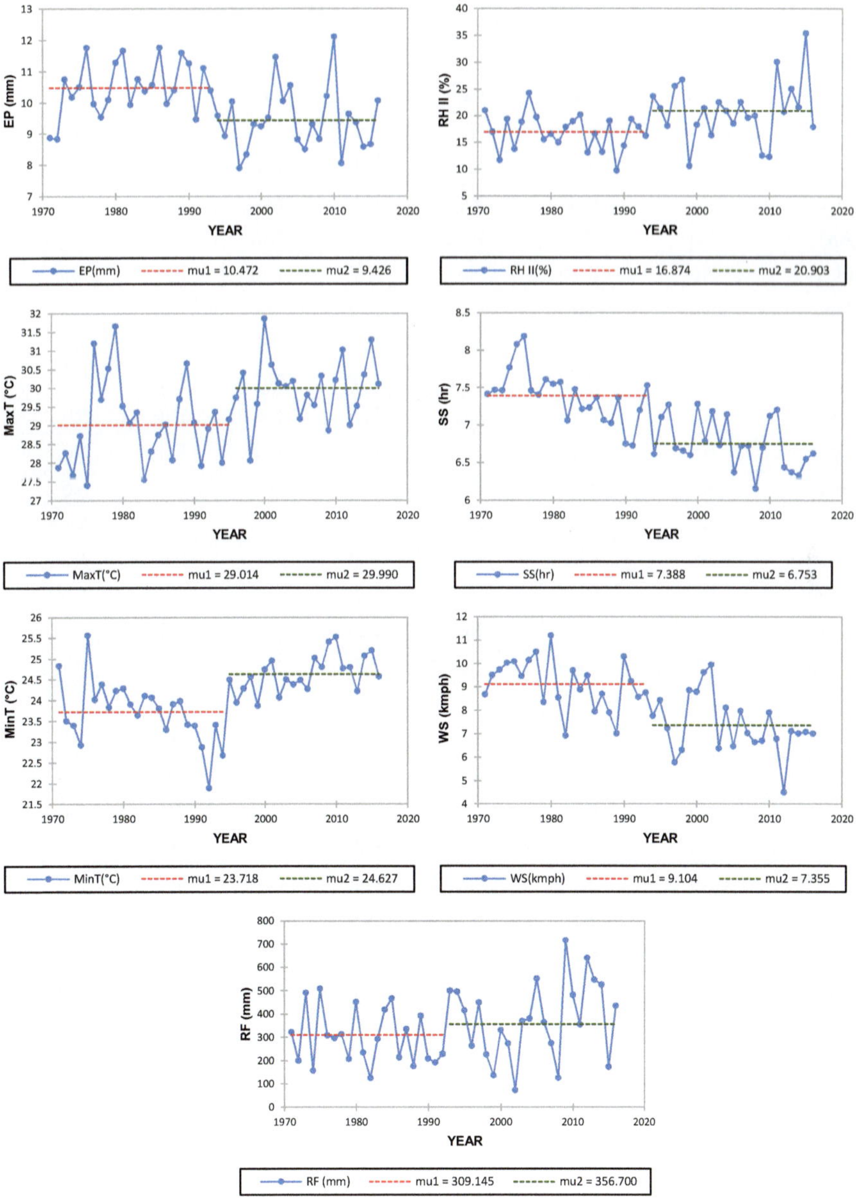


**Fig. 17.4** Result change-point test of all climatic parameters (where EP (mm) plot is for April series, similarly MaxT (°C)-December, MinT (°C)-August, RH (%)-April, SS (h)-Annual, WS (kmph)-May and RF (mm)-Annual Series)

**Table 17.6** Conclusion trend test of various climatic parameters for design period P1, P2 and P3

| Period | P1 (whole series) | | | | | | |
|---|---|---|---|---|---|---|---|
| | MaxT | MinT | RH | WS | EP | SS | RF |
| Jan | No | No | No | Fall | No | Fall | No |
| Feb | No | Rise | No | Fall | No | Fall | No |
| Mar | No | Rise | Rise | Fall | No | Fall | No |
| Apr | No | No | Rise | Fall | Fall | No | No |
| May | No | Rise | No | Fall | Fall | No | No |
| Jun | No | No | No | Fall | No | No | No |
| Jul | No | Rise | Rise | Fall | Fall | Fall | No |
| Aug | No | Rise | Rise | Fall | No | Fall | No |
| Sep | No | Rise | Rise | Fall | No | Fall | Rise |
| Oct | No | No | Fall | Fall | No | Fall | No |
| Nov | Rise | No | No | Fall | No | No | No |
| Dec | Rise | No | No | No | Fall | No | Rise |
| Annual | No | Rise | Rise | Fall | Fall | Fall | No |
| Rainy | Rise | Rise | Rise | Fall | Fall | Fall | No |
| Summer | No | Rise | No | Fall | No | Fall | No |
| Winter | Rise | Rise | No | Fall | No | Fall | No |
| Period | P2 (up to 1993) | | | | | | |
| | MaxT | MinT | RH | WS | EP | SS | RF |
| Jan | No | No | No | Fall | No | No | No |
| Feb | No | No | No | Fall | Rise | No | No |
| Mar | No | No | No | Fall | No | No | No |
| Apr | No | Fall | No | Fall | No | Fall | No |
| May | No | No | No | No | DF | No | No |
| June | No | Fall | No | No | DF | Fall | No |
| Jul | No | Fall | No | Fall | No | Fall | No |
| Aug | No | Fall | Rise | Fall | Fall | Fall | Fall |
| Sep | No | Fall | No | No | No | No | Fall |
| Oct | No | Fall | No | Fall | No | No | No |
| Nov | No | No | No | Fall | No | No | No |
| Dec | No | No | Rise | No | Fall | No | No |
| Annual | No | Fall | No | Fall | No | Fall | No |
| Summer | No | Fall | No | Fall | No | No | No |
| Rainy | No | Fall | No | Fall | Rise | No | DF |
| Winter | No | No | No | Fall | No | No | No |

(continued)

**Table 17.6** (continued)

| Period | P3 (1994–2016) | | | | | | |
|---|---|---|---|---|---|---|---|
| | MaxT | MinT | RH | WS | EP | SS | RF |
| Jan | No | No | No | No | No | No | No |
| Feb | No | No | No | No | No | Fall | No |
| Mar | No | Rise | No | No | No | Fall | No |
| Apr | No | Rise | No | No | No | No | No |
| May | No | Rise | No | No | No | No | No |
| June | No | No | No | Fall | No | No | DF |
| Jul | No | Rise | No | No | No | No | No |
| Aug | No | Rise | No | No | No | Fall | No |
| Sep | No | Rise | DF | No | No | No | Rise |
| Oct | Rise | No | Fall | No | No | No | No |
| Nov | No | No | No | No | No | No | No |
| Dec | No | Rise | Fall | No | No | No | Rise |
| Annual | Rise | Rise | No | Fall | No | Fall | Rise |
| Summer | No | Rise | No | Fall | No | No | No |
| Rainy | No | Rise | No | No | No | No | No |
| Winter | Rise | Rise | No | No | No | Fall | No |

**Table 17.7** Design storm intensities using 37-year rainfall record for 17 different durations at Raipur station

| Year | 15 min | 30 min | 45 min | 1 h | … | … | … | 12 h | 16 h | 24 h |
|---|---|---|---|---|---|---|---|---|---|---|
| 1980 | 18.0 | 36.0 | 36.0 | 36.0 | … | … | … | 118.8 | 158.4 | 195.5 |
| 1981 | 6.2 | 10.0 | 14.9 | 19.9 | … | … | … | 89.0 | 89.0 | 89.0 |
| 1982 | 10.3 | 19.3 | 29.0 | 37.8 | … | … | … | 80.0 | 80.0 | 80.0 |
| 1983 | 13.0 | 24.0 | 30.0 | 30.0 | … | … | … | 114.4 | 138.3 | 138.3 |
| … | … | … | … | … | … | … | … | … | … | … |
| … | … | … | … | … | … | … | … | … | … | … |
| … | … | … | … | … | … | … | … | … | … | … |
| 2014 | 14.4 | 28.8 | 36.0 | 36.0 | … | … | … | 162.0 | 162.0 | 162.0 |
| 2015 | 14.1 | 28.1 | 42.2 | 42.2 | … | … | … | 80.0 | 105.3 | 112.8 |
| 2016 | 20.0 | 36.5 | 41.8 | 55.8 | … | … | … | 92.0 | 92.0 | 92.0 |
| **Mean** | **13.7** | **24.1** | **31.2** | **37.6** | … | … | … | **117.1** | **131.6** | **138.1** |
| **SD** | **9.9** | **15.6** | **15.7** | **17.2** | … | … | … | **54.1** | **67.9** | **71.4** |
| $\alpha$ | **7.7** | **12.2** | **12.3** | **13.4** | … | … | … | **42.2** | **52.9** | **55.7** |
| $\beta$ | **9.2** | **17.1** | **24.2** | **29.8** | … | … | … | **92.8** | **101.1** | **106.0** |

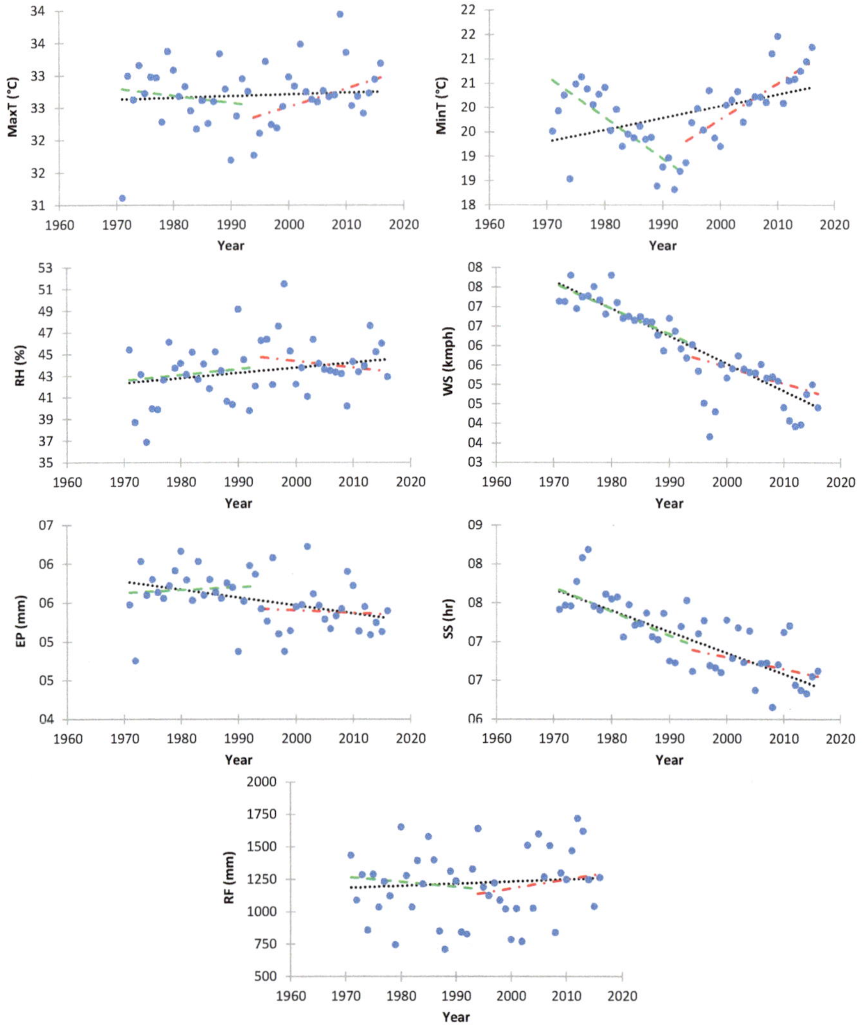


**Fig. 17.5** Annual trends in P1, P2 and P3 periods of climatic parameters that rise in trend in annual MaxT, MinT, RH and RF while fall in trend in annual WS, EP and SS

and P3 for all monthly annual and seasonal series. From above analysis, we may not lead to any conclusion as the most of series are either doubtful or showing no trend in all series. Jaiswal et al. (2017) analysed EP data for Raipur city and concluded falling trend, which supports the temperature trend of above study. This falling EP significantly shows an increasing urban land use (built-up) at Raipur city. The trend analysis of SS indicates that all the three periods P1, P2 and P3 result a significant fall in the trend of almost all monthly annual and seasonal series. P1 period shows falling trend in January to December, Annual, Rainy, summer and winter except April–May and November and their corresponding significance interval presented in Table 17.5, which may also lead to the cause of reduction in EP in Raipur city. The trend analysis of RF series for period P2 shows negative of falling trend in August and September while period P3 shows positive trend in September–November and Annual Series. Overall, positive trend has been seen in period P1 for RF at Raipur city. The result may lead to conclusion of the RF at Raipur city is increasing in trend after mid of 1993–1995. Increasing RF trend in Raipur city and reducing the number of rainy days are (Khavse et al. 2015) a cause for increasing the RF intensity in Raipur city which has been analysed and showed in frequency analysis section.

### *17.4.3 Frequency Analysis and Climate Impact Results*

After all this analysis with climatic parameters, the change-point 1993 which has been proven by trend analysis was used to draw variation in Intensity–Duration–Frequency relations. Table 17.7 shows the derived intensity values from 15 min to 24-h annual maximum rainfall data and statistic values. Using Eqs. 17.1 and 17.2, Xt has been calculated for 15, 30, 45 min, … up to 24 h durations with 2, 5, 10, 25, 50 and 100 years return period. The reason of selection of return periods for this study is to show the importance of climate uncertainty at the time of preparing IDF curves in an urban area. Although for storm water management Xt can be driven from 2, 5 or 10 year return period, while for big structure where the chances of heavy disaster involved like Dams the return period must be taken 100 year (Te Chow 1988). Maximum rainfall intensities are then calculated from Xt by dividing it from corresponding durations (Table 17.8).

The results of frequency analysis and developed IDF curve are derived for three rainfall series, namely whole series (1980–2016), series up to 1993 (1980–1993) and series after 1994 (1994–2016). The designed rainfall intensities (in mm/h) were plotted in ordinate against the durations (in min) in abscissa on logarithmic scale as shown in Fig. 17.6.

It can be clearly noticed from Table 17.8, the derived intensity values for different return periods have been significantly increased by increasing annual rainfall series, since 1980–1993 the intensity of 100-year return period rainfall was 96 mm/h, has increased if IDF will derive from 1980 to 2016 which is 179 mm/h. After our change-point application, the intensity of 100-year return period increases to 212 mm/h. This indicates changing/increasing rainfall pattern in Raipur city due to climatic uncertainties. Figure 17.7 has been plotted for individual return periods for all three series and has shown increment in intensity of rainfall if the derivable data has taken from 1994 to 2016. All three series and their indication can be seen as legend in each graph. From the analysis of IDF relation, it has been found that the derived value of intensities for difference return period of series (1980–2016) is over-estimate of the derived value of intensities of series (1980–1993). Similarly, the derived value of intensities for difference return period of series (1980–2016) is under-estimate of the derived value of intensities of series (1994–2016). The results from the same station

**Table 17.8** Derived rainfall for different return period using GEV-I for Raipur city

| Duration | | $I_T$ (mm/hr) for (1980-2016) | | | | | | $I_T$ (mm/hr) for (1980-1993) | | | | | | $I_T$ (mm/hr) for (1994-2016) | | | | | |
|---|---|---|---|---|---|---|---|---|---|---|---|---|---|---|---|---|---|---|---|
| hr | min | 2yr | 5yr | 10yr | 25yr | 50yr | 100yr | 2yr | 5yr | 10yr | 25yr | 50yr | 100yr | 2yr | 5yr | 10yr | 25yr | 50yr | 100yr |
| 0.25 | 15 | 48 | 83 | 106 | 136 | 158 | 179 | 45 | 59 | 68 | 79 | 87 | 96 | 51 | 94 | 123 | 159 | 186 | 212 |
| 0.5 | 30 | 43 | 71 | 89 | 112 | 129 | 146 | 38 | 50 | 58 | 69 | 76 | 84 | 47 | 80 | 102 | 130 | 151 | 172 |
| 0.75 | 45 | 38 | 57 | 69 | 85 | 96 | 108 | 32 | 42 | 49 | 58 | 64 | 70 | 42 | 64 | 78 | 96 | 109 | 122 |
| 1 | 60 | 35 | 50 | 60 | 73 | 82 | 92 | 29 | 39 | 46 | 55 | 61 | 67 | 39 | 55 | 66 | 81 | 91 | 101 |
| 1.5 | 90 | 30 | 41 | 48 | 57 | 63 | 70 | 26 | 33 | 38 | 44 | 48 | 52 | 33 | 44 | 52 | 62 | 69 | 76 |
| 2 | 120 | 26 | 34 | 39 | 46 | 51 | 55 | 23 | 29 | 32 | 37 | 40 | 44 | 28 | 37 | 42 | 50 | 55 | 60 |
| 3 | 180 | 21 | 27 | 32 | 37 | 41 | 45 | 19 | 23 | 26 | 30 | 33 | 36 | 22 | 30 | 34 | 40 | 45 | 49 |
| 4 | 240 | 18 | 23 | 27 | 32 | 35 | 39 | 16 | 19 | 21 | 24 | 26 | 28 | 19 | 25 | 30 | 35 | 39 | 43 |
| 5 | 300 | 15 | 21 | 24 | 28 | 32 | 35 | 14 | 17 | 19 | 22 | 24 | 26 | 16 | 22 | 26 | 31 | 35 | 38 |
| 6 | 360 | 14 | 19 | 22 | 26 | 29 | 32 | 13 | 16 | 18 | 21 | 23 | 25 | 15 | 20 | 24 | 29 | 32 | 35 |
| 7 | 420 | 13 | 17 | 20 | 24 | 26 | 29 | 11 | 15 | 17 | 20 | 22 | 24 | 13 | 18 | 21 | 25 | 29 | 32 |
| 8 | 480 | 12 | 16 | 18 | 22 | 24 | 27 | 10 | 14 | 16 | 20 | 22 | 24 | 12 | 17 | 19 | 23 | 26 | 28 |
| 10 | 600 | 10 | 14 | 17 | 20 | 23 | 25 | 09 | 13 | 15 | 18 | 21 | 23 | 11 | 15 | 18 | 21 | 24 | 26 |
| 11 | 660 | 10 | 14 | 16 | 20 | 22 | 24 | 08 | 12 | 14 | 17 | 20 | 22 | 10 | 14 | 17 | 21 | 23 | 26 |
| 12 | 720 | 09 | 13 | 16 | 19 | 21 | 24 | 08 | 11 | 13 | 16 | 18 | 20 | 10 | 14 | 17 | 20 | 23 | 26 |
| 16 | 960 | 08 | 11 | 14 | 17 | 19 | 22 | 06 | 09 | 11 | 13 | 15 | 16 | 09 | 13 | 16 | 19 | 22 | 25 |
| 24 | 1440 | 05 | 08 | 10 | 12 | 14 | 15 | 04 | 06 | 08 | 09 | 10 | 11 | 06 | 09 | 11 | 13 | 15 | 17 |

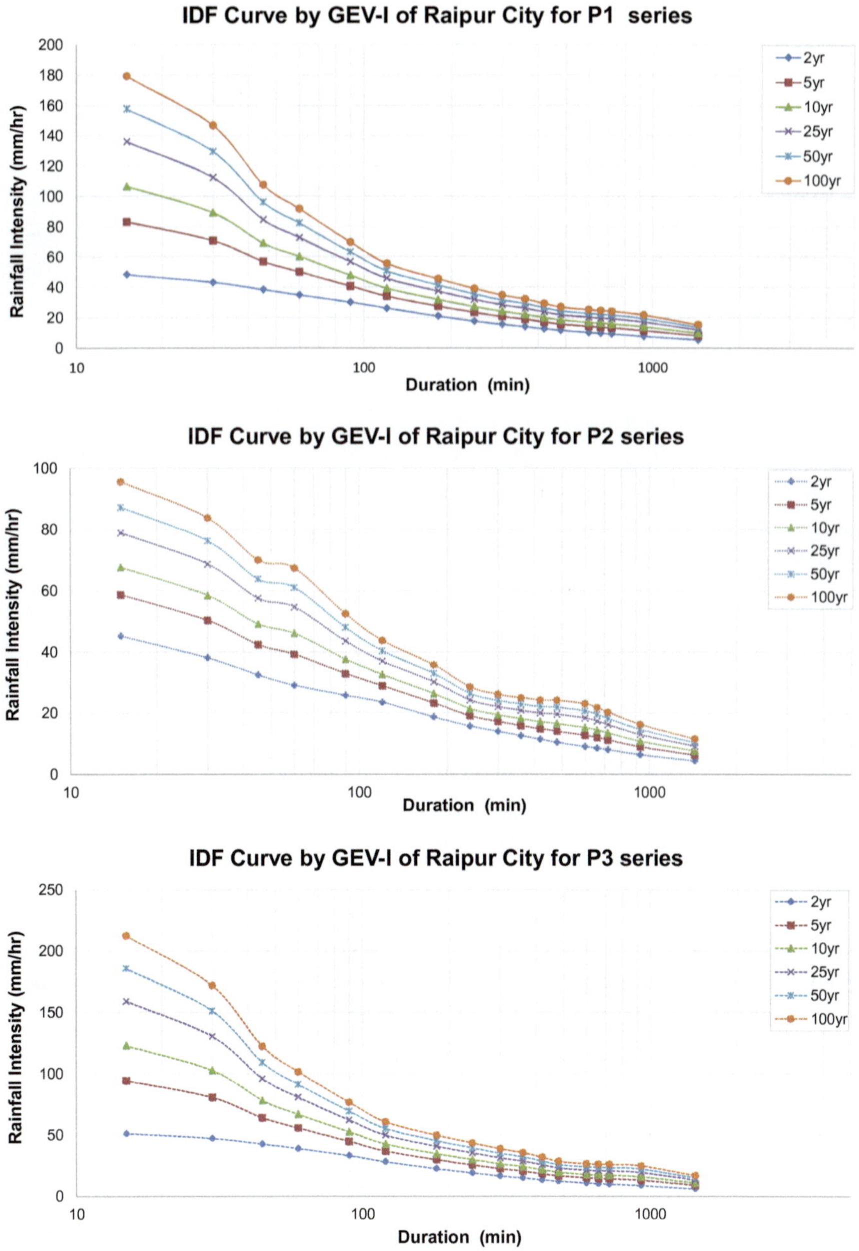


**Fig. 17.6** Designed intensity–duration–frequency relations using GEV-I of Raipur city for whole series of rainfall, series up to 1993 and series 1994–2016

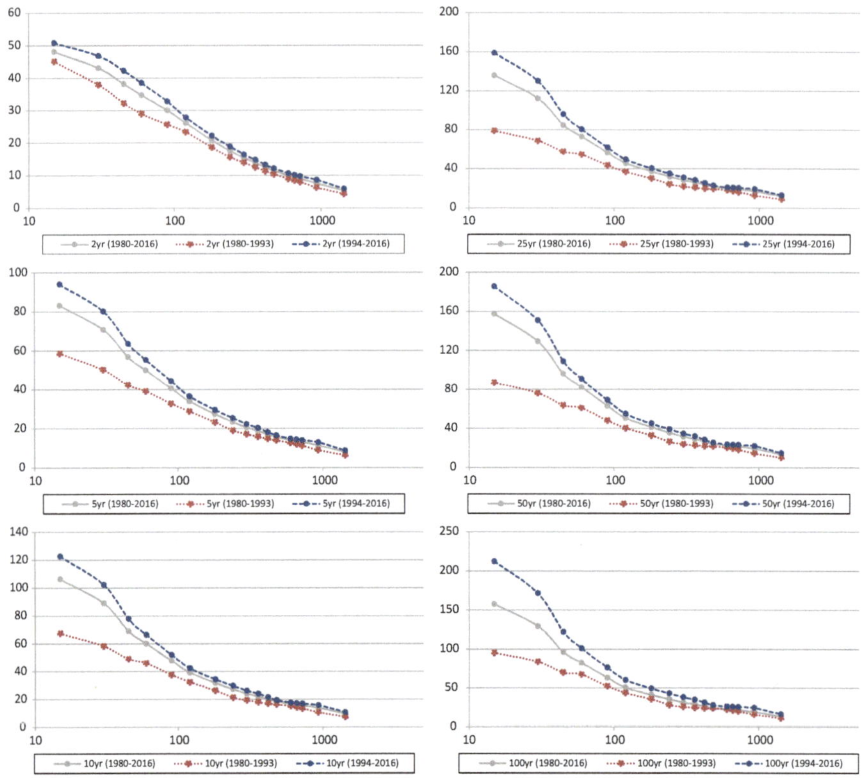


**Fig. 17.7** Comparison of intensity–duration–frequency relationships plotted between rainfall intensity (in mm) and duration (in min) with return period of 2, 5, 10, 25, 50, 100 years under climate uncertainty

but without any inconsistency showing a significant change in designed rainfall intensities, which may lead to conclusion of climate variability at Raipur city (Cheng and AghaKouchak 2014).

## 17.5 Conclusion

The assessment of variability in rainfall series under climatic uncertainty due to possible extremes weather events/climate change, climate shift and human impact is an important for water resource planner to manage the water resources at regional and local scale. Present study reveals impact of change in climate on design rainfall events so called IDF relations. Change-point test using four methods, non-parametric trend analysis, IDF curve preparation and impact of climate variability/weather extremes on IDF have been carried out in this study. The results of change-point test have been using PeT, BRT, SNHT and vNRT on various time series on MaxT, MinT, WS, RH, SS, EP and RF have been applied in Raipur city. Results of change-point test indicated a climate shift between 1990 and 2000 using critical values of test statistic. However, 1993 has repeated in the results of many change-point test, hence this year 1993 has been taken as standard change-point for Raipur city. The result attributes the effect of urbanization, rapid development of industrial and commercial activities in the region which altered natural ecology and environment in Raipur city. Application of change-point in the data series has made three different periods (P1 whole series of a variables 1971–2016, P2 covers time series up to 1993 and P3 covers time series from 1994 to 2016) to check changing pattern in the trend of climatic data. Trend analysis of P1, P2 and P3 period to check the validity of change-point test results reveals rising trend in annual MaxT, MinT, RH and RF while falling trend in annual WS, EP and SS. From the result, the temperature was the main factor which was causing the effect of changing rainfall pattern in Raipur city. The MaxT of P1 period results significant rising trend in November, December, rainy and winter, P3 period also shows rising trend in October, Annual and winter series. MinT parameter shows rising trend in March–April–May–July–August–September–December–Annual–rainy–summer and winter in P3 period. This rise in minimum and maximum temperature may cause of urbanization, increasing industrialization and climate change late after year 1993 in Raipur city. Frequency analysis has been separately applied in shorter duration rainfall data. From the analysis of IDF relation, it has been found that the derived value of intensities for different return period of series (1980–2016) was over-estimate of the derived value of intensities of series (1980–1993). Similarly, the derived value of intensities for difference return period of series (1980–2016) was under-estimate of the derived value of intensities of series (1994–2016). The frequency analysis results show reduction in number of rainy days in Raipur city, however the rainfall trend was positive. So, if annual rainfall trend is increasing and numbers of rainy days are decreasing, this will increase the possibility of getting higher intensity rainfall events in Raipur city. Also for engineering point of view, under-estimated rainfall intensities may cause a risk to selection of design rainfall intensities for building big hydraulic structure in and around the Raipur city. Selection of under-estimated rainfall intensity may also cause of limiting the scope of best management practices in storm water management in the study area. According to significant impact of climatic variables identified in this study and results of frequency analyses, it can be conveniently utilized by the water resource manager, planner and decision-makers for scientific management planning to mitigate the storm water management in Raipur city planning area.

**Acknowledgements** The authors would like to thank Deutscher Akademischer Austauschdienst (DAAD), as the research material is based upon work within the programme 'A New Passage to India'. We would like to thank Indian meteorological department for allowing us to digitize and collect continuous climatic data of Raipur city at Regional centre Nagpur. One of the author would like to thank NEERI, Nagpur for providing laboratory facility during the field work of this study.

## Appendix 1: Change-Point Detection Tests

**Pettitt's Test (PeT)** Pettitt (1979) proposes to test for no change-point against one change without hypothesis on the initial distribution. It consists of a series sweep by splitting it into two sub-samples. Comparing the distributions before and after the cut-off time enables finding a possible change-point that allows for significant differences. This non-parametric test for comparing independent samples and results the date of change-point, that can be utilized for determining the occurrence of random changes in climatic data (Gao et al. 2011; Zarenistanak et al. 2014; Zhang et al. 2004).

$$U_t = \sum_{i=1}^{t} \sum_{j=t+1}^{n} \operatorname{sign}(X_i - X_j), \operatorname{sign}(X_t - X_j) = \begin{bmatrix} -1 & \text{if}(X_i - X_j) < 0 \\ 0 & \text{if}(X_i - X_j) = 0 \\ 1 & \text{if}(X_i - X_j) > 0 \end{bmatrix} \tag{17.3}$$

$$K_n = \text{Max}|U_t| \tag{17.4}$$

The change-point of the series is located at $K_n$ (test statistic), provided that the statistic is significant. The significance probability of $K_n$ is approximated for $\rho \leq 0.05$.

**von Neumann Ratio Test (vNRT)** The method vNRT is similarly as first-order serial correlation coefficient (Bartels 1984). This test has been used and explained by Buishand (1982) and Kang and Yusof (2012). In this test, the null hypothesis is that the data is independent identically distributed random values; the alternative hypothesis is that the values in the series are not randomly distributed. '$N$' is defined as the ratio of the mean square successive (year to year) difference to the variance (Morozova and Valente 2012; Von Neumann 1941). Equation 17.5 may be used for identify break point in a data series of '$n$' number of samples.

$$N = \frac{\sum_{i=1}^{n-1} (x_i - x_{i-1})^2}{\sum_{i=1}^{n-1} (x_i - \overline{x})^2} \tag{17.5}$$

**Buishand's Range Test (BRT)** This test assumes that tested values are independent and identically normally distributed (null hypothesis). The alternative hypothesis assumes that the series has a break. This test is more sensitive to breaks in the middle of time series (Costa and Soares 2009; Morozova and Valente 2012). The adjusted partial sum is defined as:

$$S_0 = 0 \text{ and } S_k = \sum_{i=1}^{k} (x_i - \overline{x}), (1 \leq i \leq n) \tag{17.6}$$

When the series is homogeneous, then the value of $S_k$ will fluctuate around zero. The year '$k$' has shift when $S_k$ has reached a negative shift or positive shift. Rescaled adjusted range, '$R$' is obtained by Eq. 17.7 and $R/\sqrt{n}$ can be used to compare with critical values given by Wijngaard et al. (2003).

$$R = \frac{\text{Max}(S_k) + \text{Min}(S_k)}{\overline{x}} \tag{17.7}$$

**Standard Normal Homogeneity test (SNHT)** In SNHT, the null hypothesis is the same as in the Buishand Range test. This test is one of the most popular homogeneity tests in climate studies (Alexandersson 1986). The null and alternative hypotheses in this test are the same as in the Buishand test; however, unlike the Buishand test, SNHT is more sensitive to the breaks near the beginning and the end of the series (Costa and Soares 2009; Morozova and Valente 2012). A statistic $〚T〛_k$ is used to compare the average of first '$n$' years with the last of $(n - k)$ years (Gonzalez-Hidalgo et al. 2009; Zollo et al. 2012). The year '$k$' will be considered as shift if value of $T_k$ is maximum. To reject null hypothesis, the test statistic $T_0 =$ $〚\text{maxT}〛_k$ is greater than the critical value, which depends on the sample size.

$$T_k = kZ_1^2 + (n - k)Z_2^2 \tag{17.8}$$

$Z_1$ and $Z_2$ can be computed as:

$$Z_1 = \frac{1}{k}\sum_{i=1}^{k}\frac{(X_i - \overline{X})}{\sigma x} \tag{17.9}$$

$$Z_2 = \frac{1}{n-k}\sum_{i=1}^{k}\frac{(X_i - \bar{X})}{\sigma x} \tag{17.10}$$

## Appendix 2: Trend Analysis Tests

In the present study, the trend analyses for climatic parameters are estimated using application of LRT, MRT and SRT as explained here.

**Linear Regression Test (LRT)** A LRT attempts to explain the relationship between two or more variables using a straight line. Then the test data fitted in the straight line and slope of the line will be checked for notable difference from zero or not. This test has been used by Eslamian et al. (2009), Ghosh and Mujumdar (2006), Kim and Siegmund (1989), and student's $t$-test is used to confirm the hypothesis in this test. For a series of observations $X_n$, a straight line is fitted to the tested data and test static '$t$' computed, where, '$c$' is $t$ and '$m$' is intersect and slope of the line, $\sum \varepsilon_i{}^2$ is the sum of RSME (root mean square error) and $S_b$ is the standard error of slope.

Intercept

$$c = \overline{Y} - m\overline{X} \tag{17.11}$$

Slope

$$m = \frac{\sum (X_i - \overline{X})(Y_i - \overline{Y})}{\sum (X_i - \overline{X})^2} \tag{17.12}$$

$$\sum \varepsilon_i^2 = \sum (Y_i - \overline{Y})^2 - m^2 \sum (X_i - \overline{X})^2 \tag{17.13}$$

$$S_m^2 = \frac{\sum \epsilon_i^2}{(n-2)\sum (X_i - \overline{X})^2} \tag{17.14}$$

$$t = \frac{m}{S_m} \tag{17.15}$$

**Mann–Kendall Test (MKT)** MKT (non-parametric), generally used to identify monotonic trends in series of climate data or hydrological data which doesn't require the data to be normally distributed. MKT has been used widely, as it has recommended by WMO, Mitchell et al. (1966). Many other researchers use this method for determining the trend in climatic data series are Yue and Hashino (2003), Kahya and Kalayci (2004), Modarres and de Paulo Rodrigues da Silva (2007), Tabari et al. (2011). The test statistic is calculated according to:

$$S = \sum_{i=1}^{n} \sum_{j=1}^{i-1} \mathrm{sign}(X_i - X_j); \quad \mathrm{sign}(X_i - X_j) = \begin{bmatrix} 1, & \mathrm{if}(X_i - X_j) > 0 \\ 0, & \mathrm{if}(X_i - X_j) = 0 \\ -1, & \mathrm{if}(X_i - X_j) < 0 \end{bmatrix} \tag{17.16}$$

where, '$n$' is the total length of data, $X_i$ and $X_j$ are sequential samples. The null hypothesis $H_0$ for MKT is that the data comes from a population with independent realizations and the data follows a monotonic trend represents alternative hypothesis, $H_\mathrm{a}$.

The statistic '$S$' is approximately normal distributed provided that the $Z_{\mathrm{MK}}$ is employed by Eq. 17.19. The mean of '$S$' is approximately normally distributed with the mean $E(S)$ and variance can be computed using Eqs. 17.17 and 17.18, where, '$t$' is the number of the tied groups in the data set and $\Sigma t$ denotes the summation over all tie number of values.

$$E(S) = 0 \tag{17.17}$$

$$\mathrm{Var}(S) = \left[(n-1)(2n+5) - \frac{1}{n}\sum_t t(t-1)(2t+5)\right] \quad (17.18)$$

$$Z_{\mathrm{MK}} = \begin{cases} -1, & \text{if } S < 0 \\ 0, & \text{if } S = 0 \\ \frac{S-1}{\sqrt{\mathrm{Var}(S)}}, & \text{if } S > 0 \end{cases} \quad (17.19)$$

**Spearman's Rho Test (SRT)** SRT (non-parametric) measure of rank correlation (statistical dependence between the rankings of two variables). It determines how strong the relationship between two variables can be described using a monotonic function (Lettenmaier 1976; Otten 1973; Shadmani et al. 2012; van Belle and Hughes 1984). If there is no trend in the samples and observations are independent, then all rank orderings are equally likely. In SRT test, the difference between order and rank $d_l$ for all observations $x_n$ used to compute and Spearman's $\rho$, variance Var($\rho$) and test statistic 'Z' using Eqs. 17.20, 17.21 and 17.22.

$$\rho = 1 - \frac{6\sum d_i^2}{n(n-1)} \quad (17.20)$$

$$\mathrm{Var}(\rho) = \frac{1}{(n-1)} \quad (17.21)$$

$$Z = \frac{\rho}{\sqrt{\mathrm{Var}(\rho)}} \quad (17.22)$$

## References

Agrawal V (2013) Raipur as a new capital: impact on population. Int J Eng Res Sci Tech 2:41–46

Alexandersson H (1986) A homogeneity test applied to precipitation data. J Climatol 6:661–675

Ambade B (2015) Chemical composition of runoff water in Raipur city, central India. Appl Water Sci 5:1–12

Azzam R, Strohschön R, Baier K, Lu L, Wiethoff K, Bercht AL, Wehrhahn R (2014) Water quality and socio-ecological vulnerability regarding urban development in selected case studies of Megacity Guangzhou, China. In: Kraas F, Aggarwal S, Coy M, Mertins G (eds) Megacities: our global urban future. Springer, Netherlands, Dordrecht, pp 33–58

Bai J (1997) Estimation of a change point in multiple regression models. Rev Econ Stat 79:551–563

Baier K, Schmitz KC, Azzam R, Strohschön R (2014) Management tools for sustainable ground water protection in mega urban areas—small scale land use and ground water vulnerability analyses in Guangzhou, China. Int J Environ Res 8:249–262

Barrow E, Maxwell B, Gachon P (2004) Climate variability and change in Canada. Past, present and future. ACSD science assessment series. Environment Canada, Toronto, Ontario

Bartels R (1984) The rank von neumann test as a test for autocorrelation in regression models. Commun Stat Theor Methods 13:2495–2502

Bates BC, Chandler RE, Bowman AW (2012) Trend estimation and change point detection in individual climatic series using flexible regression methods. J Geophys Res 117
Bisai D, Chatterjee S, Khan A, Barman NK (2014) Long term temperature trend and change point: a statistical approach. Open J Atmos Clim Change 1:32–42
Bodhankar N (2002) Biodiversity and ecosystem analysis with reference to traditional water system, Chhattisgarh: beautiful and bountiful, p 7
Boroujerdy P (2008) The analysis of precipitation variation and quantiles in Iran. In: Proceedings of the 3rd WSEAS international conference on energy and environment. University of Cambridge, UK, pp 248–253
Buhairi MHA (2010) Analysis of monthly, seasonal and annual air temperature variability and trends in Taiz City—Republic of Yemen. J Environ Prot 01:401–409
Buishand TA (1982) Some methods for testing the homogeneity of rainfall records. J Hydrol 58:11–27
Census-of-India (2011a) District census handbook: Raipur. Census of India, pp 1–452
Census-of-India (2011b) Data on rural and urban areas, Chhattisgarh. In: Pillay RG (ed) Census of India
Chandra R, Saha U, Mujumdar PP (2015) Model and parameter uncertainty in IDF relationships under climate change. Adv Water Resour 79:127–139
Cheng L, AghaKouchak A (2014) Nonstationary precipitation intensity-duration-frequency curves for infrastructure design in a changing climate. Sci Rep 4:7093
Christ K, Baier K, Schmitz K-S, Jha R, Azzam R (2012) Water pollution and environmental problems as a result of urbanization in Hyderabad. Wasserwirtschaft 102:25–30
Costa AC, Soares A (2009) Homogenization of climate data: review and new perspectives using geostatistics. Math Geosci 41:291–305
CRISIL, IAT (2014) Revised city development plan for Raipur, capacity building for urban development project. Ministry of Urban Development Government of India
Croitoru A-E, Holobaca I-H, Lazar C, Moldovan F, Imbroane A (2011) Air temperature trend and the impact on winter wheat phenology in Romania. Clim Change 111:393–410
De Paola F, Giugni M, Topa ME, Bucchignani E (2014) Intensity-duration-frequency (IDF) rainfall curves, for data series and climate projection in African cities. Springerplus 3:133
Dhorde AG, Zarenistanak M (2013) Three-way approach to test data homogeneity: an analysis of temperature and precipitation series over southwestern Islamic Republic of Iran. J Indian Geophys Union 17:233–242
Eslamian S, Hassanzadeh H, Khordadi MJ (2009) Detecting and evaluating climate change effect on frequency analysis of wind speed. Int J Global Energy Issues 32:295–304
Farooq S, Ahmad S (2008) Urban sprawl development around Aligarh city: a study aided by satellite remote sensing and GIS. J Indian Soc Remote Sens 36:77–88
Fealy R, Sweeney J (2005) Detection of a possible change point in atmospheric variability in the North Atlantic and its effect on Scandinavian glacier mass balance. Int J Climatol 25:1819–1833
Gadgil A, Dhorde A (2005) Temperature trends in twentieth century at Pune, India. Atmos Environ 39:6550–6556
Gallagher C, Lund R, Robbins M (2013) Changepoint detection in climate time series with long-term trends. J Clim 26:4994–5006
Gao P, Mu XM, Wang F, Li R (2011) Changes in streamflow and sediment discharge and the response to human activities in the middle reaches of the Yellow River. Hydrol Earth Syst Sci 15:1–10
Ghosh S, Mujumdar PP (2006) Future rainfall scenario over Orissa with GCM projections by statistical downscaling. Curr Sci 90:396–404
GIZ, ASEM, RMC (2010) City sanitation plan Raipur. In: Asem G (ed)
GIZ, ASEM, RMC (2011) City level strategy: city sanitation plan for Raipur. In: Asem G (ed) Raipur municipal corporation. Alchemy Urban Systems (P) LTD.

Gonzalez-Hidalgo JC, Lopez-Bustins J-A, Štepánek P, Martin-Vide J, de Luis M (2009) Monthly precipitation trends on the Mediterranean fringe of the Iberian Peninsula during the second-half of the twentieth century (1951–2000). Int J Climatol 29:1415–1429

Guha S, Govil H, Mukherjee S (2017) Dynamic analysis and ecological evaluation of urban heat islands in Raipur city. SPIE, India, p 23

IPCC (2007) The fourth assessment report of the intergovernmental panel on climate change. Switzerland, Geneva

Jain SK, Kumar V (2012) Trend analysis of rainfall and temperature data for India. Curr Sci:37–49

Jaiswal RK, Lohani AK, Tiwari HL (2015) Statistical analysis for change detection and trend assessment in climatological parameters. Environ Processes 2:729–749

Jaiswal RK, Tiwari HL, Lohani AK (2017) Assessment of climate change impact on rainfall for studying water availability in upper Mahanadi catchment, India. J Water Climate Change

Jandhyala VK, Fotopoulos SB, You J (2010) Change-point analysis of mean annual rainfall data from Tucumán, Argentina. Environmetrics 21:687–697

Kahya E, Kalayci S (2004) Trend analysis of streamflow in Turkey. J Hydrol 289:128–144

Kang HM, Yusof F (2012) Homogeneity tests on daily rainfall series. Int J Contemp Math Sci 7:9–22

Karmeshu N (2012) Trend detection in annual temperature and precipitation using the Mann Kendall test–a case study to assess climate change on select states in the Northeastern United States, Department of Earth and Environmental Science. University of Pennsylvania

Karpouzos D, Kavalieratou S, Babajimopoulos C (2010) Trend analysis of precipitation data in Pieria Region (Greece). European Water 30:31–40

Khan A, Chatterjee S, Bisai D, Barman NK (2014) Analysis of change point in surface temperature time series using cumulative sum chart and bootstrapping for Asansol weather observation station, West Bengal, India. Am J Clim Chang 03:83–94

Khan R, Jhariya D (2016) Land use land cover change detection using remote sensing and geographic information system in Raipur municipal corporation area, Chhattisgarh. Int J Geo Sci Geo Inform 3:1–4

Khavse R, Deshmukh N, Manikandan N, Chaudhary J, Kaushik D (2015) Statistical analysis of temperature and rainfall trend in Raipur District of Chhattisgarh. Curr World Environ 10:305–312

Kim H-J, Siegmund D (1989) The likelihood ratio test for a change-point in simple linear regression. Biometrika 76:409–423

Kimball BF (1938) Probability-distribution-curve for flood-control studies. EOS Trans Am Geophys Union 19:460–483

Kioutsioukis I, Melas D, Zerefos C (2010) Statistical assessment of changes in climate extremes over Greece (1955–2002). Int J Climatol 30:1723–1737

Lee S, Kim SU (2014) Comparison between change point detection methods with synthetic rainfall data and application in South Korea. KSCE J Civ Eng 20:1558–1571

Lettenmaier DP (1976) Detection of trends in water quality data from records with dependent observations. Water Resour Res 12:1037–1046

Lund R, Reeves J (2002) Detection of undocumented changepoints: a revision of the two-phase regression model. J Clim 15:2547–2554

Madhuri R, Verma MK, Tripathi RK (2012) Impact of urbanisation on water resources in Raipur City using GIS and remote sensing. Int J Earth Sci Eng 05:1815–1819

Mavromatis T, Stathis D (2010) Response of the water balance in Greece to temperature and precipitation trends. Theoret Appl Climatol 104:13–24

Menne MJ, Williams CN Jr (2005) Detection of undocumented changepoints using multiple test statistics and composite reference series. J Clim 18:4271–4286

Mitchell JM, Dzezerdzeeskii B, Flohn H, Hofmeyer WL, Lamb HH, Rao KN, Wallen CC (1966) Climatic change. WMO Tech. Note 79, WMO No. 195. TP-100. Geneva, p 79

Modarres R, de Paulo Rodrigues da Silva V (2007) Rainfall trends in arid and semi-arid regions of Iran. J Arid Environ 70:344–355

Morozova AL, Valente MA (2012) Homogenization of Portuguese long-term temperature data series: Lisbon, Coimbra and Porto. Earth Syst Sci Data 4:187–213
Narain S (2017) Why urban India floods Indian cities grow at the cost of their wetlands. Centre for Science and Environment
Otten A (1973) The null distribution of Spearman's S when n = 13(1)16. Stat Neerl 27:19–20
Pettitt AN (1979) A non-parametric approach to the change-point problem. J Roy Stat Soc Ser C (appl Stat) 28:126–135
Pohlert T (2017) Non-parametric trend tests and change-point detection, CC BY-ND
Portnyagin YI, Merzlyakov EG, Solovjova TV, Jacobi C, Kürschner D, Manson A, Meek C (2006) Long-term trends and year-to-year variability of mid-latitude mesosphere/lower thermosphere winds. J Atmos Solar Terr Phys 68:1890–1901
Praharaj S, Han JH, Hawken S (2017) Innovative civic engagement and digital urban infrastructure: lessons from 100 smart cities mission in India. Proc Eng 180:1423–1432
Pranuthi G, Dubey S, Tripathi S, Chandniha S (2014) Trend and change point detection of precipitation in urbanizing Districts of Uttarakhand in India. Indian J Sci Technol 7:1573–1582
Prodanovic P, Simonovic S (2008) Intensity duration frequency analysis under changing climatic conditions. In: CD proceedings, 4th international symposium on flood defense: managing flood risk, reliability and vulnerability. Paper
Putra D, Baier K (2008) Impact of urbanization on groundwater recharge–the example of the Indonesian Million City Yogyakarta, UN Habitat—United Nations settlement programs: fourth session of the world urban forum, Nanjing, China
RMC (2010) Detailed project report, sewerage, storm, surface water drains and lake protection works for Raipur city. Meinhardt (Singapore) Pte. Ltd.
Serra C, Burgueño A, Lana X (2001) Analysis of maximum and minimum daily temperatures recorded at Fabra observatory (Barcelona, NE Spain) in the period 1917–1998. Int J Climatol 21:617–636
Shadmani M, Marofi S, Roknian M (2012) Trend analysis in reference evapotranspiration using Mann-Kendall and spearman's rho tests in arid regions of Iran. Water Resour Manage 26:211–224
Sinha MK, Verma MK, Ahmad I, Baier K, Jha R, Azzam R (2016) Assessment of groundwater vulnerability using modified DRASTIC model in Kharun Basin, Chhattisgarh, India. Arab J Geosci 9
Sinha UK, Kulkarni KM, Sharma SK, Ray A, Bodhankar N (2002) Assessment of aquifer system using isotope techniques in urban centres Raipur, Calcutta and Jodhpur, India. In: International Atomatic Energy Agency (IAEA-TECDOC-1298), Austria, pp 77–94
Smith DI (1999) Urban flood damage and greenhouse scenarios—the implications for policy: an example from Australia. Mitig Adapt Strat Glob Change 4:331–342
Strohschön R, Baier K, Azzam R (2009) Urban units as an analysis tool for mega-urban development. The case of Guangzhou, China. In: Proceedings of megacities–interactions between land use and water management. Mitteilungen zur Ingenieurgeologie und Hydrogeologie 99, pp 63–72
Tabari H, Marofi S (2010) Changes of pan evaporation in the west of Iran. Water Resour Manage 25:97–111
Tabari H, Marofi S, Aeini A, Talaee PH, Mohammadi K (2011) Trend analysis of reference evapotranspiration in the western half of Iran. Agric Meteorol 151:128–136
Te Chow V (1988) Applied hydrology. Tata McGraw-Hill Education
UN-Habitat (2011) Cities and climate change: global report on human settlements 2011. Earthscan, London
UNEP (2011) Vision for change country papers, based on the global survey on sustainable lifestyle
van Belle G, Hughes JP (1984) Nonparametric tests for trend in water quality. Water Resour Res 20:127–136
Von Neumann J (1941) Distribution of the ratio of the mean square successive difference to the variance. Ann Math Stat 12:367–395

Wakode HB, Baier K, Jha R, Ahmed S, Azzam R (2014) Assessment of impact of urbanization on groundwater resources using GIS techniques-case study of Hyderabad, India. Int J Environ Res 8:1145–1158

Ward R (1989) Hydrology of floods in Canada: a guide to planning and design. National Research Council Canada, Ottawa

Wijngaard JB, Klein Tank AMG, Können GP (2003) Homogeneity of 20th century European daily temperature and precipitation series. Int J Climatol 23:679–692

Yue S, Hashino M (2003) Long term trends of annual and monthly precipitation in Japan. JAWRA J Am Water Resour Assoc 39:587–596

Zarenistanak M, Dhorde AG, Kripalani RH (2014) Trend analysis and change point detection of annual and seasonal precipitation and temperature series over southwest Iran. J Earth Syst Sci 123:281–295

Zhang X, Zwiers FW, Li G (2004) Monte Carlo experiments on the detection of trends in extreme values. J Clim 17:1945–1952

Zollo A, Mercogliano P, Turco M, Vezzoli R, Rianna G, Bucchignani E, Manzi M, Montesarchio M (2012) Architectures and tools to analyse the impact of climate change on hydrogeological risk on Mediterranean area. CMCC research paper

# Chapter 18
# An Analysis of Textural Characteristics and Depositional Environment of Sediments in the Gumti River, Tripura, India

**I. Ahmed, N. Das (Pan), and J. Debnath**

**Abstract** Grain size analysis is an essential parameter to understand the depositional environment of sediments. In the present study, grain size analysis of bed sediments of the Gumti River has been done. Sediments samples were collected from 20 locations and processed using standard sieving method to calculate the statistical parameters like mean, standard deviation, skewness, and kurtosis. It has been found that the mean size of sediments ranges between medium to fine sand which are mainly moderate to well sorted in nature. Medium sand is mostly present in the upper and middle courses of the river Gumti, while the lower course is dominated by fine sands. The skewness value shows that most of the sediment samples are fine to very fine skewed in nature, whereas kurtosis of the samples varies from leptokurtic to paltykurtic nature of sediments. From frequency distribution curve, it has been observed that almost all the samples are unimodal in nature with peak of the curve located in size 2 to 3 phi. CM diagram has been prepared to understand the energy condition of the depositional medium and dominant mode of sediment transportation. It reveals two modes of sediment deposition, i.e., bottom and graded suspension under tractive force.

**Keywords** Sediment · Gain size analysis · Sieving · Gumti River

## 18.1 Introduction

Studies of grain size distribution provide vital information about sediment transport processes and depositional environments. Several pioneer works has been done in order to identify the relationship between grain size, transportation process, and deposition mechanism of sediments (Inman 1952; Folk and Ward 1957; Mason and Folk 1958; Friedman 1961, 1962, 1967; Sahu 1964; Visher 1969). Distribution of sediments in any particular reach of a channel is highly influenced by energy condition

I. Ahmed (✉) · N. Das (Pan) · J. Debnath
Department of Geography and Disaster Management, Tripura University, Tripura, India

 
P. Kumar et al. (eds.), *Water Resources Management and Sustainability*,
Advances in Geographical and Environmental Sciences,
https://doi.org/10.1007/978-981-16-6573-8_18

and transportation mechanism of water. The hydrological process of rolling, suspension, and saltation often controls the distribution of grain size of sediments (Inman 1949). In India, several studies have been conducted in the past to analyze the textural characteristics of river sediments (Ramesh and Subramanian 1992; Seralathan and Padmalal 1994; Prabhakara Rao et al. 2001). An analysis of river bed materials helps to identify dominant sediment type in a particular river (Mohtar et al. 2015). The downstream distribution of bed sediments is often studied to understand the regional fluvial geomorphological and hydrological settings (Zhanqiao et al. 2009). In general, grain size of river bed sediments is used to decrease downstream due to mutual friction. Besides, fining of bed sediments in lower course reflects local plummet in energy level of water to carry load. Gumti is the largest river of Tripura which was utilized as a trade route between India and Bangladesh in the past. But due to gradual sedimentation, the river is swiftly losing its navigability, and therefore, it is indispensable to study the bed sediment characteristics of the river. The present study aims to analyze the textural characteristics of sediments of the Gumti River and to identify the environment of sediment deposition.

## 18.2 Study Area

Gumti is the largest and longest river of Tripura. It is formed by confluence of two tributaries namely Raima and Sarma. The study area is extended between 23°25′16″ to 23°27′38″ N latitude and 91°15′22″ to 91°48′07″ E longitude (Fig. 18.1). A stretch of 120 km of the Gumti River, downstream from the Dumbur dam, has been selected for the present study to identify the spatial variation in the textural characteristics of sediments of the Gumti River.

## 18.3 Methodology

For the present study, 20 samples were collected from the channel of the Gumti River (starting from upstream toward downstream) in dry polythene bags for analysis in the fluvial hazard analysis laboratory. Sites for sample collection were selected on the basis of tributary confluence point, bank erosion, and sedimentation prone areas of the Gumti River where there is a chance of change in bed sediment characteristics. The exact location of each sample collection sites was marked by GPS. The samples were dried in an oven at 60 °C for 24 h in order to remove the moisture content. Sieving technique was applied to separate the grains into various size classes. Initially, 100 g of each sample were collected from the river bed and treated with 10% diluted HCl for removal of carbonate. The samples were further dried up and sieved in table sieve shaker (vibration type) for 15 min using ASTM sieve at half phi interval. Samples from each sieve were collected, weighed, and further processed in software. The Gradistat software package (Blott and Pye 2001) was

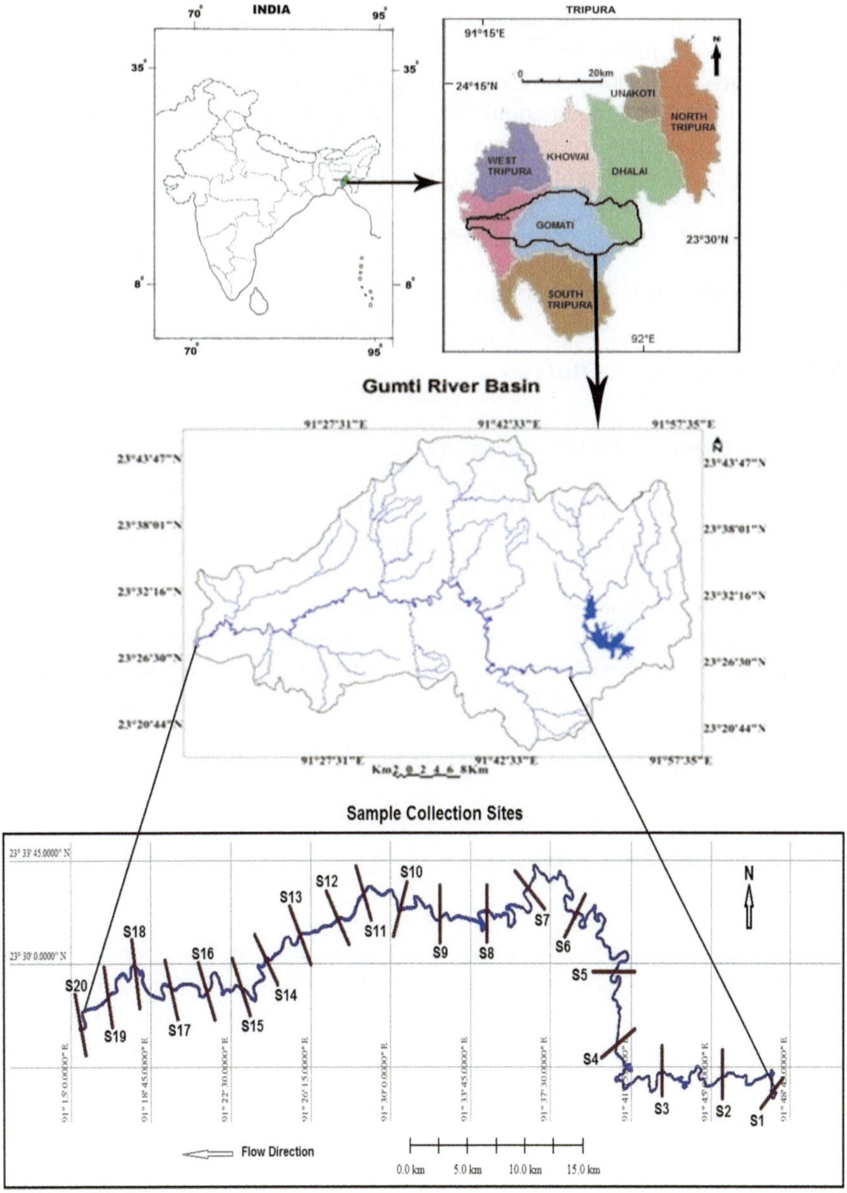


Fig. 18.1 Location of the study area

used to calculate various statistical parameters like mean, standard deviation, skewness, and kurtosis following Folk and Ward (1957) Graphic method to decipher statistically the significance of variation in sediment characteristics. Scatter plot was prepared to understand the relationship between different statistical parameters. In order to identify the depositional environment, CM diagram (Passega 1957; Passega and Byramjee 1969) was prepared using Gstat software package (Dinesh 2009).

## 18.4 Results and Discussion

### *18.4.1 Grain Size Analysis*

#### 18.4.1.1 Frequency Distribution Curves

Frequency distribution curve (FDC) represents the weight percentage of sediments of different size classes at different sections which are used to describe the nature of sediments (Karuna Karudu et al. 2013; Ganesh et al. 2013). Frequency distribution curves of sediment samples of the Gumti River indicate unimodal nature with medium to fine sand in almost all the sites. In the samples collected from upper and middle courses of the Gumti River, prominent peak of the frequency distribution curve fell in size 2Φ, whereas it became 2.5 to 3Φ in the extreme lower course (Fig. 18.2). The unimodal nature of the sediment samples indicates one phase of sediment deposition from single source with less fluctuation in energy condition.

#### 18.4.1.2 Cumulative Curves

Cumulative curves of all the samples have been prepared on a probability scale to identify the nature of sediment transportation. Visher (1969) identified different modes of sediment transportation on the basis of length and slope of distinct segments present in a cumulative curve. In the present study, three distinct modes of sediment transportation (traction, saltation, and suspension) have been identified. The curves show the dominance of saltation mode (up to 90%) though traction and suspension also played some role. The truncation points of traction and saltation of most of the sediment samples lie in 1Φ, while for saltation–suspension the points lie in 3Φ (Fig. 18.3). The low percentage of traction mode generally indicates relatively low-energy condition of the agent to transport load.

#### 18.4.1.3 Graphic Mean

Mean size of sediments is very much influenced by the source of sediment supply and the environment of deposition. The variation in energy level of the medium of

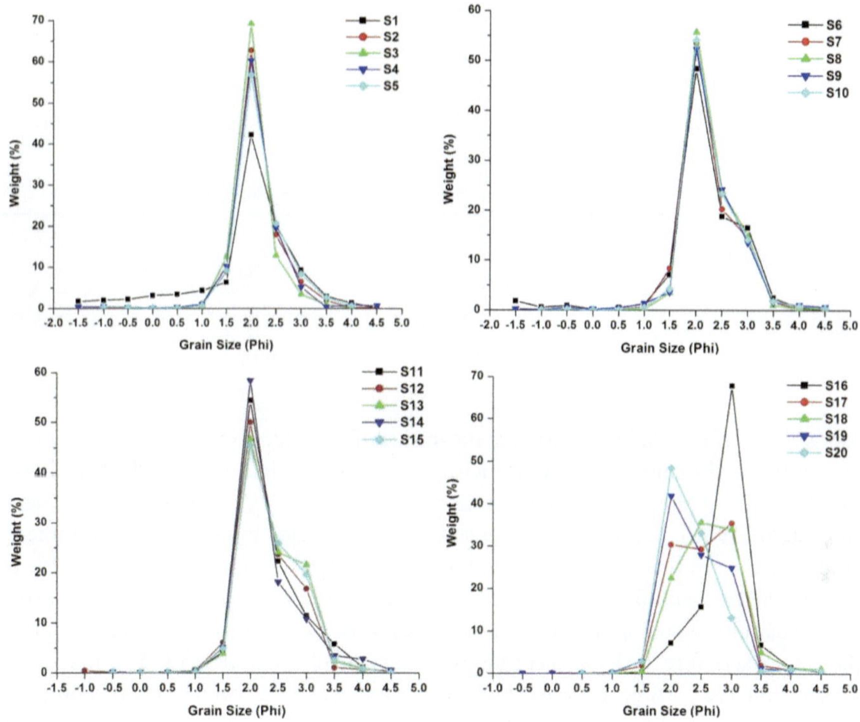


**Fig. 18.2** Frequency distribution curve of 20 sediment samples of the river bed

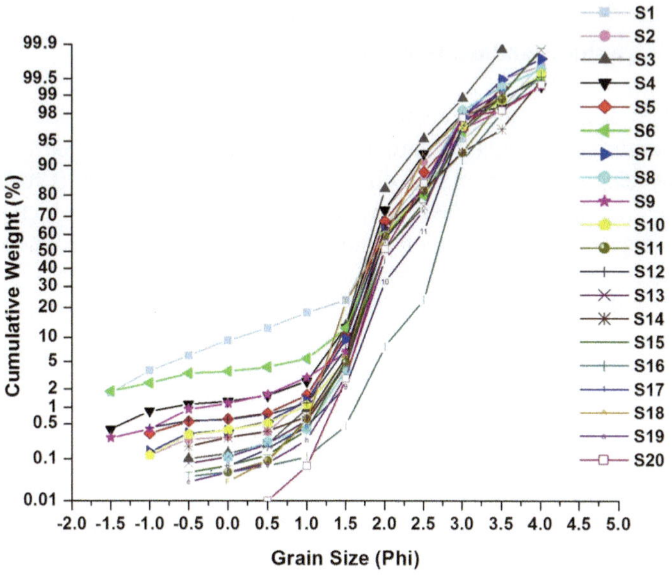


**Fig. 18.3** Cumulative frequency curve of 20 bed sediment samples of the Gumti River

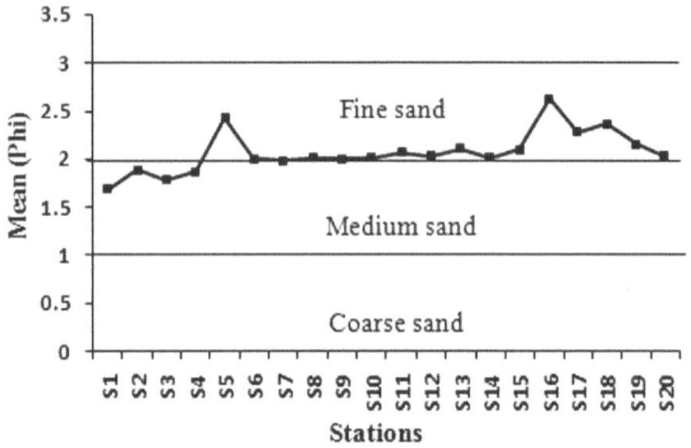


**Fig. 18.4** Mean size of 20 sediment samples of the Gumti River

deposition is reflected by variation in mean size of sediments (Sahu 1964). The mean size of bed sediments of the Gumti River varies from 1.681 to 2.625Φ (Fig. 18.4). Nearly, 65% of the samples fall under fine sand category. According to Reineck and Singh (1980), deposition of fine sediments indicates low-energy environment. Predominance of fine grained sediment is mostly observed in samples 8–20 of lower course which reveals the existence of low-energy condition of the Gumti River during the deposition of sediments due to low velocity. Besides, presence of the Maharani Barrage, just before the location of sample 8, also results in fining of sediments downstream from the barrage, because, large volume of water has been trapped behind the barrage which results in decrease of energy condition of the river downstream from the barrage.

#### 18.4.1.4 Graphic Standard Deviation

The standard deviation measures the absolute dispersion or variability of a distribution. Variation in sorting (standard deviation) value attributed to variability in the velocity of depositing current and difference in water turbulence (Chauhan et al. 2014). The graphic standard deviation reflects the environment in which sediments have been deposited. The value of standard deviation of all the sediment samples ranges between 0.331 and 0.966Φ. Maximum percent of the sediment samples (75%) falls under well sorted category (Fig. 18.5). It indicates that in maximum location sediments are deposited under uniform environmental condition with lack of fluctuation or variability in kinetic energy or velocity of the Gumti River. Whereas, moderately well sorted nature of sediment samples 6, 11, 14, 18, and moderately sorted sample 1 reflects relative variation in energy level during the deposition of sediments.

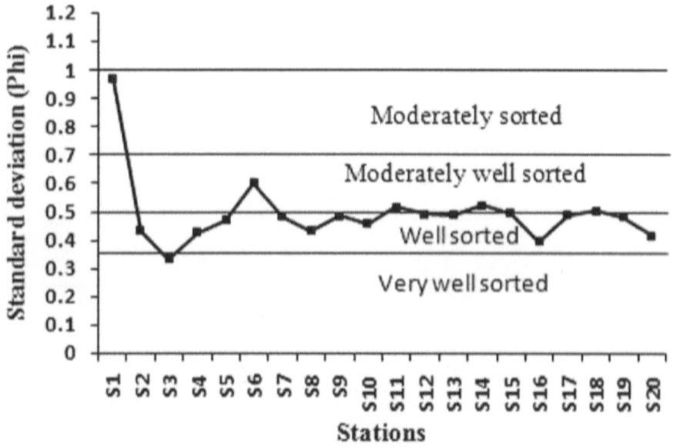


**Fig. 18.5** Variation in nature of sorting of 20 bed sediment samples of the Gumti River

#### 18.4.1.5 Graphic Skewness

Skewness indicates the difference between the manners in which items are distributed in a particular distribution compared to a symmetrical or normal distribution. In grain size analysis, skewness measures the predominance of fine or coarse sediments in any distribution. The negative value of skewness indicates coarse skewed, i.e., more material in the coarser-tail, whereas positive value denotes fine skewed, i.e., more material in fine-tail of any distribution. Besides, positive skewness indicates that sediments are deposited in sheltered low-energy environment (Duane 1964; Valia and Cameron 1977).

Sediment samples of the Gumti River show skewness value ranging between –0.306 (S1) and 0.024Φ (S18) which reflects coarse to very fine skewed nature of the distribution (Fig. 18.6). Around 85% of the sediment samples fall under positive skewness category which revealed that in most of the locations sediments have been deposited in sheltered low-energy environment.

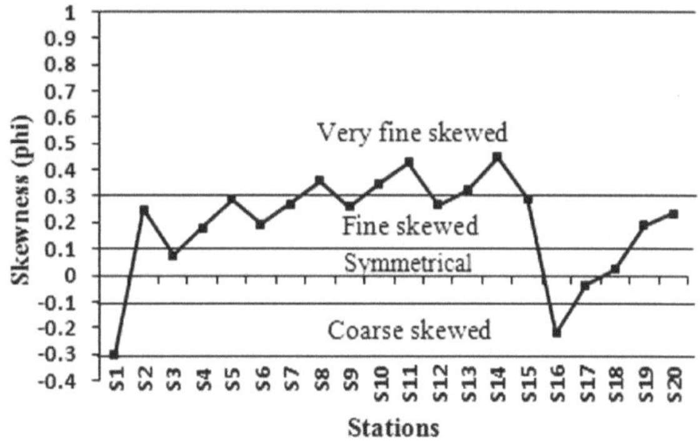


**Fig. 18.6** Variation in nature of skewness of 20 bed sediment samples of the Gumti River

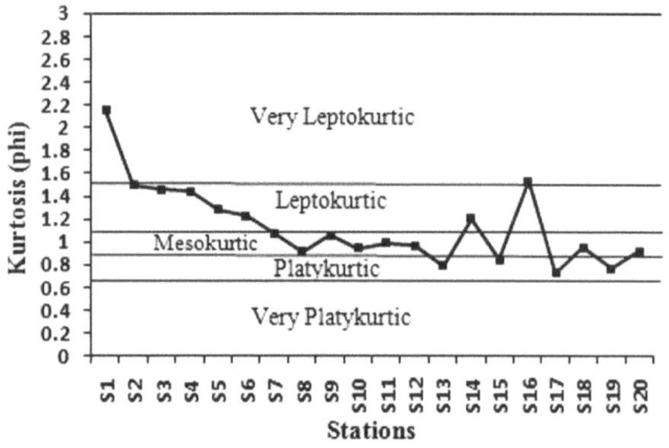


**Fig. 18.7** Variation in nature of kurtosis of 20 bed sediment samples of the Gumti River

#### 18.4.1.6 Graphic Kurtosis

Kurtosis describes the extent to which a distribution is more peaked or flat topped than a normal curve. The flow condition of the depositing medium is reflected in the changes in kurtosis value (Baruah et al. 1997; Ray et al. 2006). The kurtosis of the Gumti River sediments is platykurtic to very leptokurtic in nature (0.735 to 2.155Φ). There is a dominance of mesokurtic sediments ranging from 0.912 to 1.075Φ represented by eight samples (Fig. 18.7), while, leptokurtic and platykurtic nature is present in six and four sediment samples, respectively. The very leptokurtic nature is represented by only two samples (S1 and S16). The mesokurtic to leptokurtic character of sediments indicates the continuous addition of finer or coarser materials after the winnowing action and retention of their original characters during deposition (Avramidis et al. 2013).

#### 18.4.1.7 Bivariate/Scatter Plots

Generally, this type of plotting gives a large amount of information about an environment in which sediments are deposited. Inman (1949, 1952) and Griffiths (1951) are the earliest workers to notice the relationship between different statistical parameters in grain size analysis. Later on, several scholarly works have been done in this regard to identify sediments of different environment (Folk and Ward 1957; Friedman 1961, 1967; Moiola and Weiser 1968). Mean versus standard deviation plots give an idea about the nature of sediments. When sand mode increases and gravel diminishes, the mean size becomes finer and the sorting started to improve (Folk and ward 1957). Thus, fine sands used to be better sorted in nature. In the present study, also, it appears that in most samples fine sand is well sorted (Fig. 18.8), which indicates low energy condition of the Gumti River. In general, the ideal fractions are nearly symmetrical, but the mixing produces either positive or negative skewness which depends on the proportion of size classes in the admixture (Srivastava and Mankar 2009). The plots mean versus skewness reveals that the sediments become more positively skewed

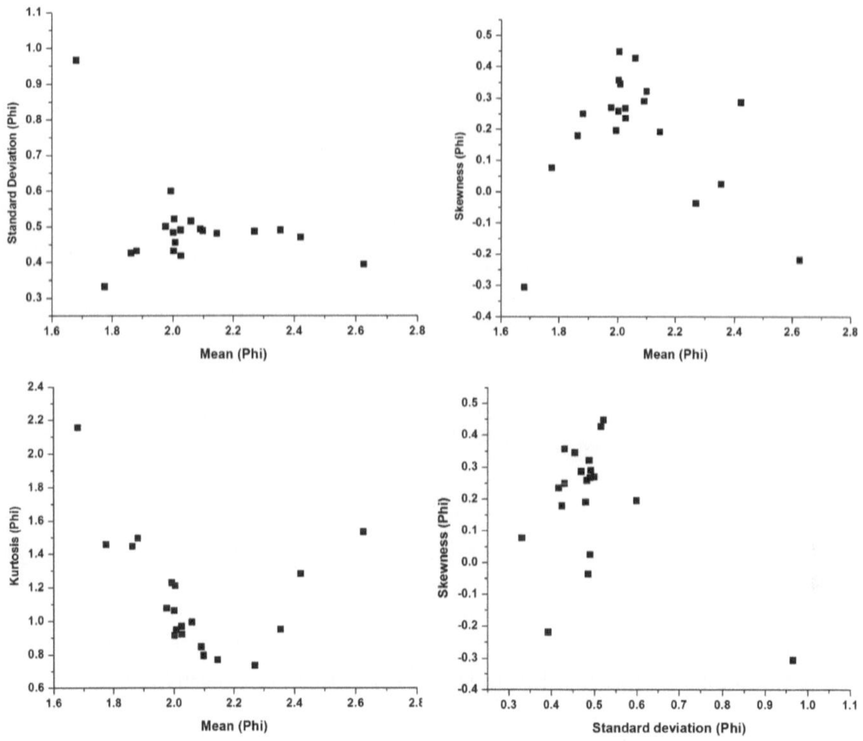


**Fig. 18.8** Bivariate plotting of statistical parameters of sediment samples

with decrease in grain size (Fig. 18.8). Thus, in the present study, positively skewed fine sand predominates indicating low-energy environment.

The relation between mean size and kurtosis is complex and theoretical (Folk and Ward 1957). The model plot suggested by Folk and Ward (1957) signifies that the sorting in peak and tails, i.e., the index of kurtosis is affected by the mixing of two or more size classes of sediments. In the present study, fine sand dominates in most of the samples with some mixture of medium sand which makes the sorting moderate to well in the tail, hence platykurtic to very leptokurtic nature in the sediment samples exhibit. The maturity of sand is revealed by fine size and dominant platykurtic nature of sediments (Rao et al. 2001) which has been observed in samples 13, 15, 17, and 19. The plots between skewness versus standard deviation (Fig. 18.8) reveals that most of the well sorted sediment samples are positively skewed in nature. Besides, the plots produce a scatter trend probably due to either unimodal sediments with good sorting or equal mixing of two modes.

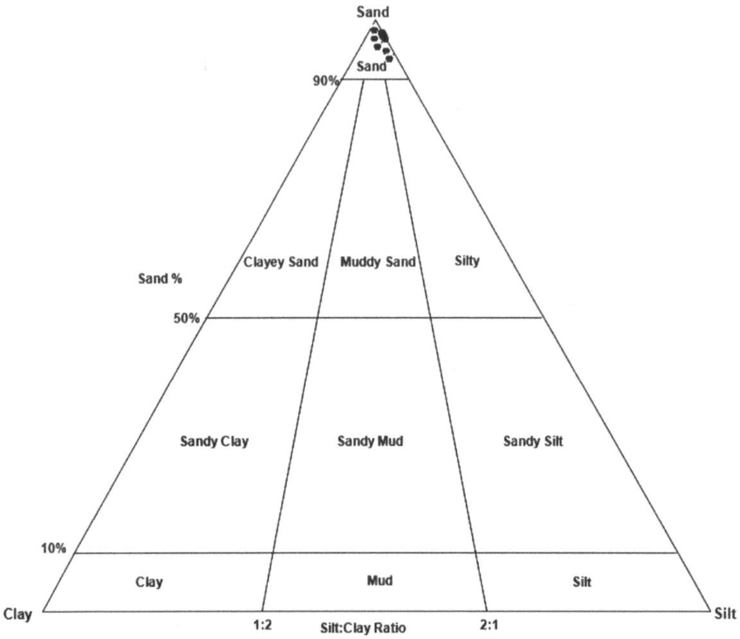


**Fig. 18.9** Triangular diagram showing distribution of sediments in the Gumti River bed

### *18.4.2 Triangular Diagram*

Graphical representation of datasets is very important for better understanding of any scientific enquiry. One of the most common graphical techniques used by the sedimentologists is the plot of sand, silt, and clay on a triangular diagram. It is very much helpful for rapid classification and comparison of sediment samples (Poppe et al. 2004).

In the present study, classification of sediments has been done by plotting the percentage of sand, silt, and clay on a triangular diagram suggested by Folk and Ward (1957). Triangular plotting of 20 bed sediment samples of the Gumti River reveals that the sediments fall under sand category (Fig. 18.9).

### *18.4.3 CM Pattern*

CM pattern is very effective means of analyzing transportation mechanism, and the environment in which sediment has been deposited with respect to size, range, and energy level of transportation (Passega 1957, 1964; Passega and Byramjee 1969). In the present study, CM pattern is used to identify the environment in which sediments of the Gumti River has been deposited. Parameter C represents one percentile value (maximum grain size) of distribution and indicates competency of the transporting agent, while parameter M represents median grain size of sediment samples. The values of C and M have been obtained from cumulative curves. The relationship between these two parameters is affected by bottom turbulence. CM diagram helps to understand the hydrodynamic force worked during the deposition of sediments. The values of C and M of all the sediment samples, collected from the Gumti River, range between 400 to 3500 and 150 to 300 μ, respectively.

The CM plot shows that sediments of the Gumti River are deposited by bottom and graded suspension under tractive current (Figs. 18.10 and 18.11). Most of the sediment samples fall under graded suspension and no rolling condition, which indicate

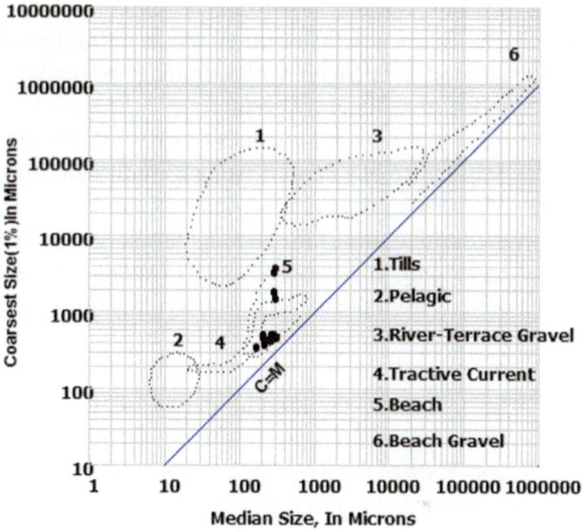


**Fig. 18.10** The basic CM pattern of the Gumti River bed sediments

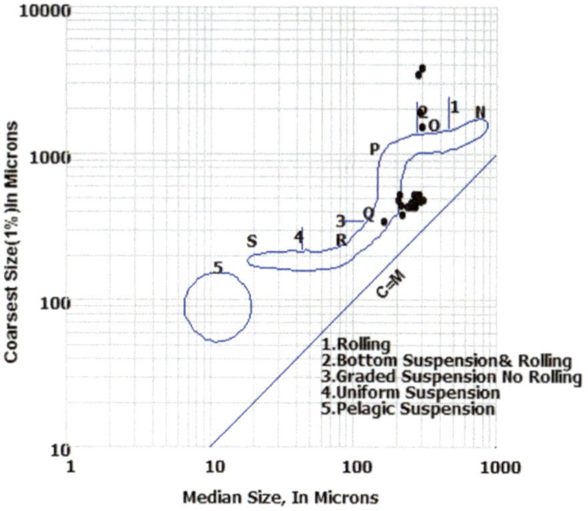


**Fig. 18.11** The tractive current of CM pattern of the Gumti River bed sediments

comparatively low energy level, while few of the samples collected from upstream reach fall under bottom suspension condition (Fig. 18.11).

## 18.5 Conclusion

The grain size analysis of 20 bed sediment samples of the Gumti River provides considerable information about the properties of sediments and their depositional environment which helps to understand the various processes affecting river erosion and deposition of sediments. The following important conclusions can be drawn from this analysis:

(i) The frequency distribution curves of almost all the sediment samples show unimodal character with dominant peaks lie in size 2Φ. It indicates single mode of sediment deposition.
(ii) The cumulative frequency curves of all the sediment samples show the dominance of saltation mode of sediment transportation.
(iii) The mean grain size of bed sediments of the Gumti River ranges between medium to fine sand. The regulation of river water through Maharani Barrage has highly influenced the size of sediments, as fine sand predominates downstream from the barrage, which indicates low-energy condition of the river in this part.
(iv) In general, the sediments are well sorted in nature which indicates lesser fluctuation in the kinetic energy of the depositing agents. Around 85% of sediment samples are positively skewed in nature which shows low-energy condition of the River Gumti. The kurtosis values of all the samples show very leptokurtic to platykurtic nature.
(v) Scatter plots of different statistical parameters indicate that in most of the cases fine sands are well sorted in nature. Besides, positively skewed nature predominates in most of the fine sand sediment samples.
(vi) The CM pattern indicates that the bed sediments of the Gumti River are deposited by bottom and graded suspension under tractive current.

The overall study reveals that very low energy condition prevailed in the Gumti River during the pre-monsoonal period, especially, in the lower reach, which is one of the prominent reasons for sedimentation problem in the river. Moreover, human intervention in the form of Maharani Barrage has largely affected the grain size distribution of the Gumti River bed sediments.

## References

Avramidis P, Samiotis A, Kalimani E, Papoulis D, Lampropoulou P, Bekiari V (2013) Sediment characteristics and water physicochemical parameters of the Lysimachia Lake, Western Greece. Environmen Earth Sci 70:383–392

Blott SJ, Pye K (2001) GRADISTAT: a grain size distribution and statistics package for the analysis of unconsolidated sediments. Earth Surf Proc Land 26:1237–1248

Baruah J, Kotoky P, Sarma JN (1997) Textural and geochemical study on river sediments: a case study on the Jhanji River. Assam. J Indian Assoc Sedimentol 16(1):195–206

Chauhan R, Ramanathan AL, Adhya TK (2014) Patterns of seasonal variability in Granulometric characteristics of Bhitarkanika Mangrove-estuarine Complex, East Coast of India. Indian J Mar Sci 43(6):1077–1084

Dinesh AC (2009) G-Stat-A Software in VB6 for grain size statistical analysis, CM digrams, trend diagrams, etc. designed and developed by A.C.Dinesh, Geologist, Marine wing, GSI, Mangalore

Duane DB (1964) Significance of skewness in recent sediments, Western Pamlico Sound North Carolina. J Sediment Petrol 34(4):864–874

Folk RL, Ward WC (1957) Brazos River Bar: a study in the significance of grain size parameters. J Sediment Petrol 27(1):3–26

Friedman GM (1961) Distinction between dune, beach and river sands from their textural characteristics. J Sediment Petrol 31(4):514–529

Friedman GM (1962) On sorting, sorting coefficients and the log normality of the grain-size distributions of sandstones. J Geol 70:737–753

Friedman GM (1967) Dynamic processes and statistical parameters compared for size frequency distribution of beach and river sands. J Sediment Petrol 37(2):327–354

Ganesh B, Naidu AGSS, Rao MJ, Karudu TK, Avatharam P (2013) Studies on textural characteristics of sediments from Gosthani River Estuary - Bheemunipatnam, A.P., East Coast of India, J Ind Geophys Union, 17 (2):139–151

Griffiths JC (1951) Size versus sorting in some Caribbean sediment. J Geol 59(3):211–243

Inman DL (1949) Sorting of sediments in the light of fluid mechanics. J Sediment Petrol 19:51–70

Inman DL (1952) Measures for describing size of sediments. J Sediment Petrol 22:125–145

Karuna Karudu T, Jagannadha Rao M, Ganesh B, Avatharam P, Naidu AGSS (2013) Studies on textural characteristics of Erra Kalva River, West Godavari District, Andhra Pradesh, East Coast of India. Int J Geomat Geosci 4(2):280–295

Mason CC, Folk RL (1958) Differentiation of beach, dune and aeolian flat environments by size analysis, Mustang Island, Texas. J Sediment Petrol 28:211–226

Mohtar WHMW, Nawang SAB, Porhemmat M (2015) Statistical characterisation of grain size distribution in fluvial sediment of Kelantan Rivers. Jurnal Teknologi 74(3):103–109

Moiola RJ, Weiser D (1968) Textural parameters: an evaluation. J Sediment Petrol 38(1):45–53

Passega R (1957) Texture as a characteristics of clastic deposition. Am Association Petrol Geol 41:1952–1984

Passega R (1964) Grain size representation by CM patterns as a geological tool. J Sediment Petrol 34:830–847

Passega R, Byramjee R (1969) Grain-size image of clastic deposits. Sedimentology 13:233–252

Poppe LJ, Eliason AH, Hastings ME (2004) A visual basic program to generate sediment grain-size statistics and to extrapolate particle distributions. Comput Geosci 30:791–795

Prabhakara Rao A, Anil Kumar V, Yugandhara Rao A, Ravi GS, Krishnan S (2001) Grain size parameters in the interpretation of depositional environments of coastal sediments between Bendi Creek and Vamsadhara River, east coast of India. J Indian Assoc Sedimentol 20(1):109–116

Ramesh R, Subramanian V (1992) Textural characteristics of the Krishna River sediments, India Geo J 28(4):449–455

Rao PA, Anilkumar V, Yugandhar Rao A, Ravi GS, Krishnan S (2001) Grain size parameters in the interpretation of depositional environments of coastal sediments between Bendi creek and Vamsadhara River, East Coast, India. J Indian Assoc Sedimentol 20(1):106–116

Ray AK, Tripathy SC, Patra S, Sarma VV (2006) Assessment of Godavari estuarine mangrove ecosystem through trace metal studies. Environ Int 32:219–223

Reineck HE, Singh IB (1980) Depositional sedimentary environments with reference to terrigenous clastics, 2nd edn. Springer, Berlin, Heidelberg, New York

Sahu BK (1964) Depositional mechanisms from the size analysis of clastic sediments. J Sediment Petrol 34:73–83

Seralathan P, Padmalal D (1994) Textural studies of surficial sediments of Muvattupvzha River and central Vembanad estuary, Kerala. J Geol Soc India 43(2):179–190

Srivastava AK, Mankar RS (2009) Grain size analysis and depositional pattern of upper Gondwana sediments (Early Cretaceous) of Salbardi Area, Districts Amravati, Maharashtra and Betul, Madhya Pradesh. J Geol Soc India 73:393–406

Valia HS, Cameron B (1977) Skewness as paleo-environmental indicator. J Sediment Petrol 47(2):784–793

Visher GS (1969) Grain size distribution and depositional process. J Sediment Petrol 39:1074–1106

Zhanqiao W, Zhongyuan C, Maotian L, Jing C, Yiwen Z (2009) Variations in downstream grain-sizes to interpret sediment transport in the middle-lower Yangtze River, China: a pre study of Three-Gorges Dam. Geomorphology 113:217–229

# Chapter 19
# Morphometric Analysis and Geohydrological Inference of Bhilangna Drainage Basin, Uttarakhand (India) Using Remote Sensing and GIS Techniques

**Sourav Bhadwal, Naresh Kumar Verma, Sanjit Kumar, Manish Kumar, and Pankaj Kumar**

**Abstract** Morphometric analysis is the method of studying landforms using quantitative techniques. The Bhilangna river of Uttarakhand Himalaya has been selected for analysis which provides a perennial source of irrigation and holds a significant place in the economic scenario of Uttarakhand. In this study, the quantitative analysis of morphometric parameters has been examined under linear, areal and relief aspects. The basin has elongated in shape and sloping down from north-east to south-west direction sharing with gentle slope (0°–5°) to very steep slope (up to 79°) which comprise 88.3% of total area. The stream order values run from 01 to 06. The drainage density has been found to be moderate in the basin, which indicate permeability of the surface strata. The bifurcation ratio ($R_b$) varies 0.76 to 2.33. The hypsometric integral of Bhilangna drainage basin is 0.42 which indicate that the basin is passing through late mature stage of development. Therefore, the present study not only enhances the detailed knowledge of morphological and a geohydrological characteristic of drainage basin, but it also helps in better understanding and sustainable utilisation of water resources. Results can be useful for policy formation, plan development and sustainable management of the region.

**Keywords** Morphometric analysis · Geohydrological · Bhilangna drainage · Hypsometric integral

S. Bhadwal · S. Kumar · M. Kumar (✉)
Department of Geography, School of Basic Sciences, Central University of Haryana, Mahendergarh, Haryana, India
e-mail: manish.ks@cuh.ac.in

N. K. Verma
Special Centre for National Security Studies, Jawaharlal Nehru Institute of Advanced Study, Jawaharlal Nehru University, New Delhi 110067, India

P. Kumar
Department of Geography, Delhi School of Economics, University of Delhi, Delhi 110007, India



P. Kumar et al. (eds.), *Water Resources Management and Sustainability*,
Advances in Geographical and Environmental Sciences,
https://doi.org/10.1007/978-981-16-6573-8_19

## 19.1 Introduction

Water covers 71% of Earth surface from which 96.5% as ocean, 1.7% as groundwater, 1.7% as glacier and ice caps, 0.001% as in water vapour, clouds (formed of solid, liquid and suspended particles in air) and precipitation, from this we can say that percentage share of fresh water is very small to fulfil the demand of population that increasing very fastly, and we know that the water is vital source for all living being. Country like India (which is diverse in physiography as well as climate) is an agricultural country defined by the climate regime. In last few decades, Himalayan region could not escape from rapid unplanned urbanisation as the other parts of country due to factors like increasing migration, tourism and religious activities are the key reasons behind the water shortage problem; it is known that sources of water for mountainous region are springs and rivers, and these sources are snow and glacier-fed; the increasing impact of climate change will affect the quality and quantity of water availability from these sources. Ultimately, all the pressure is occurred on ground water which makes ground water more critical. So, there is an urgent need to access the water resources because it plays a very important role in sustainability and economic development.

The word morphometry simply used to denote structure measurements. The morphometry is the mathematical study or analysis of the earth surface and landforms. The word morphometry can be divided into two parts morpho mean 'form' and metry means 'measurement'. The meaning of morphometry is a 'measurement of landform or surface'. 'River basins comprise a distinct morphologic region and have special relevance to drainage pattern and geomorphology' (Strahler 1954). It provides the quantitative description of the basin geometry to understand elements as slope, structural controls, geological and tectomorphic history of drainage basin (Strahler 1964). 'The morphometric analysis is significantly used for various small river basins and sub-basins in the various parts of the world' (Horton 1945). It is important to understand the landform process, physical and erosional characteristics. The prioritisation of watershed is considered as one of the most important aspects of playing and development of natural resources for water conservation measures (Javed et al. 2011). The morphometric analysis of Bhilangna river basin is carried out through measurement of relief gradient of channel network, aerial, linear and slope of the basin (Nautiyal 1994; Nag and Chakrborty 2003). This advanced methodology of quantative investigation of basin morphology was contributed by Horton who is one the main pioneers in river morphometry field. There are many other geomorphologists, mainly Strahler et al. (1956, 1957), Scheidegger (1965), Shreve (1967), Gregory and Walling (1973), and books by Blooms (2002); Pinter and Keller (1995) have proliferate the morphometric investigation.

In the era of remote sensing and GIS techniques, more precise data can be generated for precise morphometric analysis. The remote sensing technique satellites provide the synoptic view of a larger area with greater accuracy. Many researchers have used remote sensing data integrated with GIS generated more accurate data on morphometric parameters (Kusre 2016; Nongkynrih and Husain 2011; Kulkarni

2015; Altaf et al. 2013; Yadav et al. 2014; Rao et al. 2010) and concluded that remote sensing has emerged as a powerful tool and useful in analysing the drainage morphometry. Digital elevation model (DEM) when processed in GIS software provides 3D evaluation of the earth surface. The digital elevation method (DEM) of the study area was produced in this study to extract out various morphometric parameters like drainage density, drainage basin and drainage order and drainage frequency values of the basin. Hydrological and spatial analysis by using satellite information through GIS and remote sensing will be easier to identify and differentiate between drainage (Pirasteh et al. 2010). The present study is carried out to analyse various morphometric parameters of Bhilangna drainage basin, i.e. aerial, linear relief and hypsometry of the area of interest.

## 19.2 Study Area

Bhilangna river is a right bank tributary of Bhagirathi river, a river which is source-stream of Ganga river. Total length of Bhilangna river is 80 km. Bhilangna river originates at the Khatling glacier at the elevation of 3717 m above sea level. This river is approximately 50 km south of Gaumukh which is considered as the source of Bhagirathi and Ganga river. Bhilangna meets its tributaries Balganga Ghyansali. The Bhilangna drainage basin lies in the Tehri district of Garhwal Himalaya in Uttrakhand. Its latitudinal and longitudinal position are 30.3743 N to 30.88042 N and 78.78055 E to 78.89864 E having an area of 791.9 km$^2$ (Fig. 19.1). District Tehri is surrounded by Uttarkashi in north side, Rudraprayag in east side whilst Dehradun and Garhwal districts in west and south side of study area. The climate of the Bhilangna drainage basin is tropical monsoon type. The climate comprises summer season from April to June, whereas monsoon season are from June to September and winters from December to March. During winter's, the temperature falls down remarkably and minimum temperature goes down to 0 °C. In summer, the temperature varies between 19 °C and 33 °C. The air temperature in the region varies from 4.6 °C to 36.5 °C. The annual precipitation ranges from 1483.8 to 2122.3 mm with the mean of 1700 mm. The relative humidity in the region fluctuates 36 and 75% precipitation occurs in valley whilst snow at higher ridges.

## 19.3 Geology

Geology of various aspects of Bhilangna valley of Garhwal Himalaya have been carried out by various researchers (Saklani 1972; Tewari 1972; Valdiya 1980; Rao and Patil 1982; Islam and Thakur 1988). The rock of the Bhilangna valley has been classified into two tectonic units, viz

*Garhwal Group*: The rocks of the Garhwal groups are metabasic with intercalations of Phyletic and Chlorite Schist, Quartzite.

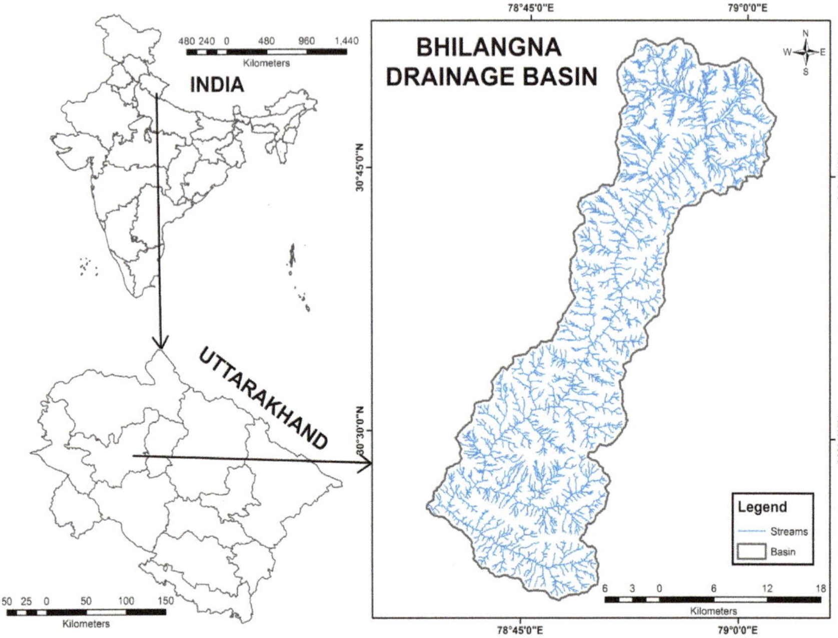


**Fig. 19.1** Study area map of Bhilangna drainage basin

*Central Crystalline Group*: consist of deform Granitic Gneisses, Porphyritic Gneiss, Talc schist, Mica Schist and Quartzo—Feldspathic Schist.

The MCT I (Chail Thrust) divides Garhwal group from the overlying outer crystalline.

The MCT II (Jutogh Thrust) separates the outer crystalline from the central crystal.

These rocks have suffered polyphasal deformation showing development of generation of folds, foliation and lineations. They have undergone regional metamorphism varying from green Schist to upper amphibolite facies. The rocks in Tehri Garhwal area are of two wide geo-structural zones, viz. lesser and central Himalaya. The lesser Himalaya is stretched upto south of study area, whereas northern part of study area is covered by central Himalaya. Lesser Himalaya a large portion of this region which comprise of different groups like Jaunsar group, Blaini-Krol group and Tal group. Generally, the rocks of the lesser Himalaya zone provide us information about weathring and upliftment occur in these regions.

## 19.4 Geohydrological Inferences

A geo-hydrological-based morphological study plays a key role in planning, designing of the area and development of river engineering structures. Quantitative analysis of morphometric elements like relief, areal and linear has got prodigious importance in various field like river geomorphic evaluation, soil and water management and natural resource management. River morphometry is one of the branches of fluvial geomorphology which deals with the shape and the hydrological characteristics of a basin area. Morphometric analysis of a catchment provides a both quantitative as well as qualitative analysis of the river basin, which is an important for characterisation of basin. Watershed (basin) is a natural hydrological unit from which surface run off flows takes a particular channel. The both quantitative and qualitative analysis method of river basin was first used by Horton and then modified by Strahler. The morphometric analysis of Bhilangna drainage basin presents that basin values of relief and elongation are moderate. The linear pattern of basin suggests the erosional characteristics of the study area because of vertical as well as lateral erosion in downward area. Circulatory ratio, elongation ratio as well as values of form factor shows that the basin is in elongated shape, which indicate the low run off and average peak of flow.

Drainage network is dendritic type, which suggests lack of structural control. The morphometric parameters ascertained using GIS helped us to understand various parameters such as infiltration, run off, terrain ruggedness, slope and hill shade. studies carried out using high resolution data provides more accurate results for demarcation of basin area and areal planning and management.

## 19.5 Research Methodology

The present study has been derived from secondary data which is collected from district census handbook, published and unpublished sources. The methodology adopted in this study is based on analytical investigation. The boundary of Bhilangna drainage basin has been delineated with the help of survey of India (SOI) top sheet. All the satellite images were mosaiced and re-projected into UTM projection (Table 19.1).

## 19.6 Extraction of Drainage Basin

The Bhilangna drainage basin is extracted from the ALOS PALSAR (Hi-Res corrected terrain data), digital elevation model (DEM) data by the Japan Aerospace Exploration Agency (JAXA) and Alaska Satellite Facility (ASF) having spatial resolution of 12.5 m. The basin area of Bhilangna was then extracted using geoprocessing technique in ArcGIS. The pour point and DEM are two input parameters necessary for the extraction purpose. A pour point is drawn on the cells having highest flow accumulation after this we filled sink DEM, then find the flow direction of that particular area in order to generate a drainage network and finally. The watershed was carved out by giving pour point where water flow out of an area.

**Table 19.1** Data used for morphometric analysis of Bhilangna drainage basin, Uttrakhand, India

| Type of data/software | Details | Source |
|---|---|---|
| ALOS PALSAR | Resolution-12.5 m | Alaska satellite facility and JAXA |
| | Frequency-L-Band | |
| | Instrument type-Radar imagery | |
| | Beam mode-FBS, FBD and PLR | |
| SOI TOPOSHEETS | Top sheets No. 53 J/13, 53/N, | Survey of India (SOI), Dehradun |
| | 53 J/10, 53 J/14, 53 J/11 and 53 J/15 | |
| | Scale-1:50,000 | |

## 19.7 Stream Network Extraction

For the extraction of the stream network of the study area, various geoprocessing tools, i.e. first start with filling DEM in GIS environment then find the flow direction after this calculate flow accumulation then we create a stream network grid with the aid of stream classification method of Strahler (1964), are used. According to Strahler classification, when two same order streams join, they form next higher order stream continue till the major stream is created with the highest order. The highest stream order in the study area was computed as sixth. This has been done in ArcGIS software. To examine the basin morphometry of Bhagirathi river, various parameters such as stream order, stream length, stream number, stream length ratio, basin length, bifurcation ratio, basin area, relief ratio, circularity ratio, elongation ratio, drainage density, stream frequency, drainage texture and form factor have been analysed using formulae shown in Table 19.2.

## 19.8 Results and Discussion

The morphometric parameters of Bhilangna river catchment have been extracted cum examined, and the results are shown in Table 19.5). It is one of the major tributaries of *Bhagirathi River*, which is also the source-stream of *Ganga River*. It originates form the 'Khatling Glacier' in Himalaya. The total catchment area of the Bhilangna basin is 791.9 $km^2$. Dendritic nature of the basin is the result of the general topography of the area.

**Table 19.2** Methodology used for morphometric parameters computation

| Morphometric analysis | | Formula | References |
|---|---|---|---|
| Aspect | Parameters | | |
| Linear | Stream order (u) | Hierarchical rank | Strahler (1969) |
| | Stream number (Nu) | Total no of stream segments of 'u' order | Horton (1945) |
| | Stream length (Lu) | Total length of stream of 'u' order | Horton (1945) |
| | Mean stream length ($L_{sm}$) | $L_{sm}$ = Lu / Nu, where, $L_u$ is stream length and $N_u$ = stream number | Strahler (1964) |
| | Bifurcation ratio ($R_b$) | $R_b$ = Nu / Nu + 1 where, Nu + 1 is No. of stream segments of next higher order | Schumm (1956) |
| | Mean bifurcation ratio ($R_{bm}$) | $R_{bm}$ = mean of bifurcation ratios of all orders | Schumm (1956) |
| Areal | Drainage density ($D_d$) | $D_d$ = Lu/$A$, where $A$ is basin area in ($km^2$) and $L_u$ = stream length | Horton (1945) |
| | Form factor ($F$) | $F$ = A/$L^2$ where $A$ = basin area, $L$ = basin length | Horton (1945) |
| | Stream frequency ($F_s$) | $F_s$ = N/A; where, $N$ = total number of streams of a given basin, $A$ = total area of basin ($km^2$) | Horton (1945) |
| | Constant of channel maintenance ($C$) | $C = 1/D_d$, where $D_d$ = drainage density | Strahler (1964) |
| | Circularity Index ($R_c$) | $R_c = 4\pi A/P^2$, where $A$ = basin area, $P$ = basin perimeter | Miller (1953) |
| | Drainage texture ($T$) | $T = Dd \times F_s$ Where, $D_d$ = drainage density, $F_s$ = stream frequency | Smith (1950) |
| | Elongation ratio ($R_e$) | $R_e = 2/L \times (A/\pi),^{0.5}$ where $A$ = area of basin and $L$ = basin length | Schumm (1956) |
| Relief | Relief ratio ($R_r$) | $R_r$ = (H/L), where $H$ = relative relief (m), $L$ = length of basin (m) | Schumm (1956) |
| | Basin relief ($R$) | $R = H - h$, where $H$ = highest relief, $h$ = lowest relief | Schumm (1961) |
| | Ruggedness index ($R_i$) | $R_i = D_d \times H/1000$,where $D_d$ = drainage density, $H$ = relative relief | Schumm (1956) |
| | Hypsometric integral (HI) | HI = ($R_a/H$)/(a/A), where $R_a$ = absolute relief, $H$ = highest relief in the area, $a$ = area of absolute relief, $A$ = total basin area | Strahler (1952) |

### 19.8.1 Linear Aspects of Bhilangna Drainage Basin

*Stream Order* ($N_u$): Stream order concept was devised by Horton in 1932. Stream ordering is characterized as a measure of the position of a stream in the hierarchy of tributaries (Langbein and Leopold 1964), and the stream of Bhilangna drainage basin has been demarcated according to the Strahler's system of stream ordering the two same order streams join to form a next order stream, and the cycle proceeds till the major stream is demarcated with the highest order. With the help of GIS, the number of streams of various orders and the total number of stream in the basin are counted autonomously (Table 19.3). Total 129,367 streams were demarcated in Bhilangna drainage basin, of which 51.47% (66,591) is first order; 23.86% (30,868) is second order; 12.77% (16,525) is third order; 5.47% (7078) is fourth order; 2.78% (3601) is fifth order, and 3.63% (4704) comprises sixth order (Table 19.3). High value of first order streams indicates the chances of flash flood which is followed by heavy downpour in lower channels (Chitra et al. 2011). The stream order map of Bhilangna basin is shown in Fig. 19.2.

*Bifurcation Ratio* ($R_b$): Bifurcation ratio ($R_b$) plays important role in drainage basin analysis because it links hydrological regime of river basin under topological and climatic condition. Bifurcation ratio is the ratio of stream no. of given order ($N_u$) to the stream number of next order ($N_{u+1}$). Value of bifurcation ratio helps to have an idea about the basin shape and run off behaviour. An elongated basin generally has high bifurcation ratio, whereas circular basin generally has low bifurcation ratio. The higher bifurcation ratio suggests that the area is tectonically active. Higher bifurcation ratio shows strong structural control on drainage pattern, and low bifurcation ratio indicates less structural disturbances. In present study, bifurcation ratio ranges from 0.76 to 2.33.

*Stream Length* ($L_u$): The stream length is estimated from mouth of the river to the drainage basin. Based on Horton's law of stream length (Table 19.2), geometrical similarity is maintained in the basin by increase in order of streams. The stream length of various orders is mentioned in Table 19.3. In most of the cases, the total length of stream segments is high in first order streams and decreases with an increase in the stream order. It is obtained by stream length of an order by stream number of stream order. The mean stream length increases with the increase of stream order. Somehow, Bhilangna drainage basin contradicts the basic principle (The mean stream length of fourth order is slightly less than third order). From this, we can conclude that basin shows variable lithology with asymmetry in nature, and the river basin is found along major structural lineament. The $L_{sm}$ values of Bhilangna basin vary from 0.0143 to 0.0148 km (Table 19.3).

*Stream Length Ratio* ($R_l$): According to Horton, stream length ratio is the ratio between mean stream length of given order ($L_u$) to mean stream length of next lower order ($L_{u-1}$), continuously till last order of river (Table 19.2) change in stream length ratio from one order to another order indicate late youth stage of geomorphic development (Pareta et al. 2011). The $R_L$ values of Bhilangna basin ranges from 0.42 to 1.32 (Table 19.3) and are greater dependent on the topography and the slope.

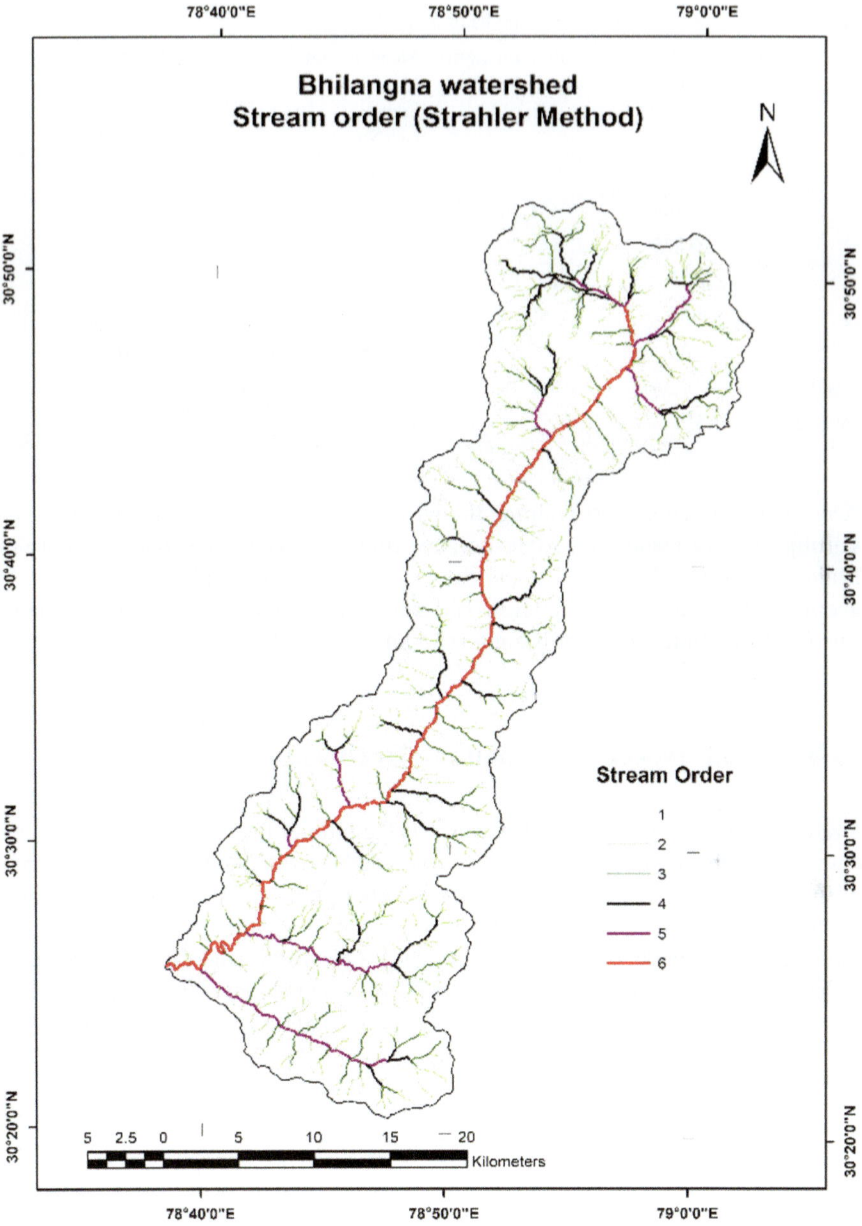


**Fig. 19.2** Stream order of Bhilangna basin

**Table 19.3** Linear aspects of the Bhilangna drainage basin

| Stream order (u) | Stream number (Nu) | Stream length Lu in km | Mean stream length Lsm = LU/Nu | Stream length ratio RL = LU/LU-1 | Bifurcation ratio Rb = Nu/Nu + 1 |
|---|---|---|---|---|---|
| 1 | 66,591 | 953.88 | 0.0143 | | |
| 2 | 30,868 | 451.20 | 0.0146 | 0.47 | 2.15 |
| 3 | 16,525 | 241.79 | 0.0146 | 0.53 | 1.86 |
| 4 | 7078 | 103.12 | 0.0145 | 0.42 | 2.33 |
| 5 | 3601 | 52.75 | 0.0146 | 0.51 | 1.96 |
| 6 | 4704 | 69.64 | 0.0148 | 1.32 | 0.76 |
| Total | 129,367 | 1872.38 | 0.0874 | 3.290 | 9.06 |
| Average Rb = 1.521 | | | | | |

According to Gregory and Walling, the basin length ($L_b$) is the length of the basin starting from the headwaters to the point of confluence (i.e. Balganga a tributary of Bhilangna river). Bhilangna river origin near the Khatling glacier at an altitude of about 4000 m above mean sea level. The length of basin indicates elongated shape. The $L_b$ of the Bhilangna river basin is 60.5 km (Table 19.5).

## 19.9 Areal Aspects of Bhilangna Drainage Basin

*Stream Frequency ($F_s$)*: Stream frequency is the ratio of the total number of stream entities of all the orders in the basin to the total catchment of the basin. The occurrence of stream segments relies upon the nature and types of rocks, vegetation cover, range and recurrence of precipitation and soil penetration potential. The value of stream frequency of Bhilangna basin is 27.22 (Table 19.5). The value of stream frequency is dependent more or less on the amount of rain and the topography of the region.

*Drainage Density (D)*: Drainage density ($D$) is expressed as total length of the streams of all orders per unit area (Horton 1945), and it is expressed as kilometre per square kilometre. The spatial distribution of drainage density is grouped into four classes, i.e. extremely low (0.25–1.58), low (1.58–2.22), moderate (2.22–2.87) and high (2.87–4.27). Strahler (1964) noted that low value drainage density is favoured where basin relief is low and whereas higher values are associated with high drainage density. The drainage density value of the Bhilangne river basin is 2.36 km/km$^2$, which shows moderate permeability of the surface strata. The drainage density map of Bhilangna basin is shown in Fig. 19.3.

*Drainage Texture (T)*: According to Smith (1950), drainage texture ($T$) is a after effect of stream frequency and drainage density. The '$T$' relies on basic lithology, infiltration capacity and relief aspect of the landscape. The drainage texture of the

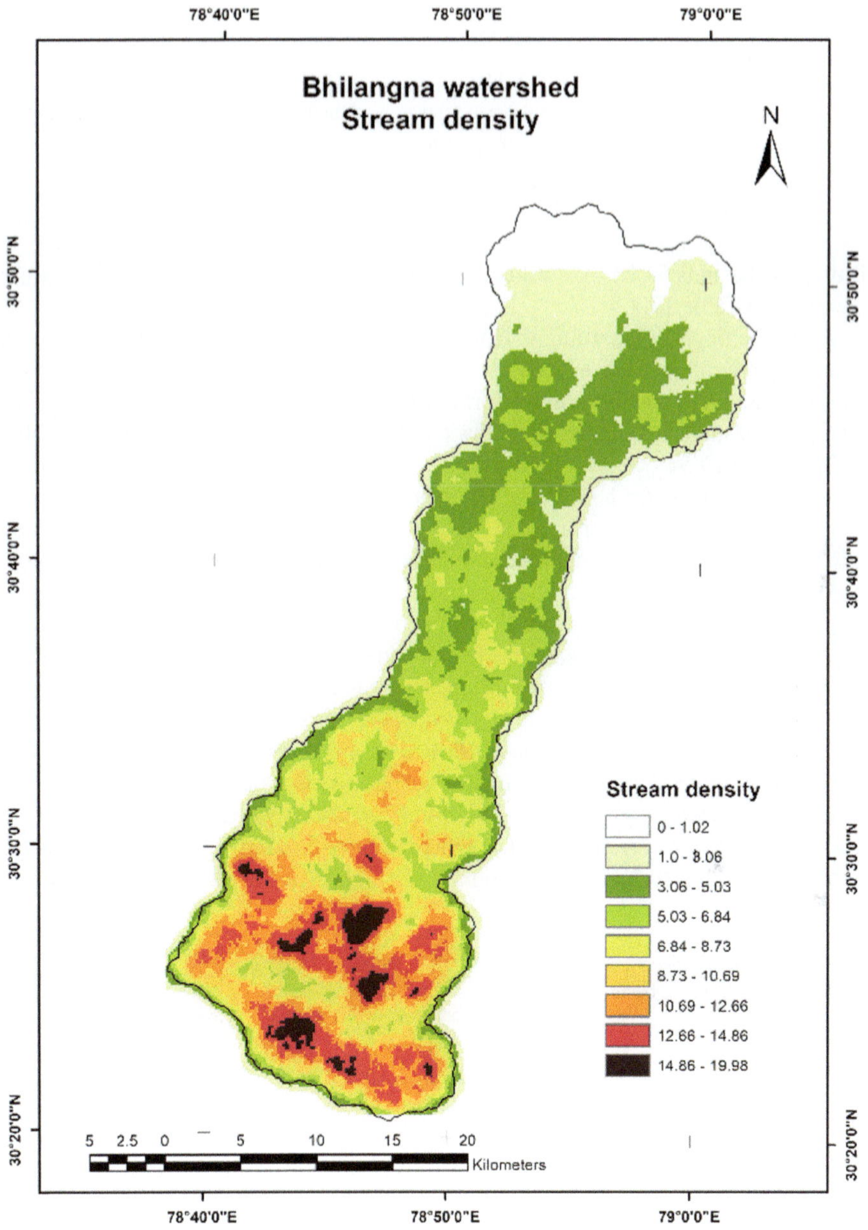


**Fig. 19.3** Drainage density of basin

**Table 19.4** Drainage texture ($T$) classification based on Smith (1950)

| Range | Texture |
|---|---|
| Below 4.0 | Course |
| From 4.0 to 10.0 | Intermediate |
| From 10 to 15.0 | Fine |
| Above 15 | Ultrafine (bad land topography) |

Bhilangna basin is 64.23. According to Smith's, characterisation of drainage texture are given in Table 19.4.

*Form Factor (F)*: According to Horton (1932) form factor as the ratio of the basin area to square root of the basin length. Long and narrow basins express larger lengths and subsequently constitutes lower form factor values. The Bhilangna river basin has an elongated shape with lesser peak flows for longer duration owing to lower *Ff* value (0.21) (Table 19.5).

*Elongation Ratio* ($R_e$): According to Schumm, ratio of elongation is characterized as the ratio of diameter of a circle having homogenous regions of the basin and maximum basin length. It measures the shape of the river basin, and it shows its dependency on geology and climatic conditions. The values of elongation ratio varies from zero, i.e. (highly elongated shape) to one (circular shape) (Bali et al. 2012). Higher elongation ratio indication active denudational process with high infiltration capacity and lesser run off in the basin, whereas lower *Re values* indicate higher elevation of the basin vulnerable to high headward erosion along structural lineaments (Reddy et al. 2004). The *Re value* of Bhilangna drainage basin is 0.52, which is an indication of steep slopes with elongated shape and moderate relief.

*Circularity Ratio* ($R_c$): According to Miller, circularity ratio is the proportion of the area of a basin to the area of a circle having the same circumference as the perimeter of the basin (Table 19.2). The low, medium and high values of the circulatory ratio are

**Table 19.5** Areal aspects of the Bhilangna drainage basin

| S. No. | Areal aspect | Bhilangna drainage basin |
|---|---|---|
| 1 | Perimeter ($P$) kilometre | 192.8 km |
| 2 | Basin area ($A$) kilometre square | 791.9 $km^2$ |
| 3 | Form factor ($F$) | 0.21 |
| 4 | Circularity ratio ($R_c$) | 0.26 |
| 5 | Elongation ratio ($R_e$) | 0.52 |
| 6 | Constant of channel maintenance ($c$) | 0.42 |
| 7 | Stream frequency ($F_s$) | 27.22 |
| 8 | Drainage texture ($T$) | 2.36 |
| 9 | Drainage density ($D_d$) | 64.23 |
| 10 | Basin length ($L_b$) | 60.5 km |

of the young, mature and old phases of the existence pattern of the basin. Bhilangna basin is in the young phase of its development having a circulatory ratio of 0.26, which indicates that basin is elongated in shape with permeable subsoil material.

*The Constant of Channel Maintenance (C)*: According to Schumm, the constant of channel maintenance is defined as that this phenemonon is a inverse of drainage density. The constant of channel maintenance ($C$) is dependent on the slope of basin, nature of rock, vegetation inclusion and the time of disintegration. Generally, higher the value of $C$, i.e. the constant of channel maintenance indicates higher permeability of rocks in the basin and vice versa (Pakhmode et al. 2003; Rao 2009; Kumar et al. 2011). The $C$ value of Bhilangna basin is 0.42 (Table 19.5) that indicate an average 0.42 $km^2$ of catchment area is needed to keep 1 km length of stream channel.

## 19.10 Relief Aspects of Bhilangna Drainage Basin

*Basin Relief* ®: The basin relief is defined as the difference in elevation between the top and the bottom of the valley floor of a basin. Basin relief is an significant factor to have an idea about the denudational attributes of the basin which plays a essential role in landform development, drainage development, surface and sub-surface water flow, permeability and erosional properties of the landscape (Magesh et al. 2011). In this study, we found that highest relief value of Bhilangna drainage basin is 6563 m and such a high value of basin indicates the gravity of water flow, low infilteration and high run off condition in early youth stage of Bhilangna river. The DEM of Bhilangna basin is shown in Fig. 19.4

*Relief Ratio ($R_r$)*: Relief ratio (Schumm, 1956) is defined as the proportion of maximum relief to horizontal distance along the longest dimension of a basin parallel to the main drainage line and it measures the overall steepness of the river basin. The value of relief ratio in the Bhilangna basin area is 0.095 which shows that the intensity of denudational process operating on the slope of the basin.

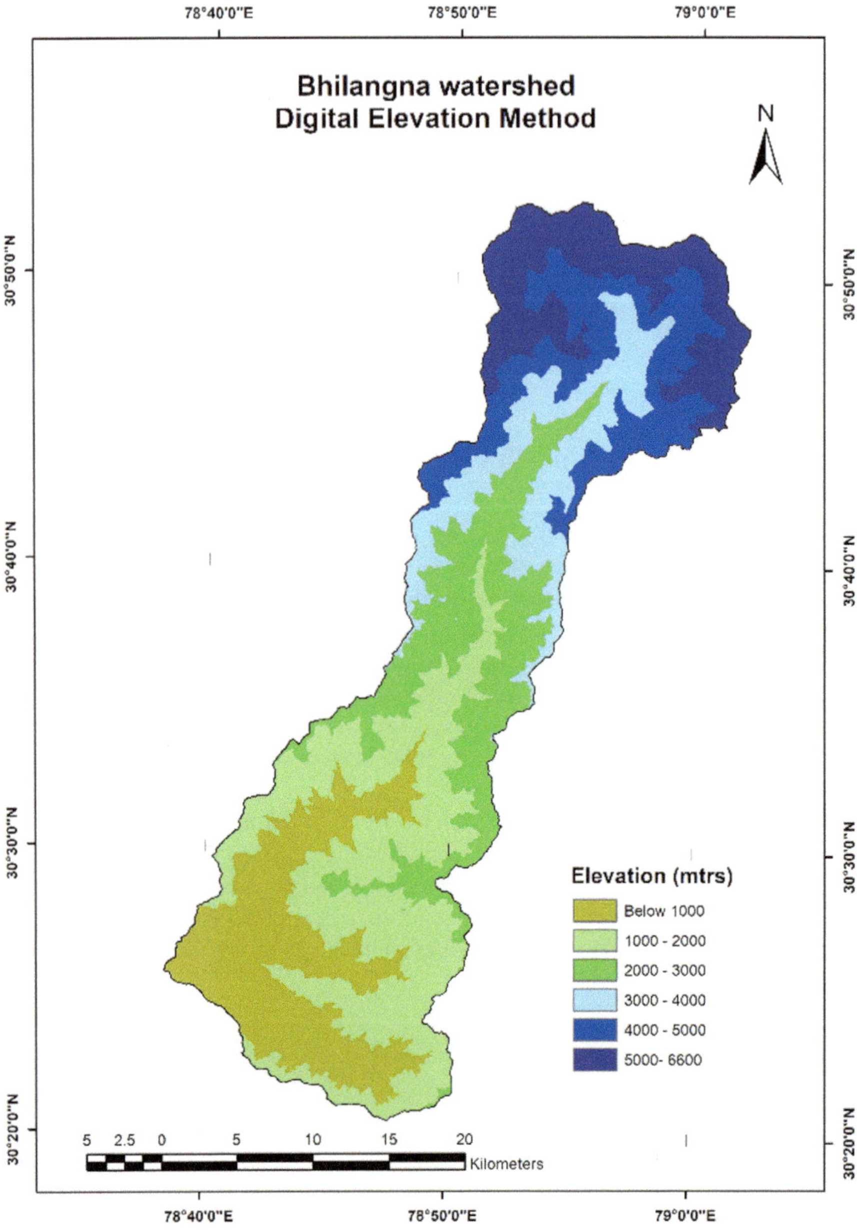


**Fig. 19.4** DEM map of Bhilangna basin

*Slope* (ɵ) is an indication of level of inclination of physical feature, topographical landforms to the horizontal surface. Slope analysis is one of the important parameters in morphometric studies. The degree of slope of Bhilangna drainage basin ranges from 0° to 79.68°. The slope map the Bhilangna catchment is shown in Fig. 19.5. Greater slope gradient will result in rapid run off with potential soil loss.

*Aspect* is referred the direction which terrain faces. Aspect has influences vegetation pattern and type, precipitation type patterns. The aspect map of Bhilangna drainage is presented in Fig. 19.6. The terrain mainly faces south-east direction, as a result higher moisture content and lower evaporation conditions prevails so does the higher vegetation index.

*Terrain Ruggedness Index (TRI)*: Terrain ruggedness index (TRI) was introduced by Riley to show the difference of elevation between two adjacent cells of digital elevation model. It calculate value from a centre cell and the eight cell surrounding it. Then, it squares each value of eight cell to make them positive and average the square, and by taking the square root of this average, we derived terrain ruggedness index. Figure 19.7 shows the ruggedness value that ranges from 0 to 0.9699. Different colours are used to indicate the different values of the TRI. Green colour shows flat area, and it don't have ruggedness. Blue colour shows the highest ranges of ruggedness. Ruggedness value is highest at the ridge area and lowest at the flat area.

A *Hypsometric Integral* (HI) indicates the properties of the area of the surface at diff elevations above or depth below a given datum in a region in simple words we can say that graph which shows the land area that exists at various elevations by plotting relative area against relative height. It expresses the stages of watershed development Strahler (1952) related the hypsometric integral of above 0.70, 0.70–0.30 and below 0.30 to the youth, mature and old stages. Therefore, the hypsometric integral of Bhilangna river basin is 0.41 which indicate that the basin is passing through late mature stage of development (Fig. 19.8).

## 19.11 Conclusion

Catchment area being one of the key elements of landform study has been analysed in this paper. Its elemental distributional pattern, stream frequency, stream density, circulatory ratio and the basin parameters have been inferred and analysed quantitatively. This study covers 15 different types of morphometric parameters and their impact on soil and topography. In this study using GIS technique is more appropriate, precise and more efficient in term of economy for extracting drainage network rather than using traditional method which take more time and quite difficult to made. This paper designates quantitative morphometric parameters of Bhilangna river. The Bhilangna river has well drainage N having stream order range from 1 to 6 that covering an area of 791.2 $km^2$. If we talk about stream order and stream number, there is strong relationship between them nearly a straight line in the lower order represents the flow direction of channels from the greater elevation to low lands. Mean bifurcation ratio of 1.52 indicates that the basin form is less affected by the geology of

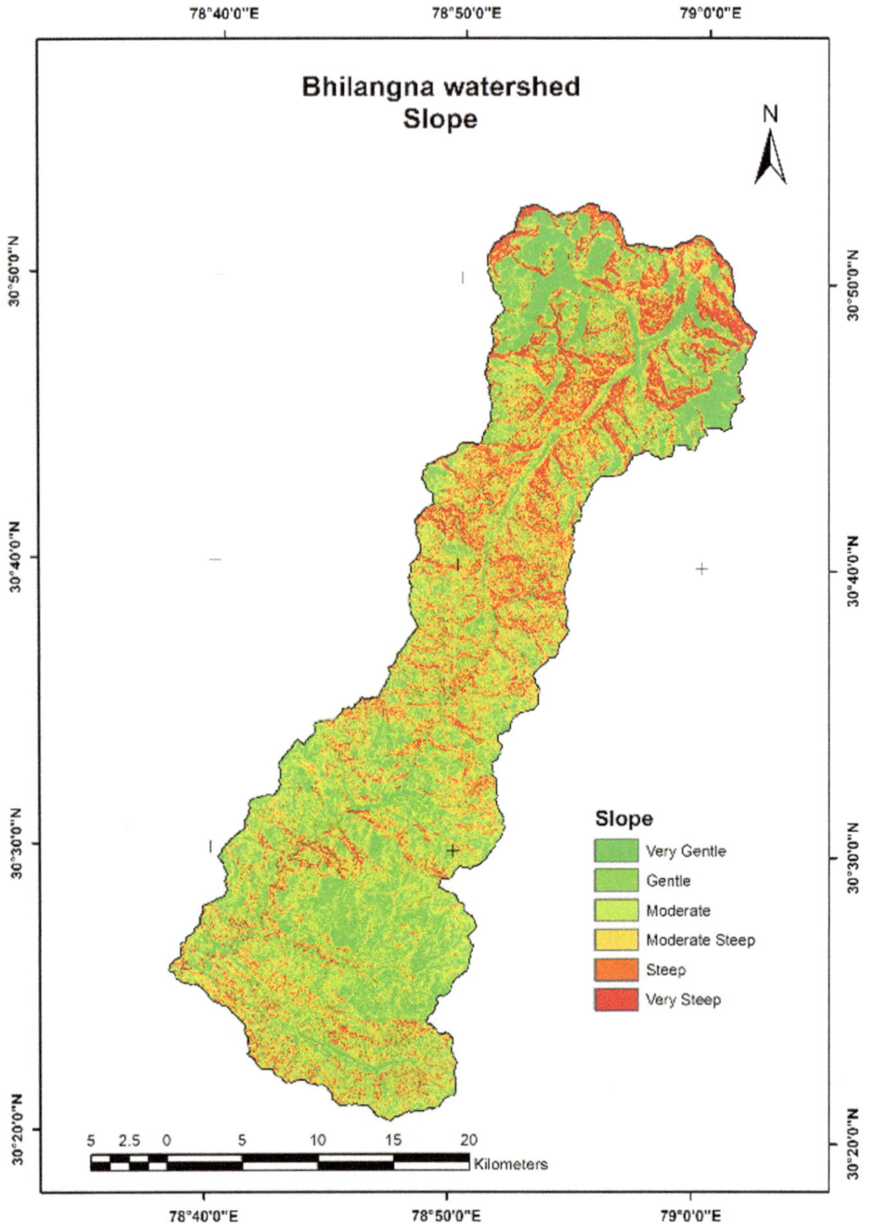


**Fig. 19.5** Slope map of Bhilangna basin

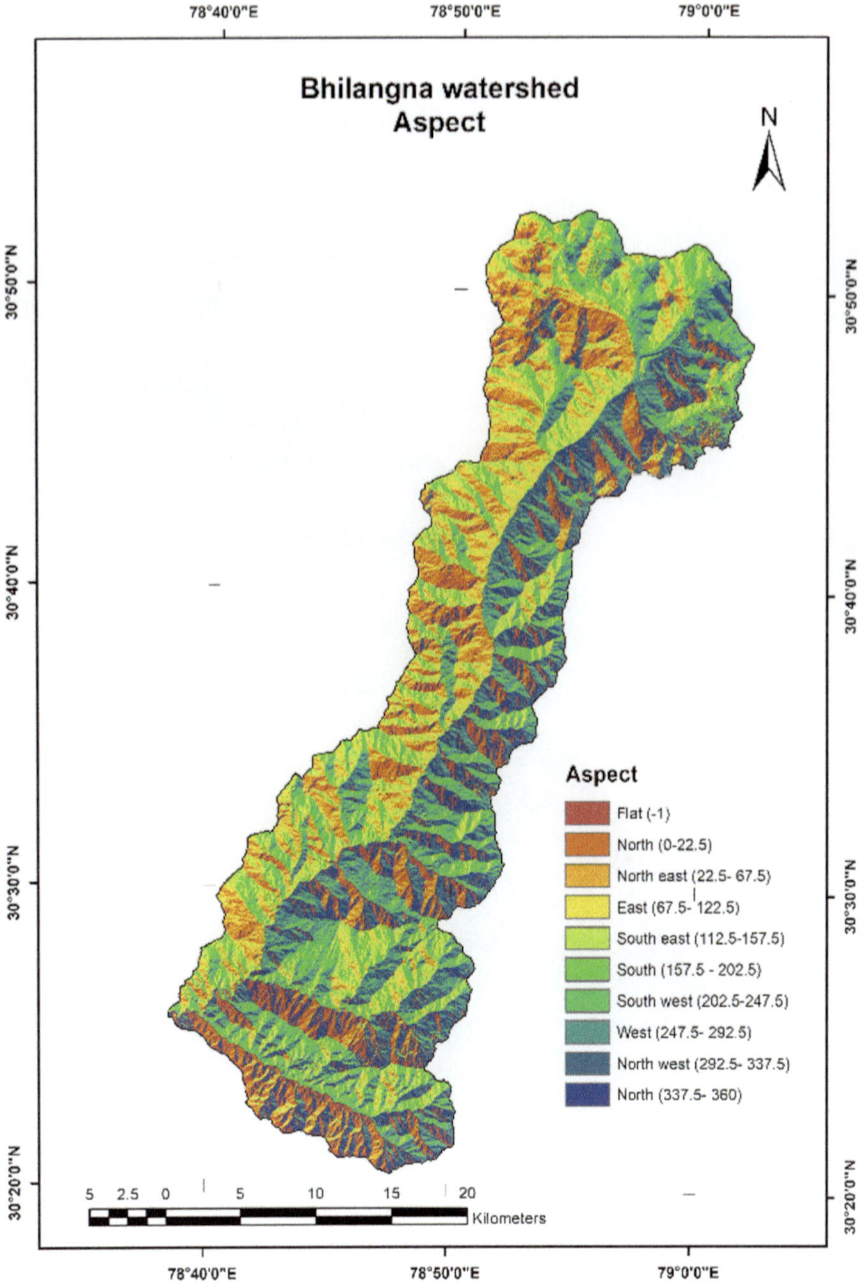


**Fig. 19.6** Aspect map of Bhilangna basin

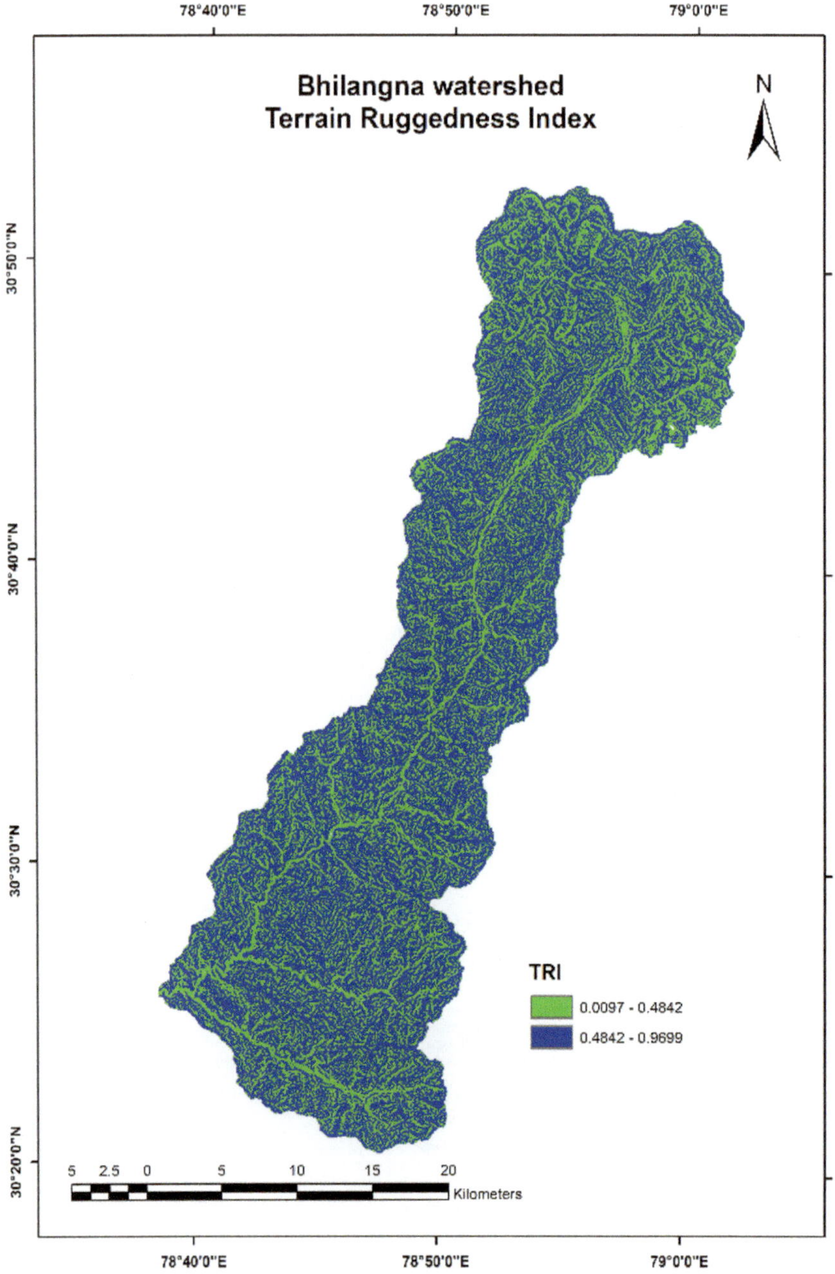


**Fig. 19.7** TRI of Bhilangna basin

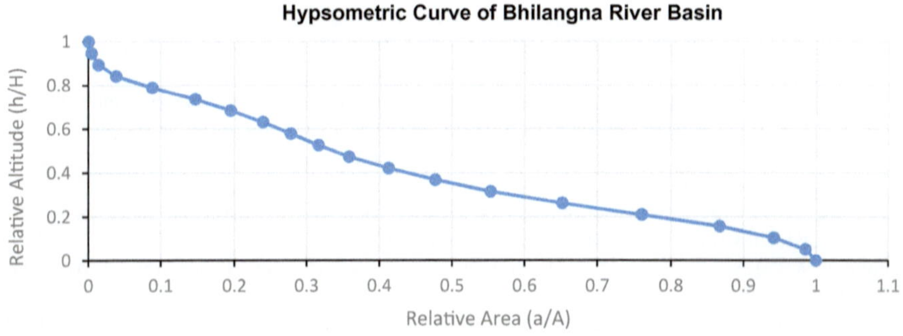


**Fig. 19.8** Relative hypsometric curve

the area and also indicates great ground water discharge potential. Drainage density of the basin is 2.36 which indicate the higher value and moderate permeability of the basin. Circularity ratio is 0.26 which is indication of elongated basin with permeable subsoil material. The value of ruggedness range from 0 to 0.9699 indicates the basin has moderate to high relief.

Hence, from the study of aerial aspects, i.e. value of form factor, relief ratio, and elongation ratio indicates that river basin is elongated with gentle to moderate slope and flatter peak for longer duration. So, floods can be managed easily in this study area. Through relief aspect, value of ruggedness number implies that river basin is prone to soil erosion. From this, it is concluded that morphometric analysis has important role in water and soil conservation, also on controlling floods. As we know, Tehri dam is built on this river, and durability of any dam is depend upon sedimentation deposition of a river. So, this study will be useful for the planners and decision makers in their purpose of basin development and management studies.

## References

Altaf F, Meraj G, Romshoo SA (2013) Morphometric analysis to infer hydrological behaviour of lidder watershed, Western Himalaya, India. Geogr J 2013(178021):14 pages

Angillieri MYE (2008) Morphometric analysis of Colangüil river basin and flash flood hazard, San Juan, Argentina. Environment Geol 55(1):107–111

Asfaw D, Getachew W (2019) Quantitative analysis of morphometry on Ribb and Gumara watersheds: implications for soil and water conservation. Int Soil Water Conserv Res 7.2: 150–157

Chitra C, Alaguraja P, Ganeshkumari K, Yuvaraj D, Manivel M (2011) Watershed characteristics of Kundah sub basin using remote sensing and GIS techniques. Int J Geomatics Geosci 2(1):311

Gregory KJ, Walling DE (1973) Drainage basin form and process

Islam R, Thakur VC (1988) Geology of Bhilangna Valley, Garhwal Himalaya. Geosci J 9(2):143–152

Horton RE (1932) Drainage-basin characteristics. Eos Trans Am Geophy Union 13(1):350–361

Horton RE (1945) Erosional development of streams and their drainage basins; hydrophysical approach to quantitative morphology. Geol Soc Am Bull 56(3):275–370

Jain AO, Thaker T, Chaurasia A, Patel P, Singh AK (2018) Vertical accuracy evaluation of SRTM-GL1, GDEM-V2, AW3D30 and CartoDEM-V3.1 of 30-m resolution with dual frequency GNSS for lower Tapi Basin India. Geocarto Int 33(11):1237–1256

Javed A, Khanday YM, Rains S (2011) Watershed prioritization using morphometric & land use/land cover parameter: a remote sensing & GIS based approach. J Geol Soc India 78:63–75

Kulkarni MD (2015) The basic concept to study morphometric analysis of river drainage basin: a review. Int J Sci Res 4(7):2277–2280

Kusre BC (2016) Morphometric analysis of Diyung watershed in northeast India using GIS technique for flood management. J Geol Soc India 87(3):361–369

Langbein WB, Leopold LB (1964) Quasi-equilibrium states in channel morphology. Am J Sci 262(6):782–794

Miller VC (1953) A quantitative geomorphic study of drainage basin characteristics in the Clinch Mountain Area Virginia and Tennessee. Columbia University, New York

Nag SK, Chakraborty S (2003) Influence of rock types and structures in the development of drainage network in hard rock area. J Indian Soc Remote Sens 31(1):25–35

Nautiyal MD (1994) Morphometric analysis of a drainage basin using aerial photographs: a case study of Khairkuli Basin, District Dehradun, UP. J Indian Soc Remote Sens 22(4):251–261

Nongkynrih JM, Husain Z (2011) Morphometric analysis of the Manas river basin using earth observation data and Geographical Information System. Int J Geomat Geosci 2(2):647–654

Pakhmode V, Kulkarni H, Deolankar SB (2003) Hydrological-drainage analysis in watershed programme planning: a case from the Deccan basalt, India. Hydrol J 11(5):595–604

Pinter N, Keller EA (1995) Geomorphological analysis of neotectonic deformation, northern Owens Valley, California. Geol Rundsch 84(1):200–212

Rao NK, Latha SP, Kumar AP, Krishna HM (2010) Morphometric analysis of Gostani river basin in Andhra Pradesh State, India using spatial information technology. Int J Geomat Geosci 1(2):179

Reddy GPO, Maji AK, Gajbhiye KS (2004) Drainage morphometry and its influence on landform characteristics in a basaltic terrain, Central India–a remote sensing and GIS approach. Int J Appl Earth Obser Geoinf 6(1):1–16

Saklani PS (1972) Lithostratigraphy and structure of the area between the Bhagirathi and Bhilangana rivers, Garhwal Himalaya. Him Geol 2:344–355

Scheidegger AE (1965) The algebra of stream-order numbers. US Geological Survey, Washington, DC, Professional Paper B, 525

Schumm SA (1961) Effect of sediment characteristics on erosion and deposition in ephemeral-stream channels (vol 352). US Government Printing Office

Shreve RL (1966) Statistical law of stream numbers. J Geol 74(1):17–37

Valdiya KS (1980) The two intracrustal boundary thrusts of the Himalaya. Tectonophysics 66(4):323–348

Yadav SK, Singh SK, Gupta M, Srivastava PK (2014) Morphometric analysis of upper tons basin from Northern Foreland of Peninsular India using CARTOSAT satellite and GIS. Geocarto Int 29(8):895–914

# Chapter 20
# Geospatial Analysis of Kosi River Course from 1998 to 2018

**Rajan Kumar, Netrananda Sahu, and Martand Mani Mishra**

**Abstract** Kosi River is one of the major left bank tributaries of the River Ganga and is rising at an altitude of over 7500 m above MSL (mean sea level) in the Himalayan kingdom of Nepal and Tibet. Ancient Sanskrit name for Kosi pronounced as Kaushiki in Vedas is one of the most sacred ancient rivers of India. In this paper, we have taken two years in 1998, and 2018 to study the channelization of the stream through group pixel formation. Unsupervised classification and stroller stream order methodology has been used in this study to analyze the stream return of the river during the years 1998 and 2018. The longitudinal profile of the river Kosi has been generated in this study by using the tool provided in the software Erdas Imagine. It has been observed through the study that there is a noticeable decrease in the sinuosity of the river and an increase in the channel belt areas.

**Keywords** Kosi · Channel · Strahler stream order · Sinuosity

## 20.1 Introduction

Next to Indus and Brahmaputra in the Indian sub-continent, Kosi is the largest river (total length 720 km) in Nepal and has got its name from the ancient Sanskrit word Kausika (Regmi 2013; Sahana and Patel 2019) (https://www.researchgate.net/profile/Rajiv_Sinha6/publication/200004867_GIS_in_Flood_Hazard_Mapping_a_case_study_of_Kosi_River_Basin_India/links/0046353bd6f9186023000000/GIS-in-Flood-Hazard-Mapping-a-case-study-of-Kosi-River-Basin-India.pdf, last accessed 2020/04/13; https://www.researchgate.net/profile/N_Chen3/publication/258806557_On_the_water_hazards_in_the_trans-boundary_Kosi_River_basin/links/567790a908ae125516ec53b0/On-the-water-hazards-in-the-trans-boundary-Kosi-River-basin.pdf, last accessed 2020/09/07). The Kosi River in north Bihar plains, precisely eastern India presents a challenge in terms of long and periodic flood hazards (Chen et al. 2017; Kattelmann 2003; Neupane et al. 2015;

R. Kumar · N. Sahu (✉) · M. M. Mishra
Department of Geography, Delhi School of Economics, University of Delhi, Delhi 110007, India



P. Kumar et al. (eds.), *Water Resources Management and Sustainability*,
Advances in Geographical and Environmental Sciences,
https://doi.org/10.1007/978-981-16-6573-8_20

Sinha 2009). Despite a long history of flood control management in the basin for more than six decades, the river continues to bring a lot of unhappiness and misery through extensive flooding. It is the dynamic nature of the Kosi River that has even attracted attention for over a century (Agarwal and Bhoj 1992; Gole et al. 1966; Wells and Dorr 1987; Arogyaswamy 1971) and a variety of mechanisms have been suggested ranging from tectonic tilting, nodal avulsions, discharge peakedness, and auto cyclic processes is also supported in the argument of Leier et al. (2005; Kecerdasan and Ikep 1993).

## 20.2 Material and Methods

### *20.2.1 Study Area*

The Kosi River Basin is a sub-basin of the Ganga basin situated as the left bank tributary of the main Ganga River (Fig. 20.1). Most of the upper catchment of the basin lies in Nepal and Tibet at great heights of the Himalayan and Kanchenjunga range. The total drainage area of the Kosi River is 74,030 km$^2$ out of which 11,410 km$^2$ lies in India and the rest 62,620 km$^2$ lies in Tibet and Nepal (RiverBasinMap.jpg (2520×1637), http://fmis.bih.nic.in/RiverBasinMap.jpg, last accessed 2020/05/17). The location of the basin lies between 25° 20′ 30″–29° 07′ 48″ North latitude and 85° 22′ 19″–88° 55′ 44″ East longitude. It is bounded by the side ridges on the left side separating it from the Brahmaputra River, while River Ganga forms its southern boundary. The topography of the basin is very steep in upper reaches and mild in lower reaches as shown in Fig. 20.1.

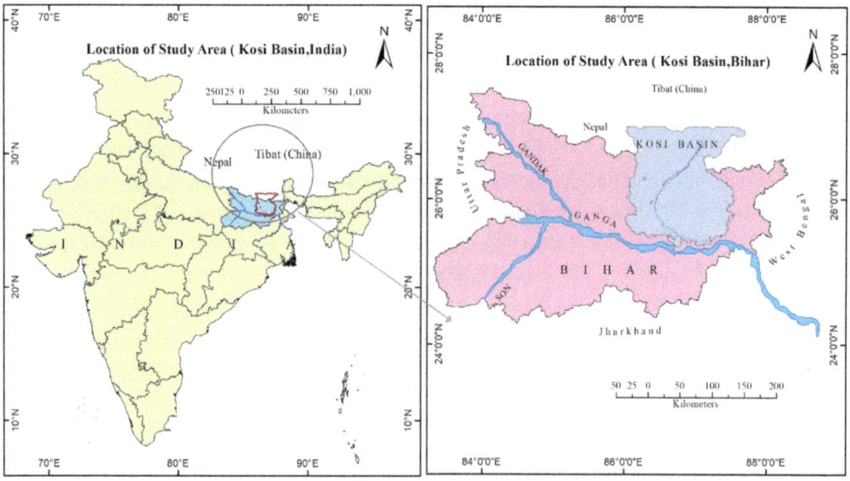


**Fig. 20.1** Index map of the Kosi River Basin, in India

### 20.2.2 Methodology

Unsupervised classification methodology is used in this study to achieve the objective. Unsupervised classification is termed as computer-automated classification based on pixels of satellite data. It uses clustering algorithms for the classification of data and classes of the image which is then specified by the user. The spectral pixels calculate numeric information from data and present it in visual form. Image processing software groups pixels based on spectral similarity and create classes for mapping. It is fairly quick and easy to run.

In this study channelization of the stream for the study years, 1998 and 2018 is done through group pixel formation. First of all the images were classified in four classes under Iso cluster unsupervised classification and then again the images were reclassified in two classes after the identification of the channel. The channel of each year is extracted converting it into vector data and then irrelevant data is deleted. The channel of each year is superimposed on each other and divided in reach (Fig. 20.2). A total stream is divided into five reaches as per location in the basin and this data is exported in Google Earth Pro and location validated (Fig. 20.3).

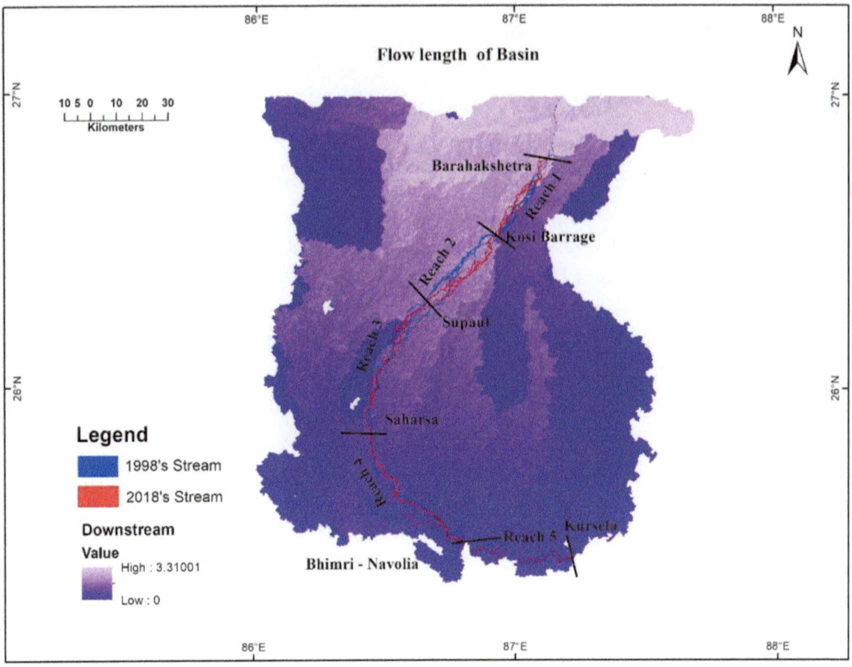


**Fig. 20.2** Flow length of the basin

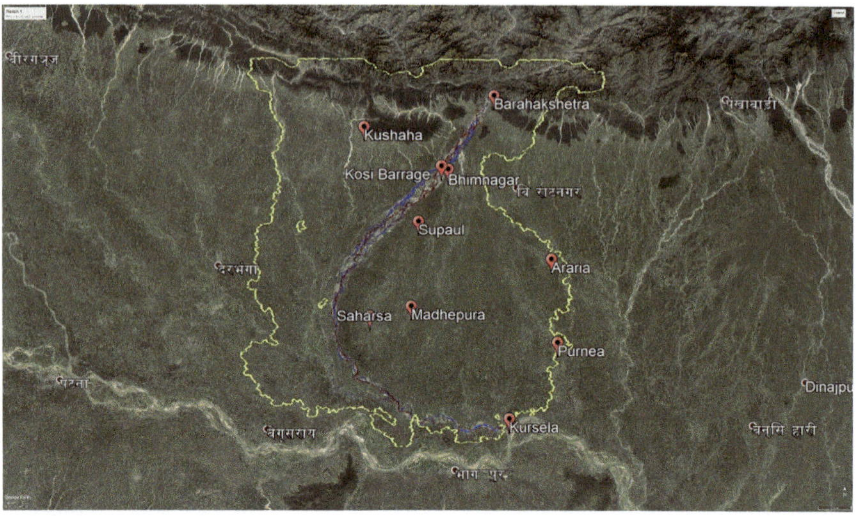


**Fig. 20.3** Ground validation of both the streams on Google Earth Pro

## 20.3 Result and Discussion

### *20.3.1 Ordering of River Network*

The ordering of the river network consists of the hierarchy of streams that are denoted as mathematical branches of numerical measures. This concept was first discussed by Horton in 1945 and later by Strahler in 1952 and 1947 (Horton 1932, 1945; Mahala 2020; Agarwal and Bhoj 1992; Strahler 1957). This concept is also termed as Strahler Stream Order. The ordering of the river network or stream network is connected with a segment (Fig. 20.4). Each segment is linked with downstream which defines the flow of the stream in a particular order. The all fingertip stream is termed as the first order. When two fingertip streams i.e. the first-order stream join at conjunction termed as second-order stream and when two second-order streams join at confluence they formed a third-order stream and so on of higher orders streams. In the case of the first-order streams, i.e., fingertip tributary joins at the confluence of second-order it remains second-order and similarly if the third-order stream is joined on the conjunction of first order or second order and so on.

The ordering of the network in the drainage basin includes three types of networks. The first is the active channel network in which water flows regularly, the second is the valley network that includes not only the active channel network but also valleys in the basin that no longer carry active flow and the third is the dry channel network that carries water after heavy rainfall.

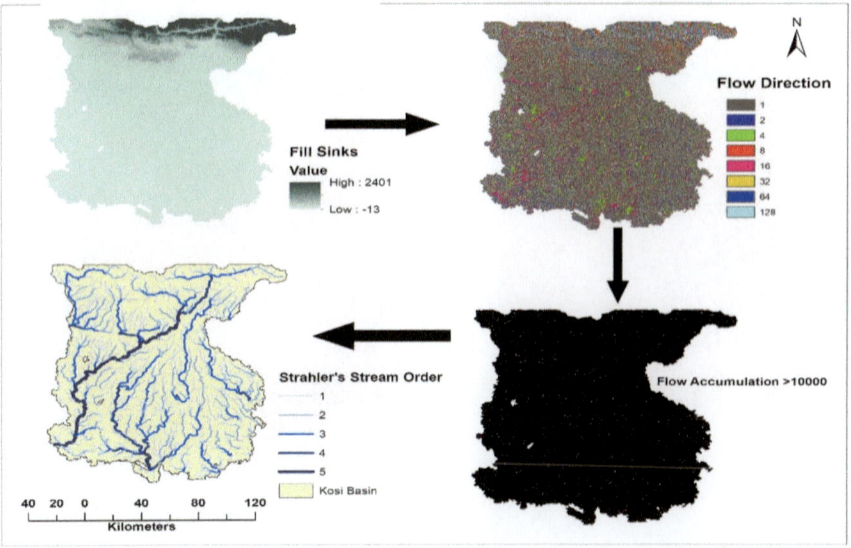


**Fig. 20.4** Stream order network delineation process

In this study, Strahler Stream Order is computed for channelizing the correct order of Kosi Basin which the threshold value of flow accumulation is taken greater than 10,000 for visualizing the active channel and valley network of Kosi Basin.

The longitudinal profile which is drawn through thalweg point (this is the line of fastest flow for any stream) is shown in Fig. 20.5 for the entire stream of 1998 from Barahkshetra to Kursela. The longitudinal profile speaks a lot about changes in elevation with distance. Kosi River which flows from Nepal and enters India at Triveni which is at a higher elevation shows little variation in channel movement. It can be interpreted from the above profile that the river in the first 120 km has a very sharp gradient with Sharpe edges at certain locations. Now the major change in the sinuosity of the river is found to be after 140 km or from Supaul to be precise it is from Saharsha perpendicular to Purina (old course used to flow from Purina, 120 km east of Saharsha) that river starts showing its meandering pattern.

The abrupt change in sinuosity at Saharsha is due to the congested streamflow of Sapt Kosi into a single stream Kosi which results in a sudden outbreak of river capture along megafan axis, which is devastating. This could also be explained with the help of the superelevation of the course with respect to the floodplain, while these topographic indices provide quantitative information about channel shift and avulsion potential. The likelihood or probability of an avulsion event at any point will be finally governed by the position of the channel with respect to embankment built and planform dynamics over time.

The longitudinal profile of any stream gives a brief idea of sinuosity which in terms helps to study the down-valley gradient of the channel concerning cross valley gradient. Figure 20.6 shows how Kosi River has produced a steep gradient for the

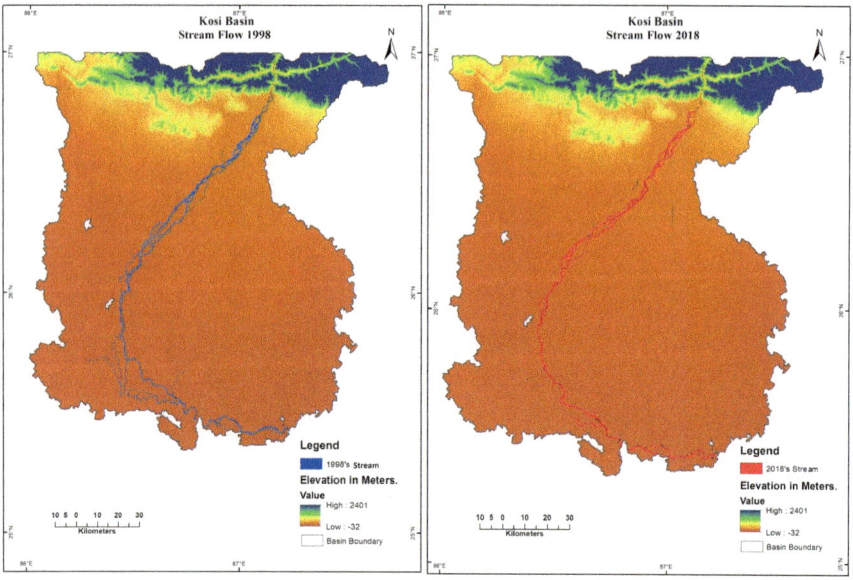


**Fig. 20.5** Stream movement in the year 1998 and 2018

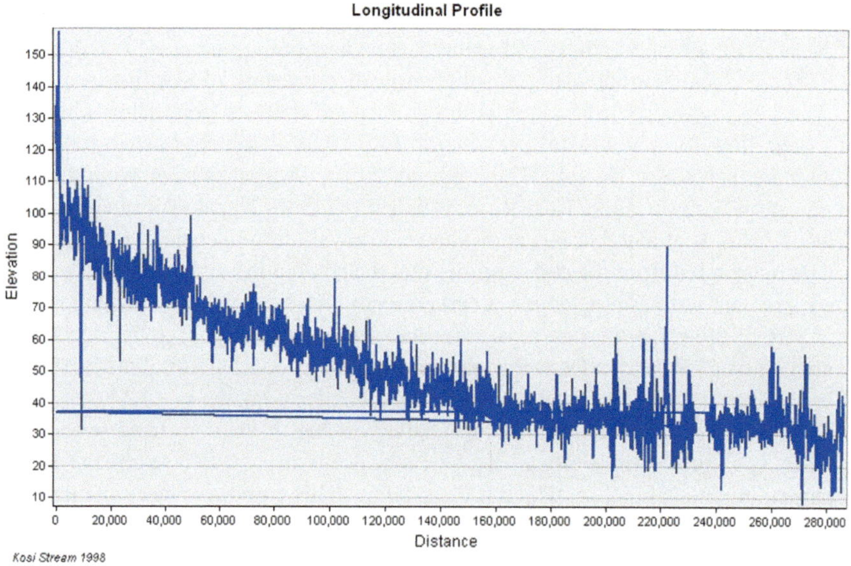


**Fig. 20.6** Longitudinal profile of the Kosi River in the year 1998

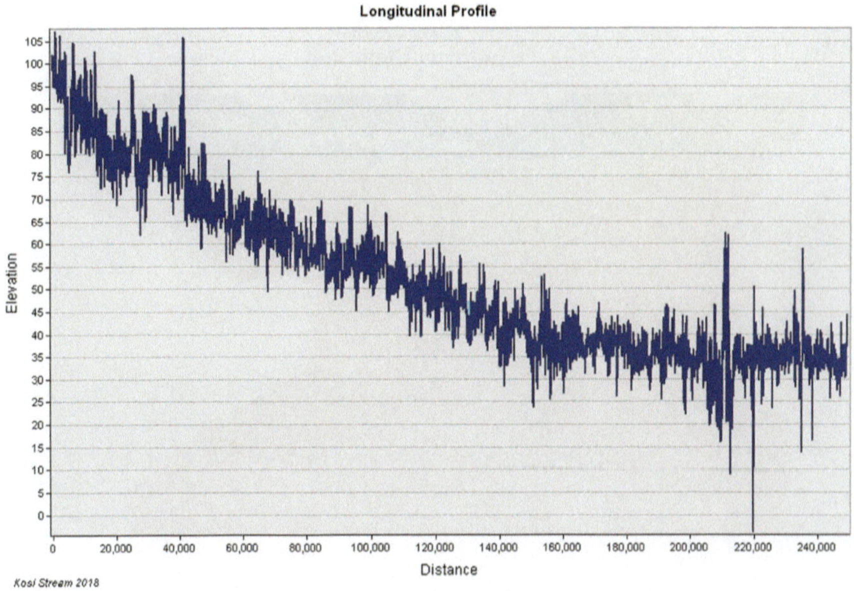


**Fig. 20.7** Longitudinal profile of the Kosi River in the year 2018

first 150 km distance and decreased further towards Kursela in Katihar district. Since 1998 the abrupt changes in river course have led to an increase in steepness of the initial reaches of the Kosi River which can be understood with the concave nature of the graph.

Towards 2018 the aggradational process has increased at the lower portion of the river and meandering is a very common phenomenon. This increase in the aggradational process is also associated with the changes in channel width and an increase in the bar area. Figure 20.7 which is drawn on thalweg point of the river behavior is the reflection of variation in a bar area or braid channel ratio in terms of aggradation and degradation process, which will have a noticeable bearing on channel movement. Therefore both morphological and topographical data should be integrated to provide better insight into the avulsion process.

The decrease in sinuosity in the upper reaches of the river is mainly due to a decrease in a down-valley gradient of the course in respect to the cross valley gradient (Fig. 20.8). The decrease in sinuosity has inched more profoundly since August 2008 which saw the most disastrous flood in a decade. The 2008 flood had its severe impact due to the river acquiring one of its old channels on the east embankment.

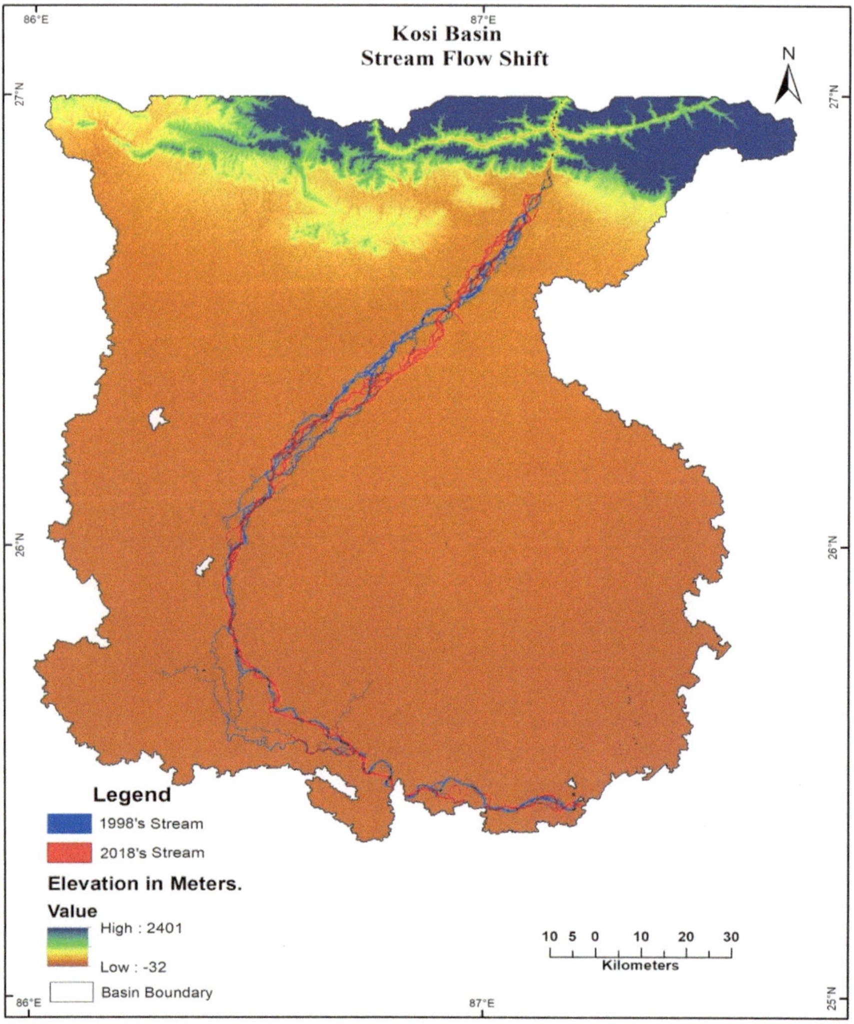


**Fig. 20.8** Superimposition of both the streams of 1998 and 2018

## *20.3.2 Reach Demarcation and Its Analysis*

Topographic analysis has been carried out with the help of the data from USGS ASTER-GDEM, which has the elevation data from the year 2011. A longitudinal profile of the Kosi has been generated for the study period using a spatial profiling tool in Erdas Imagine software.

The Kosi channel upstream and downstream of the river is divided into several reaches or transects with the location marked at a certain interval, and the cross valley profile (Fig. 20.9) is calculated for all the five transects/reaches.

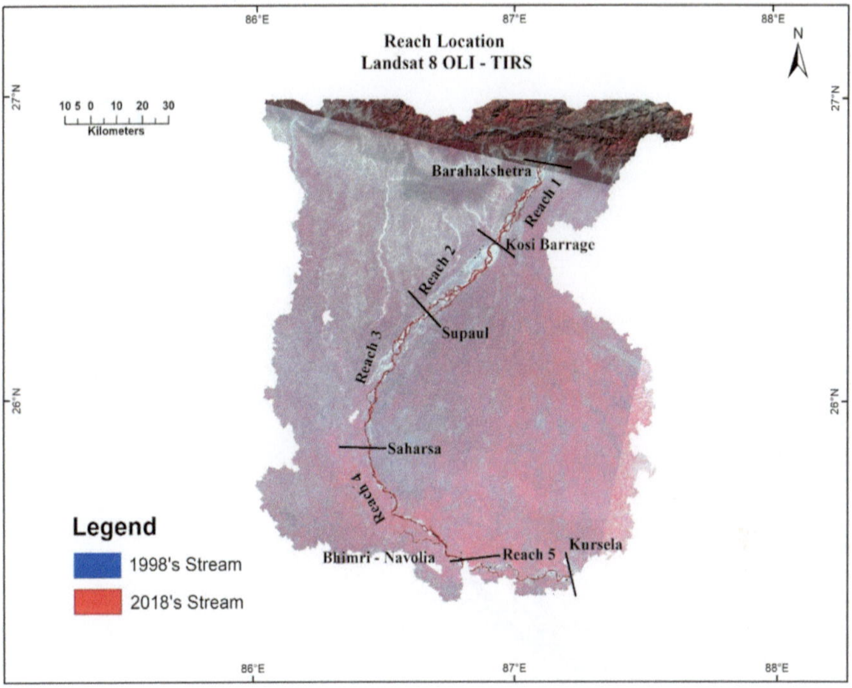


**Fig. 20.9** Demarcation of reach location

Figure 20.10 which is a portion of the mainstream known as reach one is showing the channel movement of rivers in both the years 1998 and 2018 together from Barahkshetra up to Kosi barrage or Hanuman Nagar barrage. The stream demarcated in red color is of 2018 and the other one from blue color is of 1998. It is very much clear from Fig. 20.10 that the stream of 2018 has changed its course over time; the direction of channel movement in this part of the stream is marked towards the west from the previous position of the river.

Figures 20.11, 20.12, 20.13 and 20.14 shows a microanalysis of the reach 1, i.e., Barahkshetra to Kosi River. The graph in red color is of 1998 stream and the one which is in blue is of recent stream drawn from 2018 river course. Comparing the cross profile of 1998 with 2018 gives a clear picture of how the river is changing its course very often along the embankment itself. There has been a sharp increase in steepness of elevation in 2018 with respect to 1998, the cure in the red color is smooth up to three widths and that of blue color shows a very sharp fall in elevation for a width of just 1.5 km. This is also evident from the westward movement of the river in the upper reaches.

Figures 20.15, 20.16, 20.17, 20.18, 20.19, 20.20, 20.21, 20.22, 20.23, 20.24, 20.25, 20.26, 20.27, 20.28, 20.29, 20.30, 20.31, 20.32, 20.33 and 20.34 shows graphs of longitudinal and cross profile for all remaining four transects for the years 1998 and 2018. It shows a noticeable decrease in sinuosity along a few transects or reaches (such as 2, 4, and 5). The changes in the braid channel ratio could be much more

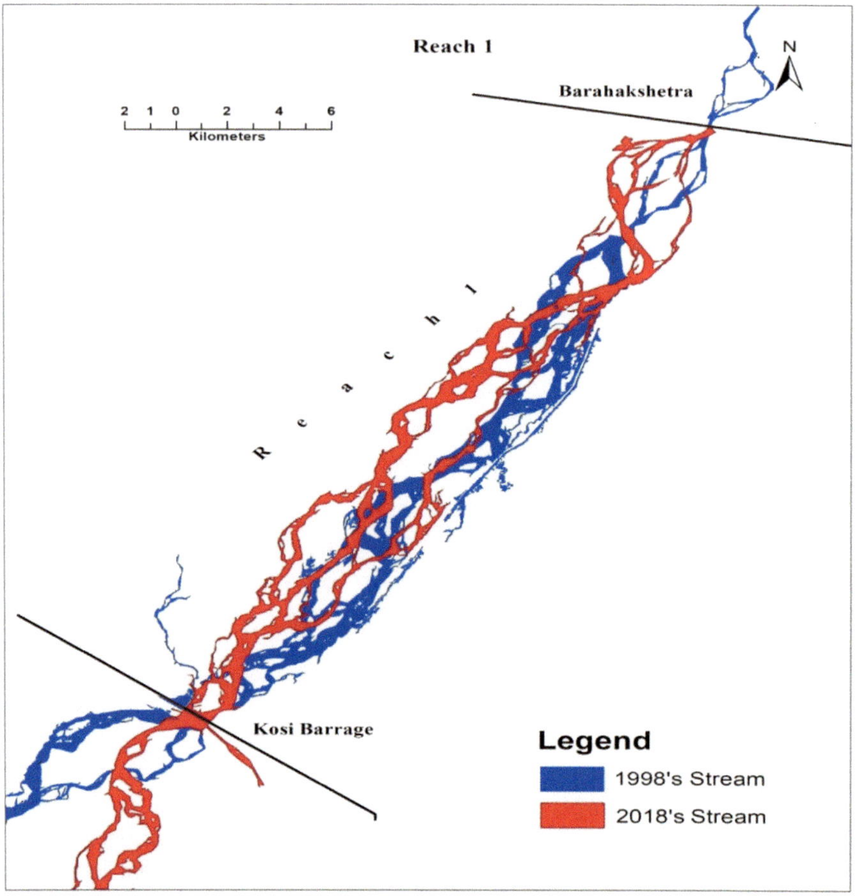


**Fig. 20.10** Reach 1 of the Kosi River from Barahkshetra to Kosi barrage

**Fig. 20.11** Longitudinal profile of reach 1, 1998

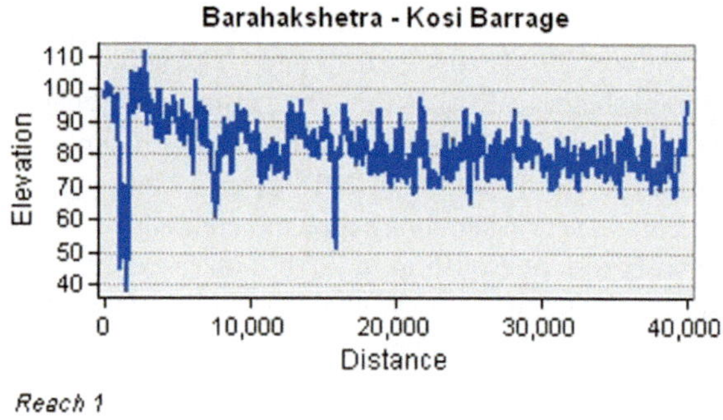

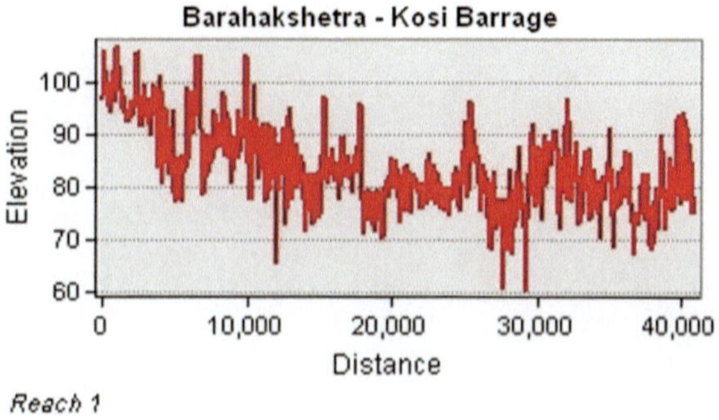


**Fig. 20.12** Longitudinal profile of reach 1, 2018

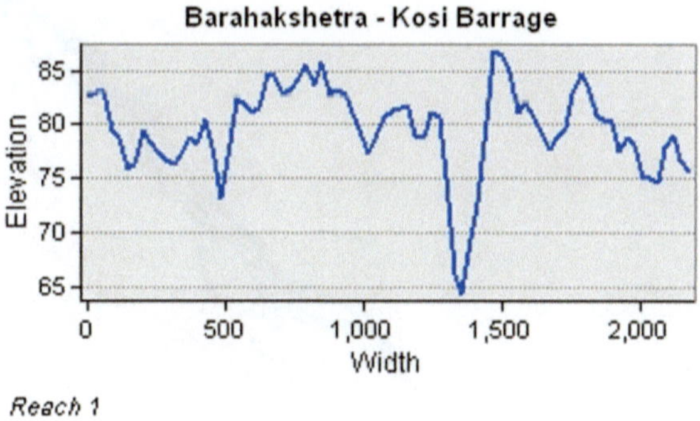


**Fig. 20.13** Cross profile of reach 1, 1998

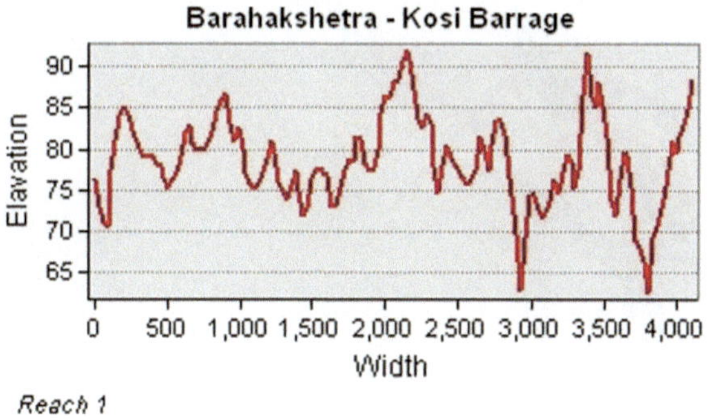


**Fig. 20.14** Cross profile of reach 1, 2018

remarkable. Almost for a total length of reaches up to Hanuman Nagar barrage show a decrease in braid channel ratio (Friend and Sinha 1993; Sinha et al. 2019; Sinha and Friend 1994). The other reaches, i.e., 1 and 3 show an increase. Now, after examining the bar area for upstream and downstream it is found that the bar area for downstream has increased around Saharsha up to Kursela and for upstream the increase is not much. This is a manifestation of a narrow path of the reach resulting in a much-reduced capacity of the channel to carry sediment caused by reduction in velocity and slope changes altered by the construction of the Kosi barrage.

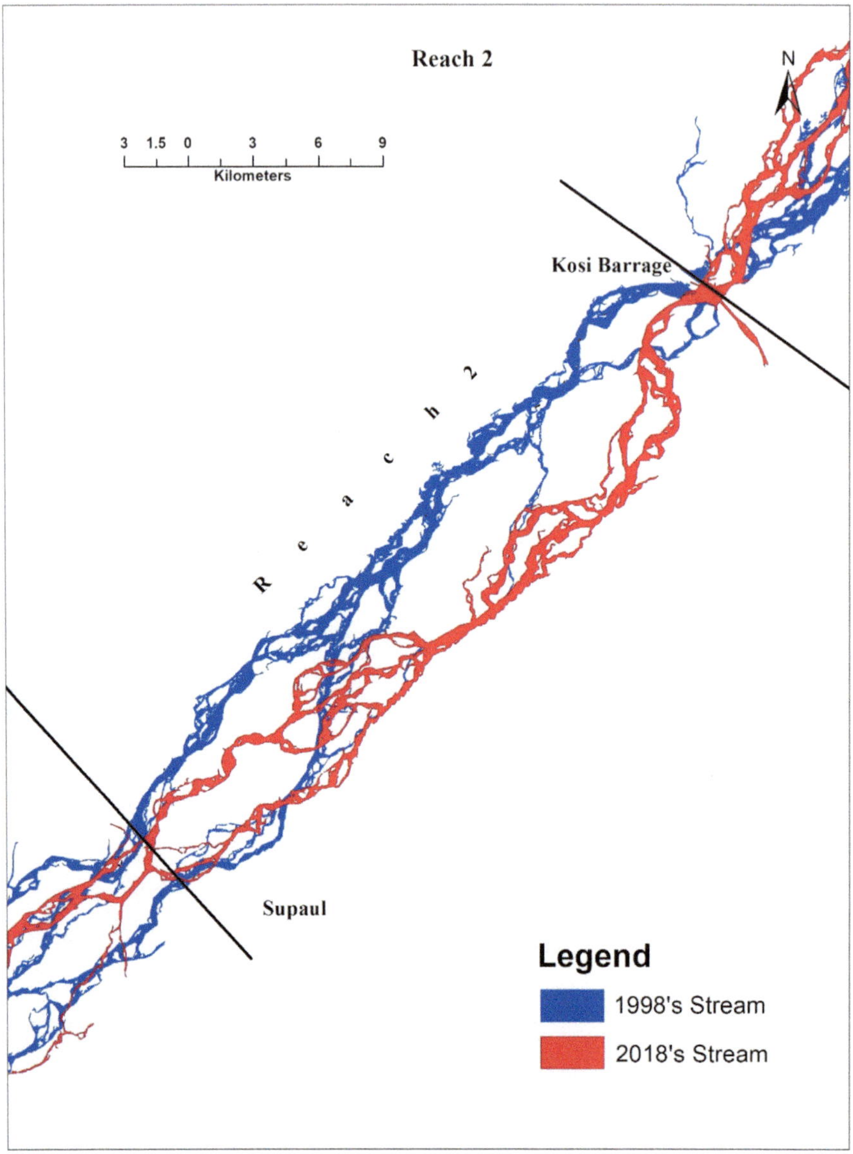


**Fig. 20.15** Reach 2 of the Kosi River from Kosi barrage to Supaul

At the upper transects of the barrage there is the settling of sediments due to temporary storage of water, on the other hand at the downstream there is channel widening in the reaches/transects downstream (barring those reaches which are adjacent to barrage and are protected) results in reduced flow depth and sedimentation. The downstream channel reaches are showing no clear trend. This is related to the policy of barrage operation and is mainly controlled by the gate opening at different time intervals.

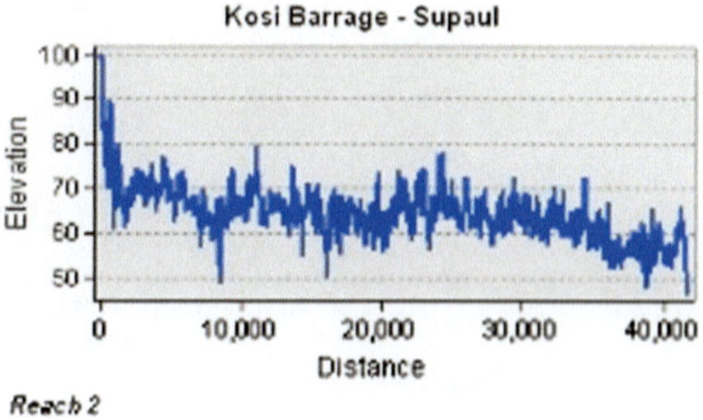


**Fig. 20.16** Longitudinal profile of reach 2, 1998

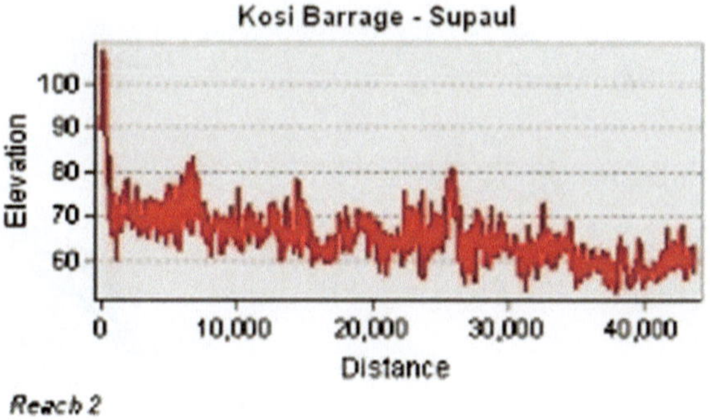


**Fig. 20.17** Longitudinal profile of reach 2, 2018

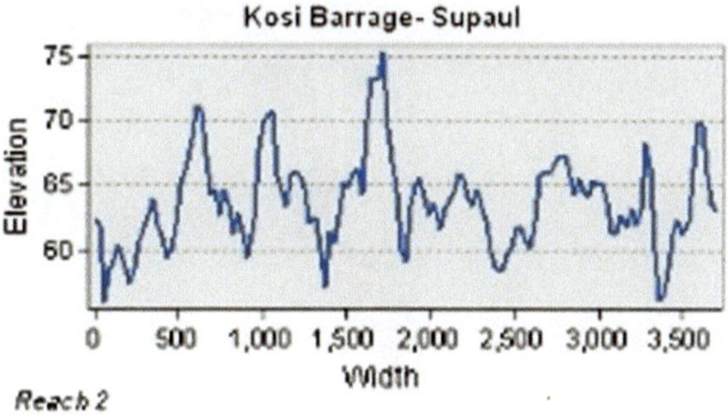


**Fig. 20.18** Cross profile of reach 2, 1998

## 20.4 Conclusion

There is extensive aggradation work of the river at lower reaches around Bhimri where sinuosity of the river is observed to be maximum in 1998 which shows an increase in channel belt areas. The movement of the deepest point that is thalweg line with reference to both the eastern and western embankments for the period 1998–2018 is shown in Figs. 20.9, 20.10, 20.11, 20.12, 20.13, 20.14, 20.15, 20.16, 20.17, 20.18, 20.19, 20.20, 20.21, 20.22, 20.23, 20.24, 20.25, 20.26, 20.27, 20.28, 20.29,

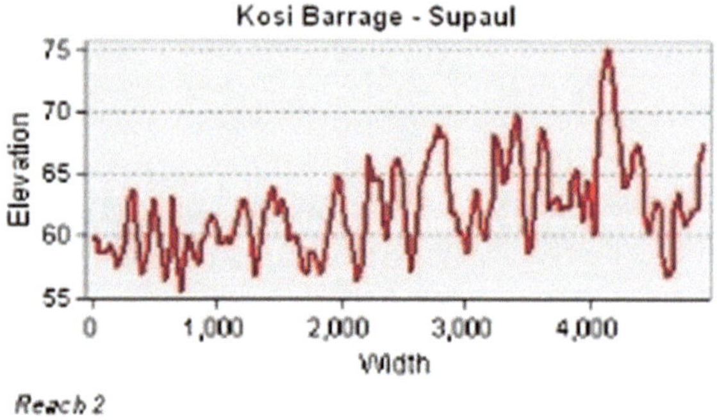


**Fig. 20.19** Cross profile of reach 2, 2018

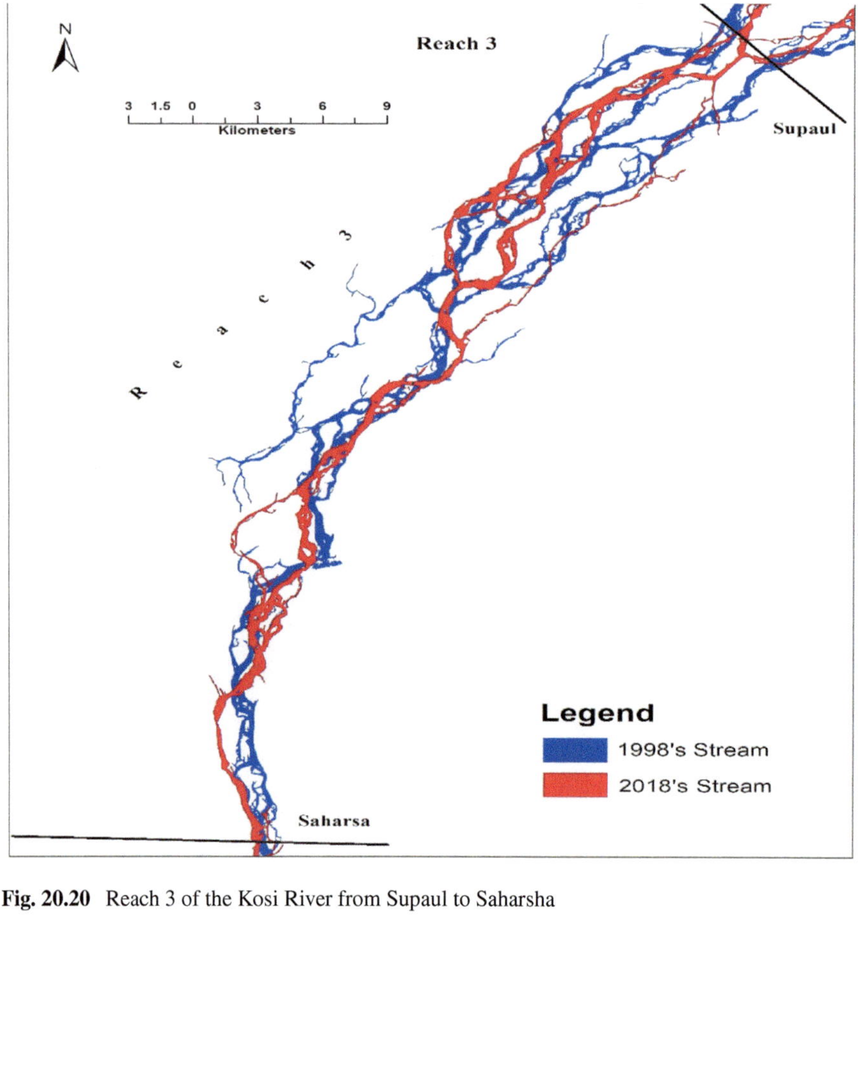


**Fig. 20.20** Reach 3 of the Kosi River from Supaul to Saharsha

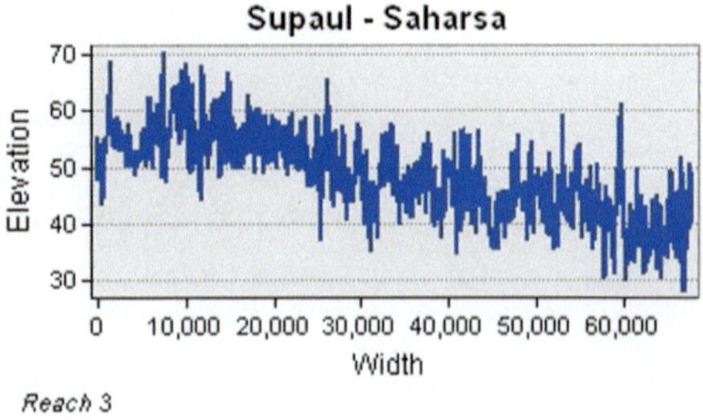


Fig. 20.21 Longitudinal profile of reach 3, 1998

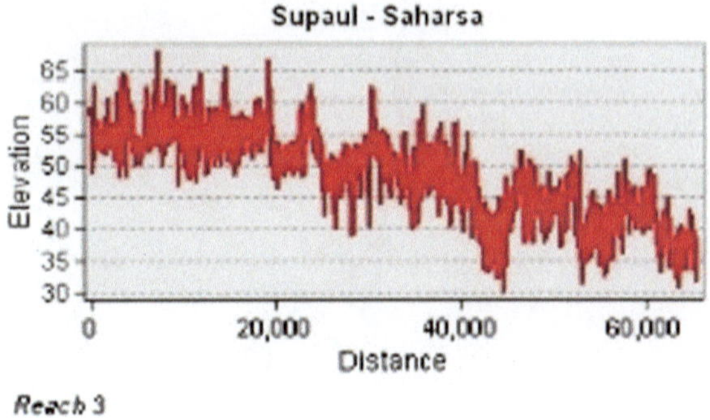


Fig. 20.22 Longitudinal profile of reach 3, 2018

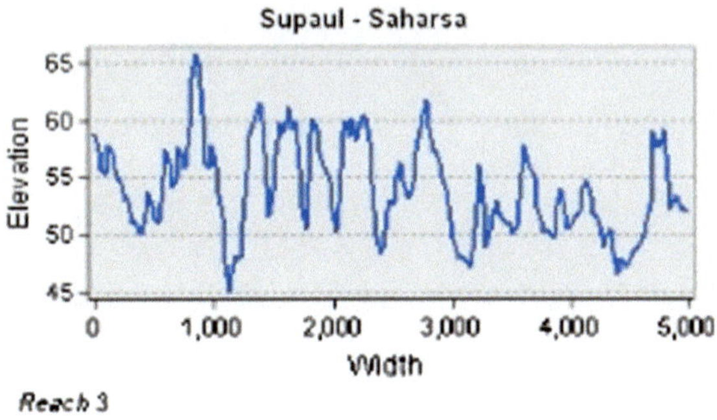


Fig. 20.23 Cross profile of reach 3, 1998

20.30, 20.31, 20.32, 20.33, and 20.34. The line was close to the eastern embankment in 1998 in most parts of the study reach. The maximum dispositioning in thalweg line is found to be at Bhimri-Novolia which is parallel to the Ganga River before meeting at Kursela. While in 2018 the thalweg line is seen moving towards the western embankment which is depicted red in color at reaches 2, 4, and 5.

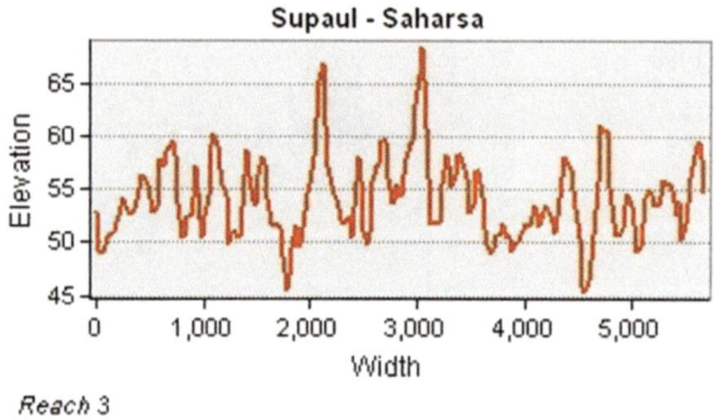


**Fig. 20.24** Cross profile of reach 3, 2018

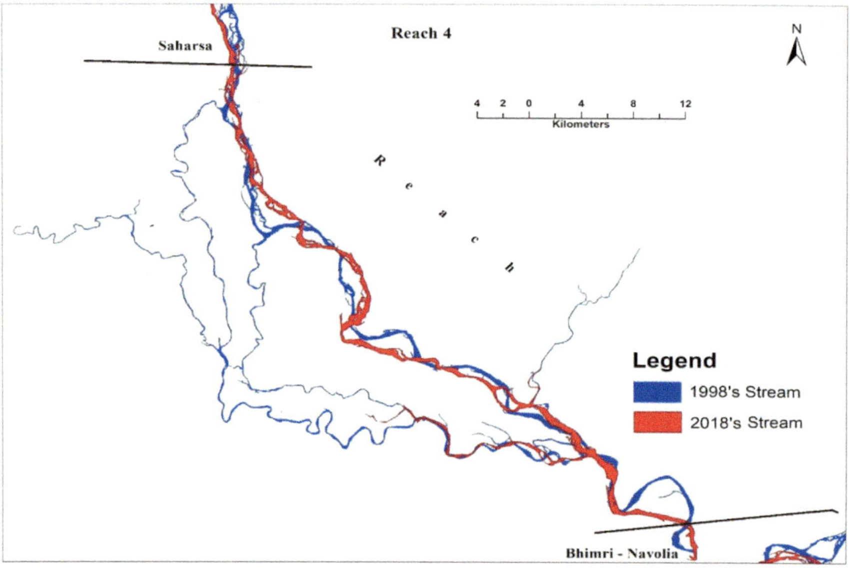


**Fig. 20.25** Reach 4 of the Kosi River from Saharsha to Bhimri

**Fig. 20.26** Longitudinal profile of reach 4, 1998

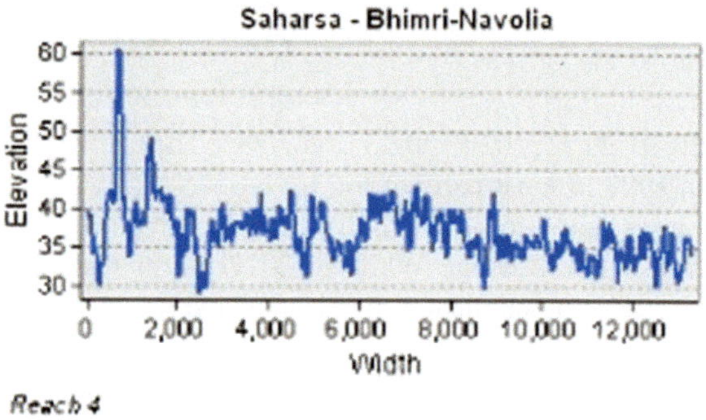

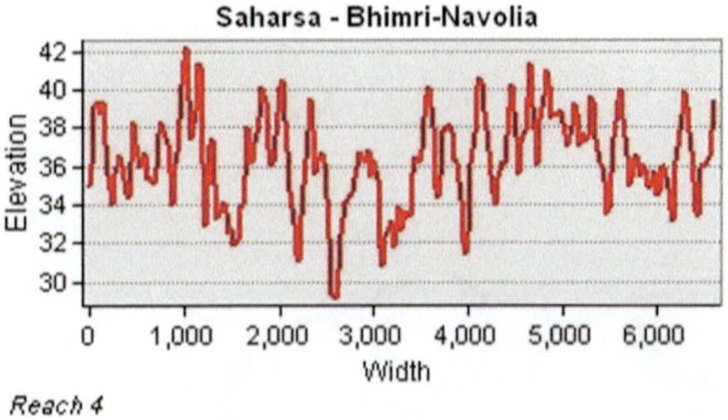


Fig. 20.27 Longitudinal profile of reach 4, 2018

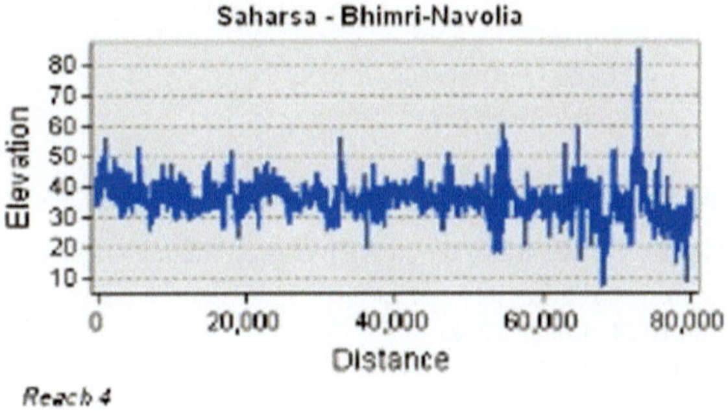


**Fig. 20.28** Cross profile of reach 4, 1998

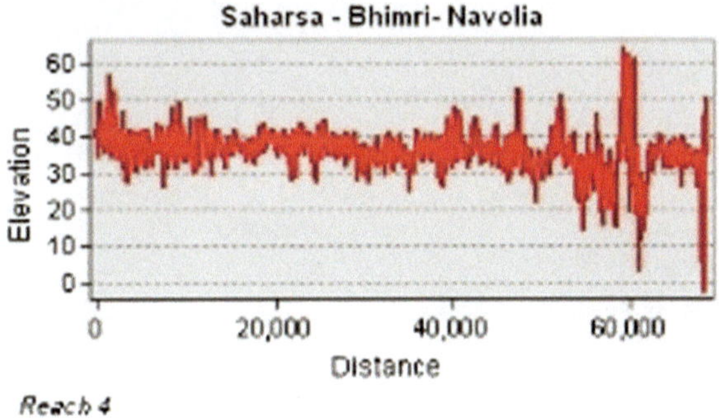


**Fig. 20.29** Cross profile of reach 4, 2018

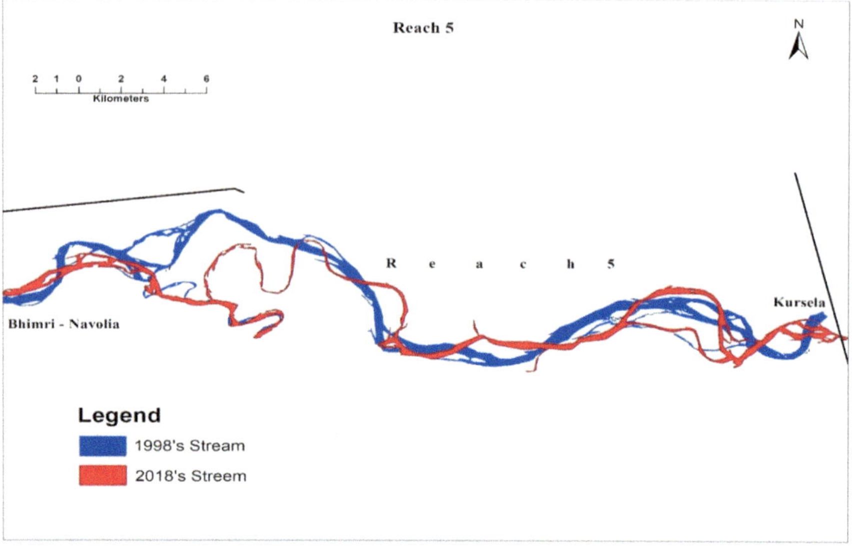


**Fig. 20.30** Reach 5 of the Kosi River from Bhimri to Kursela

**Fig. 20.31** Longitudinal profile of reach 5, 1998

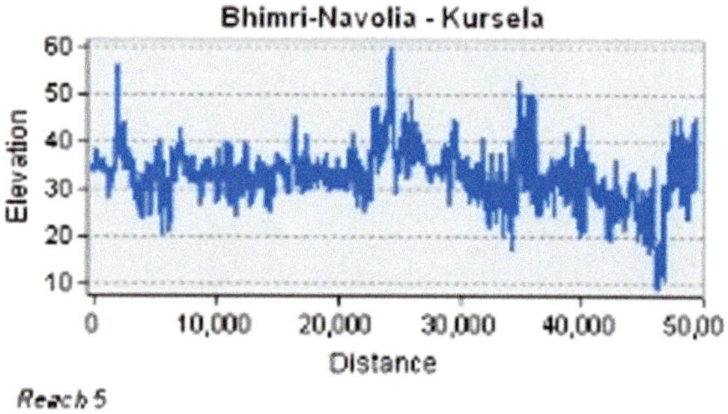


**Fig. 20.32** Longitudinal profile of reach 5, 2018

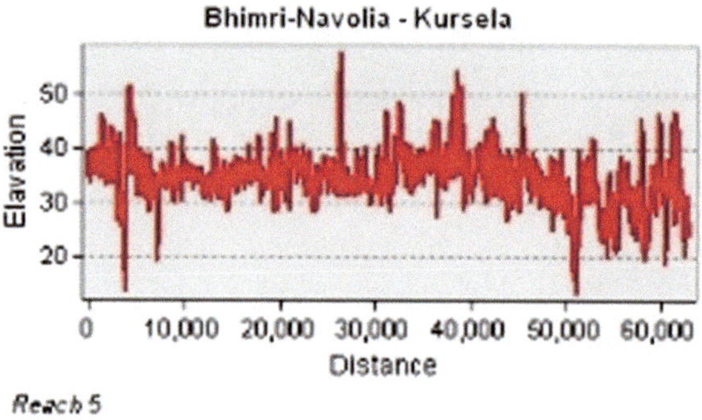

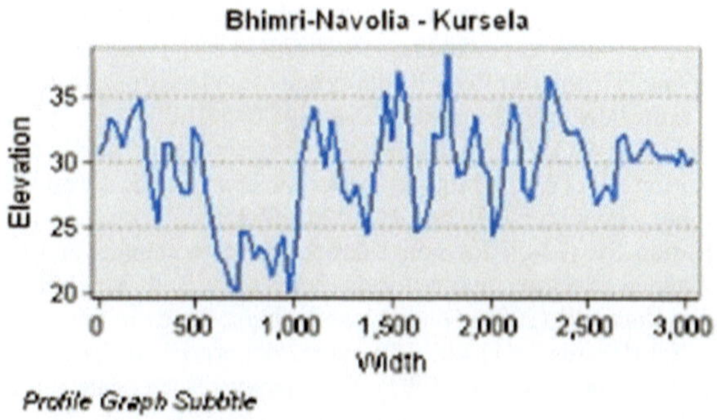


**Fig. 20.33** Cross profile of reach 5, 1998

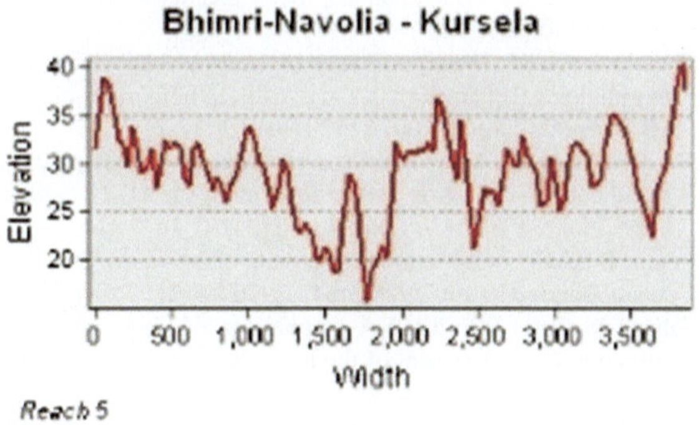


**Fig. 20.34** Cross profile of reach 5, 2018

**Acknowledgements** We are grateful to acknowledge the support provided by the Government of Bihar. ArcGIS Version 10.2 and Erdas Imagine are used to draw maps.

**Author Contributions** N.S. and R.K. conceptualized the central idea. R.K. carried out the analysis, wrote the paper, and drew all the figures. N.S. supervised the work. N.S. and M.M.M. edited and commented on the manuscript. All authors have read and agreed to the publication.

**Competing Financial Interest** There is no competing financial interest.

**Funding** This research received no external funding.

## References

Agarwal RP, Bhoj R (1992) Evolution of Kosi river fan, India: structural implications and geomorphic significance. Int J Remote Sens 13(10):1891–1901. https://doi.org/10.1080/01431169208904238

Arogyaswamy RNP (1971) Some geological factors influencing the behaviour of the Kosi. Rec Geol Surv India. Google Scholar, https://scholar.google.com/scholar_lookup?title=Some_geological_factors_influencing_the_behaviour_of_the_Kosi&journal=Records_of_Geological_Survey_of_India&volume=92&issue=2&pages=42-52&publication_year=1971&author=Arogyaswamy%2CR.N.P. Accessed 12 June 2020

Chen N et al (2017) Water hazards in the trans-boundary Kosi river basin. In: Geography. Springer, pp 383–408. https://doi.org/10.1007/978-981-10-2890-8_17
Friend PF, Sinha R (1993) Braiding and meandering parameters. Geol Soc Spec Publ 75(1):105–111. https://doi.org/10.1144/GSL.SP.1993.075.01.05
Gole C et al (1966) Inland delta building activity of Kosi river. cedb.asce.org
Horton RE (1932) Drainage-basin characteristics. Eos Trans Am Geophys Union 13(1):350–361. https://doi.org/10.1029/TR013i001p00350
Horton RE (1945) Erosional development of streams and their drainage basins; hydrophysical approach to quantitative morphology. GSA Bull. pubs.geoscienceworld.org
Kattelmann R (2003) Glacial lake outburst floods in the Nepal Himalaya: a manageable hazard? Nat Hazards 28(1):145–154. https://doi.org/10.1023/A:1021130101283
Kecerdasan I, Ikep P (1993) https://www.researchgate.net/profile/Rajiv_Sinha6/publication/237817264_Braiding_and_meandering_parameters/links/547e6c240cf2d2200ede9a0e/Braiding-and-meandering-parameters.pdf. Accessed 18 Sept 2020
Leier AL et al (2005) Mountains, monsoons, and megafans. Geology 33(4):289–292. https://doi.org/10.1130/G21228.1
Mahala A (2020) The significance of morphometric analysis to understand the hydrological and morphological characteristics in two different morpho-climatic settings. Appl Water Sci 10(1):1–16. https://doi.org/10.1007/s13201-019-1118-2
Neupane N et al (2015) Interlacing of regional water policies, institutions and agreements with livelihoods and disaster vulnerabilities in the HKH region: a case study of Kosi river basin. https://doi.org/10.1007/978-4-431-55242-0_14
Regmi B (2013) Long-term management of Kosi river basin, pp 381–390. ISBN 9783642291067. https://doi.org/10.1007/978-3-642-29107-4_21
Sahana M, Patel PP (2019) A comparison of frequency ratio and fuzzy logic models for flood susceptibility assessment of the lower Kosi river basin in India. Environ Earth Sci 78(10):1–27. https://doi.org/10.1007/s12665-019-8285-1
Sinha R (2009) The great avulsion of Kosi on 18 August 2008. JSTOR. https://www.jstor.org/stable/24112012?seq=1#metadata_info_tab_contents. Accessed 14 Jan 2020
Sinha R, Friend PF (1994) River systems and their sediment flux, Indo-Gangetic plains, Northern Bihar, India. Sedimentology 41(4):825–845. https://doi.org/10.1111/j.1365-3091.1994.tb01426.x
Sinha R et al (2019) Basin-scale hydrology and sediment dynamics of the Kosi river in the Himalayan foreland. J Hydrol 570:156–166. https://doi.org/10.1016/j.jhydrol.2018.12.051
Strahler AN (1957) Quantitative analysis of watershed geomorphology. Eos Trans Am Geophys Union 38(6):913–920. https://doi.org/10.1029/TR038i006p00913
Wells NA, Dorr JA (1987) Shifting of the Kosi river, northern India. Geology 15(3):204–207. https://doi.org/10.1130/0091-7613(1987)15%3c204:SOTKRN%3e2.0.CO;2

# Chapter 21
# Internet of Things and Machine Learning Application for a Remotely Operated Wetland Siphon System During Hurricanes

**Vivek Verma, Krishna Sai Vutukuru, Sai Srinivas Divvela, and Surya Srikar Sirigineedi**

**Abstract** Wetlands have been traditionally used to store water during rainfall events. However, during extreme events like floods and storm surge (during hurricanes), wetlands may exceed their storage capacity leading to overflow of water that sometimes results in flooding. This amplifies the problem as the purpose of wetlands is not fulfilled, and it also becomes difficult to visit the wetland site to improve upon the existing capacity. The flooding occurs because the wetlands are not used effectively. Therefore, an interconnected wetland system can be employed among the available wetlands for flood mitigation. This study discusses the latest methods to operate these wetlands remotely using machine learning (ML) techniques integrated with the Internet of Things (IoT). IoT is used to establish a secure connection between the siphon system and the control center. ML is used to check the robustness of the structural siphon system. The software interface is used to operate the siphon system manually or automatically. This provides an economical solution to access the wetland side and operate the siphon system remotely. This study provides direction to engineers/researchers with improving the horizon of such techniques for water resources applications.

**Keywords** Wetlands · Artificial intelligence · Machine learning · Internet of Things (IoT) · Hurricane · Siphon system · Flood control · Remote system architecture · Raspberry Pi

---

V. Verma (✉) · K. S. Vutukuru
Department of Civil and Environmental Engineering, Florida International University, Miami, FL, USA

S. S. Divvela · S. S. Sirigineedi
Department of Electrical and Computer Engineering, Florida International University, Miami, FL, USA



P. Kumar et al. (eds.), *Water Resources Management and Sustainability*, Advances in Geographical and Environmental Sciences,
https://doi.org/10.1007/978-981-16-6573-8_21

## 21.1 Introduction

The vision of machines communicating with each other dates to early 1800s. Machines have been providing direct communications since the invention of the telegraph (first landline) in the 1830s. In the early 1900s, first radio voice transmission (wireless telegraphy) took place hence providing a pathway to develop the Internet of Things (IoT). With the past century inventions of computers, the Internet, and Global Positioning System (GPS, hence satellites), the rapid advancement of IoT happened. Satellites and landlines provide necessary communication for much of today's IoT. The actual term "Internet of Things" was introduced by Kevin Ashton in 1999 during his work at Procter and Gamble. IoT is a system-defined when a large number (even in billions) of things/objects connected to the Internet. These objects can also interact/function autonomously. IoT's goal is to provide internet connectivity to a large number of small devices such as tracking devices, sensors, and others to collect and share data. This connection helps in decision-making, monitoring, and forecasting based on the collected data from a system. With the advent of IoT, the application of the Internet into various fields such as healthcare, automobile, agriculture, home automation, remote sensing, and others has been rapidly improving.

The IoT systems, in general, are developed for a specific application. However, recently, it has been diversified to build the IoT systems for different applications with barely any modifications needed other than replacing the sensors. The systems are developed to an extent where the IoT systems can be built and deployed with minimum programming knowledge. It is interesting to note that there are almost six billion devices (Table 21.1), which is more than the human population. From Table 21.1, the primary user of IoT is the utility industry, and the secondary user is security device industry, in the form of intruder detection and web cameras.

**Table 21.1** IoT market (in billions of units)

| Segment/Year | 2018 | 2019 | 2020 |
|---|---|---|---|
| Utilities | 0.98 | 1.17 | 1.37 |
| Government | 0.40 | 0.53 | 0.70 |
| Building automation | 0.23 | 0.31 | 0.44 |
| Physical security | 0.83 | 0.95 | 1.09 |
| Manufacturing and natural resources | 0.33 | 0.40 | 0.49 |
| Automotive | 0.27 | 0.36 | 0.47 |
| Healthcare providers | 0.21 | 0.28 | 0.36 |
| Retail and wholesale trade | 0.29 | 0.36 | 0.44 |
| Information | 0.37 | 0.37 | 0.37 |
| Transportation | 0.06 | 0.07 | 0.08 |
| Total | 3.96 | 4.81 | 5.81 |

*Source* Gartner (2019)

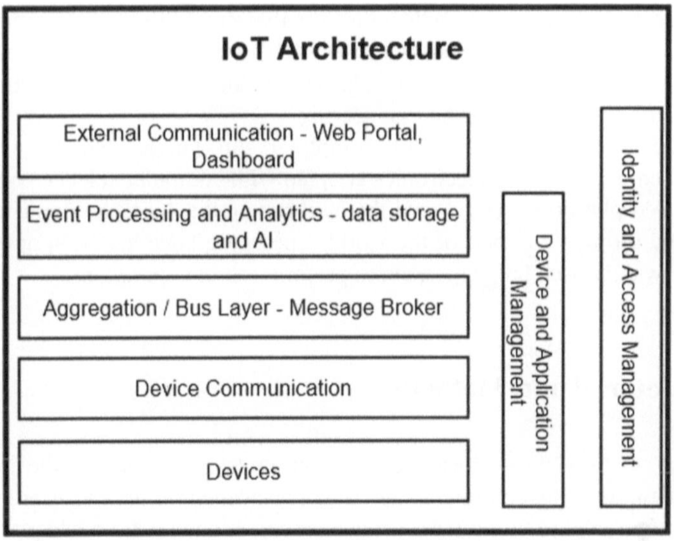


**Fig. 21.1** IoT architecture basic framework

It is predicted that building automation—like connected lighting—and automotive (connected cars) and healthcare (monitoring of chronic conditions) industries would be the highest users of IoT systems (Gartner 2019).

There are five fundamental layers of an IoT system, as shown in Fig. 21.1:

(1) Sensors/devices: Sensors or devices are the layers that help you to collect live data from the device environment. This device may have a simple one sensor or can hold multiple sensors like a smartphone.
(2) Device communication: This layer helps devices in sending all the collected data to cloud infrastructure. Wi-Fi and WAN are some examples of such communication devices.
(3) Aggregation layer: This is an intermediate layer that gathers traffic from the above devices. Typically, a message system (that depends on the protocol: like MQTT) will generally be present here. The collected data get accumulated and would be forwarded to respective destinations.
(4) Event processing and analytics: The data received from sensors will be analyzed and processed here, and data will be stored.
(5) External communication: This is the last step of IoT architecture. During this step, the collected, analyzed, processed, and stored information would be provided to the end-user. It is usually done using a windows app, a phone, or an interface.

With the advent of modern technology, such as 4G and 5G, internet penetration has been possible in the remotest of the areas. This book chapter discusses the amalgamation of modern technologies such as 4G connection using Raspberry Pi in the Azure IoT platform. After that, application of the machine learning and artificial intelligence on the proposed architecture for the remote operation siphon system which can be employed in various purposes such as maintaining water quality, flood mitigation, transfer of water from one place to another and other applications in the field of water, and environmental engineering.

The outcome of this study is a multidisciplinary effort in the field of civil and environmental engineering and electronics coupled with computer science engineering. An integrated platform is developed that will run on the 4G technology, which can be accessed through any part of the world. The entire code has been developed in one of the widely used languages, c#, to perform remote operations.

## 21.2 Material and Methods

IoT system need not always be a one-way communication. Based on the complexity of the system, two-way communication might be needed for the end-user to perform specific actions. Not providing two-way communication in individual systems might create dangerous effects. For instance, to have a controlled temperature and humidity inside an enclosed space, the user should be able to control the HVAC system with the click of a remote button or even a mobile phone.

From the above sections, it is evident that the data collection is a critical step for the application of IoT. The sensors are usually used in many fields to obtain data. Such data obtained from the sensors are used to create a decision architecture. Once a decision architecture is laid using an IoT system for a particular field using sensor data, the same IoT system can be used in various other field applications just by changing the sensors. Changing these sensors would provide the user with new data, and this can help in the decision system. For example, a decision to trigger water provision into an agricultural field at a specific location can be based on two parameters of moisture in the soil and ambient temperature. The values of moisture and temperature can be obtained by using respective sensors. These sensors can be connected, and a decision framework based on data from both these sensors can be laid. Once an IoT system is set up for this application, a similar framework can be used for wetlands (Verma et al. 2020a, b). In this case, the water level becomes one parameter, and water quality becomes another parameter. Since, it is a similar decision framework just by replacing temperature and moisture sensors using water level and water quality sensors.

Data collection can be done via Raspberry Pi. It is an ultra-tiny computer that was initially developed with the goal of education. It is made for the use of everyday use regardless of computer expertise hence making it an easy and versatile multipurpose device. Raspberry Pi is a multipurpose low-cost advanced reduced instruction set computer machines (ARM) developed by the University of Cambridge. There are several versions of Raspberry Pi in use, and they run on the most popular operating systems like Android, Linux, and Windows 10 IoT core in our experiment; we have used the Raspberry Pi 3 B+.

Middleware, as the name suggests, acts as an intermediate software and manages data, i.e., collects the data from various sensors, integrates it, provides a platform to analyze and communicate the decision to the external sensors. This whole process makes middleware the smartest component of IoT. The middleware platform is typically developed for managing several sensors and analyzing data from these sensors. The essential aspects of developing middle software are to ensure that this is user-friendly without compromising security as multiple systems are involved. Infiltration into the middleware is the most crucial security concern as middleware has data and access to all the systems used in the project. The current Internet architecture cannot be efficiently used as the communication protocol of the hypertext transfer protocol (HTTP) in the presentation layer uses too many resources. So, the existing method of developing middleware platform is product specific; hence, developing a common middleware platform is suggested by literature (da Cruz et al. 2018). The capability of IoT can be increased by facilitating the operation and creation of IoT systems across IoT, cloud, and edge infrastructures with vast heterogeneity, scalability, and dynamicity. Also, after the development of the IoT system, the actual application and deployment need additional resources and can be done remotely. This is made possible by the use of cloud computing. The problems and solutions like trustworthiness, security, and reliability can be tackled by using recently invented DEPO4IOT (Nguyen et al. 2019) which includes use of general purpose language (GPL) instead of domain specific language (DSL), introducing cloud computing into modern IoT systems.

Secure communication of sensitive data is typically carried using cryptography; however, these cryptography techniques utilize large number of resources. Hence, IoT systems cannot use conventional cryptography techniques due to their large resource utilization. Development of light-weight cryptography techniques is important step to achieve a secure IoT system (Lohachab et al. 2020). The communication between the server and hardware installed in the field is be done via Raspberry Pi. Raspberry Pi employs Azure IoT platform to make such communications. The data of these operations will be collected and analyzed using machine learning and artificial intelligence.

As shown in simple examples of Sect. 1, IoT can be used to provide a triggering mechanism for simple decision-making processes. However, having this basic framework, it can also be extended to a complex process such as auto-pilot mode of an automobile system. In case of an automobile system, different sensors with continuous feedback make this system complicated and typically these decisions are taken by experts in the field. However, with the advent of machine learning (ML) algorithms, these decisions can be obtained by feeding the data into these ML techniques. Using the techniques such as IoT, machine learning, cloud computing etc., the ability to remotely operate any system without being physically present on the site. The losses from these extreme events like hurricanes might lead to water intrusion and damage to internal components as mentioned in the literature (Vutukuru et al. 2019), and these damages can be studied in existing laboratories (Chowdhury et al. 2018). However, this integrated siphon system with remote operation can help save billions of dollars in these losses.

Usage of ML algorithms is increasing in the past decade and is expected to have an exponential increase in earth and environmental model studies (Schmidt et al. 2020). This vast increase is due to the availability of the diverse datasets. Modern day computational power and advancement of ML algorithms made it easy to apply them in the abovementioned fields. ML is proven to have predictive accuracy, but usage of ML algorithms to make decisions is hard to find. Due to recent data collection, there are large datasets available in various fields. ML can be applied to the data to analyze and make sense of it which otherwise would take a large number of resources and huge expertise. ML techniques and statistical techniques are considered to be approximate functions, but ML techniques are non-parametric with few assumptions made about the data. ML therefore needs a large amount of computational power which is easily available with advancement of computers.

The application of single ML technique as a regular tool in engineering has not had a significant progress mainly due to two reasons. Application of ML is complex than what is described in the earlier studies and engineering problems are too complex to be handled by a single ML algorithm. With better understanding of an engineering problem and application of multiple ML algorithms, there is a scope to solve the problems (Hornik 1991; Cybenko 1989). The data that are used for application of ML could be from various sources. This data often would not fit in to the ML algorithm without modification. Therefore, there is a need in preprocessing the data collected to be used as a source to the ML algorithm for concurrent results. The original data might have a large number of attributes which cannot be fed directly to a ML algorithm. The preprocessing then becomes a significant stage as this might have a huge impact on the performance of the ML algorithm. It is known that the performance levels are also dependent upon the parameters chosen as an example in neural networks which is a ML algorithm parameter selection is critical since some of them determine the topology of network and others the performance of the algorithm.

ML methods can in theory can correlate the relationship between inputs and outputs. It is possible to have less non-uniqueness in the ML methods. The research so far has proven to have high prediction accuracy than linear regression for a ML model. Due to the increasing amount of data collection from climate change and a large impact on environment with human intervention ML methods can employed to analyze. ML algorithms like artificial neural network (ANN) and (RF) algorithm can be used to study the impact. The existing research on ML algorithms is shown to have higher predictive performance than traditional statistical models like multiple linear regression. A case study on flooding events in Germany for prediction of the flooding events was done based on ML methods. Data from 30,000 flood events across Germany which was collected over the period 1950–2010 was analyzed for the magnitudes. There is a need for very deep understanding of the underlying relevant predictors for the flood patterns as they are expected to change with climate. There is a significant impact on these events due to the precipitation recent studies emphasize there is a role of antecedent soil moisture conditions in overall magnitude of the flood events. Permutation feature importance (PFI) was applied on the flood events to find the important predictors. Linear regression model and two ML models, i.e., RF and ANN were used to analyze the flood events it is seen that both the ML models outperformed linear regression which is a statistical model in terms of prediction accuracy, while RF achieved the higher prediction than ANN.

The use of ML techniques in parallel with IoT is used in hydrological models. Such models to examine forecasting include conceptual Sacramento Soil Moisture Accounting (SAC-SMA) model, data assimilation (DA), time-series deep learning (DL), catchment attributes and meteorology for large-sample studies (CAMELS), long short-term memory (LSTM) network (Feng et al. 2019). DA is mathematical technique where the observed data are combined with the output from a numerical model to achieve better estimate of an evolving system. DL as the name suggests utilizes large amount of data to directly learn response patterns. This does not require manually designed forecast models and hence making it less vulnerable to modeling structural errors. DA has a very special place in forecasting because it functions by connecting between the internal states of a process-based model and obtained variables. It updates the internal states based on this difference, and this process is reiterated until a convergence with required accuracy is achieved. While it was observed that DA had a profound place in modeling streamflow forecasting, as it works by establishing the covariance between the internal states of a process-based model and observed variables, and using the difference between the observation and the model-simulated variable(s) to update the model internal states. Some variants of the DA algorithm can also help to correct model structure equations given some prior information. DL models directly learn response patterns from massive amounts of data, without requiring manually designed features or making strong structural assumptions and are hence less influenced by model structural errors. For improving the forecasts, researchers also used LSTM-based methods that can efficiently validate discharge observation types, hence improving the overall quality of forecasts (Feng et al. 2019). Researchers also thought of comparing different versions of data integration (DI). For example, Feng et al., tested, analyzed, and compared two methods of data integration (DI): A simple method that accepts lagged observations as one of the inputs and the other where a recent observation time-series is passed as an input through a convolution neural networks (CNN) unit. CNN is a class of deep learning based on artificial neural networks.

The procedure of integrating recent observations to improve forecasts using a deep learning network is referred to as DI. The primary difference between DI and DA is attributed to the fact that DI does not utilize a dynamical model (pre-calibrated in most of the cases) as DA does (Feng et al. 2019). Instead, this method adds the data injection and prediction/forecasting into a single step. The advantage of DI over DA is that it alters internal LSTM states thereby improving future variable prediction, however it cannot predict unobserved variables.

LSTM projection models without DI showed promising results and were comparing well with the extensively calibrated operational models especially in snow-dominated and mountainous regions. However, the problem of base flow bias, hydrologically arid basins, basins with large storage changes every year, and regions with complex hydrogeology and inter basin transfers still persists. But, by applying the DI to the LSTM model, it addressed most of these problems and produced the highest forecast national-scale NSE value of 0.86 ever seen. Addition of DI also provided spatially varying and wide-spread benefits. The largest benefits of addition were recorded for basins with strong flow autocorrelation. This autocorrelation can be obtained in physical terms by either strong surface water retention or storage surface connections. The above explanation proves that integrating LSTM results in a very strong model than simpler statistical models, such as simple feedforward neural networks or auto-regressive models or simpler feedforward neural networks, for streamflow prediction of both mean and peak flow.

From the above, we see ML models are being applied in these engineering problems for better outcome. When these models are to be applied to an IoT, base live systems computation for the ML seems to be challenging. Cloud computing could be leveraged to solve this problem. Cloud is used to refer Internet, and computing is a process where computer technology is used to perform a task to achieve a goal. Cloud computing therefore means performing the computations on Internet. IoT-based systems can be hosted in cloud so as to apply the ML models on the live data collected from the fields through sensors. The data once analyzed using modeled ML algorithm can then be utilized to make decisions, trigger actuators. There are different models in cloud computing depending on computation, storage, and infrastructure services (Anawar et al. 2018). One has to choose the ideal model depending on the ML model employed as computation, and data requirements for each ML model are different.

Cloud computing is a great solution for applying ML models to IoT systems. It becomes increasingly difficult when large group of sensors are integrated due to the amount of data that is to be shared to cloud for computation. When an IoT system has to be deployed with minimal bandwidth for Internet development of such infrastructure has challenges. One of those challenges can be solved by utilizing the computational power available at the node, i.e., near the sensor. Generally, sensors are integrated to devices such as Raspberry Pi to collect and share the sensor data to cloud. As we know, while applying a ML model, the data obtained from the sensors are preprocessed for removal of noise and to fit the ML model. These preprocessing computations can be performed on Raspberry Pi which decreases the need to share huge data and huge computational power on cloud.

### *21.2.1 Server Windows App*

So far, we have discussed the client and the Azure IoT data pipeline. Now, it's time to discuss a centralized application from which all the device sensors monitoring will be done. There are many frameworks available in the market which can be used to develop these server applications either they can be done using desktop application or Web application. Some of the most popular frameworks are DOT NET, JavaFX which have been in use in industry. However, we used Win-forms DOT NET to develop our desktop application. As said earlier, we have used Win-forms Dot Net to develop our application to maintain a clear separation between the user interface and back-end data processing layer we have used model view presenter (MVP) pattern and IoT architecture as our application has to constantly communicate with the Azure IoT hub.

### *21.2.2 MVP Pattern*

This is the most popular architecture used in Win-form application development. Apart from this model-view-view model (MVVM) and model view controller to develop with Windows Presentation Framework (WPF) and ASP.NET.

### *21.2.3 Model*

The model layer involves the actual structure of data and a part of business logic of the application. This layer is decoupled from view with an extra layer in between called the presenter.

### *21.2.4 Presenter*

The presenter layer orchestrates activities between the model and view. It interacts with the model and notifies the changes to the view and vice versa it takes the commands from view and passes it to the model layer.

### *21.2.5 View*

The view layer provides the front-end graphical user interface (GUI) where the user can interact, view, or can make modifications to the system. In simple terms, this layer provides the visual representation of the presenter and model. This layer has certain events which are passed by the presenter. So, whenever data objects are updated in the presentation layer, the view will be notified to update the UI with latest information.

## 21.3 Results and Discussion

This sections first discusses about various storage units and then deals with the hardware configuration for the system. After that, software interface is discussed in detail which controls the entire system. Thereafter, communication between hardware devices and interface is discussed that is achieved using Azure IoT. Finally, this section ends with the result of the prototype test performed in the field.

Conventional storage systems for flood control include wetlands, detention ponds, and reservoirs. Wetlands are transitional zones between terrestrial and aquatic ecosystems where the water table is usually at or near the surface or the land is covered by shallow water. Wetlands provide many valuable services including fish and wildlife habitats, enhancement of water quality, recharge of groundwater table, and flood attenuation (Tsihrintzis et al. 1998; UNESCO Digital Library 1998; Juliano and Simonovic 1999). With respect to reduce flooding risks, they intercept storm runoff and could provide both short- and long-term water storage functions (Powers 2004). They retain excess water associated with heavy rainfall or snowmelt events, thus allowing for more water infiltration into the ground and reducing river flow downstream (Keddy 2010; N. R. Council 2001). However, their effectiveness is constrained due to their limited storage capacity, and the fact that some or all of this capacity may be occupied when a flood is imminent.

A detention pond is a low-lying area that is designed to temporarily hold a set amount of water while slowly draining to another location. However, the drainage is very small, and they don't have gates. Detention ponds normally are designed to release the water through an outlet control structure over a period of 2–5 days (Wet Ponds 2016). The outlet control structure of existing detention ponds could be retrofitted (e.g., increasing diameter of outlet pipes and adding actuated gates).

A reservoir is a natural or artificial storage space for fluids, which constitutes one of the most frequently used structural means of managing flood events (Guo et al. 2004). The basic function of a flood mitigation reservoir is to minimize flood peaks and water surface elevations at downstream locations. The discharge from a storage reservoir is regulated by outlet gates operated based on control rules and/or engineers' judgment. Furthermore, reservoirs are not normally operated based on a system approach but on an individual basis.

Overall, the outlet structures of conventional storage systems could be retrofitted (or added in the case of a wetland) to improve flood mitigation. For wetlands, it would be necessary to add siphons and/or trenched drains with controlled gates. For detention ponds, the retrofitting would include using larger outlet pipes equipped with controlled gates. For reservoirs, the retrofitting would include adding gates and the remote operation of the gates. In a similar way to the case of wetlands, siphons could also be used for detention ponds and reservoirs.

The schematic of a prototype of an automated and remotely operated siphon system is illustrated in Fig. 21.2. As shown in this figure, the siphon system includes four water level switches (components 1, 2 installed in the vertical section of the pipe and 3 and 4 that is installed in the storage system) to indicate the water level in the system, check valves (component 7) to prevent the backflow of the water from the siphon system, a bilge pump is employed to fill the entire pipe system with water (component 8), a remotely controlled gate (component 6) with ball valve at its end, an air vent with solenoid (component 5) to release the air pressure present in the pipe

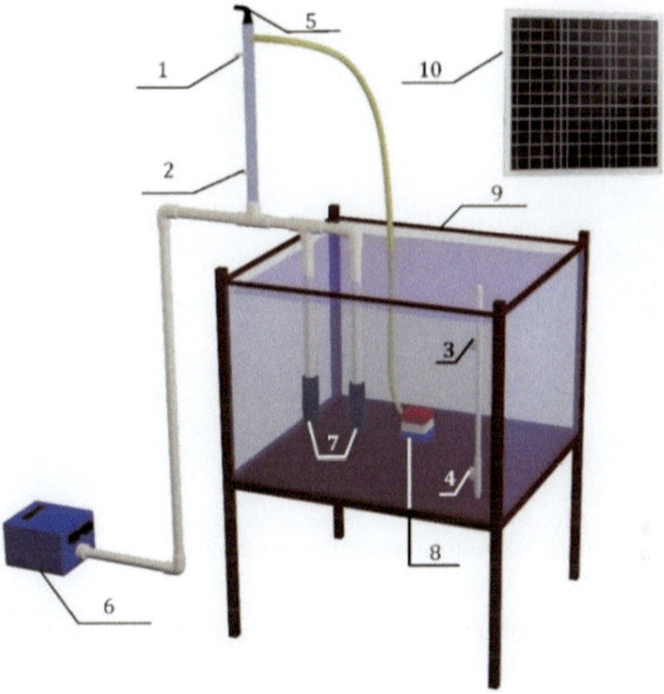


**Fig. 21.2** Hardware of siphon system in Leon and Verma (2019)

and a solar panel is installed (component 10) to recharge the batteries so that it can run by itself without using any conventional source of energy.

An interface, as shown in Fig. 21.3, is developed to remotely control all the hardware devices that are installed in the field. Right side of the interface illustrates the entire setup in a concise manner. As explained earlier, the pump is installed in the wetlands to prime the pipe so that siphon can come into action instantaneously. The priming is controlled automatically with the help of upper and lower switches installed in the vertical pipe. Air vent releases the pressure developed inside the pipe when the pump primes the system, i.e., pump and air vents operate and close at the same time. Once the system is primed whenever it receives the command to open the ball valve, water will start flowing at that instant from storage units such as wetlands and ponds to outside. Three colors, green, red, and gray, have been employed to check the status of the hardware installed. The green color resembles that the hardware device is in operation, or the water is at or above that level switch. Similarly, gray means at that point of time the device is not in use or water is below that level switch. The red color is employed when the siphon system encounters any problem. The current interface resembles twenty siphons and at least five controlled gates. All the siphons and automated gates can be controlled manually using the time interval box. It represents the time after which the siphons and or gates should

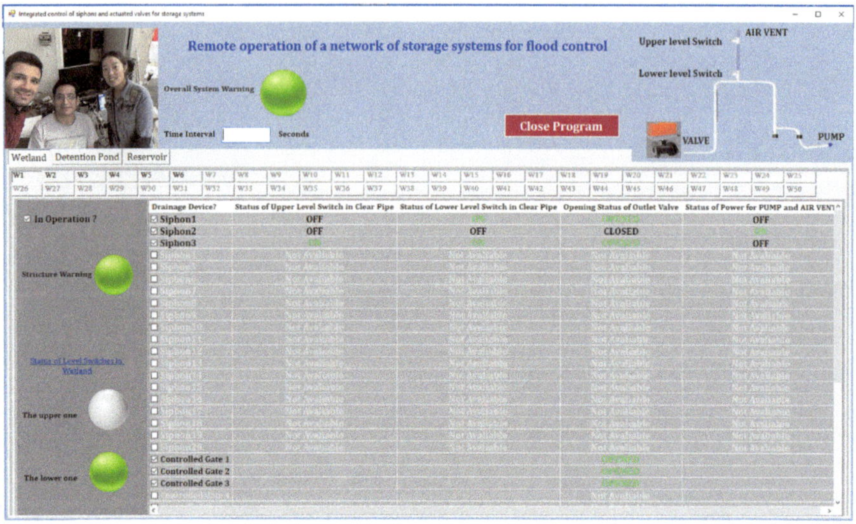


Fig. 21.3 Software interface of siphon system (Qin et al. 2019)

operate automatically either they open up or close as per the command received from the user. The storage systems which compromise of reservoir, reservoir and detention pond can be selected from the tabs in the interface on the left side of the interface. Thereafter, in this case, as wetland tab is selected thus total number of wetlands are visible, in this case it is fifty. The interface can either be closed using x mark on the top right side of the interface or by selecting close program which is present in the middle of the interface.

## *21.3.1 Azure IoT*

Azure IoT hub is a fully managed service that helps enable reliable and secure bi-directional communications between millions of devices and a solution back end. The send telemetry from a device to an IoT hub quick start shows how to create an IoT hub, provision a device identity in it, and code a simulated device app that sends device-to-cloud messages.

The below codes are illustrated with their functions about how Azure IoT is employed to obtain data of remotely operated siphon:

- SimulatedDevice.py, a modified version of the app created to send telemetry from a device to an IoT hub, which connects to your IoT hub and receives cloud-to-device messages.
- SendCloudToDeviceMessage.py, which sends cloud-to-device messages to the simulated device app through IoT hub.

The primary coding is done in python 3.7 programming language. It is important to make sure that port 8883 is open in your firewall. The device sample uses MQTT protocol, which communicates over port 8883.

The Azure IoT hub device SDK for Python is installed using the following codes in the command prompt:

```
cmd/sh
•  pip install azure-iot-device
•  Using a text editor, create a file named
SimulatedDevice.py.
•  Add the following import statements and variables at
the start of the SimulatedDevice.py file:
Python
•  import threading
import time
from azure.iot.device import IoTHubDeviceClient

RECEIVED_MESSAGES = 0
•  Add the following code to SimulatedDevice.py file.
Replace the {deviceConnectionString} placeholder value
with the device connection string for the device you
created in the Send telemetry from a device to an IoT
hub quickstart:
Python
•  CONNECTION_STRING = "{deviceConnectionString}"
•  Add the following function to print received
messages to the console:
Python
•  def message_listener(client):
     global RECEIVED_MESSAGES
     while True:
          message = client.receive_message()
          RECEIVED_MESSAGES += 1
          print("\nMessage received:")
```

```
        #print data and both system and application
(custom) properties
        for property in vars(message).items():
            print ("    {0}".format(property))

        print( "Total calls received:
{}".format(RECEIVED_MESSAGES))
        print()
```

- Add the following code to initialize the client and wait to receive the cloud-to-device message:

Python

```
•   def iothub_client_sample_run():
    try:
        client =
IoTHubDeviceClient.create_from_connection_string(CONNEC
TION_STRING)
        message_listener_thread =
threading.Thread(target=message_listener,
args=(client,))
        message_listener_thread.daemon = True
        message_listener_thread.start()

        while True:
            time.sleep(1000)

    except KeyboardInterrupt:
        print ( "IoT Hub C2D Messaging device sample
stopped" )
```

- Add the following main function:

Python

```
if __name__ == '__main__':
    print ( "Starting the Python IoT Hub C2D
Messaging device sample..." )
    print ( "Waiting for C2D messages, press
Ctrl-C to exit" )

    iothub_client_sample_run()
```

Save and close the simulated device.

Python console to send cloud-to-device message.

In this section, a Python console app is employed that sends cloud-to-device messages to the simulated device app. The device ID of the device is required that has been added in the send telemetry from a device to an IoT hub quick start. The IoT hub connection string is also required that is copied previously in get the IoT hub connection string.

In the working directory, with the help of command prompt, the Azure IoT hub service SDK for Python is installed:

cmd/sh
• pip install azure-iot-hub
• Using a text editor, create a file named SendCloudToDeviceMessage.py.
• Add the following import statements and variables at the start of the SendCloudToDeviceMessage.py file:
Python
```
• import random
import sys
from azure.iot.hub import IoTHubRegistryManager

MESSAGE_COUNT = 2
AVG_WIND_SPEED = 10.0
MSG_TXT = "{\"service client sent a message\": %.2f}"
```
• Add the following code to SendCloudToDeviceMessage.py file. Replace the {iot hub connection string} and {device id} placeholder values with the IoT hub connection string and device ID you noted previously:
Python
```
• CONNECTION_STRING = "{IoTHubConnectionString}"
DEVICE_ID = "{deviceId}"
```
• Add the following code to send messages to your device:
Python
```
• def iothub_messaging_sample_run():
    try:
        # Create IoTHubRegistryManager
        registry_manager =
IoTHubRegistryManager(CONNECTION_STRING)

        for i in range(0, MESSAGE_COUNT):
            print ( 'Sending message: {0}'.format(i) )
            data = MSG_TXT % (AVG_WIND_SPEED +
(random.random() * 4 + 2))

            props={}
            # optional: assign system properties
            props.update(messageId = "message_%d" % i)
            props.update(correlationId =
"correlation_%d" % i)
            props.update(contentType =
"application/json")
            # optional: assign application properties
            prop_text = "PropMsg_%d" % i
            props.update(testProperty = prop_text)
```

```
registry_manager.send_c2d_message(DEVICE_ID, data,
properties=props)

        try:
            # Try Python 2.xx first
            raw_input("Press Enter to continue...\n")
        except:
            pass
            # Use Python 3.xx in the case of exception
            input("Press Enter to continue...\n")

    except Exception as ex:
        print ( "Unexpected error {0}" % ex )
        return
    except KeyboardInterrupt:
        print ( "IoT Hub C2D Messaging service sample
stopped" )
•  Add the following main function:
Python
6.        if __name__ == '__main__':
7.            print ( "Starting the Python IoT Hub C2D
Messaging service sample..." )
8.
9.            iothub_messaging_sample_run()
10.
11.       Save and close SendCloudToDeviceMessage.py
file.
```

Thereafter, to listen for cloud-to-device messages as shown in Fig. 21.4, run the applications using command prompt in the working directory—shell.

Python SimulatedDevice.py.

To send cloud-to-device messages as shown in Fig. 21.5, run the following command in a new command prompt—shell.

Python SendCloudToDeviceMessage.py.

The messages received by the device is illustrated in Fig. 21.6.

**Fig. 21.4** Sample of the Python IoT hub

**Fig. 21.5** Sample of the Python IoT hub sending message

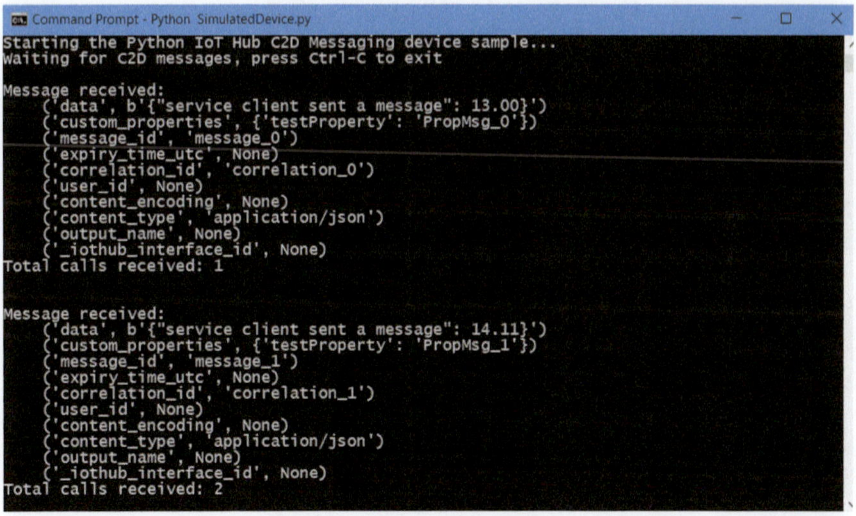


**Fig. 21.6** Summary of the Python IoT hub messages

### *21.3.2 Prototype Tests*

Figure 21.7 illustrates the siphon system installed in the field. All the hardware devices such as actuated ball valve, bilge pump, level switches, and air vent are installed in the field which are connected to the programmable logic control (PLC). The PLC collects all the data and send it to the control center where decisions are made based on the inputs received.

The pipe of the siphon system is six feet in diameter installed in the 2500 gallons container. The height of the container is about ninety-eight inch. The devices are installed in the container and the pipe to measure the water depth and discharge, respectively. The result is shown in Fig. 21.8 which illustrates the change of water depth (cm) with respect to time (s) and discharge (L/s) with respect to time. The actuated ball valve opens around 180th second, and the water starts releasing from the container. The maximum discharge was about 25 L/s which occurred around 300th second. The experiment stopped around 520th second which means that all

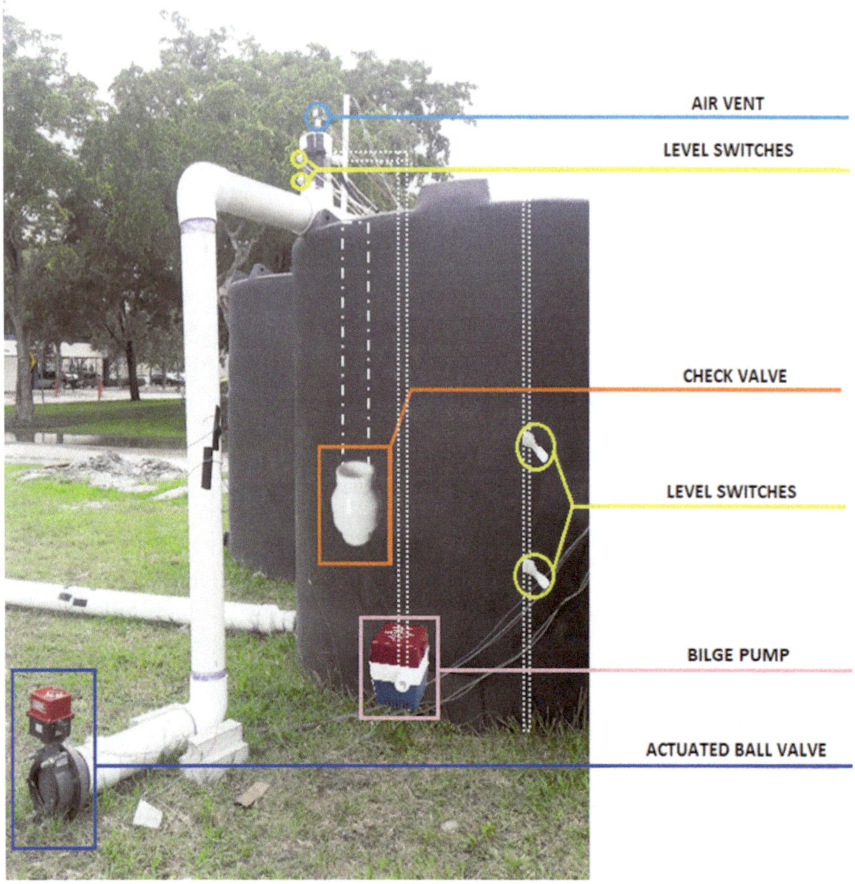


**Fig. 21.7** Siphon system setup in the field

the water is released from the container. The water depth versus time graph decreases with time as the water moves out from the container.

## 21.4 Conclusion

Water is stored conventionally using storage units such as wetlands, reservoirs, and ponds, and their management and distribution is done manually. This book chapter provides an automatic way that can upgrade the traditional and conventional methods of water utilization using latest technologies such as Internet of Things (IoT) and machine learning (ML). These technologies when combined with 4G cellular connection, the entire framework can be operated remotely using Raspberry Pi, hardware

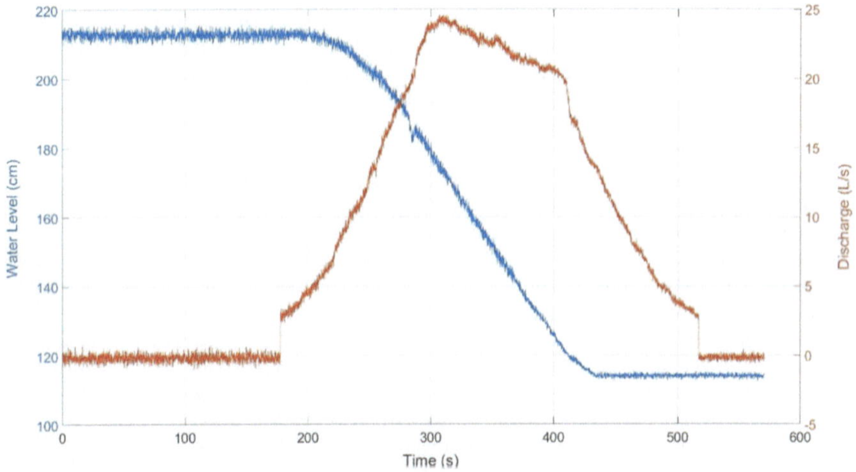


**Fig. 21.8** Water level and flow measurement data

devices installed in the field and software interface. The Azure IoT application, installed in Raspberry Pi, is employed to communicate between hardware devices and the control center. The Python code used for the communication is described in detail so that it may benefit others and can be used in other similar type of applications. The software interface updates itself at a regular interval and thus reflects the actual condition of the field in the real time. The framework once established can be controlled either manually and/or automatically according to user requirement. The book chapter also discusses several other uses of IoT and ML techniques applied in various other fields. The framework described here can be utilized for decision-making process, especially in the field of water resource engineering. The research opens new avenues for the engineers to upgrade the traditional system with the more robust, reliable, and easy to maintain system.

## References

Anawar MR, Wang S, Azam Zia M, Jadoon AK, Akram U, Raza S (2018) Fog computing: an overview of big IoT data analytics. Wireless Commun Mob Comput. https://www.hindawi.com/journals/wcmc/2018/7157192/?utm_source=twitter&utm_medium=organic_social&utm_content=EW-awareness-Multiple_portfolio-Multiple_journals-Global&utm_campaign=MECS_Org_SoMe

Chowdhury AG, Vutukuru KS, Moravej M (2018) Full- and large-scale experimentation using the wall of wind to mitigate wind loading and rain impacts on buildings and infrastructure systems

Cybenko G (1989) Approximation by superpositions of a sigmoidal function. Math Control Signals Syst 2(4):303–314. Google Scholar. https://scholar.google.com/scholar_lookup?hl=en&volume=2&publication_year=1989&pages=303-314&journal=Mathematics+of+Control%2C+Signals+and+Systems&issue=4&author=G.+Cybenko&title=Approximation+by+superpositions+of+a+sigmoidal+function

da Cruz MAA, Rodrigues JJPC, Al-Muhtadi J, Korotaev VV, de Albuquerque VHC (2018) A reference model for internet of things middleware. IEEE Internet Things J 5(2):871–883. https://doi.org/10.1109/JIOT.2018.2796561

Feng D, Fang K, Shen C (2019) Enhancing streamflow forecast and extracting insights using long-short term memory networks with data integration at continental scales. ArXiv. https://doi.org/10.1029/2019WR026793

Gartner (2019) Gartner says 5.8 billion enterprise and automotive IoT endpoints will be in use in 2020. https://www.gartner.com/en/newsroom/press-releases/2019-08-29-gartner-says-5-8-billion-enterprise-and-automotive-io. Accessed 05 Aug 2020

Guo S, Zhang H, Chen H, Peng D, Liu P, Pang B (2004) A reservoir flood forecasting and control system for China [Un système chinois de prévision et de contrôle de crue en barrage]. Hydrol Sci J 49(6). https://doi.org/10.1623/hysj.49.6.959.55728

Hornik K (1991) Approximation capabilities of multilayer feedforward networks. Neural Netw 4(2):251–257. Google Scholar. https://scholar.google.com/scholar_lookup?hl=en&volume=4&publication_year=1991&pages=251-257&journal=Neural+Networks&issue=2&author=K.+Hornik&title=Approximation+capabilities+of+multilayer+feedforward+networks.

Juliano K, Simonovic SP (1999) The impact of wetlands on flood control in the Red River Valley of Manitoba, p 75

Keddy PA (2010) Wetland ecology: principles and conservation. Cambridge University Press

Leon AS, Verma V (2019) Towards smart and green flood control: remote and optimal operation of control structures in a network of storage systems for mitigating floods, May 2019, pp 177–189. https://doi.org/10.1061/9780784482339.019

Lohachab A, Lohachab A, Jangra A (2020) A comprehensive survey of prominent cryptographic aspects for securing communication in post-quantum IoT networks. Internet Things 9:100174. https://doi.org/10.1016/j.iot.2020.100174

N. R. Council (2001) Compensating for wetland losses under the clean water act

Nguyen P et al (2019) Advances in deployment and orchestration approaches for IoT—a systematic review. In: 2019 IEEE international congress on internet of things (ICIOT), July 2019, pp 53–60. https://doi.org/10.1109/ICIOT.2019.00021

Powers TM (2004) The value of resources, an economic perspective on wetlands. In: Wetlands, pp 131–156

Qin L, Leon AS, Bian L-L, Dong L-L, Verma V, Yolcu A (2019) A remotely-operated siphon system for water release from wetlands and shallow ponds. IEEE Access 7:157680–157687. https://doi.org/10.1109/ACCESS.2019.2950270

Schmidt L, Heße F, Attinger S, Kumar R (2020) Challenges in applying machine learning models for hydrological inference: a case study for flooding events across Germany. Water Resour Res 56(5):e2019WR025924. https://doi.org/10.1029/2019WR025924

Tsihrintzis VA, John DL, Tremblay PJ (1998) Hydrodynamic modeling of wetlands for flood detention. Water Resour Manag 12(4):251–269. https://doi.org/10.1023/A:1008031011773

UNESCO Digital Library (1998) Wetlands of the humid tropics. https://unesdoc.unesco.org/ark:/48223/pf0000116084

Verma V, Bian L, Rojali A, Ozecik D, Leon A (2020a) A remotely controlled framework for gravity-driven water release in shallow and not shallow storage ponds, May 2020, pp 12–22. https://doi.org/10.1061/9780784482940.002

Verma V, Vutukuru KS, Bian L, Rojali A, Ozecik D, Leon A (2020b) Reliability and robustness evaluation of a remotely operated siphon system for flood mitigation during hurricanes, May 2020, pp 31–39. https://doi.org/10.1061/9780784482940.004

Vutukuru KS, Moravej M, Gan Chowhdury A (2019) Wind-driven rain intrusion reduction for shuttered windows

Wet Ponds 22Sept2016.pdf, chap 10 (2016) [Online]. Available https://files.nc.gov/ncdeq/Energy%20Mineral%20and%20Land%20Resources/Stormwater/BMP%20Manual/Ch10%20-%20Wet%20Ponds%2022Sept2016.pdf. Accessed 09 Aug 2020

# Chapter 22
# AHP Based Site Suitability Analysis for Water Harvesting Structure Using Geospatial Technique

**Chandra Shekhar Dwivedi, Raghib Raza, Arvind Chandra Pandey, and D. C. Jhariya**

**Abstract** Betul Chhindwara is undulating region experiences less water holding capacity due to hard rock terrain with maximum structural origin and high relief zone. Moreover, construction of major irrigation projects has limited scope due to topographical, social, financial, and environmental constraints. Hence, small structure like Check Dams, Percolation Tank, Bore Well, Nala bandh, and Farm Pond can be suitable structure for the water harvesting in various parts of Betul-Chhindwara region. The present study comprises of a four watersheds delineated in the region as Tapi River (WS 1), Kanhan river (WS 2), Pench River (WS 3), and Tawa River (WS 4) in Betul-Chhindwara district. The present study is carried out by GPS, Survey of India toposheet i.e., NF 43-8, NF 44-5, NF 43-12 and NF 44-9 on scale of 1:250,000, Aster DEM data for creation of DEM and satellite imagery of Sentinel-2. The criteria like third order stream, 2–5° slope of land, sandy-clay soil, Lineament zone, Physiography division is considered and various techniques of GIS and remote sensing are used for getting more suitable sites for rainwater harvesting. By applying above process, it was concluded that this area has limited scope of rainwater harvesting.

**Keywords** Watershed management · Geographical information system and remote sensing · Runoff estimation · Stream order · Site suitability for rainwater harvesting

## 22.1 Introduction

The water demand has significantly expanded because of the exponential ascent in human populace throughout the most recent couple of decades. FAO projects that water demand will double in the next decade, the planet is developing twice as rapidly

C. S. Dwivedi (✉) · R. Raza · A. C. Pandey
Department of Geoinformatics, School of Natural Resource Management, Central University of Jharkhand, Ranchi, Jharkhand 835205, India
e-mail: chandra.dwivedi@cuj.ac.in

D. C. Jhariya
Department of Applied Geology, National Institute of Technology Raipur, Raipur, India
e-mail: dcjhariya.geo@nitrr.ac.in

 
P. Kumar et al. (eds.), *Water Resources Management and Sustainability*,
Advances in Geographical and Environmental Sciences,
https://doi.org/10.1007/978-981-16-6573-8_22

as it estimates, and in that way water use is expected to double in the next decade (FAO 2015). Development, agriculture in an urban area, and energy demands of the growing population, as a result of mechanical advancements on land, means the country is experiencing a water scarcity. More than 1.8 billion people in the world are expected to experience water shortage by 2025, with the other two-thirds expected to deal with water pressure.

Due to different conditions, including environmental change, complex biological processes, and increasing water demand in urban areas, there are several difficulties for administrators and participants who hold the positions on the board in the field of water assets to use this tool in an integrated and holistic ways. Dry and semi-id districts are never can be uninfluenced either by the amount of precipitation. The ranchers are confronting real difficulties due to spatio-transient precipitation conveyance and erratic precipitation designs prompt climatic vulnerability and aridity circumstances. These issues influence the yield generation and expanding sustenance hazard. A feasible water asset the board in the agribusiness part is crucial for sustenance emergency and goes about as an impetus for financial improvement for the nation. Water reaping (RWH), a recharged consideration since 1980, is demonstrated as a sound way to deal with battle with water shortage issues for a huge number of centuries. Assuring the selective wellspring for enhancing waterways to address the water deficiencies issues in ASARs. RWH can be characterized as a framework for gathering, putting away, and rationing water overflow for horticulture in ASARs. RWH enabled water-related technologies to be used more effectively, particularly in desert environments, where it helps to limit groundwater extraction while maintaining crop productivity and to protect the water source and increase yield. Moreover, improve field development, support reforestation, increment nourishment generation, battle soil disintegration, and improve the usage of water assets. RWH energizes nearby groundwater saves and not just builds water accessibility, likewise upgrades work openings, improves financial most ASARs conditions (Ammar et al. 2016).

Liquidity in collection of water is ordinarily helped in determining water shortage issues, enhancing accessible water assets. While the WOCAT database is doing its research on increasing water access for rural and residential uses, we are incrementing our board's access to the World Overview of Conservation Approaches and Technologies (Mekdaschi and Liniger 2013). Be that as it may, recognizing potential locales for RWH is a significant advance toward expanding water accessibility and land efficiency in semi-bone-dry regions (Isioye et al. 2012). The reasonableness water harvesting estimations rely on a few criteria (Mahmoud and Alazba 2014) and various strategies and additionally techniques are accessible for reasonable destinations determination (Al-Adamat 2008; Ahmad 2013; de Winnaar et al. 2007). When attempting to select the best website location for a company, we must take into consideration both financial and geographic factors (Al-Adamat et al. 2010). In order to improve the water accessibility, you should investigate the factors that influence the basic strategy's ability.

When it comes to setting up competitive advantage, one needs to consider how well his or her company is able to tap into localized and niche markets, rather than locations in general (Al-Adamat et al. 2012). When using a site-appropriate risk rating scale, strategies such as prioritization and allocation are required to assemble an effective risk/opportunity mix. In the ASARs, two separate parameters (biophysical and financial) were examined. Many scientific studies come together to address biophysical parameters, for example, precipitation, waste structures, spreading, and form of property, to try to answer the question of what affects microclimate (Kadam et al. 2012; Kumar et al. 2008), and "entails integrating the biophysical sections (separation to settlement, separation to the town, to the degree of thickness of population, separation to resources) with the financial values (length of time in financial systems, enrollment in financial systems, penetration into the locality, target audience depth, financial return)" (FAO 2003; Kahinda et al. 2008; Ziadat et al. 2012; Bulcock and Jewitt 2013; Krois and Schulte 2014). These systems combine several criteria, such as remote sensing and water resources for example, with biophysical aspects such as GIS and MCA (Bamne et al. 2014; Al-Shamiri and Ziadat 2012), Hydromorphological demonstrating (HM) by remote sensing and GIS (Mahmoud and Alazba 2014), integrated with HM and/Integrated with multi-critic assessments (Krois and Schulte 2014; Khan and Khattak 2012; Elewa et al. 2012; Weerasinghe et al. 2011; Sekar and Randhir 2007). The McGuire Program on an MCA agreement was executed with the GIS project (Sayl et al. 2017; Kahinda et al. 2009; Jothiprakash and Sathe 2009; Pauw et al. 2008; Ahmed Ould Cherif et al. 2007; Mbilinyi et al. 2007). All techniques and devices utilized in past research ponder identified with site choice for RWH have a few restrictions, yet GIS/RS device is an initial step application device for reasonable destinations ID while for progressively precise outcomes and information rich districts, the reconciliation of multi criteria analysis and based on GIS HM is very suggesting techniques or additionally devices. MCAs compared to GIS, there is a good chance for high location-specific content in RWH site (Ammar et al. 2016). This investigation essentially centered on the ideal site determination for RWH in Guatemala, where the dominant part of the populace relies upon the horticultural area, which speaks to around 10% of Gross domestic product. Besides, master feelings, the physical and financial qualities were considered, and the ideal locales were distinguished in an AHP with GIS for the examination region. All things considered the investigation watershed has huge region with constrained information, so site choice procedure may end up complex while thinking about every one of these components. GIS and RS also allow site determination (preliminary resource surveys for RSH) (Ziadat et al. 2012; Bulcock and Jewitt 2013). Sayl et al. (2017) used GIS, for example, in their attempt to understand various RWH locations. They studied the hydrology and geography of various RWH to recognize differentiating options.

## 22.2 Methodology

Geographic Information Systematic Analysis (GIS) will choose the best choices for Regional Water harvesting Strategies (RWH) and utilizes an analytical hierarchy.

Most of the heavy lifting for design thinking about HP issues has already been completed in many researches examines for the recognizable proof of potential RWH destinations (Krois and Schulte 2014). In AHP strategy, the comprehensive options are calculated and evaluated to aid specialists. The main code of AHP is to take issues and create connected symbols. The original goal should be at the international and regional levels, and the induction programmed should have to follow on in progressively specific terms before they finally reach the fundamental objective. Most of the pairwise correlations in the AHP technique enable a network to estimate the loads of any model. A given target will be rated appropriately, which includes two variables: importance and relevance, supported by pairwise comparisons. We use the 9-point nonstop scale to measure all criteria. The odd numbers 1, 3, 5, 7, and 9 are alike, just slightly, but also strongly and very, and the case mod numbers 2, 4, and 8 are moderate.

This kind of framework gets built up using an approach that combines elements of both scales-up and vertical integration. Ensemble design; (i) classification; (ii) generate weightage criteria; (iii) the creation of AHP; (iv) determining the relative weights; (v) evaluating the measure.

### *22.2.1 Selection Criteria*

They chose method to portray with elements, characterize the suitable site for gathering water. From 1990s, a few investigations principally centered on various multi criteria including seepage organize, types of soil variable, slant, precipitation, and land use. RWH in any event, the level of fair funding adjusted to incorporate two important factors: biophysical and financial requirements (Ammar et al. 2016). RWH having concepts in geographic conceptually and having six classes (socio-financial aspects, soil, hydrology, geography, climate, agronomy, and soil, hydrogeology, and air) (Kahinda et al. 2008). In this examination, nine elements; Drainage, Land use/Land cover, soil texture, Geology, Geomorphology, Slope, Relief, Lineament, and Physiographic division map separations from rural terrains, and potential spillover were chosen dependent on writing survey and research of various researcher as appeared in below some points. First, multi-criteria geophysical analysis combined with GIS was used. There is not only a need for option in this decision on the basis of sweeping general statements in the references but also details that are provided in this region.

To discover reasonable locales for Check dams, Nala bunds, Bore well, Percolation tank, Farm pond, and so on for that following criteria is utilized,

A. Site Determination Criteria for Percolation Tank

- A tank can be found crossing rivers by being at a low altitude.
- Energized dirt.
- It ought to be very far under water before the planting can begin.

- Rain configuration is to be certain that the permeation tank is full when the overland measurement gets rechecked (ideally more than once).
- No sediments should collect in the catchment area to avoid silting up the tank bed.
- The reservoir needs to be downstream of the spillway or upper change chamber, with a slope of 3–5%.

B. Site Choice Criteria for Nala Bunds

- Nala bindings may be formed over greater increases may be traced to subtler slopes overshoots of second demand.
- Water-logged soil downstream of the bund would not be permitted.
- Ideally, terrains should be sorted, particularly if they have a distinctive shape.
- After a frost, the ponded stone strata in the zone should have enough penetration to provide enough of a revival by thawing.

C. Site Selection Criteria for Farm Ponds

- In ordinarily flattering soil conditions, a lake tends to have an earth side sloping 3:1 ratio on the water side and on the downstream.
- That area ought to have adequate seepage to expand the lake during at least two or three spells of good rainfall.
- Juncture with two areas of poor water supply is to be celebrated.
- It is to be used only as a permeable storage tank if it is filled in.

D. Site Selection Criteria for Bore Wells

- Useable or undeveloped land.
- Slant less than 10%.
- The dividing line between reality and fantasy.

E. Check Dam

- The incline ought to be fewer than 15%.
- The land use might be infertile, bush land, and river bed.
- The invasion rate of the dirt ought to be less.
- The kind of soil ought to be sandy mud topsoil.

These measures are employed in this investigation in order to asses' three different results for water resource management.

Analytical Hierarchy Process (AHP) consists of several criteria used to choose a position for RWH in GIS. When employed in several scientific research studies, AHP has demonstrated to be a reliable predictor of future job performance results (Krois and Schulte 2014). The Mind-Boggling Choices are created and dissected based on numerical and scientific data. Most issues have more than one dimension. The objective should be focused on the most important features while allowing lower-level factors too progressively to affect the base. In most cases, pairwise correlations are employed to calculate the loads in the AHP approach. A study's acceptability (proper conceptual acceptability) is determined by consideration of two parameters pairwise, aided by their relationships to each other. A 9-point scale is used to rank two variables. The even values 2, 4, 6, and 8 are unified when they are viewed individually, and the different quantities of 1, 3, 5, and 7, which relate in rather unique ways, are referenced to their middle-of-the-the-road importance. Figure 22.1 is showing flowchart of overall water management in the Betul-Chhindwara watershed.

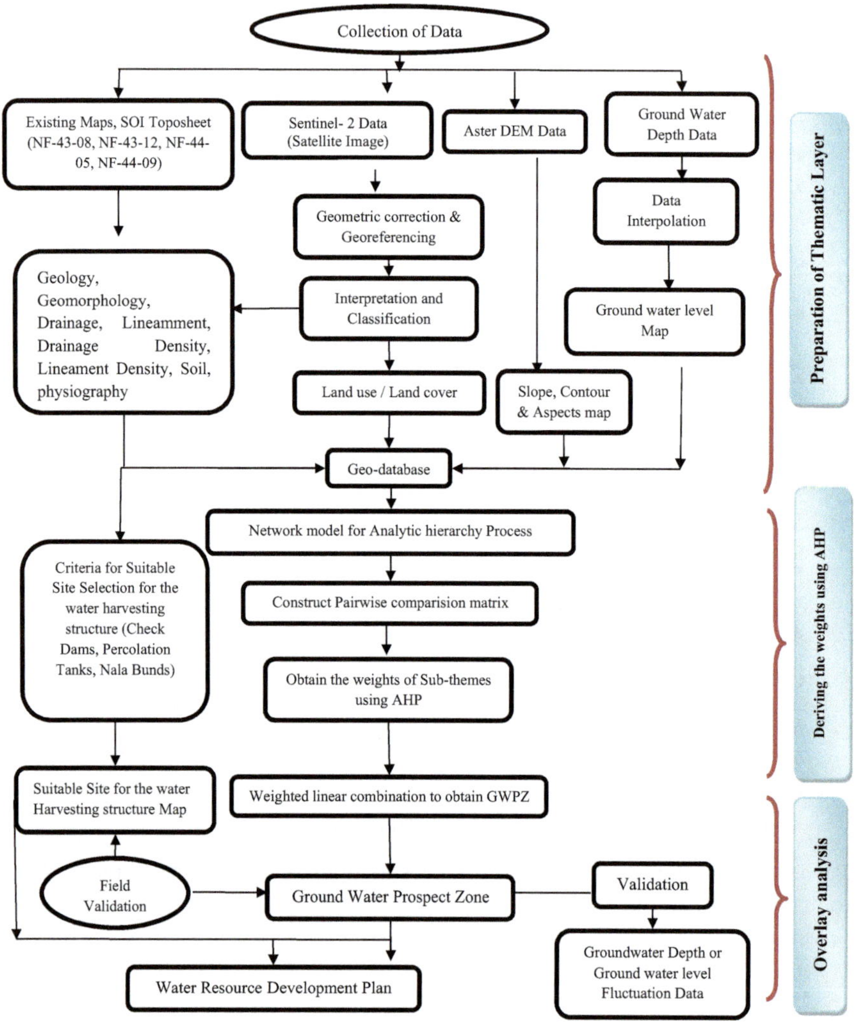


**Fig. 22.1** Flow chart of methodology used for Water Resource Development Plan

### 22.2.2 Thematic Layers Development

Advancement in topical features for computerized data's and preparing of remote detecting information, ancillary data digitization and field information for the extraction of appropriate data. To identify the likelihood of groundwater as an important zone of potential in the examination, topical layers of soil, slope, and land use and land cover, lineament, drainage, and relief were created utilizing topographic maps, topical maps, field information, and satellite picture in GIS condition. At first, the satellite picture of Sentinel-2 of 10 m spatial resolution goals relating to February 2017, ASTER DEM (Advanced Spaceborne Thermal Emission and Reflection Radiometer Digital Elevation Model) of 17 October 2011, from the Geological Survey of India (GSI) map and soil guide of Betul-Chhindwara region are geo-referenced with the assistance of Survey of India toposheet mosaic utilizing the projection UTM (Universal Traverse Mercator) and datum WGS84 (World Geodetic System 84). Often known as lineament layer density, preparation is done with the aid of lineament layer. Drainage data of 1:50,000 Topographical layers are produced from Survey of India. In this manner, seepage is refreshed with the satellite picture. Incline map was set up from ASTER DEM utilizing spatial investigator device. The soil survey and land use planning (NBSSLUP) map. Land use and Land cover map was created utilizing Sentinel-2 picture in ERDAS Imagine 9.3 and ArcGIS 10.4 software programming. For age of land use and land cover, regulated arrangement was performed utilizing K-means Classifier. K-means, a parametric choice guideline, is a very much created strategy from measurable choice hypothesis that has been connected to grouping Sentinel-2 picture. The land use and land cover map was supported by confirmed ground facts. Refer to Fig. 22.1 for the total amount of work of the method.

### 22.2.3 Deriving the Weights Using AHP/ANP

ANP is an extension of AHP for enabling decision-making by creating hierarchical criteria, classifications, and subsequently rating or ranking solutions according to their probability of success (Saaty 2004). Having five thematic classes in each theme layer suggests there are a lot of interrelated relationships among these different classes. For this reason, nine thematic relationships have been associated with ANP and class relationships formed using AHP. To create the thematic and class layers with ANP and AHP methods, there are several steps to follow:

Step 1: *Construction of model*: Following an evaluation of literature, models have been found to map groundwater availability. In constructing the model, each of the individual themes should be described and then broken down into different features and classifications to build a complex network.

Step 2: *Generation of pairwise comparison matrices*: The relative values are obtained with Saaty's 1–9 rating table (Table 22.1) (Saaty 1980). To make it simpler, the comparison matrix is laid out in Table 22.2. The development of a pairwise comparison matrix is performed based on the Saaty's nine-Spehrain thematic layer which was first used to separate out different aquifers to delineate the groundwater possibilities. By giving uncertainty with prominence of judgment by the principal Eigenvalue and their consistency index (Saaty 2004). Weights of the thematic maps of the potential groundwater were given in Table 22.4. His scale provided a method for Consistency Index (CI) and some type of consistency, called Consistency Index (CI) as variation or degree of consistency according to this equation:

$$\mathrm{CI} = \lambda \max - n/n - 1 \tag{22.1}$$

where λmax is the $n$ value by different class cardinality of eigenvalue of the pairwise comparison matrix describes the amount of variance accounted for by each pairwise similarities in relation to pairwise matrix consistency and is given by Eq. (22.2).

$$\mathrm{CR} = \mathrm{CI}/\mathrm{RI} \tag{22.2}$$

RI (Ratio Index) is the scale that measures creativity to find the value of the RI varies with $n$ values in Table 22.3.

If the CR has a value of 0.1 or less, then the inconsistency is to be permitted. As CR exceeds 10%, we must revise the subjectivity of the rating. Shown in Table 22.2 is the pairwise comparison matrix in row 0.095.

### 22.2.4 Overlay Analysis to Find Groundwater Potential Zone

GWPI is calculated by the weighted linear method (Malczewski 1999) given by Eq. 22.3:

$$\mathrm{GWPI} = \sum\sum[a_i(b_{ij}x_{ij})] \tag{22.3}$$

where $\beta ij$ = weight of the $j$th class of $i$th theme obtained by AHP and $\alpha i$ = weight of the $i$th theme obtained by ANP, $n$ = total number of thematic layers, and $m$ = total number of classes in a thematic layer, $xij$ is the pixel value of the $j$th class of the $i$th theme (Table 22.5).

Groundwater prospects of the various villages are compared with the yield data.

**Table 22.1** Saaty's 1–9 scale of relative importance

| Scale | 1 | 2 | 3 | 4 | 5 | 6 | 7 | 8 | 9 |
|---|---|---|---|---|---|---|---|---|---|
| Importance | Equal importance | Weak | Moderate importance | Moderate plus | Strong importance | Strong plus | Very strong importance | Very, very strong | Extreme importance |

**Table 22.2** Pair wise comparison matrix

| Parameter | Ground water depth | Geology | Physiography | Geomorphology | Soil | Slope | LU/LC | Lineament density | Drainage density |
|---|---|---|---|---|---|---|---|---|---|
| LU/LC | 9/9 | 9/8 | 9/7 | 9/6 | 9/5 | 9/4 | 9/3 | 9/2 | 9/1 |
| Drainage | 8/9 | 8/8 | 8/7 | 8/6 | 8/5 | 8/4 | 8/3 | 8/2 | 8/1 |
| Soil | 7/9 | 7/8 | 7/7 | 7/6 | 7/5 | 7/4 | 7/3 | 7/2 | 7/1 |
| Geomorphology | 6/9 | 6/8 | 6/7 | 6/6 | 6/5 | 6/4 | 6/3 | 6/2 | 6 |
| Geology | 5/9 | 5/8 | 5/7 | 5/6 | 5/5 | 5/4 | 5/3 | 5/2 | 5 |
| Slope | 4/9 | 4/8 | 4/7 | 4/6 | 4/5 | 4/4 | 4/3 | 4/2 | 4 |
| Lineament | 3/9 | 3/8 | 3/7 | 3/6 | 3/5 | 3/4 | 3/3 | 3/2 | 3 |
| Physiographic division | 2/9 | 2/8 | 2/7 | 2/6 | 2/5 | 2/4 | 2/3 | 2/2 | 2 |
| Relief | 1/9 | 1/8 | 1/7 | 1/6 | 1/5 | 1/4 | 1/3 | 1/2 | 1 |

**Table 22.3** Saaty's ratio index for different values of *n*

| *n* | 1 | 2 | 3 | 4 | 5 | 6 | 7 | 8 | 9 | 10 |
|---|---|---|---|---|---|---|---|---|---|---|
| RI | 0 | 0 | 0.58 | 0.89 | 1.12 | 1.24 | 1.32 | 1.41 | 1.45 | 1.49 |

**Table 22.4** Weights of the thematic maps of the potential groundwater

| Themes | Assigned weight | Normalized weight |
|---|---|---|
| LU/LC | 9 | 0.223 |
| Drainage | 8 | 0.170 |
| Soil | 7 | 0.150 |
| Geomorphology | 6 | 0.142 |
| Geology | 5 | 0.089 |
| Slope | 4 | 0.074 |
| Lineament | 3 | 0.060 |
| Physiographic division | 2 | 0.053 |
| Relief | 1 | 0.038 |

**Table 22.5** Assigned and normalized weights of the features of themes for the delineation of the groundwater potentials in the study area

| Categories | Criterion | Weight | Normalized weight | Rank | Area ($km^2$) | Percentage area % |
|---|---|---|---|---|---|---|
| Geology | Upper Gondwana | 3 | 0.14 | 0.089 | 1309.94 | 5.98 |
| | Lower Gondwana | 4 | 0.19 | | 2481.22 | 11.33 |
| | Recent alluvium | 6 | 0.28 | | 490.945 | 2.24 |
| | Deccan trap | 2 | 0.095 | | 14,309.9 | 65.35 |
| | Older alluvium/laterite | 5 | 0.24 | | 18.2458 | 0.08 |
| | Bundelkhand granite/gneiss | 1 | 0.047 | | 3285.89 | 15.0 |
| Physiography | Betul plateau | 6 | 0.214 | 0.053 | 6679.6 | 30.50 |
| | Chhindwara plateau | 6 | 0.214 | | 5185.93 | 23.68 |
| | Amarwara | 5 | 0.178 | | 2964.19 | 13.54 |
| | Gawilgarh hill | 1 | 0.035 | | 1506.36 | 6.87 |
| | Kalibhit hill and Marabdi plateau | 2 | 0.071 | | 1947.48 | 8.89 |
| | Pachmari hill | 1 | 0.035 | | 1391.14 | 6.35 |
| | Sausar upland | 3 | 0.107 | | 1176.04 | 5.39 |
| | Upper Kanhan valley | 4 | 0.143 | | 1048.05 | 4.78 |

(continued)

**Table 22.5** (continued)

| Categories | Criterion | Weight | Normalized weight | Rank | Area (km$^2$) | Percentage area % |
|---|---|---|---|---|---|---|
| Geomorphology | Structural origin | 1 | 0.33 | 0.142 | 11,778.92 | 53.79 |
| | Denudational origin | 2 | 0.66 | | 10,090.59 | 46.21 |
| Soil | Clay loam | 2 | 0.025 | 0.150 | 2856.83 | 12.95893 |
| | Sandy loam | 11 | 0.141 | | 4515.53 | 20.48299 |
| | Loam | 8 | 0.103 | | 423.479 | 1.920952 |
| | Clay | 1 | 0.019 | | 7157.8 | 32.46865 |
| | Sandy clay loam | 6 | 0.076 | | 4712.13 | 21.37479 |
| | Silt clay loam | 5 | 0.064 | | 1014.85 | 4.603483 |
| | Silty loam | 7 | 0.089 | | 144.43 | 0.655152 |
| | Silt loam clay | 4 | 0.051 | | 111.19 | 0.504371 |
| | Loamy sand | 10 | 0.128 | | 179.168 | 0.812728 |
| | Sand | 12 | 0.153 | | 209.839 | 0.951855 |
| | Sandy clay | 9 | 0.115 | | 537.747 | 2.439286 |
| | Silt clay | 3 | 0.038 | | 182.272 | 0.826808 |
| Slope | <10 | 5 | 0.416 | 0.074 | 8669.27 | 39.59 |
| | 10–20 | 3 | 0.250 | | 7287.64 | 33.28 |
| | 20–45 | 2 | 0.166 | | 3717.64 | 16.97 |
| | 45–60 | 1 | 0.083 | | 1669.73 | 7.62 |
| | 60> | 1 | 0.083 | | 552.18 | 2.54 |
| LU/LC | Dense forest | 6 | 0.272 | 0.223 | 6529.754 | 29.78 |
| | Spars forest | 5 | 0.227 | | 2705.55 | 12.34 |
| | Crop land | 4 | 0.181 | | 5523.482 | 25.19 |
| | Fallow land | 3 | 0.136 | | 5014.585 | 22.87 |
| | Water body | 2 | 0.090 | | 160.248 | 0.73 |
| | Settlement | 1 | 0.045 | | 308.236 | 1.40 |
| | Barren land | 1 | 0.045 | | 5265.788 | 24.01 |
| Lineament | Structural lineament-joint/fracture | 9 | 0.169 | 0.060 | 5866.72 | 83.70 |
| | Structural lineament-fault | 8 | 0.151 | | 7789.42 | 35.61 |
| | Structural lineament | 8 | 0.151 | | 219.398 | 3.13 |
| | Geomorphic lineament-drainage parallel | 7 | 0.132 | | 67.775 | 0.96 |

(continued)

**Table 22.5** (continued)

| Categories | Criterion | Weight | Normalized weight | Rank | Area ($km^2$) | Percentage area % |
|---|---|---|---|---|---|---|
| | Structural lineament-dyke | 6 | 0.113 | | 171.974 | 2.45 |
| | Geomorphic lineament ridge parallel | 5 | 0.094 | | 75.6263 | 1.07 |
| | Structural lineament-shear zone | 4 | 0.075 | | 49.7827 | 0.71 |
| | Structural lineament-axial trace of fold | 3 | 0.056 | | 6.59686 | 0.09 |
| | Geomorphic lineament-scrap parallel | 2 | 0.037 | | 42.2886 | 0.60 |
| | Geomorphic lineament | 1 | 0.018 | | 6.49406 | 0.09 |
| Drainage | 1 | 1 | 0.047 | 0.170 | 6324.03 | 29.09 |
| | 2 | 2 | 0.095 | | 6392.194 | 29.41 |
| | 3 | 3 | 0.142 | | 4445.615 | 20.45 |
| | 4 | 4 | 0.190 | | 3182.322 | 14.64 |
| | 5 | 5 | 0.238 | | 1389.074 | 06.39 |
| | 6 | 6 | 0.286 | | | |
| Relief | 0–4.72 | 5 | 0.416 | 0.038 | 4114.88 | 17.59 |
| | 4.72–9.19 | 4 | 0.250 | | 5452.09 | 23.32 |
| | 9.19–15.23 | 3 | 0.166 | | 7030.28 | 30.06 |
| | 15.23–23.63 | 2 | 0.166 | | 5458.28 | 23.34 |
| | 23.63–66.96 | 1 | 0.083 | | 1331.07 | 5.690 |

### 22.2.5 *Ranking Classes of Various Parameters*

The occurrence and movement of groundwater in a region are dependent on a number of different factors, such as lineament, land use and land cover, soil, Slope, drainage, relief and interrelationship between these layers (Jaiswal et al. 2003).

1. Land Use and Land Cover

Land use are interpretable by satellite images. Various types of land use and land cover pattern are identified in the study area which includes Spars Forest, Dense Forest, fallow land, Barren land, water body, Crop Land, and Settlement area (Fig. 22.2). Spars Forest and Dense Forest are poor sites for suitable site for water harvesting, and hence given lowest rank. Crop Land, fallow land, drainage are considered to have good site for water harvesting, while the water bodies, Barren land, and Settlement area have poor sites for water harvesting.

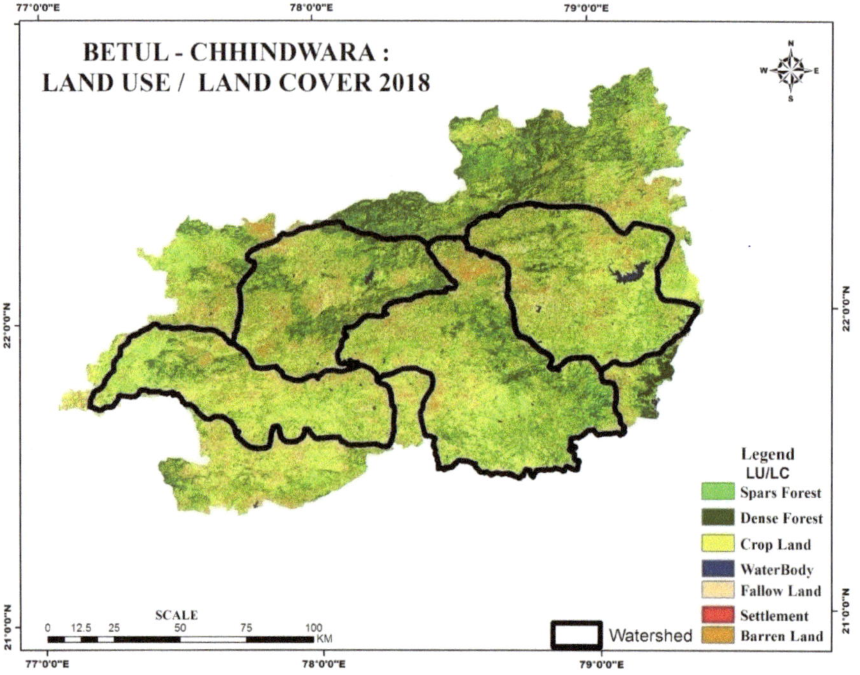


**Fig. 22.2** Land use/land cover map of 2018 year

2. Drainage

Drainage is an inverse relationship between permeability and the permeability of water in the ground; it is therefore critical in determining the characteristics of the groundwater region. Ways with high drainages are the ones that favor erosion of the surface and hence indicate high suitability and high ranks are assigned to high drainage area and vice versa. Drainage map of the study was shown in Fig. 22.3.

3. Slope

The percentage rise or fall of the research area doesn't vary much between zero and ten percent. For a class that receives a nearly constant measurement, its grade is assigned a lower rank; for a class which receives maximum values, its rank is adjusted by terrain. Slope map of study area was shown in Fig. 22.4.

4. Soil

The study area shows there are to be four major soil groups: light sandy, light loam, medium-loam, silt, and clay (Fig. 22.5). If you're using a ranking system based on soil penetration, you are assigned by based on infiltration rate. Clayey soil has low infiltration rate of absorption, hence has lower priority.

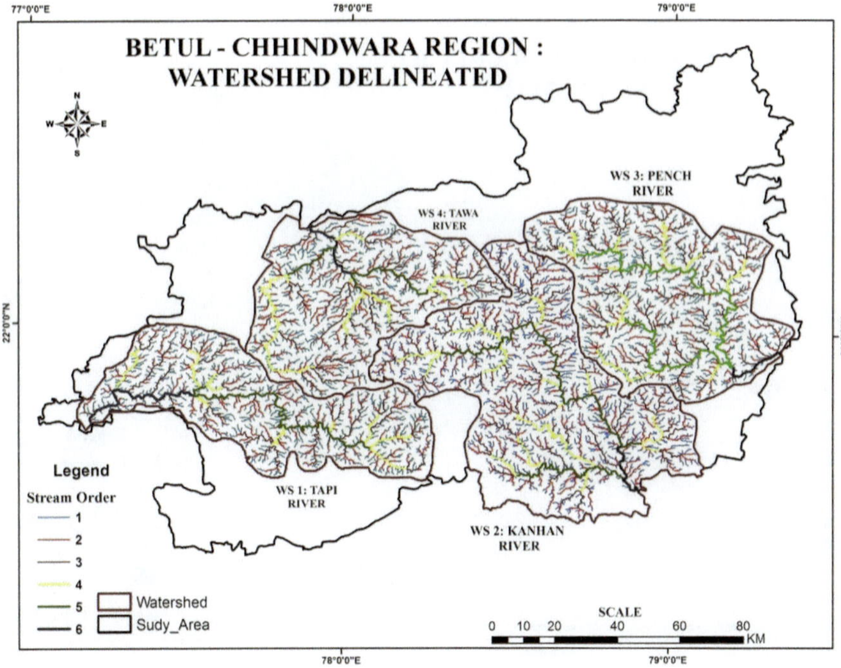


Fig. 22.3 Drainage map

5. Relief

Relief is an algorithm which is widely used to measure topographic slope positions and to automate landform classification. Relief map model with elevations and depressions representing hills and valleys, typically on an exaggerated relative scale. Many physical processes such as hilltop, valley bottom, exposed ridges, flat plain, upper and lower slope actions on landscape are correlated with relief. Low relief value is given under high suitable zone. Relief map was shown in Fig. 22.6.

6. Lineament

Structures are caused by stresses and strains over time. Fractures, faults, folds, and flaws are the relevant bedding features for enhancing permeability of the rock. The field at the center of the Earth is made up of fractures, called lineament. Lineaments are critical in groundwater assessment because they provide information about movement and storage of the water indirectly. The high Lineament density value gives the highest priority and the lowest priority to the lower one. Lineament map of study area was shown in Fig. 22.7.

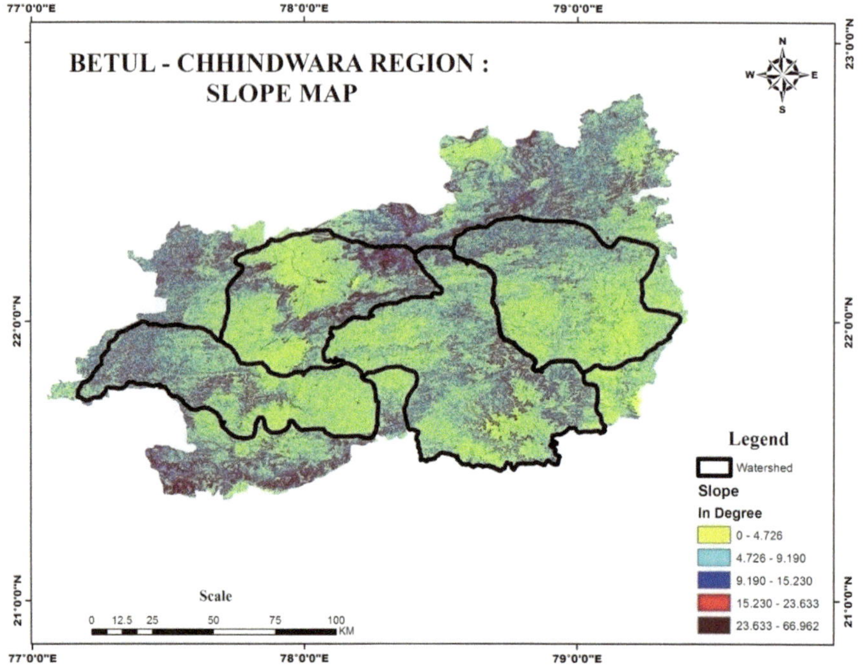


**Fig. 22.4** Slope map

7. Geomorphology

The landform assumes a fundamental job for the event what's more, dispersion of groundwater. Two sorts of geomorphic units are distinguished in the examination zone. Denudational Origin has higher water level surface, what's more, henceforth it is the best landform for high suitable zone for water harvesting then the structural origin. Geomorphology map of study was shown in Fig. 22.8.

8. Physiography Division

The Physiographic division landform assumes a fundamental job for the event what's more, dispersion of groundwater. There are eight sorts of Physiographic units are distinguished in the examination zone (Fig. 22.9). Betul and Chhindwara plateau has higher water level surface, Amarwara plateau is next to the high zone what's more, henceforth it is the best landform for high groundwater potential than the other Upper Kanhan valley to moderate important and Sausar upland and hills come to poor water harvesting zone.

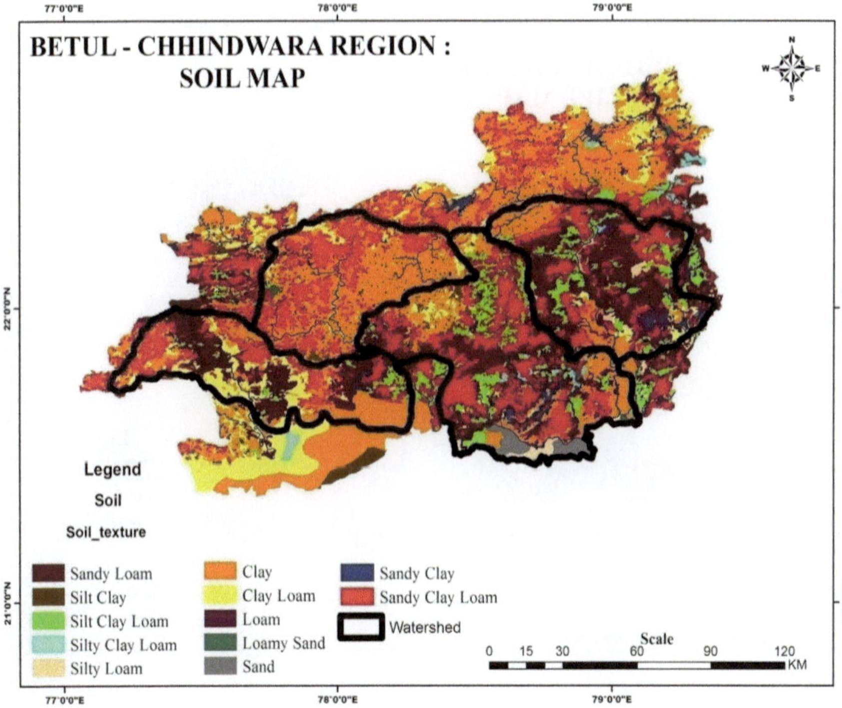


**Fig. 22.5** Soil map

9. Geology

Topographically, the territory has been isolated into different classes, i.e., Upper Gondwana rocks, Lower Gondwana rocks, Deccan trap, Bundelkhand granite and gneiss, recent alluvium, and older alluvium (Fig. 22.10). Alluvium is micaceous; fine to grained medium comprises of rock, free sand, cut, and earth and has better efficiency because of quality of water in sand what's more, rock beds, so it is given higher inclination in deciding the suitable site for water harvesting.

## 22.3 Results and Discussion

### *22.3.1 Suitable Sites for Water Harvesting Structures*

The reasonable locales for water collecting structures were related to the use of remote detecting what's more, GIS. The geology map, Physiography and geomorphology map, drainage map, land use map, soil texture, DEM, and cushion zone

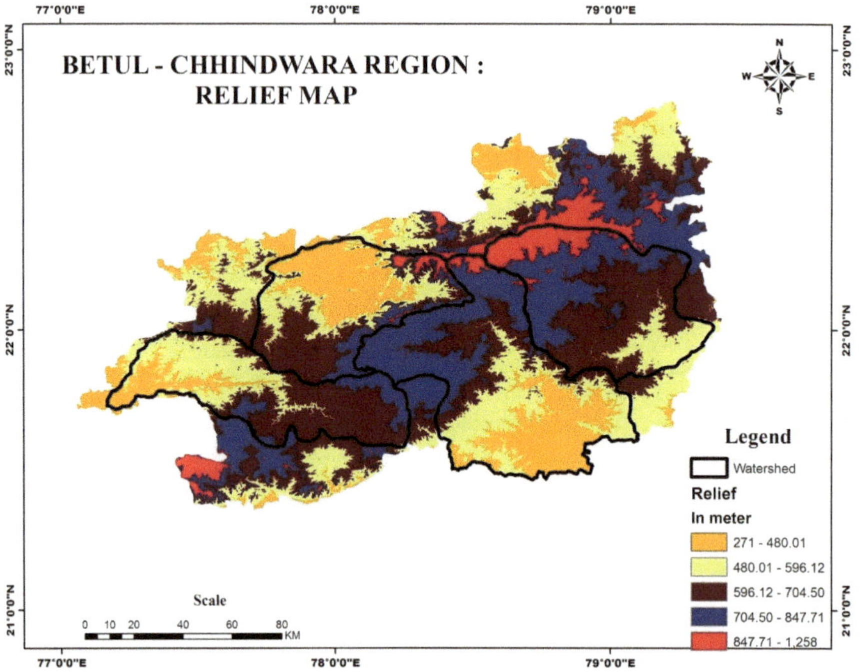


**Fig. 22.6** Ground water depth map

map were readied utilizing satellite symbolism and SOI toposheets of the Betul-Chhindwara region four watersheds. The seven classes of land use/land cover what's more, hydrological soil texture map was readied. The overlay task of land use map, hydrological soil texture guide, and Slope map was completed for choosing appropriate locales for water harvesting structures and exhibited through site reasonableness map (Figs. 22.10 and 22.11). The support zone of developed shows that every one of the areas found reasonable for water gathering structures are out of the support zone made at about 200 m from developed territory that affirms the proposed building site, which is far from the settlement.

The reasonableness of check dam locales can be affirmed as the site is situated on third request waste and fulfills the states of land use, soil type, and incline according to IMSD rules. The vast majority of the locales in Betul-Chhindwara region watersheds were observed to be appropriate for check dam yet according to ground truth and experience, 20 reasonable locales were proposed to build the check dam. Since it is situated in the reasonable land class (Spars forest, River bed), slope, (under 15%) and soil type (Silt topsoil to Sandy dirt topsoil) that fills the need of soil and water preservations and groundwater expansion. The proposed check dams could be exceptionally helpful as enhancing water system amid the dry season, and reasonable kharif/rabi harvest might be developed.

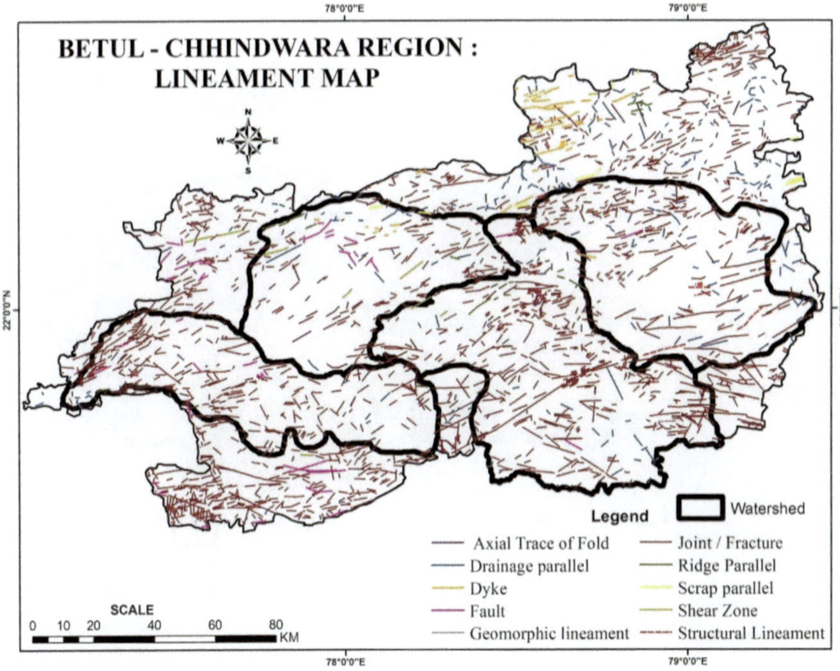


**Fig. 22.7** Lineament map

The reasonableness of destinations for permeation tanks can be affirmed, as it is situated in soil having high invasion rate and fulfilling different conditions according to IMSD rules. Figure 22.12 uncovers that few locales are observed to be appropriate for development of permeation tanks in Betul-Chhindwara region watersheds and eight destinations were proposed for the development. The proposed percolation tanks are situated in gravelly sandy mud topsoil soils, which are exceedingly appropriate for reviving groundwater. It is moreover situated in timberland land having slant under 10% which fulfills the IMSD rules. Since, it is situated in second and third request waste that spares permeation tanks from the harm because of high spillover.

The permeation tank could be the best fake groundwater energizer system in the region watersheds, which may raise the dimension of groundwater amid the stormy season. Figure 22.11 delineates that about five spots were discovered reasonable for development of check dams and percolation tanks both, be that as it may, the percolation tank was as it was proposed at these spots because of less accessibility of spillover and less expense of development. The appropriateness of other water gathering structures like Nala Bandh, Farm pond, and bore well be apportioned on the grounds that the locales did fulfill appropriateness criteria.

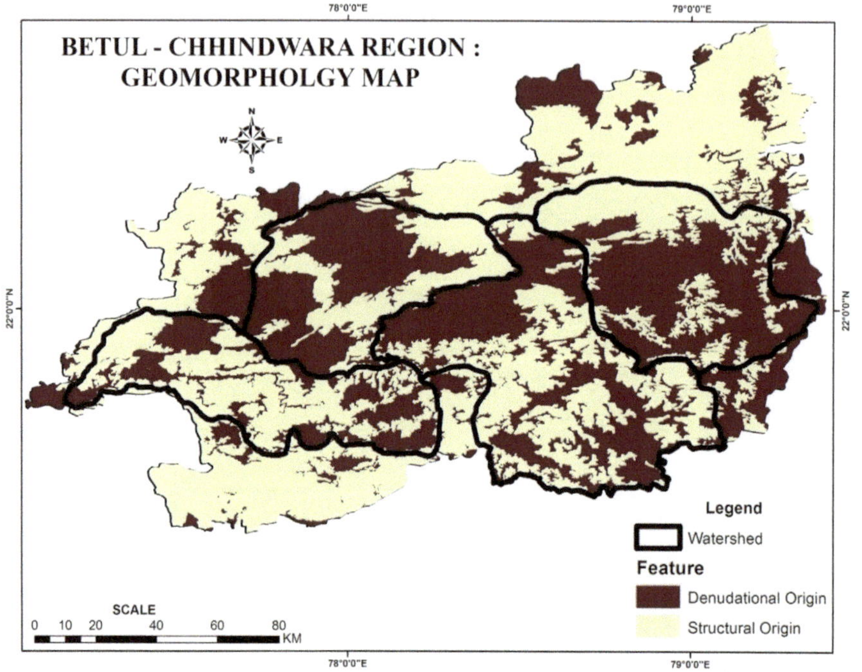


**Fig. 22.8** Geomorphology map

## 22.4 Summary and Conclusion

Water harvesting structures are critical to preserve valuable regular assets like, soil and water, which are draining step by step at disturbing rate. The Betul-Chhindwara region of Madhya Pradesh, lying in the lower regions of Satpura region, has soil and water issues. The great cultivable zones are being changed over into fallow and barren grounds due to over misuse of land and water assets. Keeping this in view, Tapi, Kanhan, Pench, and Tawa watersheds in Betul-Chhindwara region is chosen for arranging appropriate locales for development of water harvesting structures utilizing remote sensing and GIS strategies. The potential locates for water collecting structures in bet watershed were recognized through rising advancements of remote detecting what's more, GIS and 21 check dams, 8 percolation tanks, 9 nala bandh, 22 farm pond, and 16 bore well cum dug well at suitable locales are proposed. Locales are definitely not appropriate for nala bunding and farm pond due to soak incline, less soil thickness, and high spillover speed. The region spillover was found pre monsoon season (February) in this manner, an operational technique for water harvesting structures could be arranged based upon water balance investigation of the territory for legitimate the board of surface and groundwater.

The geomorphological attributes have distinctive geohydrological example of the seepage. The principle stream over through the center dimension level. The

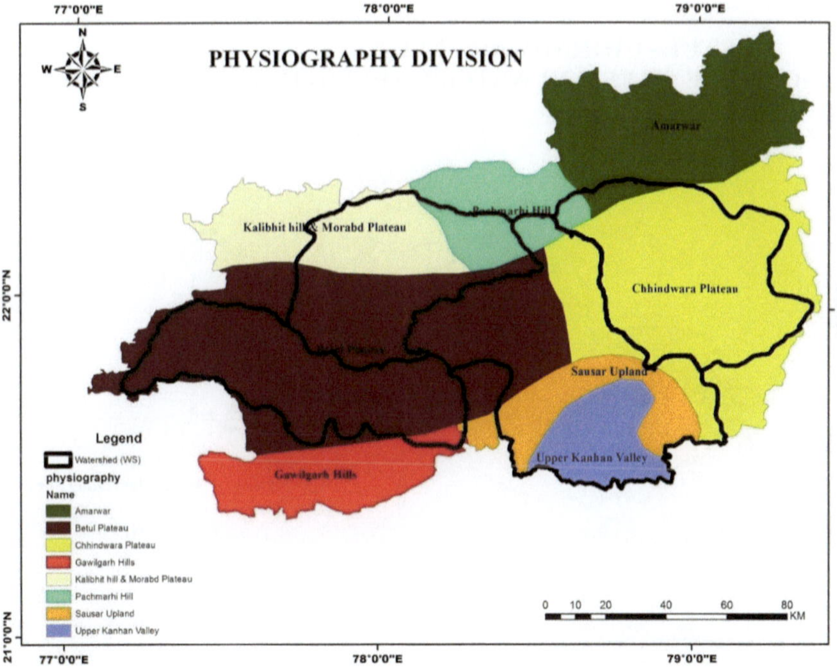


**Fig. 22.9** Physiography division map

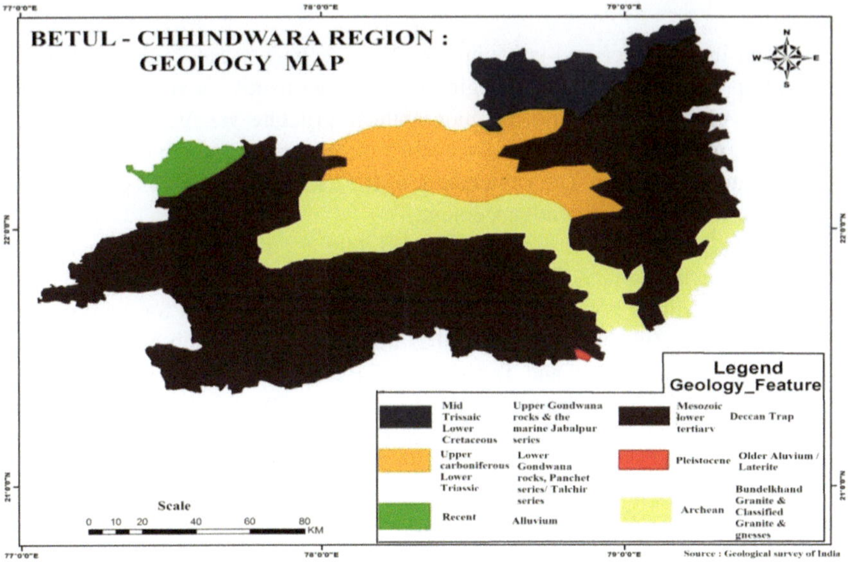


**Fig. 22.10** Geology map

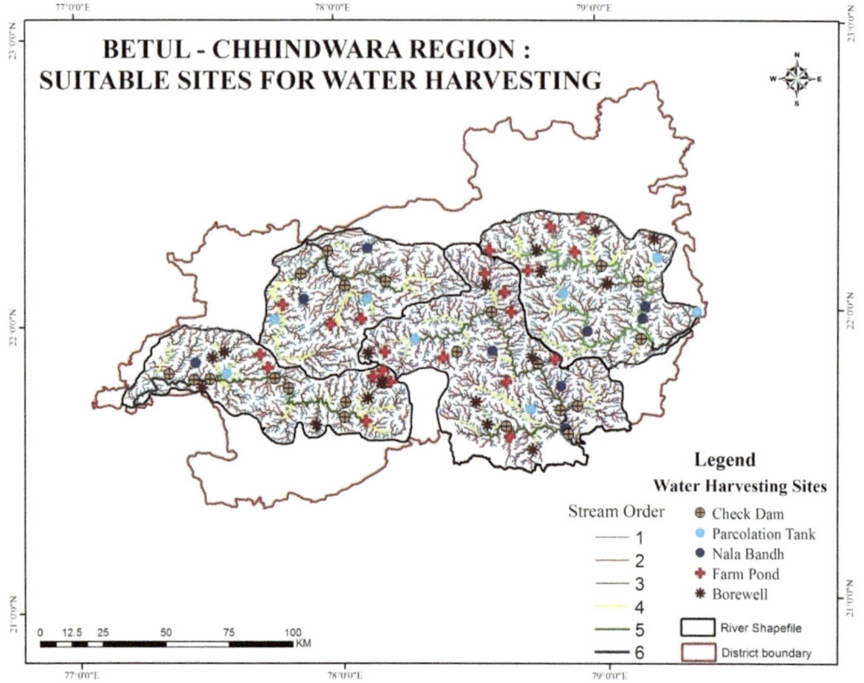


**Fig. 22.11** Suitable sites for water harvesting

barren land in the watershed is having dendritic seepage example of streams. The Deccan trap, unit of denudational beginning, has the quantity of appropriate locales for dam development, likewise the geographical structure viz AO and 1 paeochoe basaltic magma stream > 190 mt are available in Deccan trap, which is much reasonable for water harvesting. The centerpiece of the examination area is progressively appropriate for groundwater advancement.

The investigation district is having the differentiation proportion of the watered land to the unirrigated land. Watered land zone is too less to even think about developing the horticulture bore well or dug well technique is utilized for overflow estimation and sub bowl shrewd overflow has been determined. Sub-bowl one which is of Tapi, Kanhan, Pench, and Tawa river of Betul-Chhindwara region is having most elevated yield as contrast with other sub-bowls. By this it is certain that all the sub-bowls have enough yields however the downpour water isn't legitimately gathered.

Site determination for water harvesting is completed by overlying the slope, Soil texture, physiography, Geology, Geomorphology, land use/land cover, stream order, and relief support maps. The district is having the full extension for the permeation tank, checks dams, farm pond, dug well, bore well, and dug cum bore well. This examination will help in the best possible administration of the investigation area as per land use/land cover.

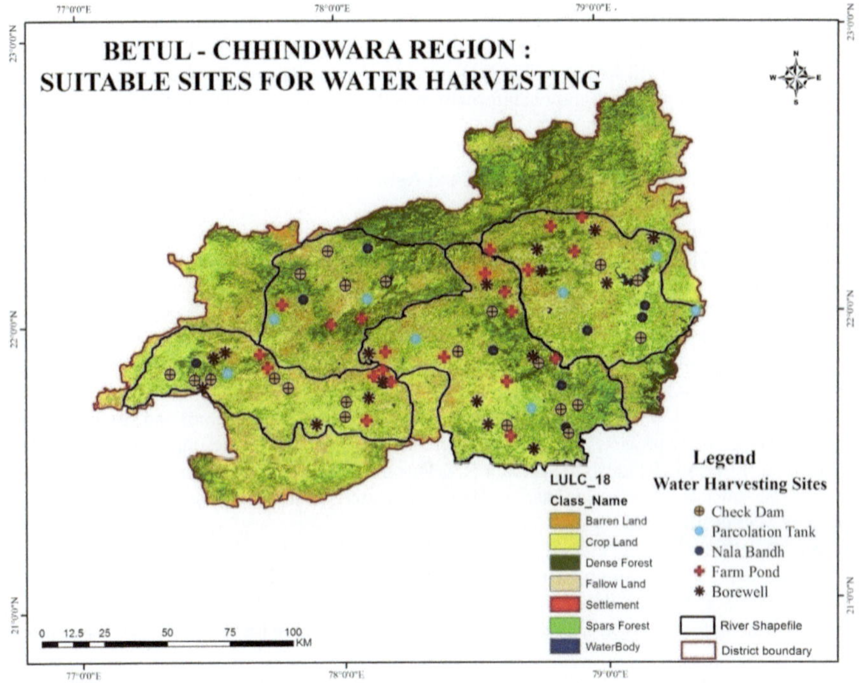


Fig. 22.12 Suitable sites for water harvesting shown on LU/LC

# References


Ahmad I (2013) Investigating of potential water harvesting sites at Potohar using modeling approach. Pak J Agric Sci 50(4):723–729

Ahmed Ould Cherif A, Nagasawa R, Hattori K, Chongo D, Perveen MF (2007) Analytical hierarchic process in conjunction with GIS for identification of suitable sites for water harvesting in the oasis areas: case study of the oasis zone of Adrar, northern Mauritania. J Appl Sci 7:2911–2917

Al-Adamat R (2008) GIS as a decision support system for siting water harvesting ponds in the basalt aquifer/NE Jordan. J Environ Assess Policy Manag 02:189–206

Al-Adamat R, Diabat A, Ghada S (2010) Combining GIS with multi-criteria decision making for siting water harvesting ponds in northern Jordan. J Arid Environ 74:1471–1477

Al-Adamat R, Alayyash S, Al-Amoush H, Al-Meshan O, Rawajfih Z (2012) The combination of indigenous knowledge and geo-informatics for water harvesting siting in the Jordanian Badia. J Geogr Inf Syst 4:366–376

Al-Shamiri A, Ziadat FM (2012) Soil-landscape modeling and land suitability evaluation: the case of rainwater harvesting in a dry rangeland environment. Int J Appl Earth Obs Geoinf 18:157–164

Ammar A, Michel R, Ouessar M, Ritsema C (2016) Identification of suitable sites for rainwater harvesting structures in arid and semi-arid regions: a review. ISWCR 4:108–120

Bamne Y, Patil A, Vikhe D (2014) Selection of appropriate sites for structures of water harvesting in a watershed using remote sensing and geographical information system. Int J Emerging Technol Adv Eng 4(11):270–275

Bulcock LM, Jewitt GPW (2013) Key physical characteristics used to assess water harvesting suitability. Phys Chem Earth Parts A/B/C 66:89–100

De Winnaar G, Jewitt G, Horan M (2007) A GIS-based approach for identifying potential runoff harvesting sites in the Thukela River basin, South Africa. Phys Chem Earth 32:1058–1067

Elewa HH, Qaddah AA, El-Feel AA (2012) Determining potential sites for runoff water harvesting using remote sensing and geographic information systems-based modeling in Sinai. Am J Environ Sci 8:42–55

FAO (2003) Land and water digital media series, 26. Training course on RWH (CDROM). Planning of water harvesting schemes, unit 22. Food and Agriculture Organization of the United Nations, FAO, Rome

FAO (2015) Available online: http://www.fao.org/ag/agp/greenercities/en/whyuph/

Isioye OA, Shebe MW, Momoh U, Bako CN (2012) A multi criteria decision support system (MDSS) for identifying rainwater harvesting site(s) in Zaria, Kaduna state, Nigeria. Creat Common 1(5):53–57

Jaiswal RK, Mukherjee S, Krishnamurthy J, Saxena R (2003) Role of remote sensing and GIS techniques for generation of groundwater prospect zones towards rural development – an approach. Int J Remote Sens 24(5):993–1008

Jothiprakash V, Sathe MV (2009) Evaluation of rainwater harvesting methods and structures using analytical hierarchy process for a large scale industrial area. J Water Resour Prot 1(6):427–438

Kadam AK, Kale SS, Pande NJ, Sankhua RN, Pawar NJ (2012) Identifying potential rainwater harvesting sites of a semi-arid, basaltic region of western India, using SCS-CN method. Water Resour Manage 26(9):2537–2554

Kahinda JM, Lillie E, Taigbenu A, Taute M, Boroto R (2008) Developing suitability maps for rainwater harvesting in South Africa. Phys Chem Earth 33:788–799

Kahinda JM, Taigbenu AE, Sejamoholo BBP, Lillie ESB, Boroto RJA (2009) GIS-based decision support system for rainwater harvesting (RHADESS). Phys Chem Earth Parts A/B/C 34(13):767–775

Khan MD, Khattak M (2012) Siting of rainwater harvesting locations in district Haripur using geographic information techniques. J Himal Earth Sci 45:2

Krois J, Schulte A (2014) GIS-based multi-criteria evaluation to identify potential sites for soil and water conservation techniques in the Ronquillo watershed, northern Peru. Appl Geogr 51:131–142

Kumar MG, Agarwal AK, Bali R (2008) Delineation of potential sites for water harvesting structures using remote sensing and GIS. J Indian Soc Remote Sens 36(4):323–334

Mahmoud SH, Alazba AA (2014) The potential of in situ rainwater harvesting in arid regions: developing a methodology to identify suitable areas using GIS-based decision support system. Arab J Geosci 8:5167–5179

Malczewski J (1999) GIS and multicriteria decision analysis. Wiley, New York, pp 392

Mbilinyi B, Tumbo S, Mahoo H, Mkiramwinyi F (2007) GIS-based decision support system for identification potential sites for rainwater harvesting. Phys Chem Earth 32:1074–1081

Mekdaschi SR, Liniger H (2013) Water harvesting - guidelines to good practices, a project report, pp 91, Centre for Development and Environment (CDE) and Institute of Geography, University of Bern; Rainwater Harvesting Implementation Network (RAIN), Amsterdam; MetaMeta, Wageningen; The International Fund for Agricultural Development (IFAD), Rome, ISBN 978-3-905835-35-9

Pauw ED, Oweis T, Youssef J (2008) Integrating expert knowledge in GIS to locate biophysical potential for water harvesting: methodology and a case study for Syria. In: ICARDA, Aleppo, p 59

Saaty TL (1980) The analytic hierarchy process: planning, priority setting, resource allocation. McGraw-Hill, New York, pp 287

Saaty TL (2004) Decision making – the analytic hierarchy and network processes (AHP/ANP). J Sys Sci Sys Eng 13(1):1–35. ISSN 1004- 3756/04/1301/1

Sayl KN, Muhammad NS, El-Shafie (2017) Robust approach for optimal positioning and ranking potential rainwater harvesting structure (RWH): a case study of Iraq. Arab J Geosci 10:413

Sekar I, Randhir TO (2007) Spatial assessment of conjunctive water harvesting potential in watershed systems. J Hydrol 334(1):39–52

Weerasinghe H, Schneider UA, Löw A (2011) Water harvest- and storage location assessment model using GIS and remote sensing. Hydrol Earth Syst Sci Discuss 8(2):3353–3381

Ziadat F, Bruggeman A, Oweis T, Haddad N, Mazahreh S, Sartawi W, Syuof M (2012) A participatory GIS approach for assessing land suitability for rainwater harvesting in an arid rangeland environment. Arid Land Res Manag 26(4):297–311

# Chapter 23
# Attaining Optimal Sustainability for Urban Wastewater Management Using Open Source Tools Like QGIS, EPANET and WATERNETGEN

**Devang Shah and Dilip Shete**

**Abstract** Nearly 3 in 10 people worldwide lack access to safe, readily available water at home, and 5.5 in 10, lack safely managed sanitation. WHO stated that the total economic return on sanitation spending is US $5.5 for every one dollar invested. Still, M.D.G. for sanitation was missed by 700 million people globally. The primary reason behind this problem is the lack of an integrated approach. In most developing countries, due to a lack of infrastructure, wastewater is not properly disposed of. The status of the water supply system is also pathetic. After a detailed study, it was found that treated wastewater usage in the residential, industrial and agricultural sectors with proper pricing is the best alternative to mitigate water stress, ensure sustainability and generate revenue. The methodology to design and implement water reuse project with economic feasibility is presented with a case study in the present article.

**Keywords** Water reuse · IRR · EPANET · WATERNETGEN · QGIS · Economic feasibility

## 23.1 Introduction

The challenge of water availability does not sensitize a layman regarding critical conditions prevailing as urban people generally enjoy sufficient water supply for daily consumption in class I and class II cities in India. In contrast, the situation is quite challenging in many underdeveloped and developing parts of India, facing moderate-to-severe water stress.

Keeping apart the Indian context, one-third of the total population lives in areas under water stress worldwide. Total water on earth is constant for millions of years

D. Shah (✉)
Civil Engineering Department, Faculty of Engineering and Technology, Parul Institute of Technology, Parul University, Vadodara, Gujarat, India

D. Shete
Faculty of Technology and Engineering, Water Resource Engineering and Management Institute, The M. S. University of Baroda, Vadodara, Gujarat, India

 
P. Kumar et al. (eds.), *Water Resources Management and Sustainability*, Advances in Geographical and Environmental Sciences,
https://doi.org/10.1007/978-981-16-6573-8_23

as the same water is changing form continuously and being used and reused millions of times as a part of the hydrologic cycle. While minor population growth has been observed for quite a long, the world's population had grown exponentially and increased more than seven-fold in the last three centuries. The last 90 years contributed to three-fold growth (Daigger 2007). The increase in industrial and agricultural needs has created water stress in most regions of the world. Water reuse is being adopted as sustainable practice in developed and highly water-scarce areas for the last four decades at a municipal scale (Leverenz et al. 2011).

In India, at least the water supply sector shows a bit satisfactory performance as 94% of the population has access to drinking water in class I and class II cities. But in the sanitation sector, the capacity exists to treat only up to 40% of sewage generated in class I and class II cities, causing 60% of generated sewage to be disposed of in an unsafe manner in natural water bodies (Central Public Health and Environmental Engineering Organization 2005). Further, using this untreated or partially treated wastewater in agriculture is prevalent in India, causing severe health hazards and soil characteristics deterioration (Mekala et al. 2008).

Further considering the nutrient value of treated wastewater, a study for coastal towns and cities has shown that wastewater valued at Rs. 1091.20 million are released into the coastal waters of India (Central Pollution Control Board 2009). Thus, if unplanned and unofficial reuse of wastewater can be converted to planned and official reuse, it can open the new horizon to utilize the wastewater's nutrient value and sell water to industries. The revenue generated this way can be utilized to construct much-needed infrastructure in the sanitation sector.

With this in view, data collection and study have been conducted to understand water reuse feasibility and optimize urban wastewater management with a case study of the Vadodara Urban Development Authority Area (VUDA) in the Gujarat state of India.

### *23.1.1 Why Water Reuse?*

Human beings' excreta and faecal discharges are nothing but food for microorganisms, which converts them into the best source of nutrients. In ancient times due to less population density, these were discharged safely to nature to return the nutrients in natural cycles. As observed from the excavations at Harappa and Mohenjo-daro, the Sindhu–valley civilization had, more than 5000 years ago, baths in many of the houses with ceramic pipes for water supply and brick conduits under the centre of the streets for drainage (Rouse and Ince 1957).

The best way adopted for disposal of drainage water in that time perhaps consisted of disposal to river or sea. As ample water was available for dilution, these discharges perhaps had served as the minimum organic load required to sustain the natural self-assimilative cycle of rivers keeping the food chain intact and fruitful. Unluckily, due to the exponential growth of population, this organic waste quantity has increased to such an extent that inland water bodies can no longer sustain this load. That is why treatment for the removal of this organic matter is given in wastewater treatment plants. The only grievance is; we are putting a lot of energy input to remove the nutrient value of this wastewater instead of directly returning it to the natural cycle for beneficial use. Here is where an engineer's intervention is required to redefine the treatment criteria and find a safe way of handling wastewater. Treated wastewater utilization in beneficial uses such as irrigation, peri-urban irrigation, watering of lawns, gardens, golf courses, green belts, etc. The nutrient value of wastewater can be adequately addressed, and load on treatment facilities can be reduced. If proper treatment technology is existing safeguarding this type of reuse patterns, the nutrient value of wastewater can be tapped. Wastewater treated as per existing treatment technologies should be sold to industries. Industries have significant ecological footprints, and they have the revenue to bear the treatment cost. Selling wastewater to industries will reduce the load on freshwater resources and generate much-required funds for developing sanitation facilities in India and all developing countries.

### *23.1.2 State of the Art*

**Sustainable Development**: The spike in population growth and high standards of living for significant portions of the human population have increased resource demands beyond the ecosystem's capacity. This brings in the concept of sustainable development as a significant factor based on evaluating enhanced water supply and wastewater management approaches.

The U.N. Report on "our common future" known as the Brundtland Report, defined sustainable development as "paths of progress that meet the needs and aspirations of the present generation without compromising future generations' ability to meet their needs" (Brundtland 1987). Many definitions of sustainable development exist, but a useful one is a balance between economic, environmental and social considerations in selecting and implementing an approach to any issue—the so-called triple bottom line (Daigger 2007).

The water available on earth today is no different in quantity from what was available thousands of years ago. The world's water resources will never change, but the human population and its demands are overgrowing. Meeting these demands will require wise investment in how we use and reuse our water (UN Water Statistics 2010). Aspiring to install a concept of sustainable development has now become broadly accepted (Wallbaum et al. 2011). A systematic research review of global urbanization research from 1991 to 2009 revealed increasing interest in urbanization studies' ecological and environmental issues (Wang et al. 2011).

**Water Stress**: There is a constant rise in recognition for water-related problems as one of the most severe and immediate environmental concerns. Water use has been increased three-fold globally since 1950, and one-sixth population of the world does not have regular access to safe drinking water. In the absence of access to a safe water supply and sanitation, the health of 1.2 billion people gets affected annually. The latest Global Environment Outlook of the United Nations Environmental Programme (UNEP) reports that about a third of the world's populations live under moderate-to-high water stress. In such countries, water consumption is more than 10% of renewable freshwater resources.

Urban water demand has been increasing steadily due to population growth, urbanization, industrial development and peri-urban agriculture. Population growth in urban areas is a significant concern for developing countries. Population growth is expected in developing countries, as the developed region's population is projected to decrease by 6% over the next 50 years. Many parts of the world face changes in climatic conditions, such as rainfall patterns, flood cycles, and droughts, which affect the water cycle.

So there is an acute need to augment the present water supply with alternative sources. Several approaches, modern and traditional, exist all over the world for efficiency improvements and augmentation. From numerous options available, wastewater reuse has become increasingly important in water management for both environmental and economic reasons.

Wastewater reuse has an ancient history of applications, primarily in agriculture, and other areas of applications, like industrial, household, and urban, are becoming more and more adopted. Among them, wastewater reuse for agriculture still represents the most significant reuse volume and is expected to increase further in developing countries. With the growing increase in applications, there exists concurrent recognition that water resource management and scheduled water cycle maintenance need up-to-date knowledge regarding basic practices, benefits and potential risks, capacity building of practitioners and planners, and appropriate policy frameworks to protect humans and the environment.

In developed countries, wastewater collection and treatment have been common practice, and wastewater reuse is practiced with proper sanitation, public health and environmental protection. The scenario is quite different in many developing countries due to the lack of appropriate infrastructure and strict wastewater treatment standards for its reuse. Unofficial wastewater reuse for irrigation is quite common in many places, causing a substantial threat for the farmers and consumers of those agricultural products due to the poor quality of wastewater. The World Health Organization (WHO) has published and updating the guidelines for wastewater reuse in agriculture. The efforts are also being made to find out easy and economical ways to localize treatment and reuse wastewater at the source.

**Global Scenario**: Environmental budgeting for requirements shows that the abstraction of water for domestic, food and industrial uses already has a huge footprint on ecosystems in several parts of the world, even though not considered "water scarce." Water will be a major constraint for agricultural development in coming decades and particularly Asia and Africa will require major institutional adjustments (Rijsberman 2006).

Currently, less than 0.5 billion people live in countries with water stress. Still, by 2050, this is expected to increase to about 4 billion, with over 2 billion in water scarcity areas (worst-case) estimates are 7 billion living in areas of water stress and

5 billion in areas of water scarcity (Daigger 2007). The world's urban population will increase from 3.5 to 4.9 billion in 2020, and maximum growth will take place in developing countries (United Nations Human Settlements Programme (UN-Habitat) 2011). So, the reuse of treated wastewater can act as a solution, an imitation of the hydrologic cycle's natural process and help in saving freshwater resources. As depicted from several pioneering studies worldwide, technological confidence has been gained for the safe reuse of reclaimed water at a municipal scale for beneficial purposes. Initially, emphasis was mainly on reuse for agricultural and non-potable reuses; the current trends prove that direct potable reuse is possible for applications that are closer to the point of generation.

There is a large gap between consumption and extraction of water, which establishes that water reuse and recycling can solve the water scarcity problem globally.

There are several well-known examples of indirect potable reuse worldwide, including facilities in Orange County, California, the NEWater facility in Singapore and Windhoek, Namibia (Miller 2006). Starting from the U.S., the practice of recycling/reuse of wastewater is a large and growing industry. Recycled water use on a volume basis is growing at an estimated 15% per year in the U.S. (Schmidt et al. 1975), systematic guidelines at the federal level (U.S. E.P.A. 2004) and state level are well developed, and a lot of research and implementation is going on towards sustainable water use (Burian et al. 2000; Draper et al. 2003; Rozos et al. 2010; Nicklow et al. 2010; May et al. 2008; Liner and DeMonsabert 2011; Mays and Schwartz 1983; Law 1996). Increasing interest in the reuse of effluent from sewage treatment plants in Australia has been observed in recent years and established to solve water shortage by various studies (Anderson 1996; Hurlimann and McKay 2007; Hurlimann 2009; Hamilton et al. 2005; Mekala et al. 2008). Development and validation of design principles for water reuse were rigorously conducted under AQUAREC Project, and remarkable progress has been achieved in E.U. countries (Joksimovic et al. 2006, 2008; Bixio et al. 2008; Hernandez et al. 2006; Urkiaga et al. 2008; Hochstrat et al. 2008; Tsagarakis 2005; Tziakis et al. 2009; Tsagarakis et al. 2004; Iglesias et al. 2010). Mediterranean region had also practiced water reuse and rapid developments regarding criteria and guidelines (Seguí et al. 2009; Hernández-Sancho et al. 2010; Shelef and Azov 1996; Marecos do Monte et al. 1996; Bahri and Brissaud 1996, 2004; Brissaud 2008; Pedrero et al. 2010). Taiwan (You et al. 1999) and Japan (Maeda et al. 1996) also started water reuse, and it has been recognized as an integral part of the water management scheme in China (Yi et al. 2011; Chu et al. 2004; Peng et al. 1995; Yang and Abbaspour 2007; He et al. 2007) and countries like Thailand (Sa-nguanduan and Nititvattananon 2011; Sujaritpong and Nitivattananon 2009).

**Wastewater Reuse Perspective in India**: Though proportion of population using an improved drinking water source in urban area is 96%, the challenges of availability as well as quality in the distribution of drinking water still persist across various areas in the country, and the proportion of population using an improved sanitation facilities is only 54% (Ministry of Environment and Forests 2011). Some major challenges cited in 11th five-year plan are regaining agricultural dynamism, providing essential basic services to the poor, protecting the environment and bridging the divide between rich and poor (Planning Commission 2006). Improved sanitation and wastewater management are central to poverty reduction and improved human health (Corcoran et al. 2010). Though the significance of wastewater reuse and recycling is accepted (Central Pollution Control Board 2009; Ministry of Water Resources Govt. of India 2012), the full potential is not utilized. To summarize, one can say that there is lack of funds to establish basic infrastructure for sanitation; which leads to pollution and informal reuse of untreated or partially treated wastewater. On other side, there is growing demand of water for domestic as well as industrial use. Projected percentage increase in industrial water use over level of year 2000 would be 61.06 if India continues to follow the path of development followed by developed countries (Jia et al. 2006). So, instead of using precious freshwater resources for industrial uses, wherever possible, marginal quality-treated wastewater should be reused, which can reduce freshwater demand, availing more water for domestic use. By proper establishment of facilities with reliability of supply, revenue generation can be achieved from industries against supplied reclaimed wastewater. Further considering nutrient value of treated wastewater, study for coastal towns and cities has showed that wastewater worth Rs. 1091.20 million are discharged into the coastal water (Central Pollution Control Board 2009). Thus, it is ironic to notice that at one end one is spending handsome amounts on fertilizers for better production and on the other hand nutrient-rich wastewater is wasted. Thus proper wastewater management implied under correct policy framework can generate revenues which can be used to build much-needed infrastructure for sanitation and to augment water supply facilities; by providing nutrient-rich wastewater for peri-urban agriculture and horticulture and cooling or process water for industries.

**Public Health Concerns**: In India, where wastewater is mainly used in agriculture, a policy framework covering the issues associated with it is lacking. The new WHO guidelines for wastewater irrigation recognize infrastructure problem in developing countries and emphasize the potential of post or non-treatment options (World Health Organization 2006). There is no standard or guideline available except CPHEEO and CPCB standards in India regarding quality of treated wastewater to be used for irrigation. These standards do not address the coliform count criteria which is most crucial for prevention of health risk to consumer or farmers. A guideline balancing between utilization of maximum nutrient value of wastewater without compromising with the health of farm workers or consumers (Haruvy 1997) should be prepared at the priority basis. This guideline should be prepared after exhaustive research and sharing outcomes of long-term research carried out in developing countries regarding critical issues like fate of volatile organic compounds, pharmaceuticals, endocrine disruptors and antibiotics etc. in wastewater reuse (Snyder and Benotti 2010; Le-Minh et al. 2010; Rodriguez et al. 2012). In addition, a number of social concerns like impaired quality of life, loss of property value, food safety, health and welfare and sustainability of land use, groundwater contamination also should be considered for formation of guidelines. Controlling potential health risks will allow urban water managers to build on the benefits from the already existing (but largely informal) wastewater reuse, those being the contribution to food security and reduction of fresh water demands (Rooijen et al. 2009). Research can be conducted to find out best suitable crops in Indian context keeping in view minimum risk to consumer health.

**Attainment of Economic Viability**: Strictly speaking in Indian context, though greater emphasis must be placed on environmental considerations, public acceptance and public policy issues rather than mere cost-effectiveness as a measure of the feasibility of a water reuse project, attaining economic viability is essential for implementation. To attain this, exhaustive research should be carried out to find out cost-effective technology solutions like improved UASB technology with simple and better operational control, use of dead ponds in town or city may as maturation pond or polishing pond for treated wastewater before reuse (Mahapatra et al. 2011) or sewage reuse after treatment in oxidation pond and duckweed pond (Ghangrekar et al. 2007) or reuse of greywater in decentralized system (Godfrey et al. 2009). The analysis tools and models developed after long-term research all over world to strike the balance between environmental and economic criteria should be used to do cost–benefit analysis (Hamilton et al. 2005; Mekala et al. 2008; Joksimovic et al. 2006, 2008; Bixio et al. 2008; Hochstrat et al. 2008; Lim et al. 2008).

Wise and immediate investment in wastewater management will generate multiple future benefit (Corcoran et al. 2010). Regarding the funds availability, the scenario is gloomy. In India, water is a highly subsidized commodity leading to market inefficiencies and hence inefficient use of the already scarce resource. A majority of urban centres (79%) show revenue deficit on water management account, that is, the revenue receipts are not sufficient to meet the revenue expenditure on the service (Central Public Health and Environmental Engineering Organization 2005). This would demand the establishment of water prices that reflect the full cost recovery principle on the one hand, and the monetarization of the environmental and social benefits of water reuse, on the other. Recycled water valuation is considered as corollary for implementation of reuse. As the present tariff for industrial users is higher and as they have ability to pay higher charges, maximum water reuse with proper price allocation should be targeted for this demand. Just to cite some examples—Chennai Metropolitan Water Supply and Sewerage Board sells 65 M.L.D. of treated wastewater to various industries at 10.75 Rs./K.L. and this demand is going to be doubled in next five years (Ravikumar 2009), Surat, Bhavnagar, Rajkot, Ahmedabad and several other cities have initiated the water reuse. The Government of Gujarat (2018) has published official guidelines for achieving targets for recycling of treated sewage (Government of Gujarat 2018). This shows the light at the end of tunnel in form of acceptance of the theory that water is an economic good and prices can be used to promote equity, efficiency and sustainability (Rogers et al. 2002). Studies have shown willingness to pay by even low-income households for improved water supply and sanitation services (Davis et al. 2008).

**Reduction in Cost of Reclaimed Water Distribution Network**: Water reclamation and reuse is not a cheap option as the infrastructural requirements are usually high, in particular because of the need to construct and/or retrofit the distribution system (Joksimovic et al. 2006). In order to reduce the cost of reuse option, optimization of wastewater distribution networks should be considered. Remarkable achievements are obtained in form of various analysis techniques and optimization models for water distribution networks (Simpson and Elhay 2011; Cheng et al. 2011; Kang and Lansey 2011a; Agrawal et al. 2007). Where treated wastewater is to be used for domestic non-potable consumption, provision of dual-purpose water distribution network is required. Studies conducted in this regard to validate and optimize the network (Kadu et al. 2008; Kang and Lansey 2011b; Tudor and Lavric 2011). One of the important limitation pointed out by Bhave and Pandolkar (2004) was risk of exposure to possible severe health hazard due to use of marginal quality water in dual-purpose network by poor, illiterate people and children for drinking purpose. Giving due regard to this and not considering dual-purpose supply system for domestic use in India, still requirement for distribution system is there for supplying treated wastewater to industries and irrigation fields and as distribution system is to cost up to 70%, it should be optimized. If different approach is used for this compared to optimization of water distribution networks with congested and looped layout, generally better and quick results may be achieved. This can be identified as research gap in this area as optimization of reclaimed water network for supplying reclaimed water to potential users is not yet been carried out in India. By applying known techniques to this kind of specific problem, a new technique or modification of existing one could be found out to attain optimal results.

**Policy Framework**: Review of National water policy gives a pleasing experience as most of the relevant, renowned and successful practices and criteria regarding water resource management are getting reflected, namely consideration of increasing participation of all stakeholders, proper maintenance of existing infrastructure, proper identification of economic value of water and setting price to recover the full cost, good governance, integrated water management, evolving an agricultural system which economizes on water use, generation of database and information management system based on modern techniques available and much more, but it surely lacks the directive for generation of framework guideline for water reuse and secondly the approach for setting up pricing of water seems to be indecisive and soft, also it doesn't reflect the strategy to safeguard poor people lifeline water needs. As India's economy has grown, so too has the spending power of its citizens. Real average household income in India has roughly doubled over the past two decades. With a booming economy and increasing disposal income, there is an urgent need to introduce a new cost head in the urban water bills and introduce the "polluter pays" principal for urban water consumers. The new cost head will contribute to sewage treatment before it enters to inland water bodies and pollutes them and increases chance of availing benefits of wastewater reuse also. Geographically targeted system would result in significant improvements in performance, for example, lifeline rates could be set at lower levels in slum areas while increasing prices for non-poor users (McKenzie and Ray 2009). Thus bringing pricing reform is the need of the hour and without prior pricing reform private sector engagement will be difficult. Private sector engagement is a welcome step towards full privatization of water management under build-own-operate-transfer (BOOT) contracts reducing economic burden on government. Australia has already implied privatization in water and wastewater sector since 1980, which has improved its economic performance in terms of productivity and returns to the shareholder, however, consumers have not substantially benefited from this process in terms of lower prices (Abbott et al. 2011). So, the "Service Provider" role of the government has to be gradually shifted to that of a regulator of services and facilitator for strengthening the institutions responsible for planning, implementation and management of water resources. The water-related services should be transferred to community and/or private sector with appropriate "Public Private Partnership" model.

To be effective and to be implemented, a requisite shift to sustainability requires active community engagement processes, political will and a commitment to political and administrative accountability, and measurement (Ling et al. 2009). Some of the examples of this political will are raising of finance for water and sewerage expansions through municipal bonds in 1998 by Ahmedabad Municipal Corporation (A.M.C.) (becoming the first Indian municipality to use this mode of raising capital, though it is not common in South Asia) (McKenzie and Ray 2009), formation of Gujarat Green Revolution Company and encouraging use of drip irrigation and appointing an international consultant for deciding pricing of water across the state for various purposes by Gujarat Government (2005). As in most democracies, any major reform needs to survive (indeed, be part of) the political process, while even small changes in prices require public approval, it is very much required to highlight the benefits and improved performances achieved by government and win the public opinion.

The management of water has become a complex policy issue bringing into its fold state, market and civil society. Attempts of commercialization of water are evident and there appears to be some justification in the criticism. The study started as an attempt to understand the politics of inter-relation between state and market in India for provisioning for water to the people. It shows that water is fast becoming a tradable commodity the consequences of which is seen and felt (Samanta 2009). The detailed description of the state of the art and methodologies used in this article is available with full details in Shah and Shete (2019).

## 23.2 Objective and Goal of the Work

**Goal**: To attain sustainable water reuse management by full-cost recovery, economic efficiency and economic optimization considering an idea of "negative pricing."

**Objective**: To achieve optimal sustainability for urban wastewater management for Vadodara Urban Development Authority Area.

**Criteria** for evaluating this objective are:

Economic returns and profitability
Reliability of water supply.

**Specific (Measurable) Indicators:**

Amount of water being reused
Cost of distribution network
Selling price of reclaimed water
Internal Rate of Return.

## 23.3 Methodology

### *23.3.1 Study Area*

The area around Vadodara city under jurisdiction of Vadodara Urban Development Authority is considered in the present study (Fig. 23.1).

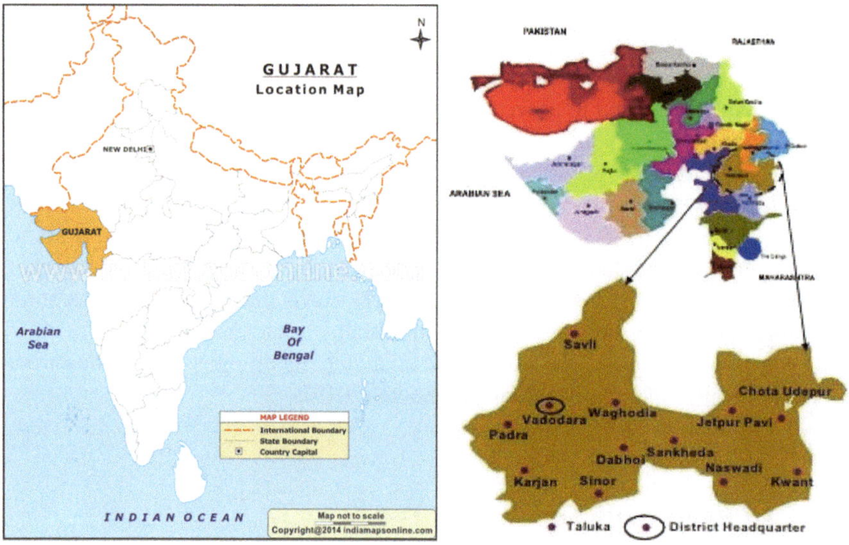


**Fig. 23.1** Study area

### 23.3.2 Data Collection

From the VMSS, data were collected for location, input and output parameters for different sewage treatment plants, cost of treatment, present and future demands, population forecasting, etc. Population was projected using various established and latest techniques, and population to be served in VUDA area was considered as 13 lakhs as obtained by auto regression method.

### 23.3.3 Specific (Measurable) Indicator

(a) Amount of water being reused

Mainly two type of uses were identified

Agricultural reuse
Industrial and residential reuse.

The criteria for water reuse as given in various standards were studied and considered. The amount of water being reused depends upon demands for industrial and residential reuse and irrigational reuse. These demands had been worked out as explained on page 11 and amount of water being reused was determined.

(b) Cost of distribution network

The major part of the cost of the distribution network depends on diameter of the pipe. Diameter depends on loss of head due to friction. Loss of head due to friction depends on friction factor.

(c) Selling price of reclaimed water

Assuming the profit over the cost of distribution of treated water should be 15%, selling price of reclaimed water was determined.

(d) Internal Rate of Return

IRR is the interest rate at which the net present value of all the cash flows both positive and negative over the period from a project equals to zero.

Based on cost analysis Internal Rate of Return was calculated.

To achieve the goal as stated earlier reclaimed water distribution network model was formulated using QGIS and EPANET.

The proposed land use map for Vadodara is obtained and superimposed on Google Earth view of VUDA area.

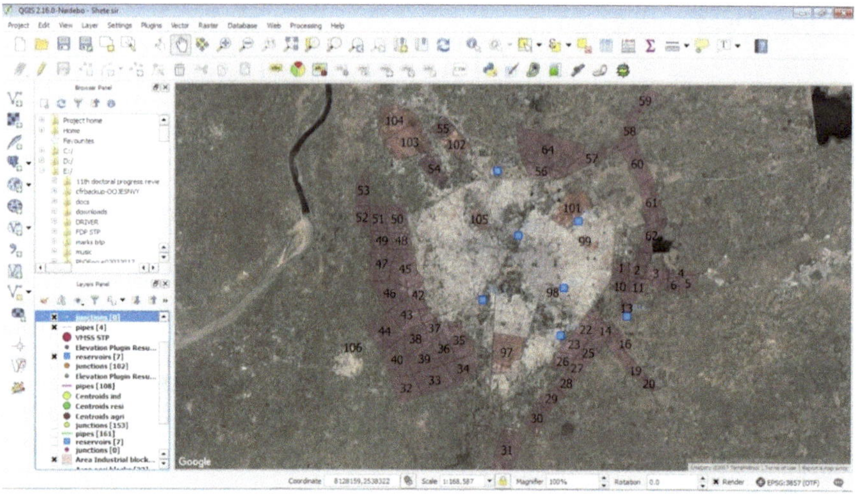


Fig. 23.2 The industrial and residential areas marked in light magenta and light burgundy colour, respectively, surrounding Vadodara city as per proposed land use map 2031 for VUDA area

### 23.3.4 Identification of Residential, Industrial and Agricultural Areas

The proposed land use map obtained from Vadodara Urban Development Authority (VUDA) was superimposed on Google earth exactly by adjusting scale and matching landmark points. After superimposing proposed land use map various areas had been identified for industrial, residential and agricultural use as per land use zones shown in the VUDA map for developing outskirt areas of Vadodara City. These areas were then marked with different layers in Quantum G.I.S. (QGIS) software as shown in Fig. 23.2.

The areas were divided into small compartments so as to create one node for laying out of proposed reclaimed water pipelines. The network of proposed reclaimed water pipeline was prepared in QGIS using GHydraulics plug-in as shown in Fig. 23.3.

The locations of sewage treatment plants, agricultural zones, industrial and residential zones were identified and the reclaimed water distribution networks from various sewage treatment plants to these areas were prepared.

With seven sources available, question of what should be the optimal proportion of reclaimed water distribution from various sources to different destination arose.

To solve this problem (Vogel Approximation Method) as a special case of linear programming with optimality checks was utilized.

Before doing distribution by V.A.M., primary distribution by visual observation based on vicinity was carried out and network was prepared.

From QGIS, details of area for industrial and residential and agricultural zones were obtained, and centroid of each area was taken to allocate node of pipe network.

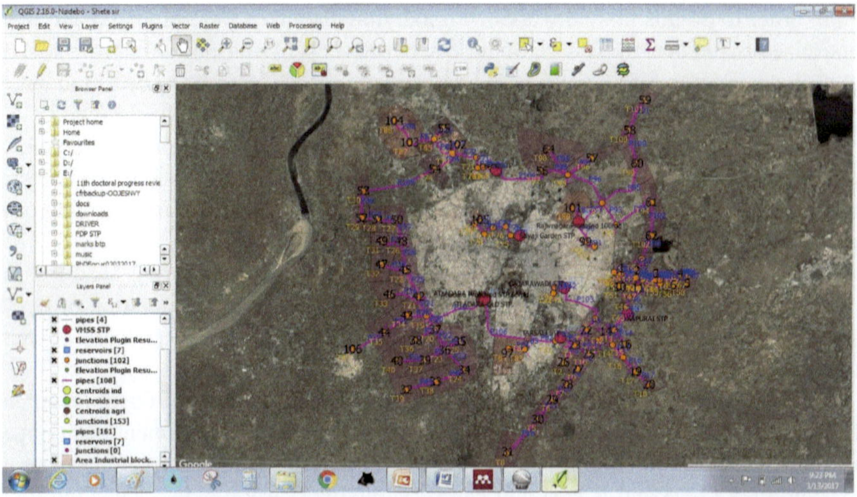

**Fig. 23.3** The industrial and residential areas and proposed reclaimed water distribution network from various sewage treatment plants in VUDA area

Elevation of all nodes and lengths of all pipes were obtained from QGIS for networks. Using GHydraulics plug-in input file in the .inp format was prepared from QGIS and imported in EPANET for further analysis.

Considering per capita supply of reclaimed water to be 100 l/day, the demand for each residential zone was found out. For determination of industrial demand, case study of Makarpura GIDC was taken. From VCCI, directory listing of industries in Makarpura GIDC was obtained. For each type of industries, demand was identified and ultimate demand for conglomeration was worked out. Agricultural demand was worked out considering banana crop and drip irrigation system.

After importing .inp file from QGIS, GHydraulics plug-in the .net file was prepared to be run in WaternetGen (an EPANET extension for sizing pipes using simulated annealing algorithm of optimization). First of all separate networks for each treatment plant were considered and design of optimum size of pipes was carried out. Then considering spatial distribution of industrial and residential zones, the integrated model was prepared. The cost of networks was worked out considering material cost, cost for excavation, cost of lowering, laying and joining, refilling the trenches, accessories, etc. complete using GWSSB schedule of rates. As there were several destinations and demand locations with demand and supply constraints, the distribution of treated wastewater from different sewage treatment plant formed an excellent transportation problem.

V.A.M. is considered to be the best method for solution of transportation problem which is a special case for linear programming. To form transportation problem, transportation tableau was required to be prepared and for that from each individual source cost/unit of transportation to each destination was required to be found out.

In order to do this first networks with individual sources were prepared and then optimized. After designing each of these networks cost of transportation from source to each destination for all sources to all destinations was required to be found out.

For this purpose, Elevated Service Reservoirs were required to be designed.

Using mass curve method E.S.R.s were designed and costs of pump and E.S.R. were found out. For calculating the cost of piping from each source to each destination, the links involved in each route were calculated and accordingly piping cost was arrived for each route.

Similarly, pumping cost was calculated for each route considering head loss in each pipe and elevation difference between concerned nodes. Calculation of O, M and R Cost was carried out using Maximum Accelerated Cost Recovery System method for depreciation.

As this was annual cost, the capital cost was also required to be brought in annual format, so total capital cost of sewage treatment plants, tertiary treatment plant, pipe, E.S.R. and pump was multiplied with capital recovery factor using interest rate at 7% and then O, M and R cost and annual capital cost were added. Knowing total supply from each source, the final demand for each node, and transportation cost for each node, the transportation model was prepared.

Finally, AMPL solver (Taha 2011) was selected for the task of solving the tableau by V.A.M. method and optimal allocation was achieved.

Using this final allocation network was prepared and the cost of all components was determined.

## 23.4 Results and Discussion

Real opportunity cost of providing the reclaimed water was calculated.

Following five scenarios were considered:

Scenario 1: Considering all costs
Scenario 2: Considering only Elevated Service Reservoir, Pump and Piping cost
Scenario 3: Considering only O, M and R Cost
Scenario 4: Considering selling prices as per Sardar Sarovar Narmada Nigam Limited
Scenario 5: Considering saved fresh water cost as per selling prices.

The optimal cost of transportation of reclaimed water from each source to different demand nodes using AMPL solver was determined. There was slight difference in cost calculations in scenarios 1 and 5. In scenario 1, selling price of saved water for irrigation purpose was considered as charged by Sardar Sarovar Narmada Nigam Limited, whereas in scenario 5 selling price of saved water was based on the cost of treated water plus 15% profit for irrigation purpose.

Tableaus for all scenarios were prepared. AMPL solver (Taha 2011) was used to solve the tableau by V.A.M. method and optimal allocation of reclaimed water from each source to different destinations and the optimal costs of transportation of

**Table 23.1** **c** Cast distribution for industrial and residential purposes and **d** cast distribution for irrigation purpose

| | |
|---|---|
| **(c)** | |
| Total ESR cost | 309,600,000 |
| Total piping cost | 871,876,597 |
| Total pump cost | 6,370,000 |
| Capital cost of piping, ESR and pump | **1,187,846,597** |
| **(d)** | |
| Total ESR cost | 407,720,000 |
| Total piping cost | 1,015,388,449 |
| Total pump cost | 10,015,543 |
| Capital cost of piping, ESR and pump | **1,433,123,992** |

Bold values show the summative or decisive values for different important parameters

reclaimed water from various sources to all destinations for various options were determined.

After selecting all the plants connected network as optimal solution, based on V.A.M. output from AMPL software, the reclaimed water distribution network was prepared.

For this final optimal network, all the costs like pipe network cost, pumping Cost, E.S.R. and pump Cost, O, M and R cost etc. were calculated that is shown in Table 23.1c, d.

For calculation of Internal Rate of Return total capital costs were calculated as follows:

(a) Capital cost treatment plants for primary and secondary treatment—Rs. **4,294,049,105**
(b) Capital cost treatment plants for tertiary treatment—Rs. **457,846,259**

**Total Capital Cost 7,372,865,953**

**(a + b + c + d) Say 7.37 × $10^9$.**

Selling price for industrial and residential reuse for scenario 1 was calculated as Table 23.2.

This selling price is similar to selling price of Sardar Sarovar Narmada Nigam Limited selling price of Rs. 40.17 Rs./KL in 2020.

Similarly, the selling price for irrigation purpose reuse was also determined and it came to be Rs. 9/K.L.

Internal rate of return was calculated using selling price of reclaimed water and all the costs to treat the urban wastewater for each of the five scenarios to ascertain the economic sustainability of the project (Table 23.3).

The optimal solution for scenario 1, Table 23.4 was subjected to sensitivity analysis by increasing and decreasing demand by 5, 10 and 15%, and % variations in AMPL optimal output were found out.

**Table 23.2** Details of selling price for industrial and residential reuse for scenario 1

| Total demand in MLD | 177.946 |
|---|---|
| The annual cost of the treatment plant, including O and M | Rs. 278,161,627 |
| The annual cost of additional treatment | Rs. 457,846,259 |
| Cost of additional treatment O and M | Rs. 1,335,605,508 |
| Annual cost of E.S.R., pump and piping | Rs. 95,724,285 |
| Total O, M and R cost for piping, E.S.R. and pump | Rs. 66,945,285 |
| Total pumping cost | Rs. 25,729,757 |
| **TOTAL COST** | Rs. **2,260,012,721** |
| **i.e. COST/MLD** | Rs. **12,700,554** |
| **i.e. COST/Kilo Liter** | Rs. 34.80 |
| Adding 15% profit | 5.22 |
| **Selling price/K.L** | Rs. **40.02** |
| **Say** | **Rs./KL 40.00** |

Bold values show the summative or decisive values for different important parameters

As % variation in AMPL optimal output varied from −0.02 to 1.23%, it was considered as negligible for practical considerations. Therefore, internal rate of return was not calculated for increase and decrease in demand by 5, 10 and 15%.

The next alternative for sensitivity analysis is variation in selling pricc. Considering increase and decrease in selling price by 5, 10 and 15%, internal rate of return was calculated for all the five scenarios, whereas 3 scenarios shown in Fig. 23.4.

Internal rate of return was also calculated considering negative pricing. At present, there are five existing S.T.P.s and one proposed S.T.P. at Rajivnagar. Looking to the future need of VUDA area, it is assumed that a new additional S.T.P. at Chhani is required. Table 23.5 represents IRR with and without negative pricing consideration for with and without 35 M.L.D. Chhani new plant.

IRR with and without considering Chhani new plant for different scenarios were calculated. Scenario 5 is the best choice under given conditions which is normally prevalent everywhere.

### Achievements with respect to objectives

Specific (measurable) Indicator

(a) Amount of water being reused

For irrigation, gardening and tree plantation purposes = 248.584 MLD.

Total water to be reused = 426.530 MLD.

(b) Cost of combined distribution network

Industrial and residential purposes = Rs. 871,876,597.

(New Chhani plant was not considered for industrial and residential purposes network because all the reclaimed water from Chhani new plant was given for irrigation purpose.)

**Table 23.3** IRR for scenario 1 considering all costs

| Sr. No. | Particulars | 0th year | 1st year | 2nd year | 3rd year | … | 28th year | 29th year | 30th year |
|---|---|---|---|---|---|---|---|---|---|
| **1** | **CAPITAL COST, Rs.** | $7.37 \times 10^9$ | | | | … | | | |
| **1.1** | **LOAN RECEIVED FROM BANK, Rs.** | $6.06 \times 10^9$ | | | | … | | | |
| **2** | **FIXED COST** | | | | | … | | | |
| 2.1 | Rate of depreciation for first half year as per MACRS, % | | | 3.75 | 7.219 | … | 0 | 0 | 0 |
| 2.2 | Rate of depreciation for second half year as per MACRS, % | | 3.75 | 7.219 | 6.677 | … | 0 | 0 | 0 |
| 2.3 | Depreciation for 1st half year, Rs. | | 0 | $1.38 \times 10^8$ | $2.66 \times 10^8$ | … | 0 | 0 | 0 |
| 2.4 | Depreciation for 2nd half year, Rs. | | $1.38 \times 10^8$ | $2.66 \times 10^8$ | $2.46 \times 10^8$ | … | 0 | 0 | 0 |
| 2.5 | Total depreciation, Rs. | | $1.38 \times 10^8$ | $4.04 \times 10^8$ | $5.12 \times 10^8$ | … | 0 | 0 | 0 |
| 2.6 | Cumulative depreciation, Rs. | | $1.38 \times 10^8$ | $5.43 \times 10^8$ | $1.05 \times 10^9$ | … | $7.37 \times 10^9$ | $7.37 \times 10^9$ | $7.37 \times 10^9$ |
| 2.7 | Amount of repayment of loan, Rs. | | $4.46 \times 10^8$ | $4.46 \times 10^8$ | $4.46 \times 10^8$ | … | $4.46 \times 10^8$ | $4.46 \times 10^8$ | $4.46 \times 10^8$ |
| 2.8 | Interest obtainable in 1st half year, Rs. | | 0 | $2.76 \times 10^6$ | $1.10 \times 10^7$ | … | $1.54 \times 10^8$ | $1.54 \times 10^8$ | $1.54 \times 10^8$ |
| 2.9 | Interest obtainable in 2nd half year, Rs. | | 0 | $5.58 \times 10^6$ | $1.66 \times 10^7$ | … | $1.57 \times 10^8$ | $1.57 \times 10^8$ | $1.57 \times 10^8$ |
| 2.10 | Total interest to be received on depreciation money, Rs. | | 0 | $8.35 \times 10^6$ | $2.76 \times 10^7$ | … | $3.10 \times 10^8$ | $3.10 \times 10^8$ | $3.10 \times 10^8$ |

(continued)

**Table 23.3** (continued)

| Sr. No. | Particulars | 0th year | 1st year | 2nd year | 3rd year | … | 28th year | 29th year | 30th year |
|---|---|---|---|---|---|---|---|---|---|
| 2.11 | Interest lost on 25% capital cost as seed money, Rs. | | $7.37 \times 10^7$ | $7.37 \times 10^7$ | $7.37 \times 10^7$ | … | $7.37 \times 10^7$ | $7.37 \times 10^7$ | $7.37 \times 10^7$ |
| **3** | Insurance, Rs. | | $5.53 \times 10^7$ | $5.53 \times 10^7$ | $5.53 \times 10^7$ | … | $5.53 \times 10^7$ | $5.53 \times 10^7$ | $5.53 \times 10^7$ |
| **4** | Cost of O and M for primary and secondary plant, Rs. | | $1.73 \times 10^8$ | $1.73 \times 10^8$ | $1.73 \times 10^8$ | … | $1.73 \times 10^8$ | $1.73 \times 10^8$ | $1.73 \times 10^8$ |
| **5** | Cost of O and M for tertiary plant, Rs. | | $1.34 \times 10^9$ | $1.34 \times 10^9$ | $1.34 \times 10^9$ | … | $1.34 \times 10^9$ | $1.34 \times 10^9$ | $1.34 \times 10^9$ |
| **6** | Maintenance and repairs of E.S.R., pump and piping, Rs. | | $1.54 \times 10^8$ | $1.54 \times 10^8$ | $1.54 \times 10^8$ | … | $1.54 \times 10^8$ | $1.54 \times 10^8$ | $1.54 \times 10^8$ |
| **7** | Pumping cost, Rs. | | $6.10 \times 10^7$ | $6.10 \times 10^7$ | $6.10 \times 10^7$ | … | $6.10 \times 10^7$ | $6.10 \times 10^7$ | $6.10 \times 10^7$ |
| **8** | Operator's salary, Rs. | | $1.44 \times 10^6$ | $1.44 \times 10^6$ | $1.44 \times 10^6$ | … | $1.44 \times 10^6$ | $1.44 \times 10^6$ | $1.44 \times 10^6$ |
| **9** | Other charges, Rs. | | $3.69 \times 10^7$ | $3.69 \times 10^7$ | $3.69 \times 10^7$ | … | $3.69 \times 10^7$ | $3.69 \times 10^7$ | $3.69 \times 10^7$ |
| **10** | CASH OUTFLOW, Rs. | | $2.34 \times 10^9$ | $2.34 \times 10^9$ | $2.34 \times 10^9$ | … | $2.34 \times 10^9$ | $2.34 \times 10^9$ | $2.34 \times 10^9$ |
| **11** | **NET CASH OUTFLOW, Rs.** | | $\mathbf{2.34 \times 10^9}$ | $\mathbf{2.33 \times 10^9}$ | $\mathbf{2.31 \times 10^9}$ | … | $\mathbf{2.03 \times 10^9}$ | $\mathbf{2.03 \times 10^9}$ | $\mathbf{2.03 \times 10^9}$ |
| **12** | **INCOME** | | | | | … | | | |
| 12.1 | Selling price of treated water for industrial and residential purpose Rs./KL | | 40.00 | 40.00 | 40.00 | … | 40.00 | 40.00 | 40.00 |
| 12.2 | Total treated water to be sold for industrial and residential purpose, K.L | | $6.50 \times 10^7$ | $6.50 \times 10^7$ | $6.50 \times 10^7$ | … | $6.50 \times 10^7$ | $6.50 \times 10^7$ | $6.50 \times 10^7$ |

(continued)

**Table 23.3** (continued)

| Sr. No. | Particulars | 0th year | 1st year | 2nd year | 3rd year | … | 28th year | 29th year | 30th year |
|---|---|---|---|---|---|---|---|---|---|
| 12.3 | Income from selling treated water for industrial and residential purpose, Rs. | | $2.60 \times 10^9$ | $2.60 \times 10^9$ | $2.60 \times 10^9$ | … | $2.60 \times 10^9$ | $2.60 \times 10^9$ | $2.60 \times 10^9$ |
| 12.4 | Total saved fresh water income as per SSNNL rates for industrial and residential purpose, Rs. | | $2.60 \times 10^9$ | $1.69 \times 10^9$ | $1.69 \times 10^9$ | … | $1.69 \times 10^9$ | $1.69 \times 10^9$ | $1.69 \times 10^9$ |
| 12.5 | Selling price of treated water for irrigation purpose, Rs./KL | | 9 | 9 | 9 | … | 9 | 9 | 9 |
| 12.6 | Total treated water to be sold for irrigation purpose, K.L | | $8.15 \times 10^7$ | $8.15 \times 10^7$ | $8.15 \times 10^7$ | … | $8.15 \times 10^7$ | $8.15 \times 10^7$ | $8.15 \times 10^7$ |
| 12.7 | Income from selling water for irrigation purpose, Rs. | | $7.34 \times 10^8$ | $7.34 \times 10^8$ | $7.34 \times 10^8$ | … | $7.34 \times 10^8$ | $7.34 \times 10^8$ | $7.34 \times 10^8$ |
| 12.8 | Total saved fresh water income as per SSNNL rate for irrigation purpose | | $4.32 \times 10^7$ | $4.32 \times 10^7$ | $4.32 \times 10^7$ | … | $4.32 \times 10^7$ | $4.32 \times 10^7$ | $4.32 \times 10^7$ |
| **13** | **CASH INFLOW** | | $5.06 \times 10^9$ | $5.06 \times 10^9$ | $5.06 \times 10^9$ | … | $5.06 \times 10^9$ | $5.06 \times 10^9$ | $5.06 \times 10^9$ |
| **14** | **NET CASH FLOW, Rs.** | **$-1.34 \times 10^{10}$** | $2.73 \times 10^9$ | $2.73 \times 10^9$ | $2.75 \times 10^9$ | … | $3.04 \times 10^9$ | $3.04 \times 10^9$ | $3.04 \times 10^9$ |
| **15** | **INTERNAL RATE OF RETURN** | | *27.47%* | | | … | | | |
| | | | **1ST YEAR** | **2ND YEAR** | **3RD YEAR** | … | **28th YEAR** | **29th YEAR** | **30th YEAR** |

Bold values show the summative or decisive values for different important parameters

**Table 23.4** Sensitivity analysis of optimal solution for different scenarios

| Description | Internal rate of return | | | | | | |
|---|---|---|---|---|---|---|---|
| | Original selling price (%) | % Increase and decrease in selling price | | | | | |
| | | 5% | 10% | 15% | −5% | −10% | −15% |
| Scenario 1 | 27.47 | 29.99 | 31.97 | 32.92 | 25.56 | 22.99 | 22.03 |
| Scenario 2 | NA | | | | | | |
| Scenario 3 | 17.76 | 19.20 | 21.19 | 23.22 | 16.31 | 14.29 | 12.18 |
| Scenario 4 | 22.37 | NA | | | | | |
| Scenario 5 | 32.88 | 36.04 | 37.95 | 38.91 | 30.97 | 27.82 | 26.86 |

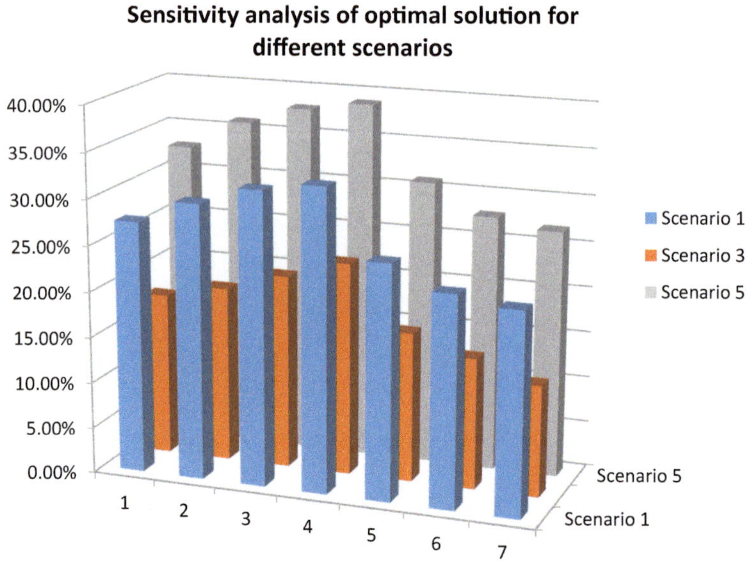


**Fig. 23.4** Sensitivity analysis of optimal solution for different scenarios

**Table 23.5** IRR with and without considering Chhani new plant for different scenarios and considering negative pricing

| Description | Considering 35 M.L.D. Chhani new plant | | Without considering 35 M.L.D. Chhani new plant | |
|---|---|---|---|---|
| | Original (%) | With negative pricing (%) | Original (%) | With negative pricing (%) |
| Scenario 1 | 26.33 | 20.10 | 27.47 | 21.01 |
| Scenario 2 | NA | | NA | |
| Scenario 3 | 16.52 | 12.77 | 17.76 | 13.83 |
| Scenario 4 | 20.91 | 16.78 | 22.37 | 18.03 |
| Scenario 5 | 32.35 | 26.13 | 32.88 | 26.42 |

**Table 23.6** Internal rate of return with and without negative pricing consideration for with and without 35 M.L.D. Chhani new plant

| Description | Considering 35 M.L.D. Chhani new plant | | Without considering 35 M.L.D. Chhani new plant | |
|---|---|---|---|---|
| | Original (%) | With negative pricing (%) | Original (%) | With negative pricing (%) |
| Scenario 1 | 26.33 | 20.10 | 27.47 | 21.01 |
| Scenario 2 | NA | | NA | |
| Scenario 3 | 16.52 | 12.77 | 17.76 | 13.83 |
| Scenario 4 | 20.91 | 16.78 | 22.37 | 18.03 |
| Scenario 5 | 32.35 | 26.13 | 32.88 | 26.42 |

Irrigation purposes:

Without New Chhani STP = Rs. 1,015,388,448.

With New Chhani STP = Rs. 1,062,799,022.

(c) Selling price of reclaimed water

Industrial and residential purposes = Rs. 40/K.L.

Irrigation purpose:

Without New Chhani S.T.P. = Rs. 9/K.L.

With New Chhani STP = Rs. 10/KL.

(d) Internal Rate of Return, Table 23.6 with Chhani New Plant.

**Criteria**

(a) Economic returns and profitability

IRR for scenario 1 considering option with and without Chhani new S.T.P. plant were 26.33% and 27.47%, respectively.

IRR for scenario 5 considering option with and without Chhani new S.T.P. plant were 32.35% and 32.88%, respectively.

As the IRR is above 11% (criteria of Asian Development Bank), the distribution of reclaimed water is economically sustainable.

(b) Reliability of water supply

As the IRR is above 11% (criteria of Asian Development Bank), the distribution of reclaimed water is sustainable. Therefore, the prospect of sustainability will make the reclaimed water distribution reliable.

As all plants are connected in selected option, during failure of any plant, water can be diverted from other plant and thus reliability of supply can be achieved.

The most important point to be noted is the cost of reclaimed water will become Rs. 29/KL in 2020 if yearly increase of 10% in anticipated till 2050. The present rate of Sardar Sarovar Narmada Nigam for non-agricultural usage is Rs. 26/KL in 2020.

That means the cost of reclaimed water is almost equal to fresh water. Further the IRR obtained is 34% which is very high and practically 12% is sufficient to make the project economically attractive. Considering 12% IRR, the cost of reclaimed water will be Rs. 7/K.L. and water for irrigation can be made available at Rs. 1/K.L. Considering 10% increase annually, it will further reduce to Rs. 3/K.L. for industrial purpose and Rs. 0.5/K.L. for irrigational purpose.

If the cost of saved fresh water by using reclaimed water is not considered than also the cost of reclaimed water for industries is coming to be Rs. 21/K.L. for industrial purpose and Rs. 2/K.L. for irrigational purposes.

Thus, it is clearly established that water reuse from sewage treatment plants for industrial, residential and irrigational purpose is not only beneficial for environmental purpose but also economically attractive and can cater as source of revenue promoting infrastructure growth by P.P.P. model.

## 23.5 Conclusion

As Specific (measurable) Indicators fulfil the criteria, the objective can be achieved. Thus, it can be stated that reuse of reclaimed water to VUDA area is sustainable. After calculating IRR, sensitivity analysis was carried out by increasing and decreasing the selling price by 5, 10 and 15% for scenario 1 and scenario 5. In all cases, IRR remained well above 11%. So, even if there is decrease of 15% in selling price the reuse option is profitable. The sensitivity analysis reveals that scenario 2, 3 and 4 are not profitable. Scenario 5 is the most profitable option therefore selling price of reclaimed water for industrial and residential purposes should be Rs. 40/K.L. and for irrigation purpose Rs. 10/K.L. Considering 12% IRR, the cost of reclaimed water will be Rs. 7/K.L. and water for irrigation can be made available at Rs. 1/K.L. Considering a 10% increase annually, it will further reduce to Rs. 3/K.L. for industrial purpose and Rs. 0.5/K.L. for irrigational purposes. The successful and reliable designing of reclaimed water can be carried out using QGIS, EPANET and such open-source software with minimum time and site work. From these conclusions, it is no longer appropriate to consider treated municipal wastewater as a "waste" that requires "disposal," but rather, it should be used as a resource that can be put to beneficial use.

## References

Abbott M, Wang WC, Cohen B (2011) The long-term reform of the water and wastewater industry: the case of Melbourne in Australia. Util Policy 19(2):115–122

Agrawal ML, Gupta R, Bhave PR (2007) Reliability-based strengthening and expansion of water distribution networks. J Water Resour Plan Manag 133(6):531–541

Anderson JM (1996) Current water recycling initiatives in Australia: scenarios for the 21st century. Water Sci Technol 33(10–11):37–43

Bahri A, Brissaud F (1996) Wastewater reuse in Tunisia: assessing a national policy. Water Sci Technol 33(10–11):87–94
Bahri A, Brissaud F (2004) Setting up microbiological water reuse guidelines for the Mediterranean. Water Sci Technol 50(2):39–46
Bhave P, Pandolkar A (2004) Dual-purpose water supply system. J Indian Water Work Assoc 36(4):281–286
Bixio D et al (2008) Water reclamation and reuse: implementation and management issues. Desalination 218(1–3):13–23
Brissaud F (2008) Criteria for water recycling and reuse in the Mediterranean countries. Desalination 218(1–3):24–33
Brundtland G (1987) Our common future. World Commission on Environment and Development. [Online]. Available: http://www.regjeringen.no/upload/SMK/Vedlegg/Taler-og-artikler-av-tidligere-statsministre/Gro-Harlem-Brundtland/1987/Presentation_of_Our_Common_Future_to_UNEP.pdf. Accessed 23 Feb 2012
Burian SJ, Nix SJ, Pitt RE, Durrans SR (2000) Urban wastewater management in the United States: past, present and future. J Urban Technol 7(3):33–62
Central Pollution Control Board (2009) Status of water supply, wastewater generation and treatment in class-I cities & class-II towns of India
Central Public Health and Environmental Engineering Organization (2005) Status of water supply, sanitation and solid waste management in urban areas. Waste Management
Cheng W, He Z, Asce AM (2011) Calibration of nodal demand in water distribution systems. J Water Resour Plan Manag 137(1):31–40
Chu J, Chen J, Wang C, Fu P (2004) Wastewater reuse potential analysis: implications for China's water resources management. Water Res 38(11):2746–2756
Corcoran E, Nellemann C, Baker E, Bos R, Osborn D, Savelli H (2010) Sick water? The central role of waste-water management in sustainable development. United Nations Environment Programme, UN-HABITAT, GRID-Arendal, Norway. www.grida.no. Retrieved from http://www.unep.org/pdf/SickWater_screen.pdf
Daigger GT (2007) Wastewater management in the 21st century. J Environ Eng 133(7):671–680
Davis J, White G, Damodaron S, Thorsten R (2008) Improving access to water supply and sanitation in urban India: microfinance for water and sanitation infrastructure development. Water Sci Technol 58(4):887–891
Draper AJ, Jenkins MW, Kirby KW, Lund JR, Howitt RE (2003) Economic-engineering optimization for California water management. J Water Resour Plan Manag 129(3):155–164
Ghangrekar MM, Kishor N, Mitra A (2007) Sewage reuse for aquaculture after treatment in oxidation and duckweed pond. Water Sci Technol 55(11):173–181
Godfrey S, Labhasetwar P, Wate S (2009) Greywater reuse in residential schools in Madhya Pradesh, India—a case study of cost-benefit analysis. Resour Conserv Recycl 53(5):287–293
Government of Gujarat (2018) Policy for reuse of treated wastewater. GWSSB, India, p 32
Govt. of Gujarat (2005) Gujarat green revolution company. [Online]. Available: http://ggrc.co.in/frmhomepage.aspx
Hamilton AJ et al (2005) Position of the Australian horticultural industry with respect to the use of reclaimed water. Agric Water Manag 71(3):181–209
Haruvy N (1997) Agricultural reuse of wastewater: nation-wide cost-benefit analysis. Agric Ecosyst Environ 66:113–119
He H, Zhou J, Wu Y, Yu Q, Zhang W, Xie X (2007) Estimating water quality pollution impacts based on economic loss models in urbanization process in Xi'an, China. J Urban Plan Dev 133(3):151–160
Hernandez F et al (2006) Feasibility studies for water reuse projects: an economical approach. Desalination 187(1–3):253–261
Hernández-Sancho F, Molinos-Senante M, Sala-Garrido R (2010) Economic valuation of environmental benefits from wastewater treatment processes: an empirical approach for Spain. Sci Total Environ 408(4):953–957

Hochstrat R, Wintgens T, Melin T (2008) Development of integrated water reuse strategies. Desalination 218(1–3):208–217
Hurlimann A (2009) Water supply in regional Victoria Australia: a review of the water cartage industry and willingness to pay for recycled water. Resour Conserv Recycl 53(5):262–268
Hurlimann A, McKay J (2007) Urban Australians using recycled water for domestic non-potable use—an evaluation of the attributes price, saltiness, colour and odour using conjoint analysis. J Environ Manage 83(1):93–104
Iglesias R, Ortega E, Batanero G, Quintas L (2010) Water reuse in Spain: data overview and costs estimation of suitable treatment trains. Desalination 263(1–3):1–10
Jia S, Yang H, Zhang S, Wang L, Xia J (2006) Industrial water use Kuznets curve: evidence from industrialized countries and implications for developing countries. J Water Resour Plan Manag 132(3):183–191
Joksimovic D, Kubik J, Hlavinek P, Savic D, Walters G (2006) Development of an integrated simulation model for treatment and distribution of reclaimed water. Desalination 188(1–3):9–20
Joksimovic D, Savic DA, Walters GA, Bixio D, Katsoufidou K, Yiantsios SG (2008) Development and validation of system design principles for water reuse systems. Desalination 218(1–3):142–153
Kadu MS, Gupta R, Bhave PR (2008) Optimal design of water networks using a modified genetic. J Water Resour Plan Manag 134(2):147–160
Kang D, Lansey K (2011) Revisiting optimal water distribution system design: issues and a heuristic hierarchical approach. J Water Resour Plan Manag
Kang D, Lansey K (2011) Dual water distribution network design under triple-bottom-line objectives. J Water Resour Plan Manag 520
Law IB (1996) Rouse Hill—Australia's first full scale domestic non-potable reuse application. Water Sci Technol 33(10–11):71–78
Le-Minh N, Khan SJ, Drewes JE, Stuetz RM (2010) Fate of antibiotics during municipal water recycling treatment processes. Water Res 44(15):4295–4323
Leverenz HL, Tchobanoglous G, Takashi A (2011) Direct potable reuse: a future imperative. J Water Reuse Desalin 1(1):2–10
Lim S-R, Park D, Park JM (2008) Analysis of effects of an objective function on environmental and economic performance of a water network system using life cycle assessment and life cycle costing methods. Chem Eng J 144(3):368–378
Liner B, DeMonsabert S (2011) Balancing the triple bottom line in water supply planning for utilities. J Water Resour Plan Manag 137(4):335–342
Ling C, Hanna K, Dale A (2009) A template for integrated community sustainability planning. Environ Manage 44(2):228–242
Maeda M, Nakada K, Kawamoto K, Ikeda M (1996) Area-wide use of reclaimed water in Tokyo, Japan. Water Sci Technol 33(10–11):51–57
Mahapatra D, Chanakya HN, Ramachandra TV (2011) Assessment of treatment capabilities of Varthur Lake, Bangalore, India. Environ Technol 14(1/2/3/4):84–102
Marecos do Monte MHF, Angelakis AN, Asano T (1996) Necessity and basis for establishment of European guidelines for reclaimed wastewater in the Mediterranean region. Water Sci Technol 33(10–11):303–316
May RJ, Dandy GC, Maier HR, Nixon JB (2008) Application of partial mutual information variable selection to ANN forecasting of water quality in water distribution systems. Environ Model Softw 23(10–11):1289–1299
Mays L, Schwartz M (1983) Models for water reuse and wastewater planning. J Environ Eng 109(5):1128–1147
McKenzie D, Ray I (2009) Urban water supply in India: status, reform options and possible lessons. Water Policy 11(4):442
Mekala GD, Davidson B, Samad M, Boland A-M (2008) Wastewater reuse and recycling systems: a perspective into India and Australia

Miller W (2006) Integrated concepts in water reuse: managing global water needs. Desalination 187(1–3):65–75

Ministry of Environment and Forests (2011) Sustainable development in India: stocktaking in the run up to Rio+20

Ministry of Water Resources Govt. of India (2012) Draft national water policy. Water Policy. [Online]. Available: http://wrmin.nic.in/index1.asp?linkid=201&langid=1

Nicklow J et al (2010) State of the art for genetic algorithms and beyond in water resources planning and management. J Water Resour Plan Manag 136(4):412–432

Pedrero F, Kalavrouziotis I, Alarcón JJ, Koukoulakis P, Asano T (2010) Use of treated municipal wastewater in irrigated agriculture—review of some practices in Spain and Greece. Agric Water Manag 97(9):1233–1241

Peng J, Stevens DK, Yiang X (1995) A Pioneer project of wastewater reuse in China. Water Res 29(1):357–363

Planning Commission (2006) Towards faster and more inclusive growth—an approach to the 11th five year plan (2007–2012)

Ravikumar R (2009) Chennai reuse price. The Hindu (BusinessLine). [Online]. Available: http://www.thehindubusinessline.com/industry-and-economy/article2468215.ece#.T0IxJLlys8A.mendeley. Accessed 20 Feb 2012

Rijsberman F (2006) Water scarcity: fact or fiction? Agric Water Manag 80(1–3):5–22

Rodriguez C et al (2012) Recycled water: potential health risks from volatile organic compounds and use of 1,4-dichlorobenzene as treatment performance indicator. Water Res 46(1):93–106

Rogers P, De Silva R, Bhatia R (2002) Water is an economic good: how to use prices to promote equity, efficiency, and sustainability. Water Policy 4:1–17

Rooijen DJ, Biggs TW, Smout I, Drechsel P (2009) Urban growth, wastewater production and use in irrigated agriculture: a comparative study of Accra, Addis Ababa and Hyderabad. Irrig Drain Syst 24(1–2):53–64

Rouse H, Ince S (1957) History of hydraulics. Dover Publications, Inc., New York, NY

Rozos E, Makropoulos C, Butler D (2010) Design robustness of local water-recycling schemes. J Water Resour Plan Manag 136(5):531–538

Samanta S (2009) Politics of access to drinking water in urban areas in India: state and market interventions a case study of Hyderabad. Retrieved from http://shodhganga.inflibnet.ac.in/handle/10603/4070

Sa-nguanduan N, Nititvattananon V (2011) Strategic decision making for urban water reuse application: a case from Thailand. Desalination 268(1–3):141–149

Schmidt CJ, Kugelman I, Clements EV (1975) Municipal wastewater reuse in U.S. J Water Pollut Control Fed 47(9):2229–2245

Seguí L, Alfranca O, García J (2009) Techno-economical evaluation of water reuse for wetland restoration: a case study in a natural park in Catalonia, Northeastern Spain. Desalination 246(1–3):179–189

Shah DA, Shete DT (2019) Attaining optimal sustainability for urban wastewater management—case study of Vadodara development authority area. A thesis submitted to Gujarat Technological University as fullfillment for PhD degree. http://shodhganga.inflibnet.ac.in:8080/jspui/handle/10603/260689

Shelef G, Azov Y (1996) The coming era of intensive wastewater reuse in the Mediterranean region. Water Sci Technol 33(10–11):115–125

Simpson A, Elhay S (2011) Jacobian matrix for solving water distribution system equations with the Darcy-Weisbach head-loss model. J Hydraul Eng 137(6):696–700

Snyder SA, Benotti MJ (2010) Endocrine disruptors and pharmaceuticals: implications for water sustainability. Water Sci Technol 61(1):145–154

Sujaritpong S, Nitivattananon V (2009) Factors influencing wastewater management performance: case study of housing estates in suburban Bangkok, Thailand. J Environ Manage 90(1):455–465

Tsagarakis KP (2005) Recycled water valuation as a corollary of the 2000/60/E.C. water framework directive. Agric Water Manag 72(1):1–14

Tsagarakis K, Dialynas G, Angelakis A (2004) Water resources management in Crete (Greece) including water recycling and reuse and proposed quality criteria. Agric Water Manag 66(1):35–47

Tudor R, Lavric V (2011) Dual-objective optimization of integrated water/wastewater networks. Comput Chem Eng 35(12):2853–2866

Tziakis I, Pachiadakis I, Moraitakis M, Xideas K, Theologis G, Tsagarakis K (2009) Valuing benefits from wastewater treatment and reuse using contingent valuation methodology. Desalination 237(1–3):117–125

UN Water Statistics (2010) Population increase and water resources. In: UNEP/GRID-Arendal maps and graphics library. [Online]. Available: http://maps.grida.no/go/graphic/population-increase-and-water-resources. Accessed 23 Feb 2012

United Nations Human Settlements Programme (UN-Habitat) (2011) Cities and climate change: global report on human settlements 2011. [Online]. Available: http://www.unhabitat.org/pmss/listItemDetails.aspx?publicationID=3086

Urkiaga A, de las Fuentes L, Bis B, Chiru E, Balasz B, Hernandez F (2008) Development of analysis tools for social, economic and ecological effects of water reuse. Desalination 218:81–91

U.S. E.P.A. (2004) Guidelines for water reuse. E.P.A./625/R-04/108, USA, Sept 2004, p 273

Wallbaum H, Krank S, Teloh R (2011) Prioritizing sustainability criteria in urban planning processes: methodology application. J Urban Plan Dev 137(1):20–28

Wang H, He Q, Liu X, Zhuang Y, Hong S (2011) Global urbanization research from 1991 to 2009: a systematic research review. Landsc Urban Plan 104(3–4):299–309

World Health Organization (2006) Guidelines for the safe use of wastewater, excreta and greywater: wastewater use in agriculture, vol II. WHO Press, France

Yang H, Abbaspour KC (2007) Analysis of wastewater reuse potential in Beijing. Desalination 212:238–250

Yi L, Jiao W, Chen X, Chen W (2011) An overview of reclaimed water reuse in China. J Environ Sci 23(10):1585–1593

You S, Tseng D, Guo G, Yang J (1999) The potential for the recovery and reuse of cooling water in Taiwan. Resour Conserv Recycl 26:53–70

# Appendix A

**R.B. Singh: Publications Related to Water Resources-Management and Sustainability** (*Source* Singh 2021, Tables 6 to 9, pp. 177–179)

(A) **Papers in Peer Reviewed Journals** (15; 1997–2018)

Singh, R.B. and Verma, R.P. (1997a) Monitoring Hydrological Parameters and Sustainable Water Resource Development in the Mahanadi River Basin. *Acta Hydrotechnica*, 15/19, pp. 159–162.

Singh, R.B. (1997b) Water Resources Monitoring for Environmental Sustainability Under Increasing Uncertainty in the Himalayan Region. In: *Sustainability of Water Resources Under Increasing Uncertainty*, IAHS Red Book 240, pp. 479–485.

Singh, R.B. (2001) Sustainable Regional Water Management in the Yamuna River Basin: Case Study of Delhi Region, Pub. In: *Regional Management of Water Resources*. IAHS Red Book 268, pp. 25–32.

Singh, R.B. (2002) Sustainable Management of Water Resources in an Agriculturally Developed Region of the Indo-Gangetic Ecosystem, India, Pub. In: *FRIEND 2002-Regional Hydrology: Bridging the Gap Between Research and Practice*. IAHS Red Book 274, pp. 315–321.

Singh, R.B. (2009) Integrated Water Resources Management in the Vulnerable Indian Environment. *IAHS Red Book 330*, pp. 187–193.

Singh, R.B. and Chandana, Vaneeta (2011a) Spatial Analysis of the Yamuna Water Quality in Pre and Post Monsoon Periods. *IAHS Red Book Pub. 348*, pp. 8–13.

Sahu, Netranand, Yamashiki, Y., Takara, Kaoru and Singh, R.B. (2011b) An Observation on the Relationship Between Climate Variability Modes and River Discharges of the Citarum Basin, Indonesia. *Annuals of Disaster Preventive Res. Inst.* Kyoto University (Online Jl.), No. 54B, pp. 49–55.

Singh, R.B., Gahlot, S. and Singh, Anju (2013) Ecohydrological Perspectives of Declining Water Sources and Quality in Traditional Water Bodies in Delhi. *IAHS Red Book 361*, pp. 361–368.



P. Kumar et al. (eds.), *Water Resources Management and Sustainability*, Advances in Geographical and Environmental Sciences,
https://doi.org/10.1007/978-981-16-6573-8

Sahu, N., Singh, R.B., Kumar, Pankaj, Da Silva, R.V. and Behera, S.K. (2013) La Nina Impacts on Austral Summer Extremely High-Streamflow Events in the Paranaiba River in Brazil. *Advances in Meteorology*. https://doi.org/10.1155/2013/461693.

Sahu, N., Behera, S.K., Ratnam, J.V., Da Silva, R.V., Parhi, P., Duan, W., Takara, K., Singh, R.B. and Yamagata, T. (2014a) El Nino Modoki Connection to Extremely-Low Streamflow of the Paranaiba River of Brazil. *Climate Dynamics*. https://doi.org/10.1007/s00382-013-2006-3.

Singh, R.B. and Suraj Mal (2014b) Trends and Variability of Monsoon and Other Rainfall Seasons in Western Himalaya. *Atmospheric Science Letters*. https://doi.org/10.1002/asl2.494.

Singh, R.B., Haque, Senaul and Grover, Aakriti (2015a) Drinking Water, Sanitation and Health in Kolkata Metropolitan City: Contribution Towards Urban Sustainability, Geography, Environment. *Sustainability*, No. 4 (V. 08).

Singh, R.B. and Kumar, Ajay (2015b) Climate Variability and Water Resource Scarcity in Drylands of Rajasthan, India. *Geoenvironmental Disasters*, 2(7), pp. 1–10. https://doi.org/10.1186/s40677-015-0018-5.

Khetwani, Sagar and Singh, R.B. (2018a) Assessment of Hydrological Drought in Marathwada Region: A Spatio-Temporal Analysis. *Journal of Geography, Environment and Earth Science International*, 17(1), pp. 1–10.

Khetwani, Sagar and Singh, R.B. (2018b) Groundwater Dynamics in Marathwada Region: Spatio-Temporal Analysis for Sustainable Groundwater Resource Management. *International Journal of Conservation Science*, 9(3), July–Sept, pp. 537–548.

(B) **Papers in the Indian Journals/Books** (7; 2001–2017)

Singh, R.B. (2001) Management of Water Related Disasters in the Context of the Himalayan Mountain Region of India. *Annals, National Association of Geographers of India* (NAGI), 21(2), December, pp. 41–46.

Singh, R.B. (2004) Impact of Climate Change on Water Resources Sustainability in the Himalayan-Gangetic Region. *Annals, National Association of Geographers of India* (NAGI), 24(1), pp. 32–39.

Singh, R.B. (2006) Hydro-Informatics for Integrated Watershed Management-Combining Science and Community Participation. *Indian National Geographer*, 21(1 & 2), pp. 59–66.

Singh, R.B., Pandey, B.W. and Prasad, Abhay Shankar (2014a). Living with Flood and Sustainable Livelihood Development in Lower Brahmaputra River Basin Assam. *Golden Research Thoughts*, 3(10), pp. 1–10.

Singh, R.B., Pandey, B.W. and Prasad, Abhay Shankar (2014b) Adaptation Strategies for Flood Risk Mitigation in Lower Brahmaputra River Basin, Assam Through Integrated River Basin Management. *Transactions of the Institute of Indian Geographers*, 36(2), pp. 159–170.

Singh, R.B., Pandey, B.W. and Prasad, Abhay Shankar (2014c). Extent and Magnitude of Flood Due to Soil Erosion in Lower Brahmaputra River Basin Assam. *Annals, National Association of Geographers of India (NAGI)*, 34(2), pp. 162–179.

Singh, R.B. (2017) Ensuring Clean Water Accessibility and Mitigating Water Induced Disasters—Contributing Towards Sustainable Development Goals. *Productivity*, 57(4), Jan–March, pp. 323–331.

(C) **Papers in Foreign Books and Proceedings** (4; 1983–2015)

Singh, R.B. (1983) Ecological Approach of Irrigation System: Potentiality Constrains and Management in Traditional Society. In: *Problems of the Management of Irrigated Land in Areas of Traditional and Modern Cultivation*, ed. H.G. Mensching. Proceedings of the IGU Working Group Symp., Cairo-El-Minia (Egypt), Hamburg (Germany), pp. 123–132.

Singh, R.B. (2003) Local Knowledge and Community Participation in Water Management for Poverty Alleviation: Towards Integrating Water Science and Ethics in India. In: *Proceedings of the Asia-Pacific Association of Hydrology and Water Resources*, Eds. K. Takara and Kojiima, Disaster Prevention Inst., University of Kyoto, Japan.

Singh, R.B. and Kumar, Dilip (2014). Water Scarcity. In: *Handbook of Engineering Hydrology: Environmental Hydrology and Water Management*. Taylor & Francis Group, pp. 519–544.

Singh, R.B., Haque, S. and Grover, A. (2015) Drinking Water, Sanitation and Health in Kolkata Metropolitan City: Contribution Towards Urban Sustainability. *Geography, Environment and Sustainability*, 8(4), pp. 64–81.

(D) **Papers in the Indian Books and Proceedings** (5; 1996–2008)

Singh, R.B. and Pandey, B.W. (1996) Perspectives for Managing Rural Water Resources: A Study of Ganga-Gomati Doab. In: *Research in Geography: Land Use Changes and Sustainable Development*, ed. R.B. Singh. APH Pub., New Delhi, vol. I, pp. 307–318.

Verma, Romila P. and Singh, R.B. (1998) Dam and Water Resource Development: Myth or Reality—Case of Hirakud Dam. In: *Ecological Techniques and Approaches to Vulnerable Environment*, ed. R.B. Singh. Oxford and IBH Pub., New Delhi, pp. 223–242.

Singh, R.B. and Shouraseni Sen Roy (2005) Influence of Water Availability on Poverty in India. In: Geomorphology and Environmental Sustainability (Felicitation Volume in Honour of Prof. H. S. Sharma), ed. S.C. Kalwar, et al. Concept Pub. Co., New Delhi, pp. 319–327.

Singh, R.B. and Prakash, Nawal (2006) Integrated Watershed Management for Livelihood Security: A Case Study of Uppalgaon Watershed, Uttaranchal Himalaya. In: *Use and Misuse of Land and Water Resources*, ed. Abha Lakshmi Singh. B.R. Pub., Delhi, pp. 113–127.

Singh, R.B. and Singh, Mehtab (2008) Geographical Analysis of Water Logging Induced Land Degradation in Karnal District of Haryana. In: *Perspectives in Resource Management in Developing Countries*, vol. 3, ed. Baleshwar Thakur. Concept Pub. Co., New Delhi, pp. 196–211.